20

SOLID STATE AND SEMICONDUCTOR PHYSICS

SOLID STATE AND SEMICONDUCTOR PHYSICS

John P. McKelvey

THE PENNSYLVANIA STATE
UNIVERSITY

ROBERT E. KRIEGER PUBLISHING COMPANY
MALABAR, FLORIDA

Original Edition   1966
Reprint Edition   1982, 1984

Printed and Published by
ROBERT E. KRIEGER PUBLISHING COMPANY, INC.
KRIEGER DRIVE
MALABAR, FLORIDA 32950

Library of Congress Cataloging in Publication Data

McKelvey, John Philip.
    Solid state and semiconductor physics.

    Reprint. Originally published: New York :
Harper & Row, 1966. (Harper's physics series)
    Includes bibliographical references and index.
    1. Semiconductors.   2. Solid state physics.
I. Title.   II. Series: Harper's physics series.
[QC611.M495   1981]   537.6'22   81-19390
ISBN 0-89874-396-6   AACR2

PREFACE

This book has arisen from source material used by the author in teaching a two-term course to seniors and beginning graduate students majoring in physics, electrical engineering, metallurgy, and materials sciences. The first term is ordinarily devoted to general solid state physics, the second to a detailed study of semiconductor materials and devices, in which (in addition to much new material) many applications of principles developed in the introductory section of the course are encountered. The present work, then, is essentially a text for such a course. We hope its scope is sufficiently broad and its treatment of fundamentals sufficiently detailed and understandable that it will also serve as a general reference book for scientists and engineers who are engaged in research and development work involving semiconductor materials and devices.

In view of the diverse educational backgrounds of those who must interest themselves in this material, it was felt that the inclusion of self-contained chapters on quantum mechanics and statistical mechanics would be of great assistance. It should be noted, however, that these chapters are not intended to serve as comprehensive treatments of these topics, but only as brief introductory essays which suffice, hopefully, to provide enough working knowledge to enable the reader to follow the subsequent development and appreciate the physical significance of all that follows. There is, of course, no substitute for a really profound appreciation of quantum and statistical phenomena in understanding solid state physics.

The treatment throughout is from the point of view of the physicist, although enough technical detail on such subjects as materials technology and crystal growth, semiconductor device fabrication, and device characteristics and circuitry is included to round out these aspects of the over-all picture and indicate what sources must be consulted in order to obtain more detailed information about those subjects. The material on p-n junction theory and semiconductor device analysis is intended to be quite complete and to present a comprehensive and detailed discussion of analytical techniques which are useful in this rather important but somewhat neglected area. The choice of introductory topics in general solid state physics, in Chapters 1–8, includes only those subjects which are of central importance to semiconductor physics. The introductory section, therefore, is selective rather than comprehensive, although the material presented is about sufficient for a single-term intermediate solid state physics course. Clearly, it would have been impossible, in a book of manageable dimensions, to include also a treatment of dielectrics, magnetism, color centers, resonance experiments, and other such topics which are ordinarily included in general solid state physics texts.

The level of presentation is intermediate; the author has tried to go far beyond what can be accomplished in a qualitative way without any real use of quantum theory and statistical mechanics, while avoiding the formidable mathematical involvements of a full quantum-theoretical treatment. The line of approach is pragmatic rather than axiomatic, and a determined attempt has been made to present a physical as well as mathematical understanding of all the subject matter. Fundamental principles, rather than technical details, are emphasized throughout. The particle approach, as embodied in the free-electron transport theory, has been relied upon heavily; the justification of this approach in the light of the quantum theory has been emphasized in Chapter 8. No mathematics beyond vector analysis and ordinary differential equations is required. Although partial differential equations and orthogonal function expansions are encountered on several occasions, the mathematical tools needed are developed on the spot.

It is quite impossible for me to express adequately my gratitude to all of my associates who have contributed to this work. Specifically, however, I should like to thank Drs. D. R. Frankl, P. H. Cutler, J. Yahia, and H. F. John for helpful comments, discussions, and suggestions, and Dr. F. G. Brickwedde for generously providing secretarial services in a time of dire need. My former students, Drs. J. C. Balogh, M. W. Cresswell, and E. F. Pulver have also rendered much assistance in reading, criticizing, and correcting parts of the manuscript. Many of the students in my classes have pointed out errors in the manuscript notes and have made suggestions for improvements concerning various individual topics. I am unable to recall all of these contributions in specific terms, but I am grateful for them nevertheless. I must also thank Miss Frances Fogle, Mrs. Marion Shaw, and Miss Eileen Berringer for their invaluable services in typing the manuscript.

J. P. MCKELVEY

University Park, Pa.
April, 1966

CONTENTS

SOLID STATE AND SEMICONDUCTOR PHYSICS

CHAPTER 1

SPACE LATTICES AND CRYSTAL TYPES

1·1 CONCEPT OF "SOLID"

Generally speaking, we apply the term *solid* to rigid elastic substances, that is, to substances that exhibit elastic behavior not only when subjected to hydrostatic forces, but also under tensile or shear stresses. There are, of course, materials that display both elastic and plastic or viscous behavior, so that this classification is not a completely rigorous one. We shall nevertheless adopt it as a criterion of what a solid substance is, recognizing that there is a class of substances that exhibits both solid and fluid behavior.

Materials that may be regarded as solids by this definition can generally be divided into two categories, amorphous and crystalline. In amorphous substances, the atoms or molecules may be bound quite strongly to one another, but there is little, if any, geometric regularity or periodicity in the way in which the atoms are arranged in space. Such substances are usually viscoelastic and can be regarded as supercooled liquids. A two-dimensional representation of an amorphous material is shown in Figure 1.1.

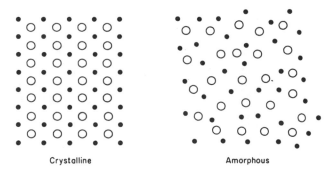

Crystalline   Amorphous

FIGURE 1.1. A schematic representation in two dimensions showing the difference in atomic arrangement between a crystalline solid and an amorphous solid.

Crystalline substances, on the other hand, are characterized by a perfect (or *nearly* perfect) periodicity of atomic structure; this regularity of structure provides a very simple conceptual picture of a crystal and simplifies the task of understanding and calculating its physical properties. For this reason, crystalline solids are better

1

understood physically than amorphous solids and liquids. In this book we shall restrict ourselves largely to a study of the physical properties of perfect or nearly perfect crystalline solids. Sometimes the presence of a relatively small number of imperfections, such as impurity atoms, lattice vacancies, or dislocations, in an otherwise perfectly periodic crystal may cause striking changes in the physical behavior of the material. To the extent that these effects may be important, we shall be concerned with them also.

Finally, we should realize that macroscopic samples of crystalline solids such as metals, ceramics, ionic salts, etc., are not always single crystals, but are often composed of an array or agglomerate of small single crystal sections of various crystal orientations separated from one another by "grain boundaries," which can be regarded as localized regions of very severe lattice disruption and dislocation. We shall treat primarily the properties of single crystal specimens, but we shall try to understand the nature of grain boundaries and to point out in what important ways they may be expected to influence the physical properties of macroscopic crystalline samples.

1·2   UNIT CELLS AND BRAVAIS LATTICES

Figure 1.2 depicts the lattice of a two-dimensional crystal that will be used as an example in explaining certain basic crystallographic terms. Referring to this figure, parallelogram $ABCD$ may be chosen as a *unit cell* of the lattice; it is determined by

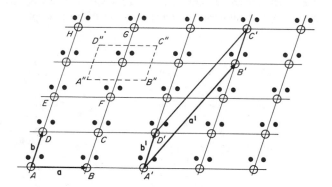

FIGURE 1.2. Unit cells and basis vectors in a two-dimensional lattice.

the *basis vectors* **a** and **b**. All translations of the parallelogram $ABCD$ by *integral* multiples of the vectors **a** and **b**, along the **a** and **b** directions, will result in translating it to a region of the crystal exactly like the original one. The whole crystal may thus be reproduced simply by reproducing the area $ABCD$ translated along the **a** and **b** directions by all possible combinations of multiples of the basis vectors **a** and **b**. In other words, every lattice point in the crystal can be described by a vector **r** such that

$$r = h\mathbf{a} + k\mathbf{b}, \qquad (1.2\text{-}1)$$

where h and k are *integers*. It is clear that this procedure can easily be extended to define unit cells and basis vectors in three-dimensional crystal lattices. We are led thus to the following definitions:

> *Unit Cell*: A region of the crystal defined by three vectors, **a**, **b**, and **c**, which, when translated by any integral multiple of those vectors, reproduces a similar region of the crystal,
>
> *Basis Vectors*: A set of linearly independent vectors **a**, **b**, **c**, which can be used to define a unit cell,
>
> *Primitive Unit Cell*: The smallest unit cell (in volume) that can be defined for a given lattice,
>
> *Primitive Basis Vectors*: A set of linearly independent vectors that defines a primitive unit cell.

Again, according to these definitions, it is clear that every lattice point in a three-dimensional crystal lattice can be described by a vector of the form

$$r = h\mathbf{a} + k\mathbf{b} + l\mathbf{c} \qquad (h,k,l \text{ integers}). \qquad (1.2\text{-}2)$$

It should be noted that the unit cell can be defined in more than one way; for example, $A'B'C'D'$ and $A''B''C''D''$ are other possible choices for the unit cell in Figure 1.2. All three of these unit cells are primitive units cells for this lattice, and there is thus a corresponding ambiguity in the choice of a set of primitive basis vectors. Either (**a**,**b**) or (**a**′,**b**′) would serve equally well. The larger cell *EFGH* in the figure is an example of a unit cell that is *not* a primitive cell.

It can be shown that there are just 14 ways of arranging points in space lattices *such that all the lattice points have exactly the same surroundings*. These 14 point lattices, which are called *Bravais lattices*, are shown in Figure 1.3. For each of these lattices, an observer viewing the crystal from one of the lattice points would see exactly the same arrangement of surrounding lattice points no matter which lattice point he chose as a point of vantage. It might appear at first sight that there should be other possible Bravais lattices; for example, no face-centered tetragonal arrangement is included in the 14 Bravais lattices in Figure 1.3. The reason for this is that such a face-centered tetragonal structure is equivalent to a body-centered tetragonal lattice in which the side of the base of the unit cell is $1/\sqrt{2}$ times what it is for the face-centered arrangement. The proof of this statement, as well as the related question of why the face-centered and body-centered *cubic* structures are distinct, is left as an exercise for the reader.

These 14 lattices may be grouped into seven crystal systems, each of which has in common certain characteristic symmetry elements. The symmetry elements that we chose to specify these seven systems are as follows:

> (1) *n-fold rotation axis*: rotation about such an axis through an angle $2\pi/n$ radians leaves the lattice unchanged. Here n may have the values 1, 2, 3, 4, and 6. Five-fold rotational symmetry in a crystal lattice is impossible.
>
> (2) *Plane of symmetry*: one half of the crystal, reflected in such a plane, passing through a lattice point, reproduces the other half.

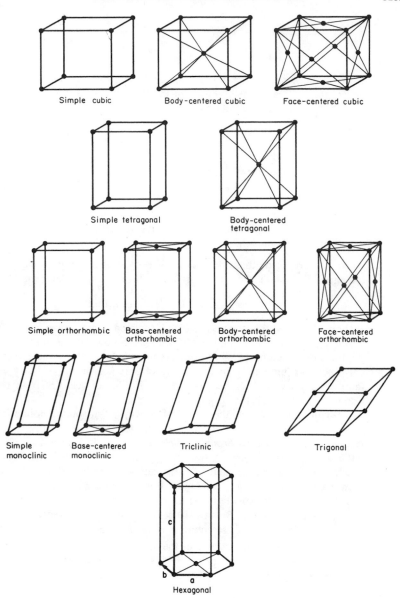

Simple cubic   Body-centered cubic   Face-centered cubic

Simple tetragonal   Body-centered
tetragonal

Simple orthorhombic Base-centered   Body-centered   Face-centered
                  orthorhombic   orthorhombic   orthorhombic

Simple   Base-centered   Triclinic   Trigonal
monoclinic monoclinic

Hexagonal

FIGURE 1.3. The fourteen Bravais lattices.

(3) *Inversion center*: a lattice point about which the operation $\mathbf{r} \to -\mathbf{r}$ (where $\mathbf{r}$ is a vector to any other lattice point) leaves the lattice unchanged.

(4) *Rotation-Inversion axis*: rotation about such an axis through $2\pi/n$ radians ($n = 1,2,3,4,6$) followed by inversion about a lattice point through which the rotation axis passes leaves the lattice unchanged.

Other choices of symmetry operations are possible, but they may all be shown to be equivalent to linear combinations of these four. The seven crystal systems, with their characteristic symmetry elements and unit cell characteristics are listed in Table 1.1. The notation used for dimensions and angles in the unit cell is depicted in Figure 1.4.

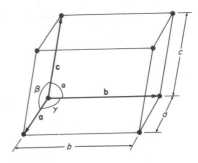

FIGURE 1.4. Notation for angles and dimensions within the unit cell.

TABLE 1.1.

The Seven Crystal Systems

System	Characteristic Symmetry Element*	Bravais Lattice	Unit Cell Characteristics
Triclinic	None	Simple	$a \neq b \neq c$ $\alpha \neq \beta \neq \gamma \neq 90°$
Monoclinic	One 2-fold rotation axis	Simple Base-centered	$a \neq b \neq c$ $\alpha = \beta = 90° \neq \gamma$
Orthorhombic	Three mutually perpendicular 2-fold rotation axes	Simple Base-centered Body-centered Face-centered	$a \neq b \neq c$ $\alpha = \beta = \gamma = 90°$
Tetragonal	One 4-fold rotation axis or a 4-fold rotation-inversion axis	Simple Body-centered	$a = b \neq c$ $\alpha = \beta = \gamma = 90°$
Cubic	Four 3-fold rotation axes (cube diagonals)	Simple Body-centered Face-centered	$a = b = c$ $\alpha = \beta = \gamma = 90°$
Hexagonal	One 6-fold rotation axis	Simple	$a = b = c$ $\gamma = 120°$ $\alpha = \beta = 90°$
Trigonal (Rhombohedral)	One 3-fold rotation axis	Simple	$a = b = c$ $\alpha = \beta = \gamma \neq 90°$

* There may, of course, be other symmetry properties in individual cases; only the ones peculiar to each particular crystal system are listed here.

It should be remembered that the lattice points of a space lattice do not generally represent a single atom, but rather a group of atoms or a molecule. In addition to the symmetry properties of the lattice points themselves, then, one must consider the symmetry properties of the molecules or groups of atoms at each lattice point about the lattice points themselves in order to completely enumerate all possible crystal structures. When this is done, it is found that there are 230 basically different repetitive patterns in which such elements can be arranged to form possible crystal structures. It is beyond the scope of this book to discuss and enumerate all these possibilities in detail. For a more complete discussion of these topics, the reader is referred to one of the standard works on crystallography listed at the end of this chapter.

1·3   SOME SIMPLE CRYSTAL STRUCTURES

It is clear from Figure 1.3 that there are three possible cubic lattices, namely the simple cubic, the body-centered cubic (b.c.c.) and the face-centered cubic (f.c.c.) lattices. The cubic unit cell shown in Figure 1.3 for the simple cubic lattice is also a *primitive* unit cell, because it contains the irreducible minimum of one atom in each such cell. This result is arrived at by noting that in the simple cubic cell there are 8 atoms at the 8 corners of the cell, each atom being shared equally among the 8 unit cells that adjoin at each corner. We may say, then, that there are 8 corner atoms, and that $1/8$ of each belongs to this particular cell, making a total of one atom per unit cell. The cubic cell for the f.c.c. structure contains 8 corner atoms, shared among 8 cells, and 6 face-center atoms, each shared between 2 cells, giving a total of $8(1/8) + 6(1/2)$ or 4 atoms in the cubic cell. In the cubic cell for the b.c.c. structure there are 8 corner atoms, each shared between 8 cells, and a central atom belonging exclusively to the cell in question, giving a total of $8(1/8) + 1$ or 2 atoms in the cubic cell. Since for simple structures containing atoms of one kind only, the primitive unit cell usually contains only one atom, we are led to suspect that neither the f.c.c. nor the b.c.c. cubic cells are primitive cells. This is indeed the case; the primitive cells for the f.c.c. and b.c.c. structures are illustrated in Figure 1.5. Each primitive cell contains only one atom. As usual, these primitive cells are not the only ones that can be constructed, there being many possible choices of sets of primitive basis vectors for either structure.

The primitive cell for the f.c.c. structure illustrated in Figure 1.5 is readily seen to be a special case of a trigonal structure with $\alpha = \beta = \gamma = 60°$. For this one particular value of α the trigonal structure possesses cubic symmetry and reduces to the f.c.c lattice. It is, of course, possible to ignore the cubic cell of the f.c.c. lattice entirely, and regard the structure as trigonal with $\alpha = 60°$, using only the primitive cell of Figure 1.5. It is usually convenient, however, to take the larger cubic cell as the basis of the crystal rather than the primitive cell, because it permits one to use an orthogonal (x,y,z) coordinate system for the lattice, which is much simpler than an oblique system, and because the lattice has all the symmetry properties associated with the simple cubic structure. In the case of the b.c.c. structure, similar considerations dictate the use of the larger cubic cell as the basis of the crystal whenever possible. For some purposes, however, it is absolutely necessary that the primitive cell be used; we shall point these out as they arise.

When equal spheres are packed into a container, the closest packing may be achieved by constructing a bottom layer in which each sphere is surrounded by 6 neighboring ones, and then putting exactly similar layers on top of this one, using the triangular interstices between spheres in the lower layer to hold spheres that are added to the upper one. It is evident that in a close-packed arrangement each sphere has just 12

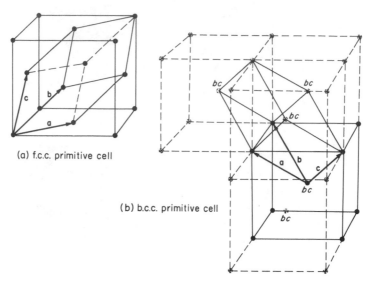

(a) f.c.c. primitive cell

(b) b.c.c. primitive cell

FIGURE 1.5. Primitive unit cells for the face-centered and body-centered cubic lattices.

nearest neighbors—6 surrounding it in the same layer, 3 in the layer above, and 3 in the layer below. Any structure that is *close-packed* in this sense is characterized by the property that each atom has 12 nearest neighbors; conversely no structure in which each atom has *more* than 12 nearest neighbors is possible.

One such close-packed structure is achieved by packing a lower layer A, then a second layer B in the manner described above, then a third layer in which the atoms are added directly over the atoms in layer A, then a fourth layer whose atoms are placed directly over the atoms in layer B, and so on, forming an array of layers $ABABABAB\ldots$, as illustrated in Figure 1.6. This structure has hexagonal symmetry, and is referred to as the *hexagonal close-packed structure* (h.c.p.). Alternatively, it is possible to put the atoms of the third layer not directly above those of the first but in a third set of positions, as shown in Figure 1.6, forming a third layer C whose atoms are directly above neither those in layer A nor those in layer B. The fourth layer is now added, using the positions corresponding to layer A, and an array of layers $ABCABCABC\ldots$ formed. This second possible close-packed structure is simply the f.c.c. structure, the normal to the close-packed layers being the cube diagonal. From Figure 1.3 it is easy to see that each atom in the f.c.c. structure has just 12 nearest neighbors, at a distance of $\sqrt{2}/2$ times the cube edge, whereby the structure must be close packed. In contradistinction, in the b.c.c. lattice, each atom has only 8 nearest neighbors, at a distance of $\sqrt{3}/2$ times the cube edge; if the lattice points in the b.c.c.

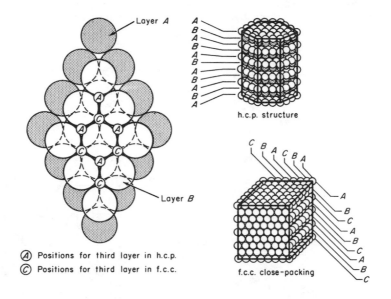

Layer A

A
B
A
B
A
B
A
B
A
B
A

h.c.p. structure

C B A C B A

A
B
C
A
B
C
A
B
C

f.c.c. close-packing

Layer B

Ⓐ Positions for third layer in h.c.p.
Ⓒ Positions for third layer in f.c.c.

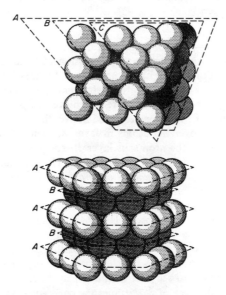

A B C

A
B
A
B
A

FIGURE 1.6. Arrangement of close-packed layers in the cubic (f.c.c.) close-packed and hexagonal close-packed lattices. [After L. V. Azaroff, *Introduction to Solids*, McGraw-Hill (1960), p. 60, with permission.]

structure were envisioned as expanding spheres, they would touch along the diagonal before they met along the cube edge, and the resulting structure would not be close packed.

The NaCl and CsCl structures are shown in Figure 1.7. The NaCl structure has alternating Na and Cl atoms at the lattice points of a simple cubic lattice. The

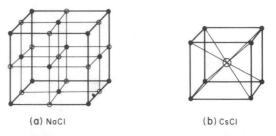

(a) NaCl   (b) CsCl

FIGURE 1.7.   The NaCl and CsCl structures.

Na atoms lie on the lattice points of a f.c.c. lattice, as do the Cl atoms. The CsCl structure is basically body-centered, with Cs atoms at the body-center positions and Cl at the cube corners (or vice-versa). A large number of ionic crystals crystallize in one or the other of these lattices.

The diamond and zincblende structures are illustrated in Figure 1.8. The lines connecting the atoms in these lattices represent covalent electron-pair bonds, which are tetrahedrally disposed about each atom. In both structures each atom has four

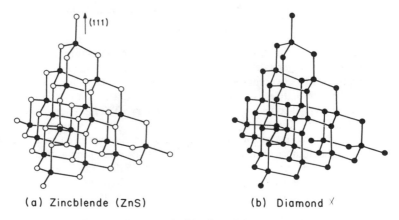

(a) Zincblende (ZnS)   (b) Diamond

FIGURE 1.8.   The zincblende and diamond lattices.

nearest neighbors. In diamond, the structure can be regarded as two f.c.c. lattices, interpenetrating, displaced from one another along the cube diagonal by 1/4 the length of that diagonal. The lattice thus has cubic symmetry. The above description of the diamond lattice can be better understood by reference to Figure 1.9, which shows how the diamond lattice can be described by a cubic unit cell. Silicon, germanium,

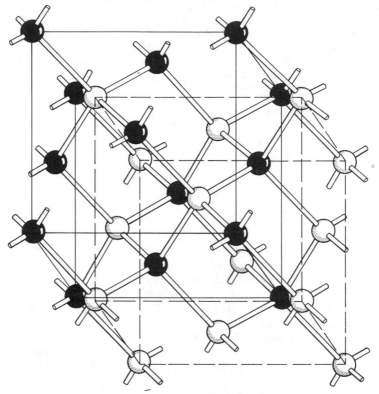

FIGURE 1.9.   A diagram of the zincblende lattice showing the outlines of the cubic unit cell.

and α- (gray) tin also crystallize in the diamond structure. The zincblende lattice is is very closely related to the diamond lattice. In zincblende the structure can be represented as two interpenetrating f.c.c. lattices, just as for diamond, except that one of the f.c.c. lattices is composed entirely of Zn atoms, the other entirely of S atoms. This lattice, of course, also has cubic symmetry. The III-V semiconducting compounds InSb, GaAs, GaSb, InP, GaP, etc., crystallize with this structure.

The "closeness" of packing of the atoms in any given structure can be inferred from the number of nearest neighbors that surround each atom. This number, which is often referred to as the *coordination number* of the crystal, can range from a maximum of 12 in close-packed structures down to 4 in diamond and zincblende, whose structures are relatively "open."

1·4   CRYSTAL PLANES AND MILLER INDICES

In a crystal with a regular periodic lattice, it is often necessary to refer to systems of planes within the crystal that run in certain directions, intersecting certain sets of atoms. Thus the close-packed layers of the f.c.c. and h.c.p. lattices form very definite and important systems of crystal planes. The orientation of such systems of planes within the crystal is specified by a set of three numbers called *Miller indices*, which may be determined as follows:

(1) Take as the origin any atom in the crystal and erect coordinate axes from this atom in the directions of the basis vectors,
(2) Find the intercepts of a plane belonging to the system, expressing them as integral multiples of the basis vectors along the crystal axes,
(3) Take the *reciprocals* of these numbers and reduce to the smallest triad of integers h, k, l having the same ratio. The quantity (hkl) is then the Miller indices of that system of planes.

For example, Figure 1.10 shows a plane whose intercepts are twice the lattice distance a, three times the lattice distance b and four times the lattice distance c. The Miller indices of the family to which this plane belongs is obtained by taking the reciprocals

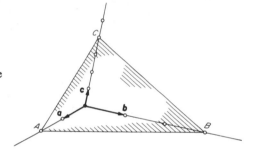

FIGURE 1.10.   A (643) lattice plane in a crystal lattice.

of these numbers, that is, $\frac{1}{2}$, $\frac{1}{3}$, $\frac{1}{4}$, and reducing to the smallest possible triad of integers having the same ratio. This can be done by multiplying each of the reciprocals by 12 in this case, giving 6, 4, 3. The Miller indices are written by simply enclosing these three numbers in parentheses; in this example the plane is a member of the (643) family of planes or, more simply, a (643) plane.

There may be a number of systems of planes whose Miller indices differ by permutation of numbers or of minus signs, yet which are all crystallographically equivalent so far as density of atoms and interplanar spacing are concerned. For example, in an *orthogonal* lattice the planes (hkl), $(\bar{h}kl)$, $(h\bar{k}l)$, $(hk\bar{l})$, $(\bar{h}\bar{k}l)$, $(\bar{h}k\bar{l})$, etc., obtained by assigning various combinations of minus signs to the Miller indices, are all equivalent in this sense. (It is conventional to write the minus signs above rather than before the Miller indices.) Likewise, in a *cubic* lattice, all the planes represented by permutations of the three Miller indices among themselves, such as (hkl), (khl), (lhk), etc., as well as those obtained by taking various combinations of minus signs, are all crystallographically equivalent. When referring to the *complete set* of crystallographically equivalent planes of which (hkl) is a member, it is customary to enclose the Miller indices in curly brackets, thus: $\{hkl\}$.

If a plane is parallel to one or two of the vectors **a**, **b**, **c**, one or two of the intercepts will be at infinity; the corresponding Miller indices are then zero. For example, if the plane shown in Figure 1.10 is revolved about the line AB until it is parallel to **c**, the intercepts become 2, 3, ∞, and the Miller indices are then (320). If it is then *further* revolved about a line perpendicular to the plane of **a** and **b** passing through point A, until it is parallel to **b** as well, the intercepts then become 2, ∞, ∞, and the Miller indices are (100). The reader should familiarize himself with the {100} (cube face planes), the {110}, and the {111} families of planes for the cubic lattices.

The indices of a *direction* in a crystal may be expressed as a set of integers which has the same ratios as the components of a vector in that direction, expressed as multiples of the basis vectors **a**, **b**, **c**. Thus the direction index of a vector $h\mathbf{a} + k\mathbf{b} + l\mathbf{c}$ is simply [hkl]. Square brackets enclosing the three indices denote direction indices. Directions with different direction indices may, of course, be equivalent crystallographically in the same manner as crystal planes with different Miller indices; it is clear, particularly, that the normals to crystallographically equivalent planes must be crystallographically equivalent directions. The complete set of crystallographically equivalent directions of which [hkl] is a member is expressed by enclosing the direction indices in angle brackets, thus: ⟨hkl⟩. In *cubic* crystals, a direction with direction indices [hkl] is normal to a plane whose Miller indices are (hkl), but this is not generally true in other systems. This is another good reason for referring the b.c.c. and f.c.c. structures to cubic rather than primitive unit cells. The proof of this result is left as an exercise.

The positions of points in a unit cell are specified in terms of fractional parts of the basis vector magnitudes along corresponding coordinate directions, taking the origin always at a corner atom. For example, the coordinates of the center point of a unit cell are $(\frac{1}{2},\frac{1}{2},\frac{1}{2})$, and the coordinates of the face centers are $(\frac{1}{2},\frac{1}{2},0)$, $(0,\frac{1}{2},\frac{1}{2})$, $(\frac{1}{2},0,\frac{1}{2})$, $(\frac{1}{2},\frac{1}{2},1)$, $(1,\frac{1}{2},\frac{1}{2})$, and $(\frac{1}{2},1,\frac{1}{2})$.

1·5   SPACING OF PLANES IN CRYSTAL LATTICES

From Figure 1.10, one might be tempted to conclude that the spacing between neighboring planes of the (hkl) system would simply be equal to the length of the normal to the plane defined by the given intercepts and the origin. This is not generally true, however, the reason being that in determining the location of an (hkl) plane from given intercept distances, the origin could have been taken to be *any* lattice point of the crystal, whereby an (hkl) plane must pass through every lattice point of the crystal. This situation is represented pictorially in Figure 1.11 for a set of "planes" in a two-dimensional lattice. The system of planes is that determined by intercepts $3a$ and $2b$ along the **a**- and **b**-axes, respectively. The Miller indices are therefore (23) (or (230) if the planes are thought to exist parallel to a **c**-axis coming out of the plane of the paper in a three-dimensional crystal). The heavily outlined planes are the ones that are determined by the intercept distances ($3a,2b$) and the choice of the origin at O. However, the origin could just as well have been chosen at the lattice point numbered "1," in which case, using the same set of intercept distances, the lightly drawn planes marked "1" are defined. Likewise, the origin could have been located at lattice points 2, 3, 4, or 5, in which cases, still using the same set of intercept distances, the sets of planes numbered 2, 3, 4, and 5 are defined. But now there is a plane of this family going through every atom of the crystal, 5 additional planes being interspersed between each pair of original heavily outlined planes. The intercepts of *adjacent* planes along the **a**- and **b**-axes are now seen to differ by $a/2$ and $b/3$, respectively. For a set of Miller indices (hk), these intercepts would differ by a/h and b/k.

In order to prove this, consider first the area $OACB$ in Figure 1.11. There are hk lattice points belonging to this area, since it consists of hk unit cells each having one lattice point. The region $ABCD$, which is clearly equal in area to $OACB$ (both can be regarded as superpositions of two triangles congruent to OAB) thus also contains hk lattice points. This area also contains just hk lattice planes (six in the above

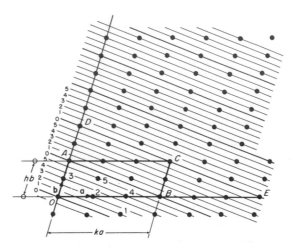

FIGURE 1.11. A representation of all the (23) lines in a two-dimensional lattice. If a **c**-axis normal to the plane of the paper exists, then this system of lines may be regarded as traces of (230) planes (which are normal to the page) in the plane of the paper. Note that one and only one line (or plane) of the system passes through each lattice point.

example), one for each of the hk lattice points within this area. Note that each plane can intersect only one lattice point within $ABCD$, except for planes intersecting lattice points on the boundaries, which can intersect two lattice points, one half or one quarter of each of which may belong to $ABCD$, depending on whether the lattice points thus intersected represent edge or corner atoms. In $ABCD$ there are thus hk planes of the system intersecting the **b**-axis in a distance hb, the difference between the **b**-intercepts of adjacent planes being $hb/hk = b/k$. Likewise, by considering area $ABEC$, which is again equal in area to $OACB$, and which likewise contains hk lattice points and hk planes, we see that there are hk planes intersecting the **a**-axis in a distance ka, whereby the difference between the **a**-intercepts of adjacent planes of the system is a/h. In order to simplify the geometrical concepts, this result was developed for a two-dimensional crystal (or for planes parallel to the **c**-axis of a three-dimensional crystal). The same argument, however, can be extended in a very straightforward manner, to a set of (hkl) planes in a three-dimensional crystal, the result being that *the distances between the intercepts of adjacent planes of a system with Miller indices (hkl) along the **a**-, **b**-, and **c**-axes are a/h, b/k, and c/l, respectively.* The development of the argument for a three-dimensional crystal will be left as an exercise.

The actual spacing d between adjacent planes can be calculated by taking any lattice point as an origin, erecting axes in the **a**-, **b**-, and **c**-directions, and finding the perpendicular distance between this origin and the plane whose intercepts are a/h,

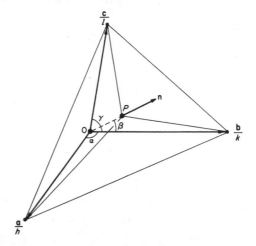

FIGURE 1.12.　Direction angles of the normal to the (hkl) plane.

b/k, c/l. Referring to Figure 1.12, it is evident that

$$d = OP = \frac{a}{h} \cos \alpha = \frac{b}{k} \cos \beta = \frac{c}{l} \cos \gamma, \qquad (1.5\text{-}1)$$

where α, β, γ are the angles between the normal to the plane and the **a**-, **b**-, and **c**-axes, respectively. If **n** is the *unit* normal vector to the plane, however, $\mathbf{n} \cdot \mathbf{a} = a \cos \alpha$, etc., and

$$d = \frac{\mathbf{n} \cdot \mathbf{a}}{h} = \frac{\mathbf{n} \cdot \mathbf{b}}{k} = \frac{\mathbf{n} \cdot \mathbf{c}}{l}. \qquad (1.5\text{-}2)$$

In an *orthogonal* lattice, taking the x-coordinate axis along **a**, the y-axis along **b** and the z-axis along **c**, the equation of the (hkl) plane whose intercepts are a/h, b/k, c/l is

$$f(x,y,z) = \frac{hx}{a} + \frac{ky}{b} + \frac{lz}{c} = 1. \qquad (1.5\text{-}3)$$

If $f(x,y,z)$ = const. is the equation of a surface, then ∇f is a vector normal to that surface, and the unit normal **n** is given by

$$\mathbf{n} = \frac{\nabla f}{|\nabla f|} = \frac{(h/a)\mathbf{i}_x + (k/b)\mathbf{i}_y + (l/c)\mathbf{i}_z}{\sqrt{\dfrac{h^2}{a^2} + \dfrac{k^2}{b^2} + \dfrac{l^2}{c^2}}}, \qquad (1.5\text{-}4)$$

and the spacing d between adjacent (hkl) planes is simply

$$d = \frac{\mathbf{n} \cdot \mathbf{a}}{h} = \frac{1}{\sqrt{\dfrac{h^2}{a^2} + \dfrac{k^2}{b^2} + \dfrac{l^2}{c^2}}}, \qquad (1.5\text{-}5)$$

according to (1.5-2). It should be noted at this point that $\mathbf{a}$, $\mathbf{b}$, and $\mathbf{c}$ are *not* unit vectors; in this example, $\mathbf{a} = a\mathbf{i}_x$, $b = b\mathbf{i}_y$, and $c = c\mathbf{i}_z$, where $\mathbf{i}_x$, $\mathbf{i}_y$, and $\mathbf{i}_z$ are the usual triad of unit vectors along the x-, y-, and z-directions, respectively. In nonorthogonal lattices the problem of finding the spacing between adjacent planes explicitly in terms of lattice spacings and angles is more complex, and will not be considered here, although the general expressions (1.5-1) and (1.5-2) are correct in both orthogonal and nonorthogonal systems.

1·6   GENERAL CLASSIFICATION OF CRYSTAL TYPES

In order to give the reader an initial overall view of the properties and characteristics of various types of crystalline materials, we shall briefly review some of the more prominent features of several types of crystals. It is, of course, possible to classify crystals in many ways, for example, according to crystal lattice, electrical properties, mechanical properties, or chemical characteristics. For the purpose at hand, however, it is most convenient to adopt a scheme of classification based on the type of inter-action responsible for holding the atoms of the crystal together. According to this scheme of classification, solids generally fall into one of four general categories; ionic, covalent, metallic, or molecular. We shall discuss each of these four classes in turn, recognizing, however, that they are not absolutely distinct, and that some crystals may at the same time possess characteristics associated with more than one of these general classes.

Ionic crystals are crystals in which valence electrons are transferred from one atom to another, the final result being a crystal that is composed of positive and nega-tive ions. The source of cohesive energy that binds the crystal together is the electro-static interaction between these ions. The electronic configuration of the ions is essentially an inert gas configuration, the charge distribution of each ion being spherically symmetric. Chemical compounds involving highly electropositive atoms and highly electronegative atoms, for example, NaCl, KBr, LiF, tend to form ionic crystals in the solid state. Ionic crystals usually have relatively high binding energies, and as a result, fairly high melting and boiling points. They are quite poor electrical conductors at normal temperatures, and are usually transparent to visible light, while exhibiting a single characteristic optical reflection peak in the far infrared region of the spectrum. The crystals are often quite soluble in ionizing solvents such as water, the solutions being highly dissociated into free ions. Ionic crystals usually crystallize in the relatively close-packed NaCl and CsCl structures.

Covalent crystals are crystals in which valence electrons are shared equally between neighboring atoms rather than being transferred from one atom to another as in ionic crystals. There is thus no net charge associated with any atom of the crystal.

The elements of columns III, IV, and V of the periodic table often enter into covalent combinations. A typical example of a covalent crystal is diamond, in which each carbon atom shares its four valence electrons with its four nearest neighbors, forming *covalent electron-pair bonds*. The spins of the two electrons in such electron pair bonds are antiparallel. These electron-pair covalent bonds in diamond are the same as the covalent carbon-to-carbon linkage which is found so frequently in organic compounds. They are stongly directional in character, that is, the electrons tend to be concentrated along the lines joining the adjacent atoms, and these lines tend to be disposed tetrahedrally about any atom. This natural tetrahedral disposition of covalent bonds is satisfied by the diamond or zincblende structures, and it is found that covalent crystals very often have these structures. Covalent crystals are usually hard, brittle materials with quite high binding energies and thus high melting and boiling points. They are typically semiconductors, whose electrical conductivity is quite sensitive to the presence of tiny amounts of impurity atoms, and increases with rising temperature at sufficiently high temperatures. They are transparent to long-wavelength radiation but opaque to shorter wavelengths, the transition being abrupt and occurring at a characteristic wavelength, usually in the visible or near infrared.

There is a continuous range between covalent and ionic properties; a given crystal may possess both covalent and ionic character, with valence electrons being partially transferred and partially shared. In Table 1.2, the crystals in the first column are

TABLE 1.2.

Covalent and Ionic Crystals

Covalent (Group IV)	III-V Compounds	II-VI Compounds	I-VII Compounds (Ionic)
C (diamond)	BN	BeO	LiF
	.BP	BeS	LiCl
	AlN	MgO	NaF
Si	AlP	MgS	NaCl
	GaP	CaO	KF
	AlAs	CaS	KCl
		ZnS	LiBr
Ge	GaAs	CaSe	KBr
	InP	ZnSe	NaI
	AlSb	CdS	RbCl
	GaSb	CdSe	RbBr
	InAs	ZnTe	KF
α-Sn	InSb	CdTe	RbI
			CsBr

composed of atoms from column IV of the periodic system, and form completely covalent crystals. The compounds in the fourth column of the table are "I-VII" compounds composed of highly electropositive and electronegative atoms, and the crystals of these compounds are highly ionic. The second and third columns list compounds (II-VI and III-V compounds) which are intermediate between these two extremes and which exhibit both ionic and covalent character, the III-V compounds being, of course, more nearly covalent than the II-VI compounds. The III-V inter-metallic compounds, despite their slightly ionic character, form a series of semi-conductors whose properties are similar in many ways to those of the corresponding group IV covalent materials.

The metallic elements in the free-state form *metallic crystals* in which free elec-trons are present. The presence of these free electrons accounts for the very high electrical and thermal conductivity of metals. The high electrical conductivity is in turn directly responsible for high optical reflection and absorption coefficients that are the most characteristic optical properties of metals. The electrical and thermal properties of metals that are due to the free electrons can be explained quite success-fully by regarding the metal crystal to be a container filled with an "ideal free electron gas." The binding energy of "ideal" metals, such as the alkali metals, arises from the interaction of the free electron gas with the positive ions of the lattice, although for many other metals the picture is more complicated. The actual binding energies of metallic crystals may be quite low, as for the alkali metals, which have relatively low melting and boiling points, or quite high, as in the case of tungsten, whose melting point is very high indeed.

Molecular crystals are crystals in which the binding between the atoms or mol-ecules is neither ionic nor covalent, but arises solely from dipolar forces between the atoms or molecules of the crystal. Even when an atom or molecule has no *average* dipole moment, it will in general have an instantaneous, fluctuating dipole moment arising from the instantaneous positions of the electrons in their orbits. This instan-taneous dipole moment is the source of an electrostatic dipole field which, in turn, may *induce* a dipole moment in another atom or molecule. The interaction between the original and the induced dipole moment is attractive and can serve to bind a crystal in the absence of ionic or covalent binding. Binding forces arising from fluctuating dipole interactions in this way are called *van der Waals forces*. These forces are usually quite weak; the binding energy due to them falls off as $1/r^6$, where r is the distance between the dipoles. Molecular crystals are thus characterized by small binding energy and consequently low melting and boiling points. They are usually poor electrical conductors. Crystals of organic compounds are usually of this type, as are the inert gases He, Ne, A, etc., in the solid state.

EXERCISES

1. Show geometrically that the face-centered tetragonal structure is equivalent to a body-centered tetragonal lattice in which the side of the base of the unit cell is $1/\sqrt{2}$ times what it is for the face-centered arrangement. Why are the face-centered and body-centered *cubic* structures distinct?

2. Show that the maximum proportion of space which may be filled by hard spheres arranged in various lattices is for the

Simple cubic	$\pi/6$,
Body-centered cubic	$\pi\sqrt{3}/8$,
Face-centered cubic	$\pi\sqrt{2}/6$,
Hexagonal close-packed	$\pi\sqrt{2}/6$,
Diamond structure	$\pi\sqrt{3}/16$.

3. Discuss physically why an actual crystal cannot possess a fivefold axis of rotational symmetry.

4. (a) Draw sketches illustrating a (100) plane, a (110) plane, and a (111) plane in a cubic unit cell.

(b) How many equivalent {100}, {110}, and {111} planes are there in a cubic crystal? Regard planes (hkl) and ($\overline{hkl}$) as *identical*.

⑤ (a) How many equivalent {123} planes are there in a cubic crystal?

(b) How many equivalent {111} planes are there in an orthorhombic crystal?

(c) How many equivalent {123} planes are there in an orthorhombic crystal? Again regard (hkl) and ($\overline{hkl}$) planes as identical in all cases.

⑥ Prove that in a cubic crystal the [hkl] direction is normal to the (hkl) plane.

7. Prove that the intercepts of adjacent (hkl) planes along the a-, b- and c-axes in a three-dimensional lattice are a/h, b/k, and c/l, respectively, by an extension of the methods used for the two-dimensional case in Section 1.5.

8. Prove that for van der Waals forces, the interaction energy between atoms or molecules falls off as $1/r^6$, where r is the distance between the interacting atoms or molecules. *Hint:* Begin by considering the potential energy of an electric dipole in an external electric field.

9. Show from the results of Maxwell's electromagnetic theory for plane waves that for a very good conductor ($\sigma \to \infty$, $\kappa \to \infty$, σ = conductivity, κ = dielectric constant) we should expect high surface reflectivity and strong internal absorption.

GENERAL REFERENCES

W. H. Bragg, *An Introduction to Crystal Analysis*, G. Bell & Sons, Ltd., London (1928).

W. L. Bragg, *The Crystalline State*, Vol. I, G. Bell & Sons, Ltd., London (1955).

M. J. Buerger, *Elementary Crystallography*, John Wiley & Sons, New York (1956).

W. F. de Jong, *General Crystallography*, W. H. Freeman & Co., San Francisco (1959).

F. C. Phillips, *An Introduction to Crystallography*, 2nd edition, Longmans, Green & Co., London (1956).

F. Seitz, *Modern Theory of Solids*, McGraw-Hill Book Co., Inc., New York (1940).

J. C. Slater, *Quantum Theory of Matter*, McGraw-Hill Book Co., Inc., New York (1951).

R. W. G. Wyckoff, *The Structure of Crystals*, 2nd edition, The Chemical Catalog Co., Inc., New York (1931).

CHAPTER 2

X-RAY CRYSTAL ANALYSIS

2·1 INTRODUCTION

The use of X-ray diffraction as a technique for crystal structure analysis dates from von Laue's discovery of the X-ray diffraction effect for single crystal samples in 1912. Laue predicted that the atoms of a single crystal specimen would diffract a parallel, monochromatic X-ray beam, giving a series of diffracted beams whose directions and intensities would be dependent upon the lattice structure and chemical composition of the crystal. These predictions were soon verified by the experimental work of Friedrich and Knipping. A schematic diagram of the experimental arrangement is shown in Figure 2.1(a). The location of the diffraction maxima·was explained by W. L.

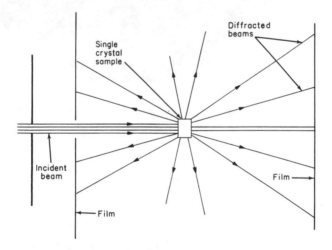

FIGURE 2.1. (a) Schematic diagram of X-ray diffraction by the Laue technique.

Bragg[1] on the basis of a very simple model in which it is assumed that the X-radiation is reflected specularly from successive planes of various (hkl) families in the crystal, the diffraction maxima being found for directions of incidence and reflection such that the reflections from adjacent planes of a family interfere constructively, differing in phase by $2\pi n$ radians, where n is an integer.

According to this idea, the path difference for successive reflections must equal an integral number of X-ray wavelengths. But this path difference, from Figure 2.2, is $2d \sin \theta$, where d is the spacing between adjacent atomic planes, as given by (1.5-2) or (1.5-5), and θ is the glancing angle between the atomic plane and the incident beam.

---

[1] W. L. Bragg, *Proc. Cambridge Phil. Soc.* 17, 43 (1912).

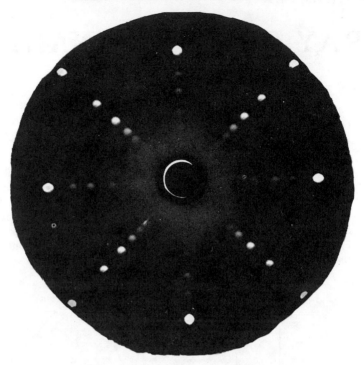

FIGURE 2.1 (*Cont'd*). (b) A Laue diffraction pattern of a lithium fluoride crystal, incident X-ray beam along a {100} direction. [Photo courtesy of H. A. McKinstry, Materials Research Laboratory, Pennsylvania State University.]

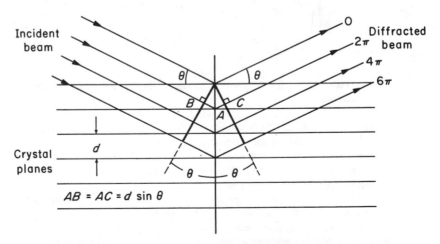

FIGURE 2.2. The Bragg picture of X-ray diffraction in terms of in-phase reflections from successive planes of a particular (*hkl*) family.

The strongly diffracted beams, then, must propagate out from the crystal in directions for which the Bragg equation

$$n\lambda = 2d \sin \theta \qquad (2.1\text{-}1)$$

is satisfied.

The experimental observation of X-ray diffraction patterns was greatly simplified by the introduction of the powder method by Debye and Scherrer[2] in 1916. In this method, as illustrated in Figure 2.3(a), a parallel, monochromatic beam of X-rays is

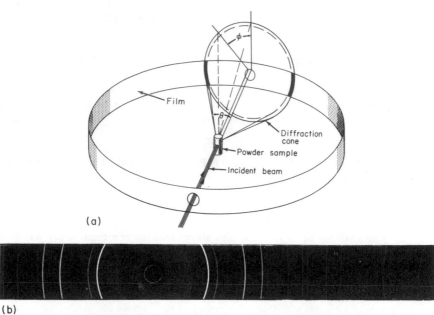

(a)

(b)

FIGURE 2.3. (a) Schematic diagram of X-ray diffraction by the Debye-Scherrer powder technique. (b) A Debye-Scherrer powder diffraction pattern from a sample of a complex scandium-zirconium oxide. [Photo courtesy of H. A. McKinstry, Materials Research Laboratory, Pennsylvania State University.]

allowed to pass through a very finely powdered specimen. Just by chance, some of the microcrystals of the powdered specimen will be oriented at the correct diffraction angle for a particular set of planes (hkl), as given by (2.1-1), and a diffracted beam will result. Since the diffraction condition can be satisfied for any possible angular orientation ϕ of the normal to the scattering planes around the incident beam axis, and since there will always be microcrystals oriented such as to produce the (hkl) diffraction for any value of ϕ, the diffracted beam will have the form of a cone whose apex angle is θ, rather than just a pencil of rays. It is customary to wrap a film strip around the inside of a cylindrical chamber, concentric with the sample, so as to intercept a certain portion of these diffraction cones, a series of arcs being produced on the film. A powder pattern made in this way is shown in Figure 2.3(b).

[2] P. Debye and P. Scherrer, *Physikal. Zeitschr.* **17**, 277 (1916).

2·2 PHYSICS OF X-RAY DIFFRACTION: THE VON LAUE EQUATIONS

X-rays can easily be produced by allowing high-energy electrons to strike a metal target anode. The X-rays so produced possess, in addition to a continuous background spectrum, a few very intense, nearly monochromatic spectrum lines, whose frequency is characteristic of the target material. These lines arise from the excitation of inner-shell atomic electrons to more highly excited states, from which they decay to the original ground state with the emission of X-ray quanta. The production of X-rays by the interaction of high-energy electrons with matter is discussed in some detail by Leighton.[3]

If a potential V_0 exists between the cathode and anode of the X-ray tube, the electrons acquire energy eV_0, where e is the magnitude of the electronic charge, from the accelerating potential as they reach the anode. The most energetic X-ray quantum which can be produced by such electrons is that for which the quantum energy $h\nu$ equals eV_0. Thus, for such a quantum,

$$eV_0 = h\nu = hc/\lambda, \tag{2.2-1}$$

where h ($= 6.62 \times 10^{-27}$ erg-sec) is Planck's constant. The shortest X-ray wavelength which can be produced is thus

$$\lambda = \frac{hc}{eV_0}. \tag{2.2-2}$$

For a voltage of 10 kilovolts, this shows that the minimum X-ray wavelength which can be excited is 1.24×10^{-8} cm., or 1.24 Ångström units. This is just of the order of interatomic distances in actual crystals, and is, according to the Bragg equation (2.1-1), just right for producing observable diffraction effects with reasonable values of d and θ. An X-ray tube in which electrons are accelerated by a potential of a few tens of kilovolts may thus be regarded as satisfactory for producing X-rays which are suitable for crystal diffraction work.

When the atoms of a crystal are exposed to electromagnetic radiation, such as X-rays, they experience electrical forces due to the interaction of the charged particles of the atoms with the electric field vector of the electromagnetic wave. The atomic electrons are therefore vibrated harmonically at the frequency of the incident radiation, thus undergoing acceleration. These accelerated charges, according to electromagnetic theory, reradiate electromagnetic energy at the frequency of vibration, that is, at the incident wave frequency. At visible light frequencies, where the incident wavelength is much larger than the interatomic distances, the superposition of the waves thus reradiated or *scattered* by the individual atoms of the crystal simply gives rise to the well-known effects of optical refraction and reflection. At X-ray frequencies, however, the incident wavelength is comparable to the interatomic spacing, and diffraction of the radiation by the atoms of the crystal can be observed.

Bragg assumed that systems of crystal planes could act to reflect X-rays specularly, provided that the condition for constructive interference between reflections from successive atomic planes is satisfied. We wish now to examine in detail the way in

---

[3] R. B. Leighton, *Principles of Modern Physics*, McGraw-Hill Book Co., Inc., New York (1959), pp. 405–421.

which X-rays scattered from different individual atoms can recombine, proving the validity of the Bragg picture, and establishing methods which can be used to extend the Bragg result in a number of ways. Let us examine the radiation scattered by two identical scattering centers separated by a distance $\mathbf{r}$. The vector $\mathbf{n}_0$ is defined to be a unit vector in the direction of the incident beam, and the vector $\mathbf{n}_1$ is taken to be a unit vector in an arbitrary scattering direction, as shown in Figure 2.4. The incident radiation is assumed to be a parallel beam, and the scattered beam is assumed to be detected at a very distant observation point. The path difference between the radiation scattered at P and that scattered at O is then

$$PA - OB = \mathbf{r} \cdot \mathbf{n}_0 - \mathbf{r} \cdot \mathbf{n}_1 = \mathbf{r} \cdot (\mathbf{n}_0 - \mathbf{n}_1). \tag{2.2-3}$$

The vector $\mathbf{n}_0 - \mathbf{n}_1 = \mathbf{N}$ is the normal to what would in the Bragg picture be called the reflecting plane, if $\mathbf{n}_1$ were a diffraction direction, as shown in Figure 2.5. It is clear from this figure, also, that the magnitude of this vector is

$$N = 2 \sin \theta. \tag{2.2-4}$$

The phase difference ϕ_r between the radiation scattered at the two points is

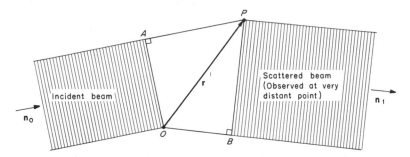

FIGURE 2.4.   Geometry of the X-ray scattering situation discussed in Section 2.2.

simply $2\pi/\lambda$ times the path difference, whereby

$$\phi_r = \frac{2\pi}{\lambda} (\mathbf{r} \cdot \mathbf{N}). \tag{2.2-5}$$

Now, in order that the direction $\mathbf{n}$ be a diffraction maximum, the scattering contribution from *every atom in the crystal* in that direction must differ in phase by an integral multiple of 2π radians. In order for this to be true, it is only necessary for the radiation from atoms separated by the *primitive* lattice vectors $\mathbf{a}$, $\mathbf{b}$, and $\mathbf{c}$ to add in phase, for then the contribution from other atoms, separated from the origin by integral combinations of these vectors will certainly add in phase. If scattering contributions from neighboring atoms were to differ in phase in any other manner, it would then always be possible to find an atom somewhere in the crystal which would contribute radiation just π radians out of phase with the contribution from a given atom; these contributions would then cancel, atom by atom, giving no diffracted beam. A consideration of an example where neighboring atoms along the $\mathbf{a}$-direction contribute radiation components in a given direction which are out of phase by π radians (or $\pi/2$, $\pi/4$, $\pi/6$, etc.,) will quickly serve to verify this assertion.

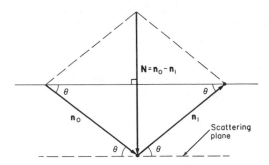

FIGURE 2.5. Geometrical relation of the incident and diffracted beams, the scattering normal, and the "scattering plane."

We need thus only require that an integral multiple of 2π be obtained in (2.2-5) when $\mathbf{r}$ equals $\mathbf{a}$, $\mathbf{b}$, or $\mathbf{c}$, that is, we must require simultaneously that

$$\frac{2\pi}{\lambda}(\mathbf{a} \cdot \mathbf{N}) = 2\pi h' = 2\pi n h$$

$$\frac{2\pi}{\lambda}(\mathbf{b} \cdot \mathbf{N}) = 2\pi k' = 2\pi n k \qquad (2.2\text{-}6)$$

$$\frac{2\pi}{\lambda}(\mathbf{c} \cdot \mathbf{N}) = 2\pi l' = 2\pi n l.$$

Here h', k', l' can be *any* three integers; in general, these three integers *may* contain a largest integer common factor n greater than unity, in which case we can write $h' = nh$, $k' = nk$ and $l' = nl$, where now h, k, and l are three integers in the same ratio as h', k', and l', but having no common factor greater than unity. If h', k', and l' do not have a common factor greater than unity, then n is simply taken to be unity. If α, β, and γ are the angles between the scattering normal $\mathbf{N}$ and the $\mathbf{a}$-, $\mathbf{b}$-, and $\mathbf{c}$-axes of the crystal, respectively, then, according to (2.2-4), $\mathbf{a} \cdot \mathbf{N} = aN \cos \alpha = 2a \sin \theta \cos \alpha$, etc., whereby (2.2-6) can be written

$$2a \sin \theta \cos \alpha = h'\lambda = nh\lambda,$$

$$2b \sin \theta \cos \beta = k'\lambda = nk\lambda, \qquad (2.2\text{-}7)$$

$$2c \sin \theta \cos \gamma = l'\lambda = nl\lambda.$$

These equations are called the Laue equations. For a given incident wavelength λ and given values of the integers h, k, l, and n, the equations determine a certain value of θ and two of the three quantities ($\cos \alpha$, $\cos \beta$, $\cos \gamma$). However, only two of the three quantitites (α,β,γ) are independent, because once the angles between a vector and two of the three coordinate axes are fixed, the direction of the vector is fixed and the third angle can be determined trigonometrically. For example, in an orthogonal coordinate system, an elementary result of analytic geometry is that $\cos^2 \alpha + \cos^2 \beta + \cos^2 \gamma = 1$. The three equations (2.2-7) thus serve to determine a unique value for θ and N, thus defining a scattering direction. The direction cosines of the scattering normal N are seen from (2.2-7) to be proportional to h/a, k/b, and l/c. However, neighboring planes whose *Miller indices* are (hkl) intersect the **a**-, **b**-, and **c**-axes at intervals a/h, b/k, and c/l; the direction cosines of the normal to the (hkl) family of planes are therefore, according to (1.5-1), *also* proportional to h/a, k/b and l/c. The scattering normal N is thus identical to the normal to the (hkl) planes, and hence the (hkl) planes may be regarded as the reflecting planes of the Bragg picture.

The Bragg equation can be shown to follow from the Laue equations by setting $h = a \cos \alpha/d$, $k = b \cos \beta/d$, $l = c \cos \gamma/d$, in Equation (1.5-1), whereby each of the three Laue equations reduces to

$$n\lambda = 2d \sin \theta,$$

where d is the distance between adjacent planes of the (hkl) system, and where the order n of the diffraction is the greatest common factor between the orders of interference h', k', and l'. It is customary to refer to an observed X-ray reflection by the numbers $(h'k'l')$ which give the order of interference between neighboring atoms along the crystal axes; thus the first order diffraction maximum for the (111) planes is referred to as the (111) reflection, the second order diffraction maximum for the same set of planes ($n = 2$, $h' = 2h$, $k' = 2k$, $l' = 2l$) as the (222) reflection, the third order as the (333) reflection, etc.

2·3 THE ATOMIC SCATTERING FACTOR

The calculations of the previous section were based on the assumption of point scattering centers at the lattice points. We now wish to take into account that the scattering, which is the result of an interaction between the atomic electrons and an X-ray beam, may take place anywhere the electrons may happen to be. More precisely, we wish to modify our calculations by considering that the X-radiation is scattered by a continuous distribution of "electron density" associated with each lattice point. This concept, as we shall see later, is in full accord with a wave-mechanical view of the process. It should be noted, however, that we are neglecting the scattering effect of the atomic nuclei, which interact much less strongly with the X-rays. We shall find that the effect of the finite extent of the electron density distribution is that the amplitude of the diffracted radiation is multiplied by a factor involving the X-ray wavelength, the glancing angle and the electron density distribution associated with the atoms.

We begin by inquiring into the ratio between the X-ray diffraction amplitude, at the Bragg angle, scattered by an element of charge $\rho(\mathbf{r})dv$, located in a volume element dv about the point $\mathbf{r}$, and that scattered by a single point electron at the lattice point, as illustrated in Figure 2.6. Here $p(\mathbf{r})$ is the electron density; $Z^{-1}p(\mathbf{r})dv$ is the probability that an electron will be found in the volume element dv. It must be required, of course, that

$$\int_v \rho(\mathbf{r}) \, dv = Z, \tag{2.3-1}$$

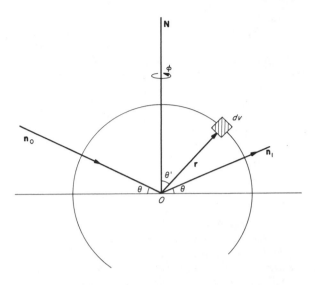

FIGURE 2.6. Geometry of the X-ray scattering situation discussed in Section 2.3. Radiation scattered at the Bragg angle by an electron density contained within the volume element dv at $\mathbf{r}$ is compared with that which would be scattered by a point electron at O.

where Z is the number of atomic electrons per atom, that is, the atomic number of the atoms of which the crystal is composed. The above integral is taken over all space, although in making actual calculations it is usually assumed that the "electron clouds" associated with different atoms in the crystal do not overlap appreciably, hence that the electron cloud of a single atom is confined to the volume of a unit cell.

The difference in phase ϕ_r between the radiation scattered at the origin and that scattered in the element dv at $\mathbf{r}$ is, from (2.2-5)

$$\phi_r = \frac{2\pi}{\lambda}(\mathbf{r} \cdot \mathbf{N}). \tag{2.3-2}$$

If the scattering amplitude from the point electron along the direction $\mathbf{n}_1$ is represented as $Ae^{i(ks-\omega t)}$, where $k = 2\pi/\lambda$ and s is a distance coordinate along the scattering direction $\mathbf{n}_1$, then the scattering amplitude along that direction from the element dv will be $\rho(\mathbf{r})dv$ times as strong (since it will be proportional to the amount of charge in that element) and out of phase by an amount given by (2.3-2). The ratio of the amplitude of the radiation scattered by the element dv to that scattered by a point electron at the origin, which we shall call df, will then be

$$df = \frac{Ae^{i(ks-\omega t)+i\phi_r}\rho(\mathbf{r})\,dv}{Ae^{i(ks-\omega t)}} = \rho(\mathbf{r})e^{(2\pi i/\lambda)(\mathbf{r}\cdot\mathbf{N})}\,dv. \tag{2.3-3}$$

If we integrate over all space, we shall find the ratio f of the scattered amplitude from the whole atom to that for a point electron at the lattice point. Thus,

$$f = \int_v \rho(\mathbf{r})e^{(2\pi i/\lambda)(\mathbf{r} \cdot \mathbf{N})} \, dv.$$ (2.3-4)

However, according to (2.2-4)

$$\frac{2\pi}{\lambda}(\mathbf{r} \cdot \mathbf{N}) = \frac{2\pi}{\lambda} Nr \cos \theta' = \frac{4\pi}{\lambda} r \sin \theta \cos \theta' = \mu r \cos \theta',$$ (2.3-5)

where $$\mu = \frac{4\pi}{\lambda} \sin \theta.$$ (2.3-6)

If the charge density of the atom is spherically symmetric, and thus a function of r only, $\rho(\mathbf{r}) = \rho(r)$ and

$$f = \int_0^\infty \int_0^\pi \int_0^{2\pi} \rho(r)e^{i\mu r \cos \theta'} r^2 \sin \theta' \, dr \, d\theta' \, d\phi.$$ (2.3-7)

It is possible to evaluate the angular parts of this integral, integrating first over ϕ, then over θ' (letting $x = \cos \theta'$, $dx = -\sin \theta' d\theta'$), finally expressing the exponential factors as trigonometric functions, obtaining

$$f(\mu) = \int_0^\infty 4\pi r^2 \rho(r) \frac{\sin \mu r}{\mu r} \, dr.$$ (2.3-8)

This quantity, the ratio of the amplitude scattered by the actual atom to that scattered by a point electron on the lattice point, is called the atomic scattering factor. As $\theta \to 0$, $\mu \to 0$, and $\sin \mu r/\mu r \to 1$, whereby

$$\lim_{\mu \to 0} f(\mu) = \int_0^\infty 4\pi r^2 \rho(r) \, dr = \int_v \rho \, dv = Z.$$ (2.3-9)

The values of $\rho(r)$ must be obtained by evaluating, quantum mechanically, the "wave functions" of the atoms in the crystal. We shall see in a later chapter precisely what is involved in this process. In practice, the wave functions for free atoms are often used in scattering factor calculations to obtain approximate results for expected intensities, rather than the actual modified wave functions which are appropriate for the atoms when present in a crystal lattice.

2·4   THE GEOMETRICAL STRUCTURE FACTOR

Up to this point we have assumed that we are dealing with unit cells having atoms only at the corners (that is, primitive unit cells). If we wish to predict the characteristics of radiation diffracted from crystals having complex unit cells containing more than one atom, such as the cubic cells for the b.c.c. and f.c.c. structures, we must account for the interaction of beams which are diffracted by the various atoms within the unit cell. If, for the $(h'k'l')$ reflection, we denote the ratio of the amplitude of the radiation scattered by the entire unit cell to that scattered by a point electron at the origin by $F(h'k'l')$, then

$$F(h'k'l') = \sum_i f_i e^{i\phi_i}. \tag{2.4-1}$$

Here f_i is the atomic scattering factor for the ith atom of the unit cell and ϕ_i refers to the phase difference between radiation scattered at the origin and radiation scattered from the ith atom of the unit cell. The sum is taken over all atoms belonging to a unit cell. The phase difference ϕ_i is given by (2.3-2), whereby (2.4-1) may be written

$$F(h'k'l') = \sum_i f_i e^{(2\pi i/\lambda)(\mathbf{r}_i \cdot \mathbf{N})}, \tag{2.4-2}$$

where $\mathbf{r}_i$ is a vector from the origin to the ith atom of the unit cell. If the fractional positional coordinates of the ith atom, as defined in Section 1.4, are (x_i, y_i, z_i), then $\mathbf{r}_i$ can be written

$$\mathbf{r}_i = x_i \mathbf{a} + y_i \mathbf{b} + z_i \mathbf{c}. \tag{2.4-3}$$

However, according to Equation (2.2-6), $\mathbf{a} \cdot \mathbf{N} = h'\lambda$, etc.; substituting (2.4-3) into (2.4-2) and using this result, we find

$$\mathbf{r}_i \cdot \mathbf{N} = \lambda(h'x_i + k'y_i + l'z_i), \tag{2.4-4}$$

whereby $\qquad\qquad F(h'k'l') = \sum_i f_i e^{2\pi i(h'x_i + k'y_i + l'z_i)}. \tag{2.4-5}$

When all atoms of the crystal are *identical*, this can be written in a particularly simple form, for then all the f_i have the same value f. Equation (2.4-5) then becomes

$$F(h'k'l') = fS \tag{2.4-6}$$

where $\qquad\qquad S = \sum_i e^{2\pi i(h'x_i + k'y_i + l'z_i)}. \tag{2.4-7}$

The total scattered amplitude is thus given by the product of the atomic scattering factor and a factor S which is governed by the geometrical arrangement of atoms within the unit cell and which is called the *geometrical structure factor*. For crystals where all atoms are not identical, no such separation is possible, and the more general form (2.4-5) must be used.

The *intensity* of the diffracted beam is proportional to the square of the amplitude F, or, more accurately, since F is complex, to the square of the absolute value of F, which is F^*F, where F^* is the complex conjugate of F. If $F(h'k'l')$ as given by (2.4-5) is written as the complex number $\alpha + i\beta$, then $F^* = \alpha - i\beta$ and

$$|F|^2 = F^*F = \alpha^2 + \beta^2, \tag{2.4-8}$$

where $$\alpha = \mathrm{Re}(F) = \sum_i [f_i \cos 2\pi(h'x_i + k'y_i + l'z_i)], \qquad (2.4\text{-}9)$$

and $$\beta = \mathrm{Im}(F) = \sum_i [f_i \sin 2\pi(h'x_i + k'y_i + l'z_i)]. \qquad (2.4\text{-}10)$$

As an example, let us discuss the case of a b.c.c. crystal in which all atoms are identical. There are two atoms in the cubic unit cell in this structure, one corner atom and one body-center atom, and we may proceed by arbitrarily assigning the corner atom whose coordinates (x_i, y_i, z_i) are $(0,0,0)$ to the unit cell and excluding the other corner atoms from consideration; the other atom in the cell is the body-center atom which belongs exclusively to the unit cell and whose cell coordinates are $(\frac{1}{2}, \frac{1}{2}, \frac{1}{2})$. According to this scheme, then, the diffraction amplitude according to (2.4-6) and (2.4-7) is, for the $(h'k'l')$ diffraction direction,

$$F(h'k'l') = f \sum_i e^{2\pi i(h'x_i + k'y_i + l'z_i)} = f[1 + e^{i\pi(h' + k' + l')}]. \qquad (2.4\text{-}11)$$

The way in which *one* corner atom was assigned to the cell and the others left out may be thought to be rather arbitrary; a more reasonable way might be proposed, in which $\frac{1}{8}$ of each corner atom is assigned to the cell, each at its respective cell co-ordinate location. This alternative way of proceeding can be shown to lead to exactly the same expression as that obtained in (2.4-11) for the geometrical structure factor. The verification of this statement is assigned as an exercise. In general, the assignment of the atoms belonging in the unit cell can be made in an arbitrary fashion, as above, as long as the correct number of atoms of each category (face, corner, edge, interior) is assigned to the unit cell.

According to (2.4-11) the geometrical structure factor $1 + \exp[i\pi(h' + k' + l')]$ vanishes for any $(h'k'l')$ reflection for which $h' + k' + l'$ is an *odd* number, since $\exp(n\pi i) = -1$ when n is odd. Therefore, in the b.c.c. structure, certain $(h'k'l')$ reflections which would be present for a simple cubic structure of the same cube edge dimension, are missing. There is, for example, no (100) reflection, although the (200) is present; there is likewise no (111) reflection, although (222) is present. This effect can be understood physically, at least for the case of the (100) reflection, by noting first that for a simple cubic structure the beams reflected from the top and bottom cube faces of the unit cell differ in phase by 2π for the (100) diffraction direction. In the b.c.c. structure having the same cube edge dimension, however, there is another plane of atoms (body centers) located parallel to and halfway between the top and bottom cube face planes of the unit cells, as shown in Figure 2.7. The density of atoms in these intermediate planes is the same as that of the top and bottom cube face planes, and they therefore give rise to diffracted beams of intensity equal to those produced by the top and bottom planes of the unit cell, but out of phase with these beams by π radians. The diffracted beams from the top planes and body-center planes thus interfere destructively in pairs, giving no net diffracted beam for this set of conditions. The (200) reflection is present, however, because for this case the top and bottom planes give rise to beams which are out of phase by 4π. The planes of body centers then contribute beams which differ by 2π in phase from the reflections from top and bottom planes, hence reinforcing them rather than cancelling. The other missing reflections for the b.c.c. structure may be accounted for physically using similar lines of reasoning.

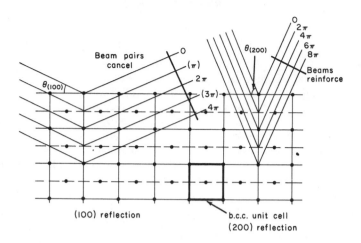

FIGURE 2.7.   Phase relations for the (100) and (200) diffraction angles in a body-centered cubic structure.

2·5   THE RECIPROCAL LATTICE

It is convenient at this point to introduce the concept of the reciprocal lattice. The
reciprocal lattice is a lattice of points which is related in a certain way to the direct
space lattice. The Bragg condition for X-ray diffraction can be expressed in a very
simple way with the aid of the reciprocal lattice, and we shall find in a later chapter
that the wave-mechanical behavior of electrons in periodic crystal lattices is also
most readily understood in terms of the reciprocal lattice of the crystal.

   If $\mathbf{a}$, $\mathbf{b}$, and $\mathbf{c}$ are the *primitive* translation vectors of a direct space lattice, the
primitive translation vectors of the *reciprocal* lattice, $\mathbf{a^*}$, $\mathbf{b^*}$, and $\mathbf{c^*}$ are defined by
the relations

$$\mathbf{a^*} \cdot \mathbf{a} = \mathbf{b^*} \cdot \mathbf{b} = \mathbf{c^*} \cdot \mathbf{c} = 1 \qquad (2.5\text{-}1)$$

and $\qquad \mathbf{a^*} \cdot \mathbf{b} = \mathbf{a^*} \cdot \mathbf{c} = \mathbf{b^*} \cdot \mathbf{c} = \mathbf{b^*} \cdot \mathbf{a} = \mathbf{c^*} \cdot \mathbf{a} = \mathbf{c^*} \cdot \mathbf{b} = 0. \qquad (2.5\text{-}2)$

Since from (2.5-2) $\mathbf{a^*} \cdot \mathbf{b} = \mathbf{a^*} \cdot \mathbf{c} = 0$, the vector $\mathbf{a^*}$ is perpendicular to the plane
determined by $\mathbf{b}$ and $\mathbf{c}$. It is therefore parallel to the vector $\mathbf{b} \times \mathbf{c}$, and can be expressed
as a multiple of that vector, that is,

$$\mathbf{a^*} = A(\mathbf{b} \times \mathbf{c}), \qquad (2.5\text{-}3)$$

where A is a scalar constant. However, from (2.5-1),

$$\mathbf{a^*} \cdot \mathbf{a} = A(\mathbf{b} \times \mathbf{c}) \cdot \mathbf{a} = 1, \qquad (2.5\text{-}4)$$

whereby, solving for A and substituting the value so obtained into (2.5-3),

$$\mathbf{a}^* = \frac{\mathbf{b} \times \mathbf{c}}{\mathbf{a} \cdot \mathbf{b} \times \mathbf{c}}. \tag{2.5-5}$$

In a similar fashion, one can show that

$$\mathbf{b}^* = \frac{\mathbf{c} \times \mathbf{a}}{\mathbf{a} \cdot \mathbf{b} \times \mathbf{c}}$$

$$\mathbf{c}^* = \frac{\mathbf{a} \times \mathbf{b}}{\mathbf{a} \cdot \mathbf{b} \times \mathbf{c}}$$

Using the same general approach, the reverse transformations

$$\mathbf{a} = \frac{\mathbf{b}^* \times \mathbf{c}^*}{\mathbf{a}^* \cdot \mathbf{b}^* \times \mathbf{c}^*}, \tag{2.5-6}$$

etc., can be obtained.

Suppose, as an example, we wish to determine the reciprocal lattice for a f.c.c. structure. We must start with the primitive unit cell of Figure 1.5(a) for which the primitive basis vectors $\mathbf{a}$, $\mathbf{b}$, and $\mathbf{c}$ can be written

$$\mathbf{a} = \frac{a}{2} (\mathbf{i}_x + \mathbf{i}_y)$$

$$\mathbf{b} = \frac{a}{2} (\mathbf{i}_x + \mathbf{i}_z) \tag{2.5-7}$$

$$\mathbf{c} = \frac{a}{2} (\mathbf{i}_y + \mathbf{i}_z),$$

where a is the edge of the cubic cell, and where the x-coordinate axis is taken to be along $\mathbf{a}$, the y-axis along $\mathbf{b}$ and the z-axis along $\mathbf{c}$. Then

$$\mathbf{a}^* = \frac{\mathbf{b} \times \mathbf{c}}{\mathbf{a} \cdot \mathbf{b} \times \mathbf{c}} = \frac{\frac{a^2}{4} (\mathbf{i}_x + \mathbf{i}_z) \times (\mathbf{i}_y + \mathbf{i}_z)}{\frac{a^3}{8} (\mathbf{i}_x + \mathbf{i}_y) \cdot [(\mathbf{i}_x + \mathbf{i}_z) \times (\mathbf{i}_y + \mathbf{i}_z)]} = \frac{1}{a} (\mathbf{i}_x + \mathbf{i}_y - \mathbf{i}_z);$$

similarly, $$\mathbf{b}^* = \frac{1}{a} (\mathbf{i}_x - \mathbf{i}_y + \mathbf{i}_z), \qquad \mathbf{c}^* = \frac{1}{a} (-\mathbf{i}_x + \mathbf{i}_y + \mathbf{i}_z). \tag{2.5-8}$$

It is clear that these vectors have the cube diagonal directions, as do the primitive vectors of the b.c.c. structure, as shown in Figure 1.5(b). In addition, it should be noted that the lattice spacing of the reciprocal lattice is proportional to $1/a$. More explicitly, it is evident that any vector $\mathbf{r}^*$ connecting two lattice points of the reciprocal lattice must have the form

$$\mathbf{r}^* = l\mathbf{a}^* + m\mathbf{b}^* + n\mathbf{c}^*, \tag{2.5-9}$$

TABLE 2.1.

Lattice Points for the f.c.c. Reciprocal Lattice

(All points are of the form $[(l+m-n),\ (l-m+n),\ (-l+m+n)]$)

	$l = -2$	$l = -1$	$l = 0$	$l = 1$	$l = 2$
$m = -2$	$(-n-4, n, n)$ $-2,-2,-2$	$(-n-3, n+1, n-1)$ $-2,0,-2$	$(-n-2, n+2, n-2)$ $-2,2,-2$		
$m = -1$	$(-n-3, n-1, n+1)$ $-2,-2,0$	$(-n-2, n, n)$ $-2,0,0$ $-1,-1,-1$ $0,-2,-2$	$(-n-1, n+1, n-1)$ $-1,1,-1$ $-2,2,0$ $0,0,-2$	$(-n, n+2, n-2)$ $0,2,-2$	
$m = 0$	$(-n-2, n-2, n+2)$ $-2,-2,2$	$(-n-1, n-1, n+1)$ $-1,-1,1$ $-2,0,2$ $0,-2,0$	$(-n, n, n)$ $0,0,0$ $-1,1,1$ $-2,2,2$ $1,-1,-1$ $2,-2,-2$	$(1-n, n+1, n-1)$ $1,1,-1$ $0,2,0$ $2,0,-2$	$(2-n, n+2, n-2)$ $2,2,-2$
$m = 1$		$(-n, n-2, n+2)$ $0,-2,2$	$(1-n, n-1, n+1)$ $1,-1,1$ $0,0,2$ $2,-2,0$	$(2-n, n, n)$ $2,0,0$ $1,1,1$ $0,2,2$	$(3-n, n+1, n-1)$ $2,2,0$
$m = 2$			$(2-n, n-2, n+2)$ $2,-2,2$	$(3-n, n-1, n+1)$ $2,0,2$	$(4-n, n, n)$ $2,2,2$

where l, m, and n are integers; substituting the expressions for $\mathbf{a^*}$, $\mathbf{b^*}$, and $\mathbf{c^*}$ from (2.5-8) into (2.5-9), it is obvious that the points of the reciprocal lattice must have (x,y,z) coordinates of the form

$$(x,y,z) = \frac{1}{a}\left[(l + m - n), \quad (l - m + n), \quad (-l + m + n)\right], \quad (2.5\text{-}10)$$

where l, m, and n may take on integer values. Table 2.1 shows a tabulation of these possible lattice points; the various boxes of the table correspond to specific values of l and m; this restricts the form of coordinates listed in each box to vary with n only according to (2.5-10) with the pertinent values of l and m inserted. In the tabulation given here any point whose x, y, or z coordinate exceeds 2 (apart from the constant factor $1/a$, which is dropped in listing the entries in the table) is discarded; the points which are retained are then sufficient to depict 8 unit cells of the reciprocal lattice. When these points are plotted, inserting the constant factor $1/a$, it is clear, as it should be from the form of the reciprocal lattice primitive vectors, that the resulting reciprocal lattice is a b.c.c. structure whose cube edge dimension is $2/a$. In a similar fashion it may be shown that the reciprocal lattice for a b.c.c. direct lattice is a f.c.c. structure.

It is a property of the reciprocal lattice that a vector $\mathbf{r^*} = h'\mathbf{a^*} + k'\mathbf{b^*} + l'\mathbf{c^*}$ from the origin to any lattice point of the reciprocal lattice is normal to the (hkl) plane of the direct lattice. Here the triad (hkl) is simply the triad $(h'k'l')$ divided through by the largest common factor n among $(h'k'l')$, just as in Section 2.2. In other words, $h'/h = k'/k = l'/l = n$. To prove this result, one first notes from Figure 2.8 that vector AC, equal to $-(\mathbf{a}/h) + (\mathbf{c}/l)$, lies in the (hkl) plane, as does vector AB, which is equal to $-(\mathbf{a}/h) + (\mathbf{b}/k)$. But, using (2.5-1) and (2.5-2),

$$\mathbf{r^*} \cdot \left[-(\mathbf{a}/h) + (\mathbf{c}/l)\right] = (h'\mathbf{a^*} + k'\mathbf{b^*} + l'\mathbf{c^*}) \cdot \left[-(\mathbf{a}/h) + (\mathbf{c}/l)\right] = -\frac{h'}{h} + \frac{l'}{l} = 0,$$

$$(2.5\text{-}11)$$

recalling that $h'/h = l'/l = n$, according to (2.2-6). In like fashion,

$$\mathbf{r^*} \cdot \left[-(\mathbf{a}/h) + (\mathbf{b}/k)\right] = (h'\mathbf{a^*} + k'\mathbf{b^*} + l'\mathbf{c^*}) \cdot \left[-(\mathbf{a}/h) + (\mathbf{b}/k)\right] = -\frac{h'}{h} + \frac{k'}{k} = 0.$$

$$(2.5\text{-}12)$$

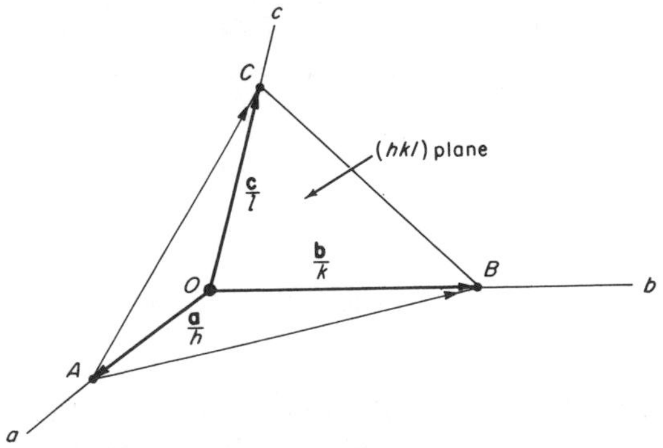

FIGURE 2.8. Vector geometry in the direct lattice for the calculations of Section 2.5.

The vector $\mathbf{r}^*$ is thus perpendicular to two linearly independent vectors, AC and AB, both of which lie in the (hkl) plane. It must thus be perpendicular to the plane itself.

Also, if $\mathbf{n}$ is a unit normal vector to the (hkl) plane of the direct lattice, we may write $\mathbf{n} = \mathbf{r}^*/r^*$, since $\mathbf{r}^*$ is known to be a vector normal to (hkl). However, if d is the distance between adjacent (hkl) planes, from (1.5-2) it must be true that

$$d = \frac{\mathbf{a} \cdot \mathbf{n}}{h} = \frac{\mathbf{a} \cdot \mathbf{r}^*}{hr^*} = \frac{\mathbf{a} \cdot (h'\mathbf{a}^* + k'\mathbf{b}^* + l'\mathbf{c}^*)}{hr^*} = \frac{n}{r^*}, \qquad (2.5\text{-}13)$$

whereby the magnitude of $\mathbf{r}^*$ is simply given by

$$r^* = n/d, \qquad (2.5\text{-}14)$$

where n is an integer defined as in equation (2.2-6).

2·6 THE BRAGG CONDITION IN TERMS OF THE RECIPROCAL LATTICE

The Bragg condition may be expressed as a relation between vectors in the reciprocal lattice. Referring to Figure 2.9, the vector AO is a vector whose length is $1/\lambda$, drawn in the direction of the incident X-ray beam, and ending at the origin of the reciprocal lattice. Note that the tail of the vector AO does not necessarily have to rest on a

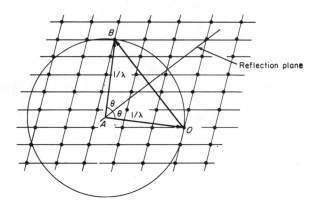

FIGURE 2.9. Vector geometry of Bragg reflection in the *reciprocal* lattice.

lattice point of the reciprocal lattice. A sphere of radius $1/\lambda$ is constructed about point A as center. Suppose now that this sphere intersects some point (h',k',l') of the reciprocal lattice at B. The vector OB then represents a vector connecting the origin of the reciprocal lattice and a point (h',k',l') of that lattice; as such it must be normal to the (hkl) plane of the direct lattice, and it must also be of length n/d, where n is the largest integral factor common to the three numbers (h',k',l'). But from the trigonometrical relations in Figure 2.9, the length of OB can also be readily expressed as $2 \sin \theta/\lambda$. Equating these two expressions for the length of vector OB, we find

$$n\lambda = 2d \sin \theta,$$

and the Bragg condition is satisfied. Referring again to Figure 2.9, it is clear that vector OB represents a normal to the reflecting planes (hkl), and that vector AB is a vector in the direction of the diffracted beam. The latter statement can be understood more easily by translating vector AB parallel with itself until its tail rests upon point O, when it will be seen that AO, OB, and AB are in the familiar relation of incident beam direction, scattering normal and diffracted beam direction. It is evident from this geometrical construction that the Bragg condition will be satisfied for a given wavelength *for each intersection of the surface of a sphere of radius* $1/\lambda$ *drawn about A with a point of the reciprocal lattice*. The appropriate Bragg angle will be given in each case by the angle between the vector AO and a plane normal to OB. When the Bragg condition is thus satisfied, the vectors AO and AB form an isosceles triangle with a lattice vector OB of the reciprocal lattice. From this disposition of the vectors, as illustrated in Figure 2.9, it can be seen that AB must be the vector sum of AO and OB, and the vector geometry of Figure 2.9 permits us to express the Bragg reflection condition in a particularly simple way.

To accomplish this, it is customary to imagine all the vectors of Figure 2.9 to be multiplied by a constant scale factor of 2π, as represented by Figure 2.10. In this figure, then, the vector **G** is simply 2π times the vector OB of Figure 2.9 and the vector **k** is 2π times the vector AO of Figure 2.9. Again the disposition of the vectors is such that vector $A'B'$ of Figure 2.10 must be the vector sum of **k** and **G**. Since the magnitude of this vector and the magnitude of the incident beam vector **k** must be equal whenever

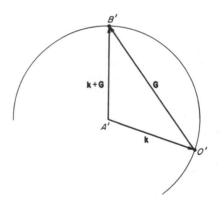

FIGURE 2.10. Vector diagram of Figure 2.9 expanded by a scale factor 2π and relabeled. [The notation follows that used by C. Kittel, *Introduction to Solid State Physics*, John Wiley & Sons, New York (1956).]

the Bragg condition is satisfied, the Bragg condition must imply that

$$(\mathbf{k} + \mathbf{G})^2 = (\mathbf{k} + \mathbf{G}) \cdot (\mathbf{k} + \mathbf{G}) = k^2, \qquad (2.6\text{-}1)$$

or, expanding the dot product and simplifying,

$$2\mathbf{k} \cdot \mathbf{G} + G^2 = 0, \qquad (2.6\text{-}2)$$

where $\mathbf{G}$ is 2π times a vector from the origin to a lattice point of the reciprocal lattice, and where $\mathbf{k}$ is a vector of magnitude $2\pi/\lambda$ along the direction of the incident X-ray beam. Equation (2.6-2) is the vector form of the Bragg equation.[4]

EXERCISES

1. In a cubic crystal, using X-radiation of wavelength 1.5 Å, a first order (100) reflection is observed at a glancing angle of 18°. What is the distance between the (100) planes of the crystal?

2. For the hydrogen atom in its lowest energy state the wave function is given by

$$\psi_{1s} = \frac{e^{-r/a_0}}{a_0^{3/2}\sqrt{\pi}}$$

where a_0 is the radius of the first Bohr orbit, $h^2/4\pi^2 m e^2 = 0.53$ Å. The electron density for such an atom is given by $\psi_{1s}^* \psi_{1s}$, whereby

$$\rho(r) = \frac{e^{-2r/a_0}}{\pi a_0^3}.$$

Using this electron density distribution function, compute the atomic scattering factor for a hypothetical crystal made up of such atoms, and plot it as a function of μ $(= 4\pi \sin \theta/\lambda)$.

3. Show that for the b.c.c. structure in which all atoms are identical, a scheme of assigning 1/8 of each corner atom at its respective cell coordinate location to the unit cell plus the body center atom in the central position leads to the same result for the geometrical structure factor for this lattice as that derived in Equation (2.4-11).

4. Find the diffraction amplitude $F(h'k'l')$ for the $(h'k'l')$ reflection of a crystal having the CsCl structure, as shown in Figure 1.7(b). Would you expect the (100) reflection to be present for this crystal? Explain.

5. Find the geometrical structure factor for the f.c.c. structure (all atoms identical). Which of the following X-ray reflections would be present, and which would be missing for such a crystal: (100), (110), (111), (200), (220), (222), (211), (221), (123)?

6. Find the geometrical structure factor for the diamond lattice. Express your result as the product of the geometrical structure factor for the f.c.c. structure times another factor. Which of the X-ray reflections mentioned in connection with Problem 5 would be present

---

[4] This construction was originated by Ewald [P. P. Ewald, *Zeitschrift für Kristallographie 56*, 129 (1921)] and is sometimes referred to as the *Ewald Construction*.

and which would be missing for this structure? *Hint:* Assign atoms to a cubic unit cell in accordance with the diagram of Figure 1.9.

7. Show that the reciprocal lattice for a simple cubic structure is another simple cubic, thus that the simple cubic structure is self-reciprocal.

8. Show that the reciprocal lattice for a b.c.c. lattice is a f.c.c. structure. Make a table of lattice points and plot them out to form a diagram of your results.

GENERAL REFERENCES

W. H. Bragg, *An Introduction to Crystal Analysis*, G. Bell & Sons, London (1928).

W. L. Bragg, *The Crystalline State*, Vol. I, G. Bell & Sons, London (1955).

R. B. Leighton, *Principles of Modern Physics*, McGraw-Hill Book Co., Inc., New York (1959).

A. Taylor, *X-Ray Metallography*, John Wiley & Sons, New York (1961).

CHAPTER 3

DYNAMICS OF CRYSTAL
LATTICES

3·1 ELASTIC VIBRATIONS OF CONTINUOUS MEDIA

We now must investigate the characteristics of elastic vibrational motion of crystal
lattices. We shall find that there are important differences between elastic waves in
lattice structures composed of discrete atoms and waves in completely continuous and
homogeneous elastic media. In order to be able to appreciate these differences clearly
we shall first briefly review some of the major features of elastic waves in homogeneous,
isotropic, linear elastic substances. We shall discuss only one-dimensional wave
motion in detail, but we shall try to indicate wherever necessary how the results may
be extended to describe two- and three-dimensional systems.

Consider an element of a homogeneous, isotropic elastic substance of length
Δx and of uniform cross-sectional area $\Delta y \Delta z$. In the absence of elastic strain, this
element extends from x to $x + \Delta x$ along the x-axis, as shown in Figure 3.1. In the

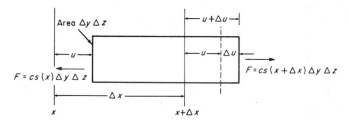

FIGURE 3.1. Displacement and extensional strain of a volume element
which originally extended from x to $x + \Delta x$. In the strained condition
the element extends from $x + u$ to $x + \Delta x + u + \Delta u$.

presence of a stress, an elastic strain is produced, the equilibrium position of the
left-hand end of the element being displaced a distance u along the x-axis, and the
right-hand end of the element being displaced a distance $u + \Delta u$. If the elastic medium
is a linear one, the applied stress and the resulting strain are related linearly by Hooke's
law. The strain s is defined as the elastic extension of the element per unit length,
that is,

$$s(x) = \lim_{\Delta x \to 0} \frac{\Delta u}{\Delta x} = \frac{\partial u}{\partial x}, \qquad (3.1\text{-}1)$$

whereas the stress is the force per unit area acting on the element. According to

Hooke's law, then, if $F(x)$ is the total force acting at point x,

$$F(x) = cs(x)\Delta y \, \Delta z, \tag{3.1-2}$$

where c is the elastic modulus of the material. Likewise, at the point $x + \Delta x$,

$$F(x + \Delta x) = cs(x + \Delta x)\Delta y \, \Delta z. \tag{3.1-3}$$

The total force on the element must be related to the acceleration of the element by Newton's law, whereby

$$F(x + \Delta x) - F(x) = c[s(x + \Delta x) - s(x)]\Delta y \, \Delta z = ma, \tag{3.1-4}$$

where m is the mass of the element and a its acceleration. Expanding $s(x + \Delta x)$ in a Taylor's series about the point x and neglecting terms of second order and higher in Δx, which will in any case vanish in the limit as $\Delta x \to 0$, we find

$$c \frac{\partial s}{\partial x} \Delta x \, \Delta y \, \Delta z = ma. \tag{3.1-5}$$

But the mass of the element is its density ρ times the volume $\Delta x \, \Delta y \, \Delta z$, the acceleration is $\partial^2 u/\partial t^2$, and the strain, according to the definition (3.1-1) is $\partial u/\partial x$. Making these substitutions, (3.1-5) becomes

$$\frac{\partial^2 u}{\partial x^2} = \frac{\rho}{c} \frac{\partial^2 u}{\partial t^2}. \tag{3.1-6}$$

This is the *wave equation* for the amplitude of elastic waves in a linear homogeneous medium as a function of x and t, where the elastic strain is confined to the x-direction. Solutions to this equation may be written in the form

$$u(x,t) = Ae^{i(\omega t - 2\pi x/\lambda)} = Ae^{i(\omega t - kx)}, \tag{3.1-7}$$

where
$$k = 2\pi/\lambda. \tag{3.1-8}$$

The solution (3.1-7) represents a sinusoidal disturbance of frequency ω and wavelength λ which is propagated along the positive x-axis. It is usually more convenient to express the results of calculations involving harmonic vibrations in terms of the *propagation constant* k than in terms of λ itself. The propagation direction of the wave is reversed by changing the sign of k. If the solution (3.1-7) is differentiated and substituted back into the wave equation (3.1-6), it will be seen that in order for (3.1-6) to be satisfied, ω and k must be related by

$$\omega = k\sqrt{c/\rho}. \quad \text{Relation} \tag{3.1-9}$$

The phase velocity of a wave is the rate of advance of a point of constant phase along the propagation direction of the wave. To determine the phase velocity, we may examine the motion of a point of constant phase, and for convenience let us choose that point to be the point of *zero* phase. Since the phase angle, according to (3.1-7) is $\omega t - kx$, the equation of motion of the zero-phase point is

$$\omega t - kx = 0$$

or,

$$x = \frac{\omega}{k} t = v_p t. \qquad (3.1\text{-}10)$$

The phase point is seen to advance at constant phase velocity v_p equal to ω/k. According to (3.1-9) the phase velocity for these waves may be expressed as

$$v_p = \omega/k = \sqrt{c/\rho}, \qquad (3.1\text{-}11)$$

whence it is clear that the wave equation (3.1-6) can be written in the form

$$\frac{\partial^2 u}{\partial x^2} = \frac{1}{v_p^2} \frac{\partial^2 u}{\partial t^2}. \qquad (3.1\text{-}12)$$

For this particular case the phase velocity is *independent* of frequency and wavelength, and depends only upon the physical constants c and ρ associated with the elastic medium.

3·2   GROUP VELOCITY OF HARMONIC WAVE TRAINS

Let us now investigate what happens when two wave trains, one of frequency ω and propagation constant k, the other of slightly different frequency $\omega + d\omega$ and propagation constant $k + dk$, are superposed. We shall assume for simplicity that the maximum amplitudes of the two wave trains are the same, in which case the two amplitudes may be written

$$u_1 = A \cos (\omega t - kx)$$
$$u_2 = A \cos [(\omega + d\omega)t - (k + dk)x]. \qquad (3.2\text{-}1)$$

The amplitude of the superposition is then

$$u_1 + u_2 = A[\cos (\omega t - kx) + \cos [(\omega + d\omega)t - (k + dk)x]] = A[\cos \alpha + \cos \beta].$$
$$(3.2\text{-}2)$$

But from elementary trigonometry,

$$\cos \alpha + \cos \beta = 2 \cos \tfrac{1}{2}(\alpha - \beta)\cos \tfrac{1}{2}(\alpha + \beta), \qquad (3.2\text{-}3)$$

and from (3.2-2) we have

$$\alpha + \beta = 2\omega t - 2kx + t d\omega - x dk \cong 2(\omega t - kx), \qquad (3.2\text{-}4)$$

and

$$\alpha - \beta = x dk - t d\omega. \qquad (3.2\text{-}5)$$

In (3.2-4) the differential terms $t\,d\omega$ and $x\,dk$ may be neglected as small in comparison with the other terms; in (3.2-5) there are no "large" terms, and the differential terms are important. Substituting (3.2-4) and (3.2-5) into (3.2-2), we find

$$u_1 + u_2 = 2A \cos (\omega t - kx) \cos (\tfrac{1}{2}x\,dk - \tfrac{1}{2}t\,d\omega). \tag{3.2-6}$$

This superposition represents a wave characterized by the original values of ω and k multiplied by a sinusoidal envelope of much longer wavelength $4\pi/dk$, called an "envelope of beats," as shown in Figure 3.2. The motion of this envelope can be

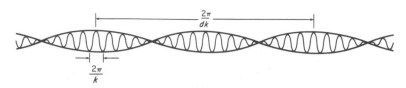

FIGURE 3.2.   Wave train formed by superposing two sinusoidal waves of frequencies ω and $\omega + d\omega$ and propagation constants k and $k + dk$.

examined, and in particular the phase velocity of the *envelope* (corresponding to the velocity of the groups or "beats" along the propagation direction) determined by the same procedure used for a simple harmonic wave in Section 3.1. The equation of motion of the zero-phase point of the envelope is thus

$$x\,dk - t\,d\omega = 0,$$

or,
$$x = \frac{d\omega}{dk}\,t = v_g t. \tag{3.2-7}$$

The envelope, or the groups of waves, are thus seen to move with a velocity $d\omega/dk$; this velocity is termed the *group* velocity v_g associated with the waves. It can be shown that the velocity with which the wave transmits energy along the propagation direction is this group velocity.[1] It can readily be seen that this is physically reasonable; for example, in the wave train of Figure 3.2, no energy can ever flow past a node, since the medium at the nodal point is absolutely motionless. The energy must thus be transmitted with the velocity with which the nodes themselves move, that is, the group velocity. For the case of a standing wave, the group velocity is zero, and it is clear that there is no net energy flow in this case.

For the elastic waves in a homogeneous medium which were discussed in Section 3.1, the phase velocity and group velocity are exactly the same, according to equation (3.1-9). This is true because ω is a *linear* function of k, whereby ω/k and $d\omega/dk$ are equal. The same result is true for electromagnetic waves in vacuo, because there again the velocity is a constant independent of frequency and wavelength. The phase velocity and group velocity may *differ* only when the phase velocity ω/k is a function

[1] L. Brillouin, *Wave Propagation in Periodic Structures*, Dover Publications, Inc., New York (1953).

of the frequency ω, thus only when ω itself is *other than a linear function* of k. This situation arises, for example, when light passes through a medium where the index of refraction (and hence the phase velocity) is a function of frequency; the phenomenon is called *dispersion*, and such a medium is referred to as a *dispersive* medium. As we shall soon see, it also arises for the case of elastic waves in a medium which is composed of discrete atoms bound together by Hooke's law forces, especially when the wavelengths involved are not much greater than the interatomic distances.

3·3   WAVE MOTION ON A ONE-DIMENSIONAL ATOMIC LATTICE

We now wish to discuss the longitudinal vibrational motion of a one-dimensional chain of identical atoms of mass m which are bound to one another by linear forces, as shown in Figure 3.3. At equilibrium the atoms will be situated on equally spaced equilibrium sites, but when vibrational motion is excited, they will execute periodic motions about these equilibrium positions, the actual displacement of the nth atom

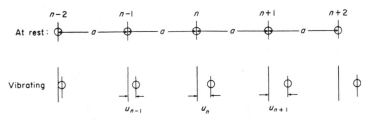

FIGURE 3.3.   Geometry of the monatomic linear lattice of Section 3.3.

from its equilibrium position being given by the amplitude u_n. It is assumed that the forces between neighboring atoms are Hooke's law forces, as if the atoms were bound together by ideal springs, and it is further assumed that the only significant force interactions between atoms are direct nearest-neighbor interactions. The direct influence of atoms $n + 2$, $n + 3$, $n + 4 \cdots$ upon atom n is thus regarded as negligible, although the behavior of atom $n + 2$, for example, may indirectly influence atom n insofar as it directly affects atom $n + 1$, which in turn acts upon n. Under these assumptions the net force acting upon the nth atom can be written, in terms of the extension of the two "springs" which bind it to atoms $n + 1$ and $n - 1$, as

$$F_n = \beta(u_{n+1} - u_n) - \beta(u_n - u_{n-1})$$

$$= \beta(u_{n+1} + u_{n-1} - 2u_n), \tag{3.3-1}$$

where β is the Hooke's law constant which expresses the proportionality between the force required to produce an atomic displacement and the displacement itself.

According to Newton's second law, this force can be expressed as the product of the mass m of the nth atom times its acceleration, whereby we obtain the equation of motion

$$m \frac{d^2 u_n}{dt^2} = \beta(u_{n+1} + u_{n-1} - 2u_n). \tag{3.3-2}$$

We are seeking periodic solutions to this equation, which we may expect to have the form

$$u_n = A e^{i(\omega t - kna)}, \tag{3.3-3}$$

in analogy with (3.1-7). In (3.3-3), however, we must represent the coefficient of k in the exponent as the *x-coordinate of the nth atom* of the chain, i.e., na. According to this scheme, we should then expect that

$$u_{n+1} = A e^{i[\omega t - k(n+1)a]}$$

and

$$u_{n-1} = A e^{i[\omega t - k(n-1)a]}. \tag{3.3-4}$$

If the solution (3.3-3) is now differentiated twice with respect to time, and, along with (3.3-4), substituted back into the equation of motion (3.3-2), the result is

$$-m\omega^2 = \beta(e^{ika} + e^{-ika} - 2)$$

$$= 2\beta(\cos ka - 1). \tag{3.3-5}$$

However, since

$$1 - \cos \theta = 2 \sin^2 \tfrac{1}{2}\theta, \tag{3.3-6}$$

this may be expressed in the form

$$\omega = \sqrt{4\beta/m} \, |\sin \tfrac{1}{2}ka|. \tag{3.3-7}$$

The absolute value signs are needed in (3.3-7) only because we must regard the frequency as an essentially positive quantity, regardless of whether k is positive or negative, that is, regardless of whether the wave is propagated to the right or to the left along the chain.

We see, thus, that there are solutions of the form (3.3-3) provided that ω is related to k by (3.3-7). It is evident that in this case ω is not a linear function of k, so that the medium we are dealing with is a dispersive one; a relation such as (3.3-7) which gives ω as a function of k is called a *dispersion relation*. Figure 3.4 shows a plot of ω vs. k, as given by (3.3-7). From this (and from (3.3-7)) it will be noted that if we restrict ourselves to values of k which are much less than π/a, and thus to values of λ *which are much greater than twice the interatomic distance* a, ω is approximately linear with k, since then $\sin \tfrac{1}{2}ka \cong \tfrac{1}{2}ka$ and

$$\omega \cong ka\sqrt{\beta/m}. \tag{3.3-8}$$

In this long-wavelength limit the phase velocity will be essentially constant, since

$$v_p = \omega/k = a\sqrt{\beta/m} = v_0. \tag{3.3-9}$$

In this same limit the group velocity is seen to be

$$v_g = d\omega/dk = a\sqrt{\beta/m} = v_0,$$

(3.3-10)

thus constant and equal to the phase velocity under these conditions. For very long wavelengths, then, the dispersion effects are negligible and the medium acts like a

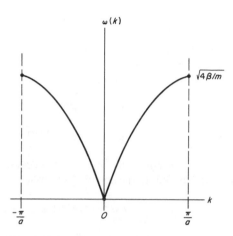

FIGURE 3.4. Dispersion relation for the mon-atomic linear lattice of Section 3.3.

continuous and homogeneous elastic medium. This is, of course, most reasonable from a physical point of view, since for such long wavelengths the "atomic" nature of the chain is of little importance insofar as the dynamical behavior of the system is concerned. As k increases, however, the dispersion effects become more important, and ω no longer varies linearly with k. Under these conditions, we find from (3.3-7) that

$$v_p = \omega/k = v_0 \left| \frac{\sin \frac{1}{2}ka}{\frac{1}{2}ka} \right|$$

(3.3-11)

and

$$v_g = d\omega/dk = v_0 \left| \cos \frac{1}{2}ka \right|,$$

(3.3-12)

where $v_0 = a\sqrt{\beta/m}$ is the long wavelength limit of both v_p and v_g, as shown by (3.3-9) and (3.3-10). Figures 3.5 and 3.6 show plots of v_p and v_g as functions of k.

It will be observed from Figure 3.6 that $v_g \to 0$ as $k \to \pi/a$ and thus as $\lambda \to 2a$; in this case the phase of vibration of neighboring atoms differ by π radians, and the character of the motion is simply a standing wave. This condition also corresponds to Bragg reflection of the elastic vibrations from successive atoms of the crystal; for this geometry the Bragg angle θ of the Bragg equation (2.1-1) is $\pi/2$, corresponding to normal incidence, in which case the Bragg equation reduces to $\lambda = 2d$ for the first order of reflection. The actual physical character of the motion is illustrated by Figure 3.7 for the long wavelength case and for the condition $\lambda = 2a$ in which the standing

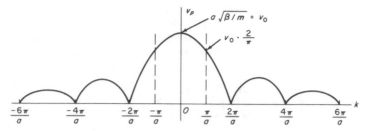

FIGURE 3.5.   Phase velocity as a function of the propagation constant k
for the monatomic linear lattice. [After L. Brillouin, *Wave Propagation
in Periodic Structures*, McGraw-Hill, New York (1946), with permission.]

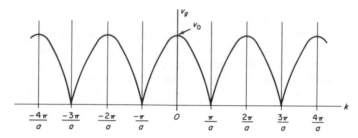

FIGURE 3.6.   Group velocity as a function of the propagation constant k
for the monatomic linear lattice.

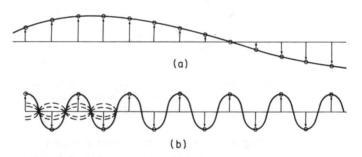

FIGURE 3.7.   Atomic displacements on the monatomic linear lattice (a)
for wavelength much larger than the interatomic spacing, (b) in the limit
where $\lambda = 2a$. The displacements are shown as transverse ones for the
sake of clarity in representation, but it should be remembered that the
actual displacements which are discussed in the text are *longitudinal*.
[After Brillouin.]

waves are set up. In these figures the atomic displacements are represented as transverse displacements for convenience of illustration, although, of course, the actual
atomic motions which are we discussing are assumed to be *longitudinal* ones in which
the actual atomic displacements are along the chain, rather than normal to it.

For any given value of k in the region $-\pi/a < k < \pi/a$ there will be a vibrational motion involving certain possible atomic displacements, as shown by Figure 3.8. However, as also shown in that figure, the *same* set of displacements of atoms as that associated with this wave can *also* be associated with a wave of larger k (thus smaller

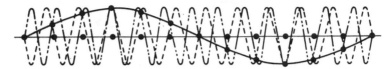

FIGURE 3.8. A single set of atomic displacements represented by several sinusoidal waves of various wavelengths. [From L. Brillouin, *Wave Propagation in Periodic Structures*, McGraw-Hill, New York (1946).]

λ), and with still another of even larger k. In other words, there may be many representations of the same atomic displacement pattern, each involving a different value of k and hence a different value of λ. These different representations can be shown to be associated with different periods of the ω vs. k plot. Each of the possible representations must have the same value of ω; therefore on a plot of ω vs. k showing several periods of the curve, such as Figure 3.9, there will be many possible values

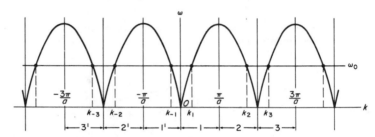

FIGURE 3.9. Several periods of the dispersion relation for the linear monatomic lattice. Note that many values of k may be associated with any given frequency within the allowed range.

of k corresponding to the same ω. In the central region, extending from $-\pi/a$ to π/a there are only two possibilities, a wave propagating to the right with propagation constant k_1, and one propagating to the left with propagation constant $k_{-1} = -k_1$. The regions marked 2 and 2' in the figure, which contain two more possible solutions, k_2 and $k_{-2} = -k_2$, can be superimposed upon the central region by translating region 2 to the left a distance $2\pi/a$ and translating region 2' to the right a distance $2\pi/a$. The character of the possible solutions in region 2 is the same as that for the possible solutions in region 1', except for a difference in k-values of $2\pi/a$; the same can be said for regions 2' and 1. Likewise, the regions 3 and 3', containing two possible solutions, for which $k = k_3$ and $k = k_{-3} = -k_3$, can be superimposed upon the central region $-\pi/a < k < \pi/a$ by translating region 3 to the left a distance $2\pi/a$ and 3' to the right a distance $2\pi/a$; the physical character of the solutions in these regions thus coincides with those in the central region except for differences in k-values of $2\pi/a$. Similar relations may be observed for higher zones. The physical fact of importance is that any possible arrangement of atomic positions which is compatible with a sinusoidal solution of *any* wavelength, however short, can be represented as, or *reduced* to a sinusoidal solution for which $-\pi/a < k < \pi/a$, and which thus belongs to the central zone. For this reason, it is not necessary in this case to consider vibrations other than those belonging to the central zone $-\pi/a < k < \pi/a$. This central region (1 and 1' in Figure 3.9) is called the *first Brillouin Zone*; the regions 2 and 2', 3 and 3', etc., for each of which the physical character of solutions is the same as that for the first zone except for a translation along the k-axis by a distance equal to an integral multiple of $2\pi/a$, are referred to as the second, third, etc., Brillouin Zones, respectively. We shall have further occasion to use the concept of Brillouin Zones in connection with the quantum mechanical behavior of electrons in periodic lattices.[2]

We must also determine how many possible *normal modes* of vibration (in other words, how many independent solutions of the wave equation which satisfy a given set of boundary conditions) can be associated with a linear chain of atoms. Suppose, for example, we consider a chain which is rigidly fixed at both ends. If the chain, which we shall take to be of length L, were a homogeneous and continuous medium, then any vibration of the form

$$u(x,t) = A \sin k_n x \sin \omega t \tag{3.3-13}$$

with
$$k_n = n\pi/L, \tag{3.3-14}$$

where n is an integer, would satisfy the boundary conditions that the displacement u at the two ends of the chain at $x = 0$ and $x = L$ must vanish, and hence would qualify as a normal mode of vibration of the system. This situation is illustrated in Figure 3.10. If the chain, however, is made up of discrete atoms, then the maximum value

FIGURE 3.10. A set of sinusoidal displacements of a transversely vibrating string, all of which satisfy the boundary condition that the displacement vanish at both ends. These various sinusoidal configurations may be regarded as *normal modes* for the system under this particular boundary condition.

of k is given by π/a, which corresponds to a minimum value of wavelength of $2a$, as shown by Figure 3.7(b). Any larger value of k or any smaller value of λ corresponds simply to a repetition of some configuration which can be described by a value of k within the central zone. The possible values of k for such a system are, according to (3.3-4) $k = \pi/L, 2\pi/L, 3\pi/L, \cdots \pi/a$. But if there are N atoms in a chain of length L, each separated from its neighbor by a distance a, then $a = L/(N-1)$, and hence the

---

[2] It should be noted that the phase velocities of equivalent waves in different Brillouin Zones differ, but this fact has no direct physical import.

permitted values of k for normal vibrations of this type are $k = \pi/L,\ 2\pi/L,\ 3\pi/L$ $\cdots (N-1)\pi/L$, giving a total of $N-1$ modes. The mode for which $k = (N-1)\pi/L$. ($\lambda = 2a$), however, must be excluded in this case, because it cannot be excited unless the end atoms move, which is forbidden by the boundary conditions. The actual number of normal modes of vibration under these boundary conditions is then $N-2$, that is, the number of atoms in the chain less two. Of course, if N is a very large number, it makes very little difference whether the number of normal modes is regarded as $N-2$ or simply N.

We should note that it is the physical boundary conditions which are imposed upon the system which limit the possible values of k to a certain finite number out of the continuum of possible values between $-\pi/a$ and π/a which are permitted by the solutions of the differential equations of motion of the system. As the above discussion shows, when we hold the ends of the chain rigidly fixed, we select a set of $N-2$ possible values of k corresponding to solutions which satisfy these boundary conditions, out of this infinite set of possible values. In any actual problem we must, however, select *some* boundary conditions on the elastic waves, and as far as the *number* of possible normal modes is concerned, it turns out that it makes little difference precisely what set of physically reasonable boundary conditions are selected. In each case there will be essentially N possible normal modes of vibration which satisfy the boundary conditions which have been picked, provided the number of atoms N is much larger than unity. This result is found also to apply to two- and three-dimensional systems.

It will be useful in many instances to select as a boundary condition the requirement that the vibrational amplitude of the Nth atom be precisely the same as that of the first atom. Physically this would correspond to a linear chain which is bent into the form of a ring, the Nth atom joining on to the first to close the chain. Boundary conditions of this sort are called *periodic* boundary conditions; the application of these boundary conditions to the solutions (3.3-3) of the equations of motion will yield the same result—that there can be essentially N solutions of the equations of motion which *also* satisfy the periodic boundary conditions. The details of proving this result are assigned as an exercise for the reader.

3·4   THE ONE-DIMENSIONAL DIATOMIC LATTICE

So far, in our examination of the dynamical properties of crystal lattices, we have concerned ourselves with the case where all the atoms of the lattice are identical. Since many common crystals are diatomic compounds, which contain atoms of two distinct chemical species, and since the dynamical characteristics of their lattices differ in a number of important ways from those of monatomic crystals, we must now address ourselves to a study of elastic vibrations in crystals of this type.

Consider now a lattice in which atoms of two species are arranged alternately, each atom being separated from its two neighbors by a distance a, as shown in Figure 3.11. The mass of the lighter atom is denoted by m, that of the heavier atom by M. The assumptions of Section 3.3 in regard to Hooke's law forces between atoms and the direct consideration of nearest neighbor interactions only are made here also. It is now necessary to write separate equations of motion for the light and heavy atoms;

this may be done in exact analogy with the procedure of Section 3.3, the result being

$$F_{2n} = m \frac{d^2 u_{2n}}{dt^2} = \beta(u_{2n+1} + u_{2n-1} - 2u_{2n}),$$

$$F_{2n+1} = M \frac{d^2 u_{2n+1}}{dt^2} = \beta(u_{2n+2} + u_{2n} - 2u_{2n+1}).$$

(3.4-1)

Again, we must seek solutions of the form $(\text{const.})e^{i(\omega t - kx)}$, representing the x-coordinate of the atom in terms of its position along the chain as in (3.3-3) and (3.3-4).

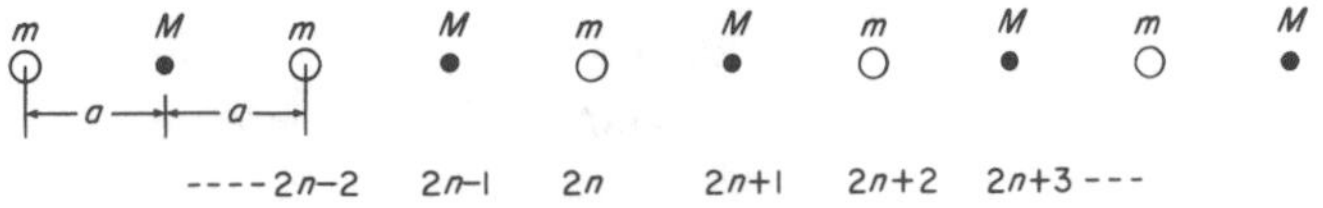

FIGURE 3.11.   Geometry of the diatomic linear lattice of Section 3.4.

Since the two kinds of atoms have different masses, their respective vibration amplitudes will not, in general, be equal, nor is it perfectly clear at this point that their frequencies are the same, although, as we shall see, this turns out to be the case. We thus assume solutions of the form

$$u_{2n} = A e^{i(\omega_1 t - 2kna)}$$

and

$$u_{2n+1} = B e^{i(\omega_2 t - [2n+1]ka)}.$$

(3.4-2)

If u_{2n} and u_{2n+1} are given by these expressions, then according to the rule that the coefficient of k in the exponent must represent the x-coordinate of the atom in question, we must also have

$$u_{2n+2} = A e^{i(\omega_1 t - [2n+2]ka)} = u_{2n} e^{-2ika}$$

$$u_{2n-1} = B e^{i(\omega_2 t - [2n-1]ka)} = u_{2n+1} e^{2ika}.$$

(3.4-3)

Differentiating the solutions (3.4-2) twice with respect to time and substituting back into the equations of motion (3.4-1), using (3.4-3) to express all displacements in terms of u_{2n} and u_{2n+1}, we find

$$-m\omega_1^2 u_{2n} = \beta[(1 + e^{2ika})u_{2n+1} - 2u_{2n}]$$

$$-M\omega_2^2 u_{2n+1} = \beta[(1 + e^{-2ika})u_{2n} - 2u_{2n+1}].$$

(3.4-4)

Solving the second of these equations for u_{2n+1}, we see that

$$u_{2n+1} = \frac{\beta(1 + e^{-2ika})}{2\beta - M\omega_2^2} u_{2n}.$$

(3.4-5)

The equations of motion require that this relation be satisfied for all values of time. However, if the values for u_{2n} and u_{2n+1} as given by (3.4-2) are substituted into (3.4-5), it is clear that this requirement can be satisfied for all values of t only if ω_1 and ω_2 are equal. Accordingly, taking

$$\omega_1 = \omega_2 = \omega, \tag{3.4-6}$$

then substituting (3.4-5) into the first of Equations (3.4-4), expressing the exponentials in terms of trigonometric functions and simplifying, it is found that

$$(2\beta - M\omega^2)(2\beta - m\omega^2) - 4\beta^2 \cos^2 ka = 0. \tag{3.4-7}$$

Rearranging this equation, collecting like powers of ω, we obtain, finally

$$\omega^4 - \frac{2\beta(m + M)}{mM} \omega^2 + \frac{4\beta^2 \sin^2 ka}{mM} = 0. \tag{3.4-8}$$

This quadratic equation in ω^2 can easily be solved by the quadratic formula to give *two* solutions for ω^2, which we shall call ω_+^2 and ω_-^2, according to whether the $+$ or $-$ sign in the quadratic formula is chosen. The result is

$$\omega_\pm^2 = \frac{\beta(m + M)}{mM} \left[1 \pm \sqrt{1 - \frac{4mM \sin^2 ka}{(m + M)^2}} \right]. \tag{3.4-9}$$

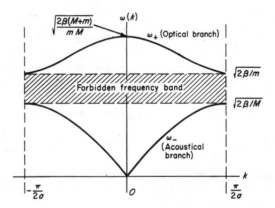

FIGURE 3.12. Dispersion relation for the diatomic linear lattice, showing acoustical and optical branches.

A plot of this result is shown in Figure 3.12. There are two branches of the ω vs. k curve, corresponding to whether the $+$ or $-$ sign is taken in (3.4-9). The upper branch, $\omega_+(k)$, is called the *optical* branch, while the lower one, $\omega_-(k)$, is called the *acoustical* branch. The significance of this terminology will become apparent in due course. For *small* values of k, $\sin ka \cong ka$, and the two roots become

$$\omega_+(0) = \sqrt{\frac{2\beta(m + M)}{mM}} \tag{3.4-10}$$

and
$$\omega_-(k) = ka\sqrt{\frac{2\beta}{m + M}} \qquad (ka \ll \pi/2). \qquad (3.4\text{-}11)$$

To obtain (3.4-11) one must expand the square root of (3.4-9) in a binomial expansion for small values of the argument.

The smallest possible wavelength of the first Brillouin Zone is twice the unit cell distance of the lattice, which here is $2a$. This gives a minimum wavelength of $4a$, corresponding to a maximum value of k of $\pi/2a$, at the boundary of the first zone. For this value of k, (3.4-9) yields

$$\omega_+ = \sqrt{2\beta/m}$$

$$\qquad (3.4\text{-}12)$$

and
$$\omega_- = \sqrt{2\beta/M}.$$

We may better understand the physical characteristics of the motion by examining what happens when $k \to 0$. For this case, the ratio of the amplitudes B/A is readily obtained from (3.4-5), the result being

$$\lim_{k \to 0} \frac{u_{2n+1}}{u_{2n}} = \frac{B}{A} = \frac{2\beta}{2\beta - M\omega_\pm^2}. \qquad (3.4\text{-}13)$$

For the acoustical branch, $\omega = \omega_-$, as given by (3.4-11), and as $k \to 0$, $\omega_- \to 0$, whereby $B/A = 1$. Hence, for the acoustical mode of vibration, the two types of atoms move in the same direction with the same amplitude, as shown in Figure 3.13(a). For

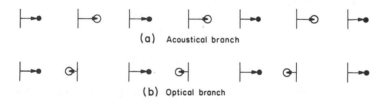

(a) Acoustical branch

(b) Optical branch

FIGURE 3.13.   Physical displacements associated with long wavelength vibrations belonging to (a) acoustical branch, (b) optical branch.

the optical branch, $\omega = \omega_+$ in (3.4-13), and using (3.4-10) it is readily seen that in this instance, $B/A = -m/M$. For the optical branch at $k = 0$, then, the vibrations of the atoms are in *opposite* directions and the amplitudes are inversely in the ratio of the masses, so that the center of mass of the unit cell remains fixed during the period of the motion, as shown in Figure 3.13(b). These characteristics are typical of the optical and acoustical branch in general. The optical mode vibrations in ionic crystals where the two types of atoms are oppositely charged, can be excited by an electric field, which tends to move the ions in opposite directions. Specifically, in ionic substances, this mode can be excited by the electric field associated with a light wave, from which the term *optical* mode vibrations is derived.

3·5   THE FORBIDDEN FREQUENCY REGION

According to Figure 3.12, there exists a band of frequencies $2\beta/M < \omega^2 < 2\beta/m$ where no solutions of the form (3.4-2) exist. It turns out that frequencies in this range simply cannot be propagated as continuous undamped harmonic vibrations of the lattice. If one attempts to excite vibrations of a frequency lying within this band, the vibrations are *attenuated* or damped by the lattice, the attenuation coefficient depending upon the frequency, the atomic masses, and the force constant.

To see this more clearly, let us write (3.4-7) in the form

$$\cos ka = \sqrt{\left(1 - \frac{m\omega^2}{2\beta}\right)\left(1 - \frac{M\omega^2}{2\beta}\right)}$$

$$= i\sqrt{\left(1 - \frac{m\omega^2}{2\beta}\right)\left(\frac{M\omega^2}{2\beta} - 1\right)} = i\delta. \qquad (3.5\text{-}1)$$

Since in the region $2\beta/M < \omega^2 < 2\beta/m$ the first factor in the top equation of (3.5-1) is positive and the second negative, it is clear that $\cos ka$ must be imaginary in this region of frequency, whereby ka itself must be a complex number. If the square root is written in the second form above, $\cos ka$ is expressed as an imaginary quantity $i\delta$, where δ is a *real* number in the frequency range of interest.

The cosine of a complex number may be expressed as the sum of real and imaginary parts by a simple calculation using Euler's theorem, as follows:

$$\cos z = \cos (x + iy) = \cos x \cosh y - i \sin x \sinh y. \qquad (3.5\text{-}2)$$

Setting this equal to $i\delta$, as required by (3.5-1), and equating real and imaginary parts of the resulting equation, one obtains

$$\cos x \cosh y = 0, \qquad (3.5\text{-}3)$$

$$\sin x \sinh y = -\delta. \qquad (3.5\text{-}4)$$

In (3.5-3), $\cosh y$ is never zero for any value of y, hence $\cos x = 0$ and $x = \pm\pi/2$, whereby $\sin x = \pm 1$. Substituting this value into (3.5-4), it is clear that

$$y = \mp\sinh^{-1} \delta, \qquad (3.5\text{-}5)$$

whence
$$z = ka = x + iy = \pm\frac{\pi}{2} \mp i \sinh^{-1} \delta \qquad (3.5\text{-}6)$$

and the solutions of the wave equation (3.4-2) become

$$u_{2n} = Ae^{i\omega t}e^{-2ikna} = Ae^{i\omega t}e^{-2ni(\pm\pi/2 \mp i\sinh^{-1}\delta)} = \pm Ae^{i\omega t \mp 2n\sinh^{-1}\delta}$$

$$= \pm Ae^{i\omega t \mp 2na\cdot a^{-1}\sinh^{-1}\delta}. \qquad (3.5\text{-}7)$$

These are attenuated or damped oscillations; the minus sign in the exponent must be chosen for the physically real case of damped waves (the plus sign represents growing waves whose amplitude increases exponentially along the chain). The attenuation constant κ, defined by writing the solution (3.5-7) as

$$u_{2n} = Ae^{i\omega t}e^{-2\kappa na} \tag{3.5-8}$$

is given by

$$\kappa = \frac{1}{a}\sinh^{-1}\delta, \tag{3.5-9}$$

where δ, in turn, is given by (3.5-1). The same general situation is found to prevail for values of ω in excess of $\omega_+(0)$ as given by (3.4-10), and also in the case of the monatomic chain for values of ω in excess of $\sqrt{4\beta/m}$; in each case it is found that the lattice simply cannot propagate frequencies in these ranges, and that any such disturbance must be damped or attenuated by the lattice.

We have so far discussed only longitudinal waves in one-dimensional lattices; it should, however, be understood that it is also possible to excite *transverse* vibrations in which the atomic displacements are perpendicular to the chain, rather than along it. The characteristics of these transverse waves are in general very similar to those of the longitudinal waves; the phenomenon of dispersion, the acoustical and optical modes of vibration for diatomic lattices, and the forbidden frequency bands all occur in much the same way as they do for longitudinal waves. Since the transverse waves involve atomic displacements normal to those associated with the longitudinal vibrations, transverse and longitudinal waves may be excited simultaneously and (to a first order of approximation) independently of one another. Likewise, since there are two possible orthogonal and independent displacement directions for transverse waves, two independent transverse oscillations may be excited simultaneously.

3·6   OPTICAL EXCITATION OF LATTICE VIBRATIONS IN IONIC CRYSTALS

In an ionic crystal such as NaCl, the sodium and chlorine atoms are ionized, the sodium atom bearing a charge e and the chlorine atom a charge $-e$. In such crystals, the electric vector of a light wave can excite optical mode vibrations, since an electric field exerts forces on the positive and negative charges which are in opposite directions. In this discussion, we shall assume that we are dealing with a diatomic linear lattice of oppositely charged atoms, extending along the x-direction, and that the light is incident in a direction perpendicular to the atomic chain in such a way that the electric vector of the wave oscillates along the x-axis. Under these conditions, at any given instant of time, all the atoms of the crystal experience the same electric field, neighboring positive and negative ions being subjected to exactly equal but opposite forces. The effect of an exciting field of this type will be to set up *forced* longitudinal vibrations in the crystal lattice, *at the frequency of the exciting source*, that is, the incident light wave. Since the electric field is the same everywhere on the lattice at a given time, the type of vibration which will be excited will be that of Figure 3.13(b), thus an optical vibration of infinitely long wavelength, for which $k = 0$.

In the presence of such a field, the force on a positive ion is $eE_0e^{i\omega_0 t}$; and on a negative ion, $-eE_0e^{i\omega_0 t}$, where ω_0 is the light frequency and E_0 is the magnitude of the electric vector of the light wave. Accordingly, a force term $eE_0e^{i\omega_0 t}$ should be added to the right-hand side of the first of the equations of motion (3.4-1) and a term $-eE_0e^{i\omega_0 t}$ to the second. If, now, solutions of the form

$$u_{2n} = Ae^{i(\omega_0 t - 2kna)}$$
$$u_{2n+1} = Be^{i(\omega_0 t - [2n+1]ka)}$$

(3.6-1)

are assumed, the vibrations thus being assumed to be at the forcing frequency, and the steps leading to Equations (3.4-4) repeated, one obtains

$$-m\omega_0^2 u_{2n} = \beta[(1 + e^{2ika})u_{2n+1} - 2u_{2n}] + eE_0e^{i\omega_0 t}$$
$$-M\omega_0^2 u_{2n+1} = \beta[(1 + e^{-2ika})u_{2n} - 2u_{2n+1}] - eE_0e^{i\omega_0 t}.$$

(3.6-2)

Since, physically, the optical vibration which is excited by the light wave is one for which $k = 0$, the above set of equations reduce to

$$(2\beta - m\omega_0^2)u_{2n} - 2\beta u_{2n+1} = eE_0e^{i\omega_0 t}$$
$$-2\beta u_{2n} + (2\beta - M\omega_0^2)u_{2n+1} = -eE_0e^{i\omega_0 t}.$$

(3.6-3)

This set of simultaneous equations may be solved for the vibration amplitudes u_{2n} and u_{2n+1}, the result being

$$u_{2n} = \frac{(-eE_0/m)}{\omega_0^2 - \omega_+^2(0)}e^{i\omega_0 t}$$

$$u_{2n+1} = \frac{(eE_0/M)}{\omega_0^2 - \omega_+^2(0)}e^{i\omega_0 t},$$

(3.6-4)

where $\omega_+(0)$ is the long-wavelength limit of the optical branch frequency, as given by (3.4-10).

According to these results, when the incident light frequency is equal to the natural frequency of the lattice for long wavelength optical mode vibrations, $\omega_+(0)$, a resonance effect should be observed, the vibration amplitudes becoming very large. We should expect the very large vibration amplitudes of the charged ions at resonance to give rise to a strong *reradiation* of electromagnetic energy at the resonant frequency. This phenomenon is actually observed for ionic crystals, a single strong peak in the optical reflectivity of the crystal occurring in the far infrared region of the spectrum, typically in the 40 to 100μ wavelength region. It is possible, knowing the elastic constants of the crystal to arrive at an estimate of the force constant β; if the atomic masses m and M are then known, the resonance frequency $\omega_+(0)$ as predicted by (3.4-10) can be calculated and compared with the optical frequency at which the reflectivity maximum is observed experimentally. The agreement found in this way is quite good for most simple ionic crystals.

If radiation from a continuous-spectrum infrared source is reflected several times by a given ionic crystal, the resulting residual radiation will be nearly monochromatic light of the lattice resonance frequency, since at other frequencies the reflectivity of the crystal will be rather small, and after several reflections such light will be strongly attenuated. This effect has led to the terminology *residual rays*, or, as the German puts it, *Reststrahlen*, referring to the reflected radiation from an ionic crystal at the reflectivity peak. The presence or absence of the characteristic Reststrahl effect serves to indicate whether a crystal is ionic or covalent, since optical mode vibrations cannot be excited in this manner in a covalent crystal in which the atoms bear no net charge. The intensity of the Reststrahl peak in a mixed ionic-covalent crystal likewise gives a quantitative estimate of the relative strength of the ionic component of the crystal binding.

Associated with the reflectivity peak, of course, is a minimum in the optical *transmission* of the crystal. Although in the simple calculation we have made, we included no provision in the form of damping terms in the equations of motion to account for energy losses from the crystal, in actual fact energy is lost. This energy loss is due not only to the reradiation of the reflected beam, but also for other reasons, the most important of which is that the *anharmonicity* of the lattice vibrations which sets in at the large amplitudes that occur near resonance causes the excitation of other modes of vibration of the lattice.

3·7   BINDING ENERGY OF IONIC CRYSTAL LATTICES[3]

In an ionic crystal, the forces which hold the crystal together arise primarily from simple electrostatic interactions between the positive and negative ions of the crystal. For this type of binding it is relatively simple to calculate the binding energy of the crystal lattice on a semiempirical basis. The original treatment of this subject is due to Madelung and Born.[4,5,6]

Let us consider as a specific example a crystal of the NaCl structure, as illustrated in Figure 1.7(a), and let U_{ij} be the potential energy of interaction between ions i and j of the crystal. The total energy of interaction between ion i and all other ions of the crystal will then be

$$U_i = \sum_j{}' U_{ij}. \tag{3.7-1}$$

The prime on the summation indicates that the term for which $j = i$ is to be *excluded* from the sum, there being obviously no binding interaction between the ion i and itself. In addition to the Coulomb energy of interaction $\pm e^2/r_{ij}$, where r_{ij} is the distance between ions i and j, one must assume that there is a *repulsive* force which becomes

---

[3] This section follows in a general way the treatment given by C. Kittel in *Introduction to Solid State Physics*, John Wiley & Sons, New York (1956).

[4] E. Madelung, Physik. Zeitschr. **11**, 898 (1910).

[5] M. Born, *Atomtheorie des Festen Zustandes*, Teubner, Leipzig (1923).

[6] M. Born and M. Göppert-Mayer, *Handbuch der Physik* **24/2**, 623 (1933).

large only for very small interionic distances. This short-range repulsive force arises, in effect, simply because when the atoms are close together, they resist any further attempt to force them to occupy the same space. If this force were not present, the crystal would *always* have a net Coulomb attractive energy between oppositely charged ions which would be greater than the Coulomb repulsive energy between ions of like charge, and the crystal would collapse to essentially zero volume! The energy of interaction between ions i and j due to this force is represented empirically by a term A/r_{ij}^n. If the exponent n is a reasonably large number, this repulsive interaction will be very small at large distances, becoming appreciable only at quite small interatomic spacings.

We may write the total energy of interaction between the two ions i and j as the sum of the Coulomb and short-range repulsive interactions in the form

$$U_{ij} = \frac{A}{r_{ij}^n} \pm \frac{e^2}{r_{ij}}. \tag{3.7-2}$$

It should be noted that the *force* arising from the short-range repulsive interaction is given by the negative derivative of the corresponding interaction energy with respect to r_{ij}, nA/r_{ij}^{n+1}. In (3.7-2) the $+$ sign applies to interactions between ions of like sign, the $-$ sign to interactions between ions of opposite charge. The constant A simply expresses the proportionality between the short-range repulsive energy and r_{ij}^{-n}. If we regard the crystal as being composed of N positive ions and N negative ions, the total binding energy of the lattice, U, may be written

$$U = NU_i. \tag{3.7-3}$$

Although the total number of ions in the crystal is $2N$, in determining the total binding energy we must count the contribution from each pair of ij interactions only once; had we written (3.7-3) with this factor of 2 on the right-hand side, we would have been incorrectly counting interactions U_{ij} and U_{ji} of (3.7-1) as separate entities. The energy U_{ij} as given by (3.7-2) is seen to be the amount of energy required to separate the ions i and j from an initial distance r_{ij} until they are infinitely far apart; the total lattice energy U is thus the energy required to convert a crystal whose nearest neighbor atoms are initially a distance r apart into separate ions, each of which is essentially infinitely far from all the others.

If, now, the dimensionless quantities x_{ij} are defined, such that

$$x_{ij} = r_{ij}/r, \tag{3.7-4}$$

where r is the distance between nearest neighbor atoms in the crystal, then (3.7-2) can be written

$$U_{ij} = \frac{A}{r^n x_{ij}^n} \pm \frac{e^2}{r x_{ij}}, \tag{3.7-5}$$

and

$$U_i(r) = \sum_j{}' U_{ij} = \sum_j{}' \frac{A}{r^n x_{ij}^n} \pm \sum_j{}' \frac{e^2}{r x_{ij}}. \tag{3.7-6}$$

However, in (3.7-6) r is now independent of the summation index j, whereby

$$U_i(r) = \frac{A}{r^n} \sum_j{}' \frac{1}{x_{ij}^n} \pm \frac{e^2}{r} \sum_j{}' \frac{1}{x_{ij}}, \tag{3.7-7}$$

which we shall write as

$$U_i(r) = \frac{B_n}{r^n} - \frac{\alpha e^2}{r}, \tag{3.7-8}$$

with

$$B_n = A \sum_j{}' 1/x_{ij}^n \tag{3.7-9}$$

and

$$\alpha = \sum_j{}' \mp 1/x_{ij}. \tag{3.7-10}$$

Figure 3.14 shows a plot of relation (3.7-8) giving U_i as a function of the nearest neighbor distance r. If the distance between neighboring atoms is large, then the

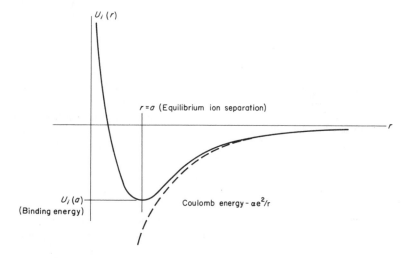

FIGURE 3.14.   Potential energy of an ion in an ionic crystal as a function of the interatomic distance r. The crystal is in equilibrium when this energy is a minimum, corresponding to an interatomic separation equal to the equilibrium lattice distance a.

short-range repulsion is negligible and the binding energy is essentially the excess Coulomb energy of attraction, which approaches zero as r goes to infinity. At shorter distances the positive short-range repulsive energy becomes important, and exceeds the negative Coulomb energy, increasing rapidly as r becomes small. There is a position of minimum energy for the system, as indicated in the diagram, and this point represents the equilibrium condition of the crystal. The value of r associated with this point is simply the equilibrium nearest neighbor interatomic distance a. At this point the Coulomb attractive force between ions is just balanced by the short-range repulsion, the net force on an atom, given by $-\partial U_i/\partial r$, being zero.

The constant α of Equation (3.7-8), which is called the Madelung constant, is a number which is determined completely by the lattice structure of the crystal. The constant B depends upon the repulsive energy coefficient A as well as the lattice structure. It can, however, be determined in terms of the equilibrium nearest neighbor distance a. To do this, one evaluates the derivative of $U_i(r)$ at $r = a$, which according to the above discussion must be zero, since the minimum point of the potential energy curve occurs at $r = a$. We have then,

$$- \frac{nB_n}{a^{n+1}} + \frac{\alpha e^2}{a^2} = 0,$$

whereby
$$B_n = \alpha e^2 a^{n-1}/n. \tag{3.7-11}$$

Substituting this value of B_n into (3.7-8), using (3.7-3) we find for the total binding energy at equilibrium

$$U_0 = NU_i(a) = - \frac{N\alpha e^2}{a} \left(1 - \frac{1}{n}\right). \tag{3.7-12}$$

In this formula, initially, we know neither α nor n. We can, in principle, however, calculate α by evaluating the sum (3.7-10) over all atoms of the crystal. In this equation, the upper sign refers to interactions between like charges, the lower to interactions between opposite ones. If, therefore, the origin is chosen as a negative ion site, the positive sign in (3.7-10) will refer to interactions with positive ions and the negative sign to interactions with negative ions. For a one-dimensional chain of atoms of alternating charge, each atom of which is separated by a distance a from its neighbors, we have, choosing a negative ion as the origin and using (3.7-10)

$$\alpha = \sum_j{}' \mp 1/x_{ij} = 2\left[1 - \frac{1}{2} + \frac{1}{3} - \frac{1}{4} + \cdots\right]. \tag{3.7-13}$$

The factor 2 is necessary because contributions from atoms to the left of the origin as well as atoms to the right of the origin must be included in the sum. From the series expansion

$$\ln(1 + x) = x - \frac{x^2}{2} + \frac{x^3}{3} - \frac{x^4}{4} + \cdots \tag{3.7-14}$$

the series of (3.7-13) can be obtained by setting $x = 1$, whence, for this case

$$\alpha = 2 \ln 2. \tag{3.7-15}$$

The evaluation of the Madelung constant proves to be a rather simple exercise for this simple example. When the same methods are applied to actual three-dimensional crystal lattices, however, it is found that the sum converges so slowly that it is a very difficult matter to evaluate α directly in this manner. It is instructive to write out the Madelung sum for nearest neighbors, next nearest neighbors, third nearest neighbors, etc., in the NaCl structure, to exhibit the nature of these convergence difficulties. There are various mathematical schemes which can be used to circumvent these difficulties, the best known of which are due to Evjen[7] and Ewald.[8] We shall not go into the details of these methods, but merely quote the results for a few of the structures of interest. It is found that for the

$$
\begin{array}{lll}
\text{NaCl structure} & \alpha = 1.7476 \\
\text{CsCl structure} & 1.7627 \\
\text{Zincblende structure} & 1.6381
\end{array}
$$

The tendency of ionic compounds to crystallize in structures which have large Madelung constants is well known. The reason for this is that a maximum value of α, according to (3.7-12) minimizes the energy of the system, giving a crystal with maximum (negative) binding energy. Highly ionic crystals thus rarely crystallize in the zincblende structure.

The constant n in (3.7-12), which is really the exponent associated with the short-range repulsive potential, can be expressed in terms of the *adiabatic compressibility* of the crystal, which is experimentally measurable. In order to do this, we may begin with the First Law of Thermodynamics, which states that

$$dQ = dU + p\,dV = T\,dS, \tag{3.7-16}$$

the symbols having their usual thermodynamic meanings. The internal energy U is assumed to consist entirely of potential energy of interaction between the ions of the crystal, as given by (3.7-12). This means that the temperature is assumed to be much less than the melting temperature on an absolute scale, for as the temperature increases the ions acquire, in addition to their potential energy, more and more *kinetic* energy of vibration, which at the melting point exceeds the potential energy of interaction between the ions and thus destroys the stability of the crystal lattice. For an adiabatic process $dQ = 0$, whereby $dS = 0$ and $dU = -p\,dV$, giving

$$p = -\left(\frac{dU}{dV}\right)_s \quad \text{and} \quad \left(\frac{dp}{dV}\right)_s = -\left(\frac{d^2U}{dV^2}\right)_s. \tag{3.7-17}$$

The adiabatic compressibility (compressibility measured under such conditions that no heat is exchanged with surroundings) may be defined as

$$K = -\frac{1}{V}\left(\frac{dV}{dp}\right)_s, \tag{3.7-18}$$

in which case, from (3.7-17)

$$\frac{1}{K} = -V\left(\frac{dp}{dV}\right)_s = V\left(\frac{d^2U}{dV^2}\right)_s. \tag{3.7-19}$$

[7] H. M. Evjen, *Phys. Rev.* **39**, 675 (1932).
[8] P. P. Ewald, *Ann. d. Physik* **64**, 253 (1921).

Also,

$$\frac{dU}{dV} = \frac{dU}{dr}\frac{dr}{dV}$$

and

$$\frac{d^2U}{dV^2} = \frac{d}{dV}\left(\frac{dU}{dV}\right) = \frac{d}{dV}\left(\frac{dU}{dr}\frac{dr}{dV}\right) = \frac{dU}{dr}\frac{d^2r}{dV^2} + \frac{dr}{dV}\frac{d}{dV}\left(\frac{dU}{dr}\right)$$

$$= \frac{dU}{dr}\frac{d^2r}{dV^2} + \left(\frac{dr}{dV}\right)^2\frac{d^2U}{dr^2}. \tag{3.7-20}$$

But the volume of the unit cell of NaCl, containing 4 Na atoms and 4 Cl atoms, as shown in Figure 1.7(a), is $8r^3$; the volume per Na atom (or per Cl atom) is thus $2r^3$, and the total volume of a large crystal containing N Na ions and N Cl ions will be found to be

$$V = 2Nr^3, \tag{3.7-21}$$

whereby

$$\frac{dr}{dV} = \frac{1}{6Nr^2} \tag{3.7-22}$$

and

$$\frac{d^2r}{dV^2} = \frac{d}{dV}\left(\frac{1}{6Nr^2}\right) = \frac{d}{dr}\left(\frac{1}{6Nr^2}\right)\frac{dr}{dV} = -\frac{1}{18N^2r^5}. \tag{3.7-23}$$

According to (3.7-20), (3.7-22), and (3.7-23), then,

$$\frac{d^2U}{dV^2} = \frac{-1}{18N^2r^5}\frac{dU}{dr} + \frac{1}{36N^2r^4}\frac{d^2U}{dr^2}, \tag{3.7-24}$$

As we have already seen, $dU/dr = 0$ at $r = a$, whence from (3.7-24) we must have

$$\left(\frac{d^2U}{dV^2}\right)_{r=a} = \frac{1}{36N^2a^4}\left(\frac{d^2U}{dr^2}\right)_{r=a}, \tag{3.7-25}$$

from which, recalling (3.7-19),

$$\frac{1}{K} = \left(V\frac{d^2U}{dV^2}\right)_{r=a} = \frac{1}{18Na}\left(\frac{d^2U}{dr^2}\right)_{r=a} \tag{3.7-26}$$

From (3.7-8), $U(r)$ can be written

$$U(r) = NU_i(r) = N\left[\frac{B_n}{r^n} - \frac{\alpha e^2}{r}\right], \tag{3.7-27}$$

whence

$$\frac{d^2U}{dr^2} = N\left[\frac{n(n+1)B_n}{r^{n+2}} - \frac{2\alpha e^2}{r^3}\right], \tag{3.7-28}$$

B_n being given by (3.7-11). Substituting for B_n from the latter equation into (3.7-28) and simplifying, letting $r = a$, we obtain finally

$$\left(\frac{d^2U}{dr^2}\right)_{r=a} = \frac{N\alpha e^2(n-1)}{a^3} = \frac{18Na}{K},$$ (3.7-29)

whence, solving for n,

$$n = 1 + \frac{18a^4}{K\alpha e^2}.$$ (3.7-30)

For NaCl, the measured value of the adiabatic compressibility K is about 4×10^{-12} cm^2/dyne, while $a = 2.8 \times 10^{-8}$ cm. Substituting these values into (3.7-30) gives the result $n \cong 8$, whereby, in this case, according to (3.7-12) we should expect $U_0 \cong -0.9N\alpha e^2/a$. It would appear, then, that approximately 90 per cent of the binding energy of the lattice is accounted for by the first term in (3.7-12), which represents the excess of Coulomb attraction energy between ions of opposite charge over Coulomb repulsion energy between ions of like charge, while the second term, representing the short-range repulsive contribution represents only about 10 per cent of the total. The fact that n turns out to be a rather large number is consistent with our original assumptions regarding the form of the short-range repulsive potential, and in fact serves to a degree to validate those assumptions.

The cohesive energy of most of the highly ionic I-VII type crystals calculated in this manner from (3.7-12) and (3.7-30) agree quite well with experimentally determined values, as shown in Table 3.1.[9] The agreement with experiment is not quite so

TABLE 3.1.
Theoretical and Experimental Binding Energies of Ionic Crystals

Substance	U_0 (calc.) kcal./mole	U_0 (exp.) kcal./mole	Repulsive exponent, n
LiCl	193.3	198.1	7.0
NaCl	180.4	182.8	8.0
KCl	164.4	164.4	9.0
RbCl	158.9	160.5	9.5
CsCl	148.9	155.1	10.5
NaBr	171.7	173.3	8.5
KBr	157.8	156.2	9.5
NaI	160.8	166.4	9.5
KI	149.0	151.5	10.5
RbI	144.2	149.0	11.0
ZnS (Zincblende)	819.0	851.0	9.0
ZnSe	790.0	845.0	9.5
PbS	705.0	731.0	10.5
PbSe	684.0	735.0	11.0

---

[9] The values quoted here are taken from F. Seitz, *Modern Theory of Solids*, McGraw-Hill, New York (1940), Tables XXIV and XXVI, pp. 80–83.

good for the less completely ionic II-VI compounds, as one might expect. We should bear in mind the fact that these calculations are to a degree empirical ones, which relate one set of experimentally determined quantities (lattice spacing and compressibility) to another (repulsive potential exponent and binding energy). A really fundamental calculation of crystal binding energy would proceed from the known wave-mechanical properties of the positive and negative ions, and would *predict*, rather than utilize, the values of the lattice constant and the compressibilities; the only measured quantities which would appear in such a formulation would be e, m, h, etc. Calculations of this type, the results of which are in fairly good agreement with experiment have been made by Löwdin.[10] A discussion of the details is unfortunately beyond the scope of the present work. The simpler Born-Madelung theory, nevertheless, is of considerable value because it verifies the simple conceptual model of ionic binding with which we started, and because it enables us to understand fairly simply the role of both Coulomb and short-range repulsive forces in determining the binding energy of ionic substances.

EXERCISES

1. For the monatomic linear lattice of Section 3.3, find the attenuation coefficient for waves of frequency greater than $\sqrt{4\beta/m}$.

2. Derive the equation of motion for transverse waves on the monatomic linear lattice of Section 3.3; compare your result with the equation of motion for longitudinal waves.

3. Find the phase velocity and group velocity for waves of both optical and acoustical branches for the diatomic lattice of Section 3.4. Plot your results as a function of k.

4. The zincblende lattice can be regarded as a series of (111) planes of alternate spacing a, $a/3$, a, $a/3$, a, $a/3$, $\cdots$, where a is one quarter of the edge of the cubic unit cell. Each plane contains atoms of only one type, so that the atomic species in successive planes alternate thus: Zn, S, Zn, S, Zn, $\cdots$. The planes are held together alternately by one covalent bond per atom, normal to the (111) plane, and three such bonds per atom at an angle to the [111] direction whose cosine is 1/3, so that the same amount of force per unit displacement is involved in stretching a single bond or a system of three bonds disposed at this angle, when an atom is displaced along the [111] direction. By considering a diatomic lattice of the type shown in the figure below, with a force constant β which is the same for any displacement, set up and

solve the equations of motion for longitudinal waves in the [111] direction in zincblende, and plot the resulting ω *vs.* k relation in the first Brillouin Zone. Where is the edge of this first zone? *Hint:* For zincblende, $b = a/3$, $m \neq M$ in the figure above.

5. Solve Problem 4 to obtain the ω *vs.* k relations for longitudinal waves in the [111] direction for the *diamond* lattice. Does the fact that diamond has only one species of atom mean that there is no "optical" branch for this structure? Compare your results generally with those of Problem 4.

6. If the elastic stiffness coefficient c for NaCl is 5×10^{11} dynes/cm², find the force constant β and calculate the Reststrahl frequency for this crystal.

---

[10] P. Löwdin, *Ark. Mat. Astron. Fysik* **35A**, Nos. 9, 30 (1947).

7. Suppose that longitudinal vibrations in a linear chain of N identical atoms are constrained to obey periodic boundary conditions, so that the motion of the first and last atom of the chain is identical. Find the number of normal modes of vibration for this system.

8. Find the Madelung constant for the linear chain of Problem 4, assuming the atoms of mass m to have a positive charge and those of mass M an equal negative charge.

9. Suppose that a homogeneous, isotropic fluid of dielectric constant κ were poured into the interatomic spaces of an ionic crystal, so that the Coulomb interaction is reduced by a factor $1/\kappa$. Assuming that the short-range repulsive potential is unchanged, calculate the new lattice spacing and the new binding energy in terms of the old spacing and the old binding energy.

10. Discuss the optical reflection characteristics of Ge, CdS, KBr, and GaAs in the far infrared region of the spectrum. Explain your predictions physically.

11. The imposition of periodic boundary conditions on a one-dimensional chain of atoms corresponds physically to the case where the head and tail of the chain are joined to form a continuous *ring* of atoms. To what physical system does a two-dimensional square net of atoms obeying periodic boundary conditions at opposite edges of the net correspond? Discuss the topological implications of a three-dimensional cubical lattice obeying periodic boundary conditions on opposite sample faces.

GENERAL REFERENCES

M. Born and M. Göppert-Mayer, *Handbuch der Physik*, Vol. **24/2**, J. Springer Verlag, Berlin (1933), p. 623.

M. Born and K. Huang, *Dynamical Theory of Crystal Lattices*, Oxford University Press, Oxford (1956).

L. Brillouin, *Wave Propagation in Periodic Structures*, Dover Publications, New York (1953).

C. Kittel, *Introduction to Solid State Physics*, 2nd Edition, John Wiley & Sons, New York (1956), Chapter 3.

F. Seitz, *Modern Theory of Solids*, McGraw-Hill Book Co., Inc., New York (1940), Chapter 2.

R. A. Smith, *Wave Mechanics of Crystalline Solids*, Chapman & Hall, Ltd., London (1961), Chapter 3.

J. M. Ziman, *Electrons and Phonons*, Oxford University Press, New York (1960), Chapters 1 and 3.

CHAPTER 4

OUTLINE OF QUANTUM MECHANICS

4·1 INTRODUCTION

In the latter part of the nineteenth century and the early years of the twentieth century scientists were confronted with a large number of experimentally observed effects which could not be properly explained on the basis of classical mechanics and classical electrodynamics. Among these poorly understood phenomena were some of the more basic thermal and electrical properties of solids. For example, the question of why some solids are very good insulators, with specific resistivities in excess of 10^{12} ohm-cm, and others are excellent conductors having resistivities of the order of 10^{-5} ohm-cm could not be understood classically at all. Starting in 1901 with Max Planck's explanation of the spectral distribution of radiation from incandescent hot bodies, however, nearly all of these puzzling experimental results were found to be consistent with the predictions of a revised scheme of mechanics, which differs in many important respects from Newtonian mechanics, but which reduces to Newtonian mechanics in the limit where the masses and energies of the particles involved become relatively large. This new and expanded scheme of mechanics, which has been particularly successful in describing events which take place on an atomic scale, is called *quantum mechanics*. Although it is manifestly impossible to present a comprehensive account of this subject here, we shall discuss some of its more important aspects, in order to obtain a working knowledge of how to solve simple dynamical problems quantum mechanically, and in order to be able to appreciate the results of calculations concerning more complex systems which are involved in explaining the physical behavior of actual crystals.

4·2 BLACK BODY RADIATION

The experimentally determined spectral distribution of radiation intensity emitted by an ideal incandescent "black body" radiator is shown in Figure 4.1. A number of attempts were made late in the nineteenth century to explain these results on the basis of classical mechanics and electromagnetic theory, by treating the radiation field as a fluid having thermodynamic properties such as temperature, pressure, entropy, etc., which were determined by the laws of classical electrodynamics and thermodynamics. This fluid was assumed to be in equilibrium with the atoms of the radiating substance, which were regarded as classical harmonic oscillators. None of these attempts was successful in describing the short wavelength part of the curves, although quite good agreement with the experimental data was obtained in the limit of long wavelengths (the Rayleigh-Jeans law). The classical calculations, which were always based upon the notion that energy could be absorbed or emitted by the atomic oscillators continuously in any amount, large or small, invariably predicted infinite spectral intensity in the short wavelength limit. This resulted in the emission of infinite total radiation energy per unit time, a result which was clearly absurd.

FIGURE 4.1. Schematic representation of the frequency distribution of radiation emitted by an incandescent black body radiator at several different values of temperature.

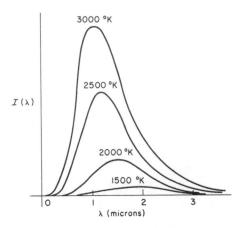

In 1901 Max Planck succeeded in explaining the form of the spectral radiation intensity curves for black body radiation by hypothesizing that the atomic oscillators could radiate or absorb energy only in discrete packets or bundles (called *photons*) whose energy was given by

$$\varepsilon = h\nu = \hbar\omega, \tag{4.2-1}$$

where
$$\hbar = h/2\pi. \tag{4.2-2}$$

Here ν is the frequency of the oscillator (or of the radiation absorbed or emitted), $\omega = 2\pi\nu$, and h is a universal atomic constant (Planck's constant) of magnitude

$$h = 6.625 \times 10^{-27} \text{ erg-sec}$$

$$\hbar = h/2\pi = 1.054 \times 10^{-27} \text{ erg-sec.}$$

The oscillators are thus pictured as existing only in certain *allowed energy states* of total energy 0, $\hbar\omega$, $2\hbar\omega$, $3\hbar\omega$, $\cdots$ $n\hbar\omega$, $\cdots$, the transitions between these allowed states being accomplished by the absorption or emission of photons of energy $\hbar\omega$.

The quantum hypothesis seemed rather bizarre to some of the scientists of that era, who were accustomed to the ideas of classical mechanics, but the agreement which was obtained thereby with the experimental data was excellent, and the value of h which Planck obtained by fitting his theory to experimental data (6.55×10^{-27} erg-sec) is only about 1 percent less than the currently accepted value. More important, however, subsequent successes of the quantum hypothesis in explaining very different phenomena left no doubt of its essential validity. In the light of present knowledge it is not difficult to accept the fact that energy as well as matter may be composed of discrete and indivisible particles.

4·3   THE PHOTOELECTRIC EFFECT

When light is allowed to fall upon the surface of a metal, electrons may be ejected from the metal into a vacuum where they may be collected by an anode, as shown in Figure 4.2(a). The maximum initial energy of these emitted *photoelectrons* may be measured by inserting a grid between the photocathode and the anode, and determining the retarding potential which must be applied between grid and cathode to reduce the photocurrent to zero, as illustrated in Figure 4.2(b). It can be demonstrated in this

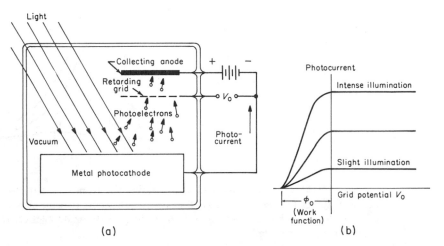

(a)

(b)

FIGURE 4.2.   (a) Schematic diagram of the apparatus used to study the photoelectric effect in metals, (b) a plot of observed photocurrent *versus* retarding potential for several levels of incident light intensity.

way that if the incident light is of frequency less than a given threshhold frequency ω_0 there is no photoelectric effect whatsoever, no matter how *intense* the light may be. For incident radiation of frequency greater than ω_0, photoelectrons are emitted in numbers proportional to the intensity of the incident illumination, causing a photocurrent proportional to light intensity to flow in the external circuit. The maximum initial energy of these electrons as measured by the retarding potential method is strictly proportional to $\omega - \omega_0$, where ω is the frequency of the incident light.

The existence of the threshold frequency cannot be explained on the basis of classical physics, and it was not until 1905 that the physical nature of the photoelectric effect was described by Einstein on the basis of Planck's quantum hypothesis. According to Einstein's explanation, the electrons inside the metal must have lower potential energy than electrons at rest in the vacuum outside the metal surface; otherwise, they would be emitted spontaneously in the dark. This fact leads to a conceptual picture of a metal such as that shown in Figure 4.3. In order to be emitted as photoelectrons, the electrons in the metal must overcome a potential energy barrier of magnitude $e\phi_0$, at least. The potential ϕ_0, called the *vacuum work function*, is a characteristic property of the emitting cathode material. According to Planck's quantum ideas, the absorption of a photon whose energy is less than $e\phi_0$ by the metal cannot excite an electron into the vacuum, but will only impart to it some kinetic energy which is finally dissipated inside the metal in the form of heat by collisions.

On the other hand, if the absorbed photon has energy $e\phi_0$, it can just excite a photoelectron into vacuum with zero kinetic energy, and if the photon energy is

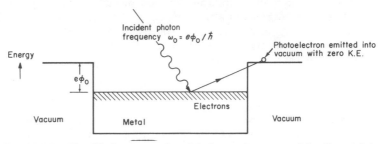

FIGURE 4.3.   Simplified conceptual model of a metal as a potential well containing free electrons upon which the Einstein theory of the photoelectric effect is based.

greater than $e\phi_0$, it can excite a photoelectron into vacuum with excess kinetic energy which may be as large as $\hbar\omega - e\phi_0$, $\hbar\omega$ being the incident photon energy. The maximum energy of the emitted photoelectrons will thus be seen to be

$$\varepsilon_m = \hbar\omega - e\phi_0 = \hbar(\omega - \omega_0), \qquad (4.3\text{-}1)$$

where the threshold frequency ω_0 must be given by

$$\omega_0 = e\phi_0/\hbar. \qquad (4.3\text{-}2)$$

This is all in agreement with the experimental facts which were described previously. In addition, Equation (4.3-1) predicts that the *slope* of the curve giving ε_m as a function of ω will be the constant value $\hbar$. When this slope is evaluated from the experimental data, it is indeed found to be constant and gives a value for $\hbar$ which is in exact agreement with the value derived by Planck from experimental data pertaining to black body radiation! The photoelectric effect thus provided, within a few years, a striking confirmation of the quantum theory in a field which was quite unrelated to the original application.

4·4   SPECIFIC HEAT OF SOLIDS

Classical physics regarded a solid crystal as an assembly of atoms held together in a periodic array by certain attractive forces. The atoms were assumed to be free to vibrate about their equilibrium positions under the constraints of the resultant forces, and to a first approximation, forces and atomic displacements would be related by Hooke's law. The effect of thermal energy, then, would be to set these atoms into vibration as harmonic oscillators about their equilibrium positions. It is an elementary result of classical kinetic theory that if the energies of an assembly of classical harmonic oscillators are distributed according to the Boltzmann law in thermal equilibrium, the average energy of an oscillator is kT for each vibrational degree of freedom, where T is the absolute temperature and k is Boltzmann's constant (equal to 1.380×10^{-16} erg/$^\circ$K.). We shall see in a later section how this result is derived. On the basis of classical physics, then, since there are three independent vibrational degrees of freedom per atom, we should expect the total internal thermal energy of a crystal composed of N identical atoms to be

$$U = 3NkT. \tag{4.4-1}$$

The *heat capacity* at constant volume, C_v, is by definition the rate of increase of internal energy per unit temperature rise, measured under conditions of constant volume, or, for this case

$$C_v = (\partial U/\partial T)_v = 3Nk. \tag{4.4-2}$$

According to the classical formula (4.4-2) the heat capacity should be independent of the temperature. If, in (4.4-2), N is set equal to Avogadro's number N_A (equal to 6.025×10^{-23} mole^{-1}), it can be seen that the *molar* heat capacity of all solid chemical elements should have the same value $3N_A k$, equal to about 6 cal $^\circ$K^{-1} mole^{-1}. This result is known as the Law of Dulong and Petit, and is in fairly good agreement with experiment for many elements *at and above room temperature*. At low temperatures,

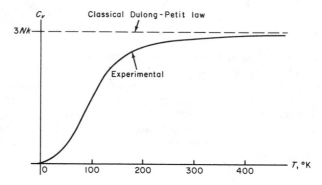

FIGURE 4.4.   Typical experimental data for the specific heat of a solid substance as a function of temperature, in comparison with the classical Dulong-Petit result.

the classical result does *not* agree with experiment, the experimental data showing that, the heat capacity approaches zero as the temperature approaches zero on the absolute scale, as shown in Figure 4.4. The same general remarks, of course, apply to the specific heat at constant volume c_v, which is defined simply as

$$c_v = (\partial U/\partial T)_v/V. \tag{4.4-3}$$

The discrepancies between the classical theory and experiment were explained in 1911 by Einstein and, more accurately, in 1912 by Debye, on the basis of the quantum theory. According to Planck's original hypothesis, a harmonic oscillator can exist only in certain discrete energy states having energy values

$$\varepsilon_n = n\hbar\omega_0, \tag{4.4-4}$$

where ω_0 is the classical oscillator frequency. This results, as we shall see later, in an average oscillator energy different from the classical result, and in heat capacities which agree quite well with the experimental data at all temperatures. Again the quantum theory had provided a simple explanation of an effect which could not be explained at all by classical methods, once more in an area quite different from that of the original application.

4·5   THE BOHR ATOM

One of the most puzzling aspects of atomic behavior to physicists of the nineteenth century was the emission and absorption spectra of the elements. The sharp, discrete spectral lines which are observed could not be understood at all in terms of classical mechanics and electromagnetic theory. In 1913, however, Niels Bohr proposed a model of the hydrogen atom, based on Planck's quantum hypothesis, which described with amazing accuracy the main features of the spectrum of atomic hydrogen.

In Bohr's model, which was founded upon the concept of the nuclear atom advanced by Rutherford in 1910, an electron of mass m and charge $-e$ is assumed to move in an orbit around a much more massive nucleus of charge $+Ze$, where Z is an integer. The introduction of the factor Z allows one to account not only for hydrogen ($Z = 1$), but also for certain other hydrogen-like ions, such as He^+ ($Z = 2$), Li^{++} ($Z = 3$), Be^{+++} ($Z = 4$), etc., which consist of a single electron and a heavy nucleus.

According to classical electrodynamics, accelerated charges always radiate energy, and a system such as that considered by Bohr, wherein the electron is always subjected to a central acceleration, should lose energy constantly by radiation, the electron spiralling gradually inward toward the nucleus. Bohr nevertheless assumed that the electron could exist in stable orbits about the nucleus without radiating energy at all, if only the *angular momentum* associated with the motion were quantized so as to have only the allowed values $n\hbar$, where n is a positive integer. Transitions between allowed steady states n and m would then be accompanied by the instantaneous emission or absorption of a photon of frequency ω_{mn} such that

$$\hbar\omega_{mn} = |\varepsilon_m - \varepsilon_n|, \tag{4.5-1}$$

where ε_m is the energy associated with the state of angular momentum $m\hbar$ and ε_n the energy associated with the state of angular momentum $n\hbar$,

Dynamically, of course, the orbit of the electron about the nucleus could be either circular or elliptical, but for simplicity we shall assume that the orbit is circular. The allowed values of angular momentum L are restricted to integer multiples of $\hbar$, so that

$$L_n = mr_n^2\omega_n = n\hbar \qquad (n = 1,2,3, \cdots), \qquad (4.5\text{-}2)$$

where r_n is the radius of the orbit with angular momentum $n\hbar$ and ω_n is the angular velocity of the electron in that orbit. For a steady orbit, the radius r_n must be such that the electrostatic force of attraction between the electron and the nucleus, Ze^2/r_n^2, is just the centripetal force mv_n^2/r_n required to hold the electron in the circular orbit, whence,

$$Ze^2/r_n^2 = mv_n^2/r_n = mr_n\omega_n^2. \qquad (4.5\text{-}3)$$

Equations (4.5-2) and (4.5-3) may now be solved as simultaneous equations for r_n and ω_n, giving

$$r_n = \frac{n^2\hbar^2}{Zme^2} \qquad (4.5\text{-}4)$$

and

$$\omega_n = \frac{Z^2me^4}{n^3\hbar^3}. \qquad (4.5\text{-}5)$$

The kinetic energy of the system, ε_k, is given by

$$\varepsilon_k = \tfrac{1}{2}mv_n^2 = \tfrac{1}{2}mr_n^2\omega_n^2 = \frac{Z^2me^4}{2n^2\hbar^2}. \qquad (4.5\text{-}6)$$

The potential energy, ε_p, is

$$\varepsilon_p = -Ze^2/r_n = -\frac{Z^2me^4}{n^2\hbar^2}. \qquad (4.5\text{-}7)$$

The total energy of the system is then

$$\varepsilon_n = \varepsilon_k + \varepsilon_p = -\frac{Z^2me^4}{2n^2\hbar^2} \qquad (n = 1,2,3, \cdots). \qquad (4.5\text{-}8)$$

The energy of the system is seen to be restricted to certain discrete values or *energy levels* corresponding to $n = 1, 2, 3, \cdots$. If $Z = 1$, the system corresponds to the hydrogen atom, and the resulting energy level diagram is as represented in Figure 4.5. Transitions between the energy levels are accomplished by the absorption or emission of a photon whose frequency is given by (4.5-1), each such transition corresponding to a possible spectrum line. The Bohr theory accounts quite successfully for the spectrum of hydrogen and hydrogen-like ions, the agreement between observed and predicted spectral frequencies being very close.

In the preceding development we have assumed the electron orbits to be circular. In general the orbits may be elliptical, and it can indeed be shown that there are regular series of allowed elliptical orbits which satisfy the quantum condition (4.5-2). These elliptical orbits, however, contribute no new energy levels, but merely represent

OUTLINE OF QUANTUM MECHANICS

orbits for which the angular momentum has a value $m\hbar$ where m is an integer less than n; the system thus has energy ε_n as given by (4.5-8), but its angular momentum, though still an integral multiple of $\hbar$, is less than the maximum value $n\hbar$ which corresponds to a circular orbit. We have also regarded the nucleus as being fixed, which is true only in the limit where the ratio of the nuclear to electronic mass becomes infinite. For finite nuclear mass, the nucleus and the electron would rotate about

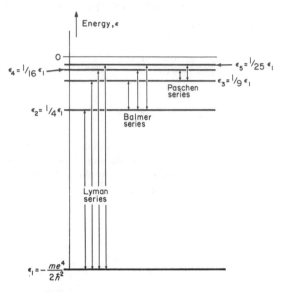

FIGURE 4.5. Energy levels of the hydrogen atom as predicted by the Bohr theory. Transitions corresponding to certain prominent spectral lines are indicated.

their common center of mass. This effect is easily taken into account, the result being that the quantity m in the preceding equations should be replaced by a *reduced mass* μ given by

$$\mu = \frac{mM}{m + M}, \tag{4.5-9}$$

M being the nuclear mass. In the case of the hydrogen atom, the reduced mass differs from the electron mass by about 1 part in 1850.

Numerous attempts were made to extend the Bohr model to explain the spectra of helium and of more complex atoms during the decade following its introduction. These attempts were fraught with difficulty because there was no simple way of determining what the orbits should be. In general, the results of these calculations were not in good agreement with experiment; nor did the Bohr theory provide any explanation for why some transitions between electronic states in complex atoms seemed to be forbidden. Nevertheless the Bohr model provided for the first time a simple, coherent and accurate explanation of the spectra of one-electron atoms, and served as a conceptual framework for visualizing certain aspects of atomic behavior which persists even to the present time.

4·6 DE BROGLIE'S HYPOTHESIS AND THE WAVELIKE PROPERTIES OF MATTER

In 1924 de Broglie suggested that with a particle of momentum p one might associate a *wave* of wavelength λ such that

$$p = \frac{h}{\lambda} = \frac{h}{2\pi}\frac{2\pi}{\lambda} = \hbar k. \qquad (4.6\text{-}1)$$

The Bohr quantum condition follows directly from this hypothesis if one assumes that the allowed orbits of the previous section accommodate an integral number of particle wavelengths as defined by (4.6-1). Since the path length of the orbit is $2\pi r_n$, we must find, according to this idea,

$$n\lambda = \frac{nh}{mv_n} = 2\pi r_n$$

or $\qquad\qquad\qquad n\hbar = mv_n r_n = L_n. \qquad (4.6\text{-}2)$

It was shown experimentally by Davisson and Germer that electrons could indeed be diffracted from crystals, giving the same pattern for a given crystal as that produced by X-rays of wavelength $\lambda = h/p$, as proposed by de Broglie. The condition for the diffraction of electrons from crystal planes is just the Bragg condition, with λ equal to h/p;

$$n\lambda = nh/p = 2d \sin \theta. \qquad (4.6\text{-}3)$$

It has since been shown that neutral atoms, protons, positive ions and other particles exhibit the same wavelike behavior, the de Broglie relation (4.6-1) being satisfied in each case.

4·7 WAVE MECHANICS

In 1926, Schrödinger developed a unified scheme of mechanics based upon the physical notions of Planck's quantum theory and utilizing de Broglie's ideas of the wavelike nature of matter. This scheme, which is called *wave mechanics*, is a very general revision of the laws of mechanics which is designed to extend that subject into the realm of atomic and nuclear phenomena. It has been very successful, and as far as one can tell at present, it is a correct way of describing the physical behavior of matter on the atomic scale, or, in fact, on any scale, since the results of wave mechanics reduce to those of classical Newtonian mechanics for the relatively large and massive systems which are observable macroscopically. Almost at the same time, an alternative system of quantum mechanics, called *matrix mechanics*, based upon the same physical ideas as wave mechanics, was devised by Heisenberg. Although quite different from Schrödinger's wave mechanics in its mathematical formulation, it was later shown to be precisely equivalent to Schrödinger's theory. We shall confine our discussions for the most part to the Schrödinger wave mechanical description, which is simpler and more easily related to physically observable situations.

In Schrödinger's formulation of quantum mechanics a complex quantity Ψ, called the *wave function* is associated with a dynamical system. This quantity is a function of three space coordinates for each particle of the system, and the time. The dynamical properties of the system are closely related to the properties of the Ψ function, and the dynamical behavior of the system can be ascertained when the Ψ function for the system is known. For a single-particle system, to which our discussion will be largely restricted, the properties of the wave function Ψ can be expressed in terms of these five postulates:[1]

1. Associated with the particle is a complex wave function $\Psi(x,y,z,t)$, where x, y, z are space coordinates and t is the time.
2. The classical expression for the total energy ε of the system (which is called the classical *Hamiltonian* of the system), given by

$$\frac{p^2}{2m} + V(x,y,z) = \varepsilon, \tag{4.7-1}$$

where p is the momentum of the particle, m its mass and $V(x,y,z)$ its potential energy, may be converted into a *wave equation* by associating certain *operators* with the classical dynamical quantities and allowing these operators to operate on the wave function as directed by (4.7-1). The operators corresponding to the pertinent dynamical quantities are

dynamical variable		associated operator	
x, y, or z	$\rightarrow$	x, y, or z	
$f(x,y,z)$	$\rightarrow$	$f(x,y,z)$	
p	$\rightarrow$	$\dfrac{\hbar}{i}\nabla$	(4.7-2)
ε	$\rightarrow$	$-\dfrac{\hbar}{i}\dfrac{\partial}{\partial t}.$	

Since the cartesian components of the ∇ operator are $(\partial/\partial x,\ \partial/\partial y,\ \partial/\partial z)$, it is

_____

[1] C. W. Sherwin, *Introduction to Quantum Mechanics*, Holt, Rinehart and Winston, New York (1959), p. 14, pp. 62–63. See also for the extension of these postulates to systems of more than one particle.

clear that the operators corresponding to the momentum components p_x, p_y, p_z are $(\hbar/i)(\partial/\partial x)$, $(\hbar/i)(\partial/\partial y)$ and $(\hbar/i)(\partial/\partial z)$, respectively. Also, since $p^2 = \mathbf{p} \cdot \mathbf{p}$, the corresponding p^2 operator should be $-\hbar^2(\mathbf{\nabla} \cdot \mathbf{\nabla})$ or $-\hbar^2\nabla^2$, where ∇^2 represents the Laplacian operator $\partial^2/\partial x^2 + \partial^2/\partial y^2 + \partial^2/\partial z^2$. Replacing the dynamical quantities p^2, $V(x,y,z)$ and ε in (4.7-1) with their corresponding operators and allowing them to operate upon the wave function Ψ, we obtain the wave equation

$$-\frac{\hbar^2}{2m}\nabla^2\Psi + V(x,y,z)\Psi = -\frac{\hbar}{i}\frac{\partial\Psi}{\partial t}. \tag{4.7-3}$$

This is Schrödinger's equation for the wave function. It is often written in the form

$$\mathscr{H}\Psi = -\frac{\hbar}{i}\frac{\partial\Psi}{\partial t}, \tag{4.7-4}$$

where $\mathscr{H}$ represents the Hamiltonian operator

$$\mathscr{H} = -\frac{\hbar^2}{2m}\nabla^2 + V(x,y,z). \tag{4.7-5}$$

3. The quantities $\Psi(x,y,z,t)$ and $\nabla\Psi$ must be finite, continuous and single-valued for all values of x,y,z and t.

4. The quantity $\Psi^*\Psi$, where Ψ^* is the complex conjugate of Ψ, is always *a real* quantity. This quantity is interpreted as a probability density, in the sense that $\Psi^*\Psi dv$ is the probability that the particle will be found in the volume element dv at time t. This is all the information about the actual location of the particle we can ever obtain from the wave function; the question of just where the particle is at a given time and what its trajectory is cannot be answered precisely, according to quantum mechanics. Of course, for a large or massive object, $\Psi^*\Psi$ will be large only within the classical boundaries of the object and will move in time as predicted by Newton's laws, but on an atomic scale it will be impossible to locate a particle precisely and to follow its trajectory in precise detail. Since $\Psi^*\Psi dv$ is a probability density and since the probability that the particle will be found *somewhere* in space is unity, we must require that the wave function be *normalized* such that

$$\int_v \Psi^*\Psi \, dv = 1, \tag{4.7-6}$$

the integral being taken over all space.

5. The average or *expectation* value $\langle\alpha\rangle$ of any dynamical variable α, with which is associated an operator $\alpha_{op.}$, is defined by

$$\langle\alpha\rangle = \int_v \Psi^*\alpha_{op.}\Psi \, dv, \tag{4.7-7}$$

the integral again being taken over all space.

The essential features of all of wave mechanics are contained in these five postulates. The remainder of our treatment of the subject will be devoted to exploring their implications and applying them to specific problems. There is no way of *proving* them except to state that their consequences, whenever they have been subjected to test by experiment, have been shown to be in agreement with observations within the limits of experimental uncertainty, which in many cases have been extremely small. We shall, of course, try to demonstrate that the results of wave mechanics are in line with some of the dictates of intuition and correspond whenever it can be expected to the results of classical mechanics.

In regard to the latter subject, we are already in a position to show that wave mechanics gives the same results as classical mechanics insofar as the average or expectation values of dynamical quantities are concerned. Let us for simplicity consider a one-dimensional situation where the dynamical system is confined to the x-axis. The expectation value of the momentum p_x, according to (4.7-7), is then

$$\langle p_x \rangle = \int_{-\infty}^{\infty} \Psi^*(x,t) \cdot -\frac{\hbar}{i} \frac{\partial}{\partial x} \Psi(x,t) \, dx, \tag{4.7-8}$$

whereby, taking a time derivative and differentiating under the integral sign on the right,

$$\frac{d\langle p_x \rangle}{dt} = \frac{\hbar}{i} \int_{-\infty}^{\infty} \frac{\partial \Psi^*}{\partial t} \frac{\partial \Psi}{\partial x} \, dx + \frac{\hbar}{i} \int_{-\infty}^{\infty} \Psi^* \frac{\partial^2 \Psi}{\partial x \partial t} \, dx. \tag{4.7-9}$$

Now the value of $\partial \Psi / \partial t$ is given by the wave equation (4.7-3), and the value of $\partial \Psi^* / \partial t$ can be expressed in a similar form by writing the wave equation in terms of Ψ^* rather than Ψ. This can be done by writing the complex function $\Psi(x,t)$ as a sum of real and imaginary parts,

$$\Psi(x,t) = u(x,t) + iv(x,t), \tag{4.7-10}$$

substituting this form for Ψ into (4.7-3), and equating real and imaginary parts on either side of the resulting expression, to obtain two equations of the form

$$-\frac{\hbar^2}{2m} \nabla^2 u + uV = -\hbar \frac{\partial v}{\partial t}$$

$$-\frac{\hbar^2}{2m} \nabla^2 v + vV = \hbar \frac{\partial u}{\partial t}. \tag{4.7-11}$$

Multiplying the second of these equations by $-i$ and adding it to the first, we get a wave equation for Ψ^* ($= u - iv$) of the form

$$-\frac{\hbar^2}{2m} \nabla^2 \Psi^* + V\Psi^* = \frac{\hbar}{i} \frac{\partial \Psi^*}{\partial t}. \tag{4.7-12}$$

In equation (4.7-9), if we express $\partial \Psi / \partial t$ by (4.7-3) and $\partial \Psi^* / \partial t$ by (4.7-12), we obtain

$$\frac{d\langle p_x \rangle}{dt} = -\frac{\hbar^2}{2m} \int_{-\infty}^{\infty} \left[\frac{\partial^2 \Psi^*}{\partial x^2} \frac{\partial \Psi}{\partial x} - \Psi^* \frac{\partial^3 \Psi}{\partial x^3} \right] dx + \int_{-\infty}^{\infty} \left[V\Psi^* \frac{\partial \Psi}{\partial x} - \Psi^* \frac{\partial}{\partial x}(V\Psi) \right] dx$$

$$= -\frac{\hbar^2}{2m} \int_{-\infty}^{\infty} \left[\frac{\partial}{\partial x} \left(\frac{\partial \Psi}{\partial x} \frac{\partial \Psi^*}{\partial x} - \Psi^* \frac{\partial^2 \Psi}{\partial x^2} \right) \right] dx - \int_{-\infty}^{\infty} \Psi^* \frac{\partial V}{\partial x} \Psi \, dx$$

$$= -\frac{\hbar^2}{2m} \left[\frac{\partial \Psi^*}{\partial x} \frac{\partial \Psi}{\partial x} - \Psi^* \frac{\partial^2 \Psi}{\partial x^2} \right]_{-\infty}^{\infty} - \int_{-\infty}^{\infty} \Psi^* \frac{\partial V}{\partial x} \Psi \, dx. \qquad (4.7\text{-}13)$$

From postulate (4) above, the integral of the quantity $\Psi^*\Psi = u^2 + v^2$ over the range ∞ to $-\infty$ must exist; this means that both u and v, as well as their derivatives must approach zero as x approaches either ∞ or $-\infty$. Thus Ψ and $\partial \Psi/\partial x$ approach zero in these limits, and the first term in (4.7-13) vanishes. But, according to postulate (5) above, the second term in (4.7-13) represents the expectation value of the quantity $-\partial V/\partial x$, which is the classical force on the particle. We have thus

$$\frac{d\langle p_x \rangle}{dt} = \langle -\partial V/\partial x \rangle = \langle F_x \rangle, \qquad (4.7\text{-}14)$$

which is simply Newton's Law of motion. It is clear, then, that insofar as the expectation values are concerned, wave mechanics is in agreement with the equations of classical mechanics.

4·8 THE TIME DEPENDENCE OF THE WAVE FUNCTION

Let us assume that the Schrödinger equation (4.7-3) can be solved by the usual mathematical technique of separation of variables. Accordingly we shall assume solutions of the form

$$\Psi(x,y,z,t) = \psi(x,y,z)\phi(t), \qquad (4.8\text{-}1)$$

where ψ depends only on the space coordinates and ϕ only upon the time. Differentiating (4.8-1) and substituting back into (4.7-3), it is easily seen that

$$-\frac{\hbar^2}{2m}\phi\nabla^2\psi + V\phi\psi = -\frac{\hbar}{i}\psi\frac{d\phi}{dt}. \qquad (4.8\text{-}2)$$

Dividing both sides of this equation by $\Psi = \phi\psi$, we find

$$-\frac{\hbar^2}{2m}\frac{\nabla^2\psi}{\psi} + V(x,y,z) = -\frac{\hbar}{i}\frac{1}{\phi}\frac{d\phi}{dt} = \varepsilon. \qquad (4.8\text{-}3)$$

In (4.8-3), the left-hand side is a function of the space coordinates alone, while the center expression is a function of time only. The only way in which this equality can hold for all values of the variables is for each expression separately to be equal to a constant, called a separation constant and denoted by ε in (4.8-3). There are really two separate equations, one for $\phi(t)$, the other for $\psi(x,y,z)$, thus

$$d\phi/dt = -\frac{i\varepsilon}{\hbar}\,\phi(t) \tag{4.8-4}$$

and

$$\nabla^2\psi + \frac{2m}{\hbar^2}\,(\varepsilon - V(x,y,z))\psi(x,y,z) = 0. \tag{4.8-5}$$

Equation (4.8-4) can easily be integrated to give

$$\phi(t) = e^{-i\varepsilon t/\hbar}. \tag{4.8-6}$$

The integration constant which would normally appear as a multiplicative factor in (4.8-6) is arbitrarily set equal to unity. There is no loss in generality sustained in doing this. According to (4.8-1) the time-dependent wave function Ψ can now be written

$$\Psi(x,y,z,t) = \psi(x,y,z)e^{-i\varepsilon t/\hbar}, \tag{4.8-7}$$

where $\psi(x,y,z)$ is a solution of (4.8-5).

Equation (4.8-5) is called the *time-independent Schrödinger equation*, and its solutions $\psi(x,y,z)$ are called *time-independent wave functions* or *stationary state wave functions*. Equation (4.8-5) can also be written as

$$\mathscr{H}\psi = \varepsilon\psi, \tag{4.8-8}$$

where $\mathscr{H}$ is the Hamiltonian operator given by (4.7-5). It is usually more convenient, when possible, to solve the time-independent equation (4.8-6) and deal as much as possible with the time-independent wave functions only. The time dependences, whenever needed, can be expressed using (4.8-7).

In order to be able to do this conveniently, we should take note of certain results which follow when the wave function is of the form (4.8-7). First of all, if the wave function has this form, then

$$\Psi^*\Psi = \psi^*e^{i\varepsilon t/\hbar}\psi e^{-i\varepsilon t/\hbar} = \psi^*\psi, \tag{4.8-9}$$

and $\Psi^*\Psi$ itself is time independent and equal to $\psi^*\psi$—thus the terminology *stationary state* wave functions for wave functions of this sort. As a consequence of this result, $\psi^*\psi$ has the same probability density interpretation with respect to the time-independent wave functions as $\Psi^*\Psi$ has with respect to the time-dependent ones. In addition, from (4.8-9) it follows that

$$\int_v \Psi^*\Psi \, dv = \int_v \psi^*\psi \, dv = 1, \tag{4.8-10}$$

and that the normalization condition for the time-independent functions (4.8-7) is thus the same as that for the time-dependent functions. Finally, if the wave functions have the form (4.8-7) and if $\alpha_{op.}$ depends only on the coordinates and not *explicitly* upon the time, the expectation value of the dynamical variable represented by the operator $\alpha_{op.}$ can be expressed as

$$\langle \alpha \rangle = \int_v \Psi^* \alpha_{op.} \Psi \, dv = \int_v \psi^* \alpha_{op.} \psi \, dv, \qquad (4.8\text{-}11)$$

since the exponential time factors cancel out just as in (4.8-9) if $\alpha_{op.}$ has no explicit time dependence. If $\alpha_{op.}$ depends explicitly upon time, then the time-dependent wave functions must in general be used to evaluate $\langle \alpha \rangle$.

It should be noted that these results are applicable only when the total wave function is of the form (4.8-7). If the total wave function Ψ is a *superposition* of functions of the form (4.8-7), as it well may be while still satisfying the time-dependent wave equation, these three results are no longer true, and the time dependent wave functions must generally be used to calculate the properties of the system.

It is easily shown, using the wave function (4.8-7) in (4.7-7), employing the time-dependent energy operator $-(\hbar/i)(\partial/\partial t)$, that the expectation value of the energy of a system in a state represented by a wave function of this type is just the value of the separation constant ε. The notation chosen for the separation constant has, of course, anticipated this result. It can also be shown from (4.8-8) that the expectation value of the Hamiltonian operator $\mathscr{H}$ for a system in a state represented by this sort of wave function (which, since $\mathscr{H}$ is a time-independent operator can be calculated using the time-independent wave function according to (4.8-11)) is just equal to ε, the expectation value of the energy.

Neither (4.8-4) nor (4.8-5) places any restriction upon the value of ε, since for *any* ε there exists a Ψ which satisfies these equations. However, the requirements of continuity, finiteness, and single-valuedness *may* select out of the infinite continuum of possible solutions only certain individual ones satisfying these conditions, which correspond to *certain discrete values* of the separation constant (thus the energy) ε. We may find in this way that only a certain set of solutions $\psi_n(x,y,z)$ which are related to an associated set of energy levels ε_n are acceptable as wave functions for the system, because only these functions satisfy the wave equation *and the boundary conditions*. These acceptable wave functions are called the *eigenfunctions* of the system (from the German *eigen*, meaning *own* or *characteristic*), and the associated energy levels are called *energy eigenvalues*. The discrete energy levels of the Planck oscillator and the Bohr atom are really eigenvalues of this sort. From (4.8-6) it is easily seen that the frequency of the wave function always obeys Planck's relation

$$\varepsilon = \hbar \omega. \qquad (4.8\text{-}12)$$

In some instances, both the wave equation and the associated boundary conditions may be satisfied for *any* value of ε, or at least for any value of ε in a given finite interval. In such cases there is a *continuous* range of permitted energies or a *continuum* of energy levels. We shall see in due time in what particular circumstances either of these possible situations arises.

4·9   THE FREE PARTICLE AND THE UNCERTAINTY PRINCIPLE

We now wish to examine the wave function of a free particle, that is, one which is subject to no forces, and which, therefore, moves in a region of constant potential. We shall again limit ourselves to a one-dimensional geometry in which motion is restricted to the x-axis. For convenience, and without loss of generality, the potential may be taken to be zero, in which case the wave equation (4.8-5) becomes

$$\frac{d^2\psi}{dx^2} + k^2\psi(x) = 0, \tag{4.9-1}$$

where

$$k = \sqrt{2m\varepsilon/\hbar^2}. \tag{4.9-2}$$

Equation (4.9-1) is a familiar differential equation which has solutions of the form

$$\psi(x) = Ae^{\pm ikx}, \tag{4.9-3}$$

where A is an arbitrary constant. These solutions satisfy all the requirements for wave functions except, in a strict sense, integrability on the interval $-\infty < x < \infty$. We shall nevertheless have to accept them, although, as we shall see, certain difficulties will arise in satisfying the normalization requirement. The time-dependent wave functions for the system, according to (4.8-7), will be

$$\Psi(x,t) = Ae^{i(\pm kx - \omega t)}, \tag{4.9-4}$$

where, from (4.9-2),

$$\omega = \varepsilon/\hbar = \frac{\hbar k^2}{2m}. \tag{4.9-5}$$

The time dependent solutions (4.9-4) represent running waves; if the + sign is chosen, the propagation direction is along the $+x$-axis, and if the − sign is chosen, the propagation direction is reversed. Accordingly, the same choice of signs for the time-independent functions (4.9-3) may be made to represent waves propagating (in time) in those respective directions. The expectation value of the momentum of the particle, according to (4.8-11), is given by

$$\langle p_x \rangle = \int_{-\infty}^{\infty} \psi^* \frac{\hbar}{i} \frac{\partial}{\partial x} \psi \, dx. \tag{4.9-6}$$

However, from (4.9-3) it is easily seen that $\partial\psi/\partial x = \pm ik\psi$, whereby (4.9-6) becomes

$$\langle p_x \rangle = \pm \hbar k \int_{-\infty}^{\infty} \psi^*\psi \, dx = \pm \hbar k = \pm \frac{h}{\lambda}. \tag{4.9-7}$$

This expression is simply a statement of de Broglie's relation (4.6-1); the de Broglie relation for free particles is thus implicitly contained in the postulates of wave mechanics. In addition, from (4.9-5) it is clear that

$$\varepsilon = \frac{\hbar^2 k^2}{2m} = \frac{\langle p^2 \rangle}{2m}, \tag{4.9-8}$$

in agreement with the classical result.

In attempting to normalize the wave function (4.9-3) in accord with the requirements of (4.8-10) it is found that the value of $A*A$ must be chosen to be infinitesimally small in order that the integral (4.8-10) not diverge. Since from (4.9-3) the probability $\psi^*\psi \, dx$ of finding the particle within an interval dx about a point x is the same at all points and equal to $A*A \, dx$, and since the *a priori* classical probability of finding a free particle within some specific finite region of an infinite line is infinitesimally small, it is not surprising that this situation should have arisen. It is, in fact, in agreement with what we should have expected physically. We shall not attempt to deal with this difficulty in a more rigorous manner, but we shall merely note that no serious trouble arises through application of the normalization condition (4.8-10) as long as we recognize the singular nature of the circumstances which govern the magnitude of $A*A$ and do not attempt to evaluate this quantity explicitly in numerical terms.

The solutions of the time-independent Schrödinger equation (4.8-5) yield wave functions of the form (4.9-3) or (4.9-4) which represent plane waves propagating along the $\pm x$-directions. Since the probability density $\Psi^*\Psi$ associated with these functions is independent of time, the solutions (4.9-3) are stationary-state wave functions. Each stationary-state wave function is associated with a unique value of energy ε, and thus, according to (4.9-8) and (4.9-7), represents a state whose propagation constant has a unique value k and whose momentum has a unique value p_x ($=\hbar k$). However, since the amplitude of the waves is constant, the waves run uniformly from ∞ to $-\infty$, and thus have *no unique location* in space. Because the boundary conditions of continuity, finiteness, single-valuedness, etc., are satisfied equally well for any value of ε or k, there is a *continuous* range of allowed values for these parameters.

It is quite possible by superposing solutions of this type, each of which corresponds to a different value of ε and k, to construct wave functions of the form

$$\Psi(x,t) = A_1 e^{i(k_1 x - \omega_1 t)} + A_2 e^{i(k_2 x - \omega_2 t)}. \tag{4.9-9}$$

Due to the *linearity* of the time-dependent Schrödinger equation, such superposition wave functions are perfectly good solutions to the time-dependent equation (4.7-3), although they do *not* satisfy the time-independent equation (4.8-3) because they are not of the form (4.8-7). The probability amplitude associated with such solutions is not independent of time, as it is for the stationary-state solutions. The process of superposition may be extended so that an infinite number of stationary-state solutions are combined into a wave function of the form

$$\Psi(x,t) = \sum_n A_n e^{i(k_n x - \omega_n t)}. \tag{4.9-10}$$

Finally, the values k_n may be chosen very close together, separated only by intervals dk, and the values A_n may assume the values of any given function of k at these points, in which case the superposition (4.9-10) may become in the limit an integral of the form

$$\Psi(x,t) = \int_{-\infty}^{\infty} A(k)e^{i(kx-\omega t)}\, dk. \qquad (4.9\text{-}11)$$

The reader may as an exercise easily verify the fact that this is still a solution to the time-dependent Schrödinger equation (4.7-3). The function $A(k)$ in (4.9-11) may be any function of k whatsoever, so long as the integral exists.

Suppose now that at some instant of time, say $t = 0$, one tries to form a wave function whose probability density $\Psi^*\Psi$ is localized in a particular region of the x-axis, for instance between the points $x = a/2$ and $x = -a/2$, as illustrated in Figure 4.6. This can be accomplished by constructing a superposition of solutions

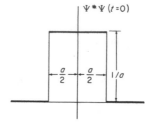

FIGURE 4.6. The probability density function for a particle which is somehow constrained to be along the region of the x-axis extending between $-a/2$ and $+a/2$.

of the stationary-state form (4.9-4), belonging to different values of k, according to (4.9-11). Let us choose

$$A(k) = C\,\frac{\sin \tfrac{1}{2}(k-k_0)a}{\tfrac{1}{2}(k-k_0)a}, \qquad (4.9\text{-}12)$$

where C and k_0 are constants. It will be seen directly that this choice for $A(k)$ is the correct one if $\Psi^*\Psi$ is to be as represented in Figure 4.6. Substituting (4.9-12) into (4.9-11) and setting t equal to zero, it is clear that

$$\Psi(x,0) = C \int_{-\infty}^{\infty} \frac{\sin \tfrac{1}{2}(k-k_0)a}{\tfrac{1}{2}(k-k_0)a}\, e^{ikx}\, dk. \qquad (4.9\text{-}13)$$

Letting
$$q = k - k_0, \qquad (4.9\text{-}14)$$

this can be written in the form

$$\Psi(x,0) = Ce^{ik_0 x} \int_{-\infty}^{\infty} \frac{\sin \tfrac{1}{2}qa}{\tfrac{1}{2}qa}\, e^{iqx}\, dq. \qquad (4.9\text{-}15)$$

Now, according to the Fourier Integral Theorem[2] an arbitrary function $f(x)$ can be expressed as a *Fourier integral* by

$$f(x) = \frac{1}{2\pi} \int_{-\infty}^{\infty} e^{iqx}\, dq \int_{-\infty}^{\infty} f(x')e^{-iqx'}\, dx'. \qquad (4.9\text{-}16)$$

---

[2] See, for example, R. V. Churchill, *Fourier Series and Boundary Value Problems*, McGraw-Hill Book Co. Inc., New York (1941), Chapter V.

If the function shown as $\Psi^*\Psi$ in Figure 4.6, that is,

$$(x) = 1/a \qquad (-a/2 < x < a/2)$$
$$= 0 \qquad (x > a/2; x < -a/2) \tag{4.9-17}$$

is expressed in integral form by using (4.9-16), it can be shown in a straightforward manner that

$$f(x) = \frac{1}{2\pi} \int_{-\infty}^{\infty} \frac{\sin \frac{1}{2}qa}{\frac{1}{2}qa} e^{iqx} \, dq = 1/a \qquad (-a/2 < x < a/2) \tag{4.9-18}$$
$$= 0 \qquad (x > a/2; x < -a/2).$$

Using this result to express the integral in (4.9-15) in simpler form, we find[3]

$$\Psi(x,0) = \frac{2\pi C}{a} e^{ik_0 x} \qquad (-a/2 < x < a/2)$$
$$= 0 \qquad (x > a/2; x < -a/2). \tag{4.9-19}$$

The normalizing constant C must now be chosen so as to satisfy (4.7-6), whereby

$$\frac{4\pi^2 C^2}{a^2} \int_{-\frac{1}{2}a}^{\frac{1}{2}a} dx = \frac{4\pi^2 C^2}{a} = 1$$

or

$$C = \sqrt{a}/(2\pi). \tag{4.9-20}$$

Then

$$\Psi(x,0) = \frac{1}{\sqrt{a}} e^{ik_0 x} \qquad (-a/2 < x < a/2)$$
$$= 0 \qquad (x > a/2; x < -a/2) \tag{4.9-21}$$

In other words, if $A(k)$ is chosen as in (4.9-12), with $C = \sqrt{a}/(2\pi)$, the wave function at time $t = 0$, given by (4.9-15) is ultimately expressable in the simple form (4.9-21). The probability amplitude $\Psi^*(x,0)\Psi(x,0)$ existing at $t = 0$ is then simply that of Figure 4.6, and the "particle" is localized within a distance $\Delta x = a$ about the origin. To accomplish this localization, however, we were forced to introduce a superposition of stationary-state wave functions corresponding to various values of k, hence to various values of *momentum* $\hbar k$. The relative amplitudes of the various momentum components in the superposition is given by the function $A(k)$, thus by (4.9-12) with $C = \sqrt{a}/(2\pi)$. A plot of these momentum contribution amplitudes *versus ka*

---

[3] Although all the superposition functions $A(k)e^{i(kx-\omega t)}$ satisfy the continuity requirement for wave functions, the actual summation as expressed by (4.9-19) does not satisfy this requirement for $x = \pm a$. This violation of the formal requirements, however, introduces no error into the calculations, since obviously the square pulse function (4.9-19) could be "rounded off" very slightly near $x = \pm a$ so as to satisfy the continuity condition without introducing any very significant alteration into the shape of the function $A(k)$.

is shown in Figure 4.7. The principal momentum contributions are from the region near $k = k_0$, and the momentum contribution falls off quite rapidly for values of k much larger than or smaller than k_0. We may regard the contributions to the super-position from values of k larger than $k_0 + \frac{1}{2}\Delta k$ and smaller than $k_0 - \frac{1}{2}\Delta k$ as negligible, where Δk is a parameter which expresses the effective width of that part of the $A(k)$

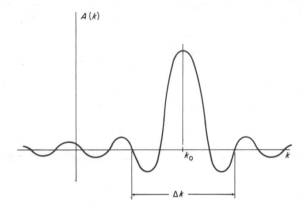

FIGURE 4.7. Momentum amplitude distribution corresponding to the probability amplitude of Figure 4.6.

curve where $A(k)$ takes on values which are not small compared to the maximum value $A(k_0)$. If we arbitrarily choose Δk such that $\Delta(ka) = 8\pi$, a position correspond-ing to the second zero of the curve on either side of $k = k_0$, which is a reasonable choice by the above criterion, we find

$$\Delta k = 8\pi/a. \qquad (4.9\text{-}22)$$

The product of the distance Δx within which the particle may be considered to be localized and the uncertainty Δk in momentum which is introduced by the super-position of states required to accomplish this localization is

$$\Delta k \cdot \Delta x = \frac{8\pi}{a} \cdot a = 8\pi,$$

or, since $k = p/\hbar$,
$$\Delta p \cdot \Delta x = 8\pi\hbar. \qquad (4.9\text{-}23)$$

Neglecting the numerical factors which result from the arbitrary choice of a cutoff point for Δk, and using the symbol $\sim$ to mean "of the order of," we may write

$$\Delta p \cdot \Delta x \sim \hbar. \qquad (4.9\text{-}24)$$

For a free particle whose momentum is characterized by a single unique value of k, the "particle" extends over all space. If the particle is required to be within a given region of space, then its wave function, which must now be regarded as a superposition of stationary-state functions related to different values of k, is characterized by a "spread" of momentum values as illustrated by Figure 4.6, such that (4.9-24) is obeyed. This is a specific illustration of an effect known as the *Heisenberg Uncertainty Principle* which is a very general consequence of the wave-mechanical description of nature. The uncertainty principle states that the position of a particle and its momentum cannot *simultaneously* be defined with arbitrary precision; any experiment which is performed to ensure that the particle will be localized within a given region of space will inevitably *introduce* an uncertainty into its momentum which will be given by (4.9-24). It should be emphasized that this effect is independent of the experimental precision with which the quantities involved can be measured; it is really a property of matter under observation.

It can be shown quite generally that uncertainty relations such as (4.9-24) exist between all pairs of conjugate dynamical variables which classically could be used to specify completely the state of motion of the system. A complete and thorough discussion of this subject is given by Powell and Craseman.[4] If in (4.9-24) we write $\Delta x = v \, \Delta t$, where v is the velocity, we obtain

$$v \, \Delta p \, \Delta t \sim \hbar. \tag{4.9-25}$$

Since for a free particle, $\varepsilon = p^2/2m$, $\Delta \varepsilon = p\Delta p/m = v\Delta p$, (4.9-25) may be expressed as

$$\Delta \varepsilon \cdot \Delta t \sim \hbar. \tag{4.9-26}$$

The Heisenberg relation (4.9-24) may thus also be written in terms of the uncertainty in energy of the wave train times its duration.

It should be noted that the behavior of the "wave packet" which was considered above may be expressed as a function of time by (4.9-11) with $A(k)$ as given by (4.9-12). It can be shown that the maximum of the probability amplitude $\Psi^*\Psi$ moves with constant velocity $\hbar k_0/m$, corresponding to a constant momentum $\hbar k_0$ as predicted by classical mechanics. In addition to this classical motion of the packet as a whole, it is also found that the packet spreads out as time elapses, in such a way that the uncertainty relation (4.9-24) is always satisfied. For wave packets representing large or massive objects which are localized within macroscopically observable distances, this spreading effect is found to be so slow as to be undetectable on a reasonable time scale.[5]

4·10 A PARTICLE IN AN INFINITELY DEEP ONE-DIMENSIONAL POTENTIAL WELL

In this example we shall examine the behavior of a particle in the one-dimensional potential well shown in Figure 4.8. For such a system we have

$$V(x) = 0 \qquad (0 < x < a)$$
$$= \infty \qquad (x < 0; x > a). \tag{4.10-1}$$

[4] J. L. Powell and B. Craseman, *Quantum Mechanics*, Addison-Wesley, Reading, Mass. (1961), pp. 69–76; pp. 182–184.
[5] E. Ikenberry, *Quantum Mechanics for Mathematicians and Physicists*, Oxford University Press, New York (1962), p. 64.

In the region $(0 < x < a)$, where $V = 0$, the time-independent Schrödinger equation (4.8-5) can again be written

$$\frac{d^2\psi}{dx^2} + k^2\psi = 0 \tag{4.10-2}$$

where

$$k = \sqrt{2m\varepsilon/\hbar^2}. \tag{4.10-3}$$

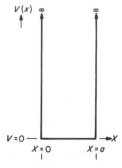

FIGURE 4.8. An infinitely deep potential "well."

In this region the particle is a free particle, and the solution to (4.10-2) could be written in the form (4.9-3). We shall find it more convenient, however, to write it in the equivalent form

$$\psi(x) = A \sin kx + B \cos kx \qquad (0 < x < a), \tag{4.10-4}$$

where A and B are arbitrary constants. Since the particle is constrained by infinitely high potential barriers at $x = 0$ and $x = a$, we shall assume that

$$\psi(x) = 0 \qquad (x < 0; \; x > a). \tag{4.10-5}$$

This assumption, which appears to be physically reasonable, will be justified further in the next section.

If it is required that $\psi(x)$ be continuous at the boundaries of the potential well at $x = 0$ and $x = a$, then (4.10-4) must reduce to zero at those two points. Substituting the values $x = 0$ and $x = a$ into (4.10-4) and setting the result equal to zero, it is clear that $B = 0$ and that

$$A \sin ka = 0. \tag{4.10-6}$$

This condition can be satisfied *only* when

$$ka = n\pi \qquad (n = 1,2,3, \cdots), \tag{4.10-7}$$

thus only for a discrete sequence of values of k, which we label k_n, where

$$k_n = n\pi/a \qquad (n = 1,2,3, \cdots). \tag{4.10-8}$$

Since k is related to the energy by (4.10-3), this condition defines a discrete set of allowed energy eigenvalues

$$\varepsilon_n = \frac{\hbar^2 k_n^2}{2m} = \frac{n^2 \pi^2 \hbar^2}{2ma^2} \qquad (n = 1,2,3, \cdots). \qquad (4.10\text{-}9)$$

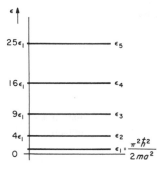

FIGURE 4.9. Allowed energy levels for a particle in the potential well illustrated in Figure 4.8.

Figure 4.9 shows an energy level diagram for this system. In view of the requirement (4.10-8), the wave functions of the system must be of the form

$$\psi_n(x) = A_n \sin \frac{n\pi x}{a} \qquad (0 < x < a),$$

$$= 0 \qquad (x < 0; x > a), \qquad (4.10\text{-}10)$$

the constants A_n being determined by the normalization requirement (4.8-10), according to which

$$\int_{-\infty}^{\infty} \psi_n^* \psi_n \, dx = A_n^2 \int_0^a \sin^2 \frac{n\pi x}{a} \, dx = 1 \qquad (4.10\text{-}11)$$

giving $$A_n = 1/\sqrt{a}. \qquad (4.10\text{-}12)$$

The time dependence of the wave functions is easily determined from (4.8-7). The functions (4.10-10) are, of course, stationary state eigenfunctions, for which $\Psi^*\Psi$ is time-independent.

For this system we find that there is a system of *discrete* energy levels as given by (4.10-9). The Schrödinger equation (4.10-2) within the region of allowed motion is the same as that for a free particle (4.9-1), for which a *continuous* range of energy values are permitted, and the solutions (4.10-10) are simply linear combinations of free-particle solutions of the form (4.9-3) pertaining to the same energy. The difference between the two cases lies solely in the *boundary conditions*, which for the potential well can be satisfied only for a discrete set of energies.

The continuity condition upon the slope of the wave functions (4.10-10) is formally violated by these functions at $x = 0$ and $x = a$. This situation arises from the singularity in the potential function $V(x)$ at those points, and causes no difficulty in the physical interpretation of the results. The behavior of the wave function at these points can be inferred as a limiting case of the results which will be discussed in the next section, confirming the calculations leading to (4.10-10).

The wave functions (4.10-10), which can be represented as

$$\psi_n(x) = \frac{A_n}{2i} \left(e^{n\pi i x/a} - e^{-n\pi i x/a} \right) \qquad (4.10\text{-}13)$$

in the region $(0 < x < a)$, are seen to be standing-wave superpositions of plane wave solutions for which $k = n\pi/a$ and $-n\pi/a$, thus superpositions of a wave propagating along the $+x$-direction and one of equal amplitude going in the opposite direction. Since both components of the superposition have the *same* energy, the linear combination (4.10-13) is a solution to the time-independent Schrödinger equation (4.10-2) and thus still represents a stationary state of the system. The corresponding physical picture is that of a particle which is reflected elastically from the walls of the potential well at $x = 0$ and $x = a$ and oscillates rapidly back and forth in the potential well. The wave function and the probability amplitude for the particle in the lowest energy state and in a higher energy state are shown in Figure 4.10. According to the quantum

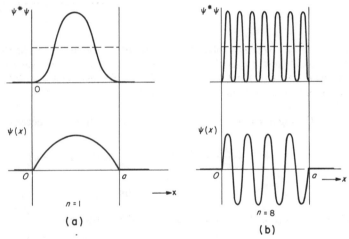

FIGURE 4.10.   Wave functions and probability amplitudes for two particular allowed energy states of a particle in the infinitely deep well.

probability amplitude, the probability of finding the particle at certain points corresponding to the zeros of $\psi^*\psi$ (which for this example simply equals ψ^2, since ψ is real) becomes vanishingly small. The quantum probability amplitude for the lower energy states is not in good agreement with the constant probability amplitude which would be expected classically and which in the figure is shown by the dashed curves. For higher energy states, however, apart from the rapid oscillation of the quantum amplitude about the classical value as an average, the agreement is good in the sense that the mean value of the quantum probability amplitude taken over any interval of appreciable length nearly equals the classical amplitude.

4·11 A PARTICLE IN A ONE-DIMENSIONAL WELL OF FINITE DEPTH

The behavior of a particle in a one-dimensional well of finite depth can be calculated along the general lines of approach which were used in Section 4.10. In this case we shall assume a potential well of the form shown in Figure 4.11, in which the depth of

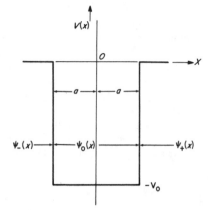

FIGURE 4.11. A potential well of finite depth.

the well is V_0. We must solve separate equations for the wave function in the regions $(x < -a)$, $(-a < x < a)$ and $(x > a)$, and we shall call the wave function in those regions $\psi_-(x)$, $\psi_0(x)$ and $\psi_+(x)$, respectively. We shall first treat the case of a particle which is classically bound to remain within the well, that is, a particle for which $-V_0 < \varepsilon < 0$.

In the regions outside the well $(x < -a; x > a)$, (4.8-5) becomes

$$\frac{d^2\psi_\pm}{dx^2} - k^2\psi_\pm(x) = 0 \qquad (x < -a; x > a) \qquad (4.11\text{-}1)$$

where

$$k = \sqrt{-2m\varepsilon/\hbar^2}, \qquad (4.11\text{-}2)$$

while inside the well, where $V(x) = -V_0$, (4.8-5) becomes

$$\frac{d^2\psi_0}{dx^2} + k_0^2\psi_0(x) = 0 \qquad (-a < x < a) \qquad (4.11\text{-}3)$$

where

$$k_0 = \sqrt{2m(\varepsilon + V_0)/\hbar^2}. \qquad (4.11\text{-}4)$$

As thus defined, under the condition $(-V_0 < \varepsilon < 0)$, k and k_0 are both real quantities. The general solutions to (4.11-1) and (4.11-3) may be written

$$\psi_+(x) = A_+e^{kx} + B_+e^{-kx} \qquad (x > a) \qquad (4.11\text{-}5)$$

$$\psi_0(x) = A_0 \sin k_0x + B_0 \cos k_0x \qquad (-a < x < a) \qquad (4.11\text{-}6)$$

$$\psi_-(x) = A_-e^{kx} + B_-e^{-kx} \qquad (x < -a). \qquad (4.11\text{-}7)$$

The boundary conditions on the wave function demand that the wave function and its derivative be continuous at $x = \pm a$ and that the wave function approach zero as $x \to \pm \infty$. We must thus require

$$\psi_0(a) = \psi_+(a) \qquad\qquad \psi_0'(a) = \psi_+'(a) \qquad\qquad (4.11\text{-}8)$$

$$\psi_0(-a) = \psi_-(-a) \qquad\qquad \psi_0'(-a) = \psi_-'(-a) \qquad\qquad (4.11\text{-}9)$$

$$\psi_+(\infty) = 0 \qquad\qquad \psi_-(-\infty) = 0, \qquad\qquad (4.11\text{-}10)$$

where the primes indicate derivatives with respect to x. It is obvious at once from (4.11-10) that $A_+ = B_- = 0$. The four remaining coefficients can be evaluated by substituting the solutions (4.11-5,6,7) into the four boundary conditions (4.11-8) and (4.11-9), whereby one obtains a set of four simultaneous equations of the form

$$A_0 \sin k_0 a + B_0 \cos k_0 a \quad - \quad B_+ e^{-ka} \qquad\qquad = 0$$

$$-A_0 \sin k_0 a + B_0 \cos k_0 a \qquad\qquad\quad -A_- e^{-ka} = 0$$

$$\qquad\qquad\qquad\qquad\qquad\qquad\qquad\qquad\qquad\qquad (4.11\text{-}11)$$

$$A_0 k_0 \cos k_0 a - B_0 k_0 \sin k_0 a + B_+ k e^{ka} \qquad\quad = 0$$

$$A_0 k_0 \cos k_0 a + B_0 k_0 \sin k_0 a \qquad\qquad -A_- k e^{-ka} = 0.$$

In order to solve the problem completely, the values of the four arbitrary constants A_0, B_0, A_-, and B_+ must be determined. The expressions (4.11-11) form a set of four *homogeneous* equations in these unknown coefficients. It is an elementary result of the theory of linear algebraic equations that such a set of equations can have no solution other than the trivial one $A_0 = B_0 = A_- = B_+ = 0$ *unless* the determinant of the coefficients of the system vanishes.[6] This means that there are no solutions of physical significance *unless*

$$e^{-2ka} \begin{vmatrix} \sin k_0 a & \cos k_0 a & -1 & 0 \\ -\sin k_0 a & \cos k_0 a & 0 & -1 \\ k_0 \cos k_0 a & -k_0 \sin k_0 a & k & 0 \\ k_0 \cos k_0 a & k_0 \sin k_0 a & 0 & k \end{vmatrix} = 0. \qquad (4.11\text{-}12)$$

To obtain this expression the quantity e^{-ka} has been factored out of the third and fourth columns of the determinant of the coefficients of (4.11-11). Since e^{-2ka} does not vanish for any real value of ka, the determinant must vanish in order that (4.11-12) be satisfied. The determinant may be expanded in minors to give the equation

$$k_0^2 \sin k_0 a \cos k_0 a + k k_0 \sin^2 k_0 a - k k_0 \cos^2 k_0 a - k^2 \sin k_0 a \cos k_0 a = 0,$$

which may be written as

$$(k \sin k_0 a + k_0 \cos k_0 a)(k_0 \sin k_0 a - k \cos k_0 a) = 0,$$

---

[6] See for example N. B. Conkwright, *Introduction to the Theory of Equations*, Ginn and Co., Boston (1941), pp. 144–145.

or, dividing by $\cos^2 k_0 a$,

$$(k \tan k_0 a + k_0)(k_0 \tan k_0 a - k) = 0. \tag{4.11-13}$$

This equation is satisfied if *either* factor vanishes. If we let

$$\mu^2 = 2mV_0/\hbar^2, \tag{4.11-14}$$

then, according to (4.11-4) and (4.11-2),

$$k^2 = \mu^2 - k_0^2. \tag{4.11-15}$$

Substituting this in (4.11-13) and equating either factor to zero, we find that (4.11-13) is satisfied if

either
$$\sqrt{\mu^2 - k_0^2} = -k_0 \operatorname{ctn} k_0 a$$
$$\tag{4.11-16}$$
or
$$\sqrt{\mu^2 - k_0^2} = k_0 \tan k_0 a.$$

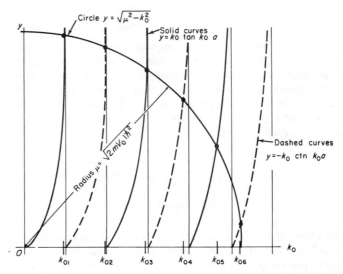

FIGURE 4.12. Diagram illustrating the solution of the transcendental equations for the energy eigenvalues associated with the potential well of Figure 4.11.

One or the other of these equations will be satisfied only for a certain set of values of k_0, which we shall represent by $k_{01}, k_{02}, k_{03}, \cdots k_{0n}, \cdots$. A set of energy eigenvalues ε_n is related to these values of k_0, according to (4.11-4), by

$$\varepsilon_n = \frac{\hbar^2 k_{0n}^2}{2m} - V_0. \tag{4.11-17}$$

The eigenvalues k_{0n} can be obtained graphically by finding the intersections of the curves $y = -k_0 \operatorname{ctn} k_0 a$ and $y = k_0 \tan k_0 a$ with the circle $y = \sqrt{\mu^2 - k_0^2}$ as shown in Figure 4.12. For each intersection point one or the other of Equations (4.11-16) is satisfied. There is thus defined a finite set of values k_{0n} and a *finite* number of discrete energy levels corresponding to these values through (4.11-17). The radius of the circle $y = \sqrt{\mu^2 - k_0^2}$ in Figure 4.11 is simply μ, which is directly related by (4.11-14) to the depth of the potential well. For a very shallow well, $\mu \to 0$, and there will be one and only one energy level in the range $-V_0 < \varepsilon < 0$. For deeper wells, the number of energy states increases, and for very deep wells may become very large. In a very deep well, as illustrated by Figure 4.12, the lower-lying levels are characterized by values of k_{0n} which are almost equally spaced, corresponding quite closely to the k_n values of the infinite well of Section 4.10, and giving rise to energy levels which, with respect to the bottom of the well, are close to those of the infinite well. The energy level diagrams for a number of finite wells are shown in Figure 4.13.

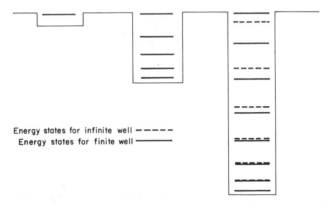

Energy states for infinite well – – – – –
Energy states for finite well ————

FIGURE 4.13. Schematic representation of the energy levels of potential wells of various depths.

The values k_{0n} for which (4.11-11) has physically meaningful solutions having been determined, the coefficients A_0, B_0, A_- and B_+ may be found. Since for the values k_{0n} which satisfy (4.11-16) the determinant of the homogeneous system (4.11-11) vanishes, the equations (4.11-11) are *no longer linearly independent* equations, and hence only the *ratios* of the coefficients can be obtained from (4.11-11) directly. The actual magnitudes of the coefficients may, however, be found by imposing the condition that

$$\int_{-\infty}^{\infty} \psi^* \psi \, dx = \int_{-\infty}^{-a} \psi_-^* \psi_- \, dx + \int_{-a}^{a} \psi_0^* \psi_0 \, dx + \int_{a}^{\infty} \psi_+^* \psi_+ \, dx = 1. \quad (4.11\text{-}18)$$

The wave functions themselves follow from (4.11-5,6,7) once the coefficients are known. The actual calculation of the coefficients is a laborious matter and will not be discussed in detail here. We shall, however, discuss some of the properties of the wave functions which are so obtained.

The wave functions for the bound energy states of this problem are illustrated in Figure 4.14. It will be noted first of all that the wave function *as a whole* is either an even function or an odd function of x. In the present example the lowest energy state is even, the next odd, the next even, and so forth, alternately. This comes about as a result of the fact that the potential $V(x)$ is itself an even function of x; it can be shown quite generally that when this is so the eigenfunctions must be either even or odd functions of x.[7]

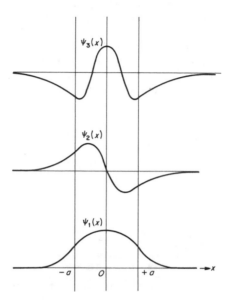

FIGURE 4.14. Schematic diagrams of the wave functions of the lowest energy states of the finite potential well of Figure 4.11.

From Figure 4.14 and from Equations (4.11-5) and (4.11-7) it can be seen that in spite of the fact that the total energy of the particle is negative and hence according to the classical picture the particle would never be able to surmount the potential barrier and appear outside the well, the wave function of the particle does indeed extend beyond the limits of the well. According to quantum mechanics, therefore, there is a definite probability that the particle will be found in the classically forbidden region beyond the actual boundaries of the well. The wave function of the particle is exponentially attenuated in this region and approaches zero far outside the well. This phenomenon, which is known as *barrier penetration*, is quite a common aspect of the quantum behavior of matter. A particle approaching a potential barrier of finite thickness and height as shown in Figure 4.15 has a certain probability of penetrating the barrier and appearing on the other side, even though this may be energetically forbidden on a classical basis. The wave function is attenuated within the barrier, and if the barrier is very high or quite thick, the attenuation becomes very strong and

---

[7] See for example E. Merzbacher, *Quantum Mechanics*, John Wiley and Sons, New York (1961), p. 53.

the probability of penetration becomes extremely small. This effect is called quantum mechanical *tunneling*, and has been observed experimentally in semiconductor devices and thin insulating films. Reflection effects are also predicted quantum mechanically for particles incident on a barrier such as that of Figure 4.15 which have *more* than enough energy to surmount the barrier classically. Quantum tunneling effects provide the generally accepted explanations for the electrical breakdown of insulators, reverse breakdown of semiconductor rectifiers and radioactive decay of α-emitting isotopes.

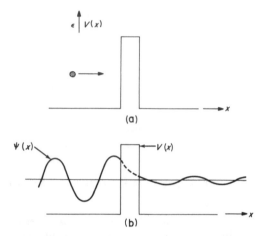

FIGURE 4.15. (a) Classical and (b) quantum mechanical pictures of a particle interacting with a potential barrier whose height is greater than the initial particle energy. The quantum picture (b) exhibits the phenomenon of *tunneling*.

So far, this discussion has been based exclusively upon the assumption that the total energy ε is negative, lying in the range $-V_0 < \varepsilon < 0$. In the case where ε is positive, it is easy to see from (4.11-1) that the solutions $\psi_\pm$ outside the well become oscillatory in nature, like the solution ψ_0 inside. In this instance, it turns out that the Schrödinger equation and all the boundary conditions can be satisfied for *any* positive value of ε. There is thus a *continuous* range of allowed energy states and corresponding eigenfunctions extending upward from $\varepsilon = 0$ in Figure 4.13. These states are referred to as the "continuum states."

To see how this comes about, let us assume that the eigenfunctions in this energy range are either even or odd functions of x; we have seen that this must be true if $V(x)$ is an even function. Then, for *even* eigenfunctions, we must have

$$\psi_0(x) = A_0 \cos k_0 x \qquad\qquad (-a < x < a)$$

$$\psi_+(x) = A_+ \cos kx + B_+ \sin kx \qquad (x > a) \qquad\qquad (4.11\text{-}19)$$

$$\psi_-(x) = \psi_+(-x), \qquad\qquad (x < -a)$$

where *now k is defined as*

$$k = \sqrt{2m\varepsilon/\hbar^2} \qquad (\varepsilon > 0). \tag{4.11-20}$$

The boundary conditions (4.11-8) and (4.11-9) still must apply, so that the values and slopes of $\psi_0(x)$ and $\psi_+(x)$ must be equal at $x = a$. From (4.11-19) this requires that

$$A_0 \cos k_0 a = A_+ \cos ka + B_+ \sin ka$$

and $$-k_0 A_0 \sin k_0 a = -kA_+ \sin ka + kB_+ \cos ka. \tag{4.11-21}$$

Dividing both equations by A_0 and solving the resulting set of simultaneous equations for A_+/A_0 and B_+/A_0, we find that

$$A_+/A_0 = \cos k_0 a \cos ka + \frac{k_0}{k} \sin k_0 a \sin ka$$

$$\tag{4.11-22}$$

$$B_+/A_0 = \cos k_0 a \sin ka - \frac{k_0}{k} \sin k_0 a \cos ka.$$

It is thus always possible to find perfectly good values for A_+ and B_+ in terms of A_0, whose value, in turn, is fixed by normalization requirements. This means that whatever value of ε or k is chosen initially, a solution for ψ_0, ψ_+, and ψ_- satisfying all the boundary conditions can be found, as asserted previously. A similar calculation can be made starting with the *odd* eigenfunctions, the same result being obtained. The general character of the wave functions for the continuum states is shown in Figure 4.16.

This state of affairs is illustrative of a much more general principle. If the total energy of the system is such that the particle is classically constrained by the potential $V(x)$ to move in a given finite region of space, then there will be a discrete set of eigenfunctions and energy levels which satisfy all the requirements for wave functions. On the other hand, for particle energies sufficiently large that a classical particle can escape from any potential minimum of the system to infinity in at least one direction, there will be a continuum of energy states and corresponding eigenfunctions.[8]

The results of Section 4.10 for the infinitely deep potential well can be obtained from the results of this section if V_0 is allowed to approach infinity. In particular, it is easily seen from (4.11-5) and (4.11-7) with $A_+ = B_- = 0$, that if k becomes indefinitely large, as it must for any state with a finite energy above the bottom of the well as $V_0 \to \infty$, the wave function outside the limits of the well approaches zero. As the well becomes very deep, the curvature of the wave function near $x = a$ and $x = -a$ where the solutions $\psi_+(x)$ and $\psi_-(x)$ are joined to the solution $\psi_0(x)$ becomes very large, although for a finite well depth, the slope is always continuous. It is only in the limit as the well depth approaches infinity that the discontinuity in the derivative of the wave function at these points, which was mentioned in Section 4.10, arises.

---

[8] A simple proof of this assertion is given by L. Pauling and E. B. Wilson, *Introduction to Quantum Mechanics*, McGraw-Hill Book Co., Inc., New York (1935).

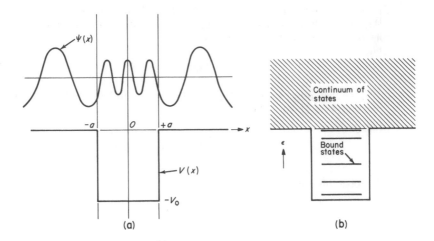

FIGURE 4.16. (a) Schematic representation of the wave function of a particle whose energy is sufficiently great that it would not in the classical picture be constrained to remain within the illustrated potential well. (b) Division of allowed energy levels for the finite potential well into a group of discrete levels corresponding to a classically bound particle and to a continuum of levels corresponding to a classically unbound particle. The wave function illustrated in (a) belongs to the latter group of levels.

$4 \cdot 12$ THE ONE-DIMENSIONAL HARMONIC OSCILLATOR

A harmonic oscillator is a particle which is bound to an equilibrium position by a force which is proportional to the displacement from that position; this force may thus be expressed as

$$F = -kx = -dV(x)/dx. \tag{4.12-1}$$

The potential $V(x)$ must then be a parabola of the form

$$V(x) = \tfrac{1}{2}kx^2. \tag{4.12-2}$$

From (4.12–1) the classical equation of motion for a particle of mass m subjected to such a force is

$$m\frac{d^2x}{dt^2} = -kx \tag{4.12-3}$$

whose solution is an oscillatory function which can be written in the form

$$x(t) = A \sin(\omega_0 t - \delta), \tag{4.12-4}$$

where

$$\omega_0 = \sqrt{k/m} \tag{4.12-5}$$

is the *classical oscillator frequency*, and where A and δ are constants. Following this classical result, let us use (4.12-5) to express the force constant k in terms of ω_0 and m, and rewrite the potential $V(x)$ as

$$V(x) = \tfrac{1}{2}m\omega_0^2 x^2. \tag{4.12-6}$$

We shall now proceed to solve Schrödinger's equation (4.8-5) using this potential function; (4.8-5) will then read

$$\frac{d^2\psi}{dx^2} + \frac{2m}{\hbar^2}\left[\varepsilon - \tfrac{1}{2}m\omega_0^2 x^2\right]\psi(x) = 0. \tag{4.12-7}$$

To solve this equation, we first let

$$\alpha = m\omega_0/\hbar \tag{4.12-8}$$

$$\beta = 2\varepsilon/(\hbar\omega_0) \tag{4.12-9}$$

$$\xi = x\sqrt{\alpha}. \tag{4.12-10}$$

Then, from (4.12-8) and (4.12-9), it is clear that

$$\frac{d^2\psi}{dx^2} + (\alpha\beta - \alpha^2 x^2)\psi(x) = 0, \tag{4.12-11}$$

and from (4.12-10), noting that $d^2\psi/dx^2 = \alpha\, d^2\psi/d\xi^2$,

$$\frac{d^2\psi}{d\xi^2} + (\beta - \xi^2)\psi(\xi) = 0. \tag{4.12-12}$$

For *large* values of $|\xi|$, such that $\xi^2 \gg \beta$, we may neglect β and write (4.12-12) as

$$\frac{d^2\psi}{d\xi^2} \cong \xi^2\psi. \tag{4.12-13}$$

This equation is satisfied approximately, for large ξ, by

$$\psi(\xi) = e^{-\xi^2/2}, \tag{4.12-14}$$

because then $d^2\psi/d\xi^2 = (\xi^2 - 1)\exp(-\xi^2/2)$, which satisfies (4.12-13) approximately if $\xi^2 \gg 1$. This suggests that the solutions $\psi(\xi)$ for (4.12-12) might be more simply expressed by writing

$$\psi(\xi) = e^{-\xi^2/2}H(\xi) \tag{4.12-15}$$

and transforming (4.12-12) to an equation for the function $H(\xi)$. This may be accomplished by simply substituting the form (4.12-15) into (4.12-12), the result being that (4.12-12) becomes

$$\frac{d^2H}{d\xi^2} - 2\xi\frac{dH}{d\xi} + (\beta - 1)H(\xi) = 0. \tag{4.12-16}$$

This differential equation may be solved by a power series technique, assuming initially that

$$H(\xi) = \sum_{k=0}^{\infty} a_k \xi^k, \tag{4.12-17}$$

whereby

$$\frac{d^2 H}{d\xi^2} = \sum_{k=0}^{\infty} (k+1)(k+2)a_{k+2} \xi^k$$

$$-2\xi \frac{dH}{d\xi} = \sum_{k=0}^{\infty} -2k a_k \xi^k \tag{4.12-18}$$

$$(\beta - 1)H(\xi) = \sum_{k=0}^{\infty} (\beta - 1)a_k \xi_k.$$

Adding these three equations and noting that the right-hand sides add to zero according to (4.12-16), we find

$$0 = \sum_{k=0}^{\infty} [(k+1)(k+2)a_{k+2} + [(\beta - 1) - 2k]a_k] \xi^k. \tag{4.12-19}$$

This equation must hold true for *all values* of ξ, and therefore the coefficient of *each power* of ξ must vanish separately. (This may be seen more clearly by evaluating the coefficients for the expansion of the function $f(\xi) = 0$ in McLaurin's power series; they are all zero.) This requirement at once establishes a *recursion relation* between the coefficients a_{k+2} and a_k, in which

$$a_{k+2} = -\frac{\beta - 1 - 2k}{(k+1)(k+2)} a_k. \tag{4.12-20}$$

In the solution of any second-order ordinary differential equation there must be two arbitrary constants. Therefore, let us regard a_0 and a_1 as fundamental arbitrary constants, to be determined from boundary conditions, in the series expression (4.12-17) for $H(\xi)$, and express all the other coefficients in terms of these two by repeated application of the recursion relation (4.12-20). It is easily seen in this way that

$$a_2 = -\frac{\beta - 1}{2!} a_0 \qquad\qquad a_3 = -\frac{(\beta - 3)}{3!} a_1$$

$$a_4 = -\frac{\beta - 5}{4 \cdot 3} a_2 = \frac{(\beta - 1)(\beta - 5)}{4!} a_0 \qquad a_5 = \frac{(\beta - 3)(\beta - 7)}{5!} a_1 \tag{4.12-21}$$

$$a_6 = -\frac{(\beta - 1)(\beta - 5)(\beta - 9)}{6!} a_0 \qquad a_7 = \frac{(\beta - 3)(\beta - 7)(\beta - 11)}{7!} a_1$$

etc.,   etc.

According to (4.12-17), then, we must have

$$H(\xi) = a_0 \left[1 - \frac{(\beta - 1)}{2!} \xi^2 + \frac{(\beta - 1)(\beta - 5)}{4!} \xi^4 - \frac{(\beta - 1)(\beta - 5)(\beta - 9)}{6!} \xi^6 + \cdots \right]$$

$$+ a_1 \left[\xi - \frac{\beta - 3}{3!} \xi^3 + \frac{(\beta - 3)(\beta - 7)}{5!} \xi^5 - \frac{(\beta - 3)(\beta - 7)(\beta - 11)}{7!} \xi^7 + \cdots \right].$$

(4.12-22)

If, in the recursion relation (4.12-20), $\beta - 1 - 2k$ should be *zero* for some value of the index k, then $a_{k+2} = 0$. But since a_{k+4} is a multiple of a_{k+2} and a_{k+6} is a multiple of a_{k+4}, etc., *all* the succeeding coefficients which are related to a_k by the recursion formula would then vanish, and one or the other of the bracketed series in (4.12-22) would *terminate* to become a *polynomial* of degree k. This phenomenon can occur only for certain *integer* values of β, in fact, from (4.12-20) only when $\beta - 1$ equals twice an integer, thus only when

$$\beta - 1 - 2n = 0$$

or $$\beta = 2n + 1 \quad (n = 0,1,2,3, \ldots).$$ (4.12-23)

It is easily seen by inspection of (4.12-22) that this termination of one or the other series to produce a polynomial does indeed occur when $\beta = 1,3,5,7, \ldots$ as predicted by (4.12-23). It turns out that the *only* solutions of (4.12-16) which lead to acceptable wave functions *via* (4.12-15) will be these polynomial solutions.

In order to understand why this is so, let us first consider the series expansion for the function e^{ξ^2}, which can be written

$$e^{\xi^2} = 1 + \frac{\xi^2}{1!} + \frac{\xi^4}{2!} + \frac{\xi^6}{3!} + \cdots = \sum_k b_k \xi^k.$$

(4.12-24)

For large values of ξ, the contributions of the initial terms of this series are negligible in comparison with those for which k is large. This is true also, of course, for the series in (4.12-22) for $H(\xi)$. For the series (4.12-24) the ratio of the coefficient b_{k+2} to the coefficient b_k is

$$\frac{b_{k+2}}{b_k} = \frac{(k/2)!}{\left[\frac{k}{2} + 1 \right]!} = \frac{1}{\frac{k}{2} + 1},$$

and for large values of k

$$\frac{b_{k+2}}{b_k} \cong 2/k.$$

(4.12-25)

The ratio of the coefficients a_{k+2} and a_k for the series solutions of (4.12-16) is given by the recursion relation (4.12-20) and, for large values of k, is readily seen to be also approximately equal to $2/k$. The series solutions for $H(\xi)$, thus, when not terminated to polynomials, behave like e^{ξ^2} for large values of ξ. According to (4.12-15), however, the actual wave functions are given by

$$\psi(\xi) = e^{-\xi^2/2}H(\xi) \cong e^{-\xi^2/2}e^{\xi^2} = e^{\xi^2/2} \qquad (4.12\text{-}26)$$

for large values of ξ. Since these functions do not remain finite at $\pm\infty$ they are not acceptable as wave functions for the physical system. The only solutions which are acceptable are the solutions which terminate as polynomials, that is, the solutions for which (4.12-23) is obeyed. If one or the other of the two bracketed series of (4.12-22) terminates as a polynomial, the solution for $H(\xi)$ then takes the form

$$H(\xi) = a_0(\text{polynomial}) + a_1(\text{infinite series})$$

or, $$H(\xi) = a_0(\text{infinite series}) + a_1(\text{polynomial}).$$

In either case, in order that the solution $H(\xi)$ lead to a physically acceptable wave function, the arbitrary constant multiplying the series solution must be chosen to be *zero*, so that the total expression for $H(\xi)$ is simply a polynomial. Expressing β in (4.12-23) in terms of ε by (4.12-9), it is clear that the only solutions which yield the polynomial expressions for $H(\xi)$ are those corresponding to a discrete set of energy eigenvalues

$$\varepsilon_n = (n + \tfrac{1}{2})\hbar\omega_0 \qquad (n = 0,1,2,3,\cdots). \qquad (4.12\text{-}27)$$

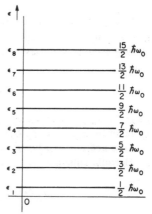

FIGURE 4.17. Energy levels of the one-dimensional quantum harmonic oscillator.

A notable feature of this system is that the lowest energy state of the system, for which $n = 0$, is characterized not by $\varepsilon_0 = 0$, but by $\varepsilon_0 = \tfrac{1}{2}\hbar\omega_0$. In other words, the lowest energy state of the harmonic oscillator is characterized not by the absence of vibrational motion but by a minimum "zero-point" vibrational energy $\tfrac{1}{2}\hbar\omega_0$. In his original theory of black body radiation, Planck assumed that the energy levels of a harmonic oscillator where given by $\varepsilon_n = n\hbar\omega_0$ rather than by (4.12-27); in that particular instance, however, the zero-point energy makes no difference in the final result. It will be noted that the potential wells of the preceding sections also exhibit this zero-point energy effect. The energy level diagram for the harmonic oscillator is illustrated in Figure 4.17. The equal spacing of levels is a unique property of the parabolic potential well.

The polynomial solutions which represent the eigenfunctions of the system can be obtained from (4.12-22) by considering β to take on the values $1, 3, 5, \cdots 2n + 1, \cdots$ in accord with (4.12-23). For each such value one or the other of the two bracketed series will terminate to a polynomial, all the coefficients beyond a certain point being zero. It is customary to choose the arbitrary constant a_0 or a_1, as the case may be, which multiplies the resulting polynomial, such that the coefficient of the highest power of ξ in the polynomial is $2^n (= 2^{(\beta-1)/2})$. When this is done, a set of polynomial solutions $H_n(\xi)$ of the form

$$H_0(\xi) = 1$$

$$H_1(\xi) = 2\xi$$

$$H_2(\xi) = 4\xi^2$$

$$H_3(\xi) = 8\xi^3 - 12 \tag{4.12-28}$$

$$H_4(\xi) = 16\xi^4 - 48\xi^2$$

$$H_5(\xi) = 32\xi^5 - 160\xi^3 + 120\xi$$

$$\text{etc.,}$$

is obtained. These polynomials are well known to mathematicians and are called the *Hermite polynomials*; Equation (4.12-16) is known as the Hermite equation. The Hermite polynomials may be defined as

$$H_n(\xi) = (-1)^n e^{\xi^2} \frac{d^n}{d\xi^n} (e^{-\xi^2}). \tag{4.12-29}$$

It can be shown that the definition (4.12-29) leads to the polynomials which satisfy the Hermite equation (4.12-16). The proof is assigned as an exercise.

The actual wave functions $\psi_n(x)$ are obtained from the polynomial solutions of the Hermite equation by (4.12-15) and (4.12-10), the result being

$$\psi_n(x) = N_n e^{-\xi^2/2} H_n(\xi) = N_n e^{-\alpha x^2/2} H_n(x\sqrt{\alpha}), \tag{4.12-30}$$

where α is given by (4.12-8) and where the normalization constant N_n is chosen such that

$$\int_{-\infty}^{\infty} \psi_n^* \psi_n \, dx = 1.$$

This condition can be shown[9] to require that

$$N_n = \sqrt{\sqrt{\frac{\alpha}{\pi}} \cdot \frac{1}{2^n n!}}. \tag{4.12-31}$$

---

[9] D. Bohm, *Quantum Mechanics*, Prentice-Hall, Englewood Cliffs, N.J. (1951), p. 305.

The time-dependent wave functions $\Psi_n(x,t)$ are obtained from the expressions of (4.12-30) by multiplying by a factor $e^{-i\varepsilon_n t/\hbar}$, in accord with (4.8-7).

Since we shall wish to compare the properties of the quantum oscillator with those of its classical counterpart, we must again briefly consider some characteristics of the classical oscillator and, in particular, we must compute the probability distribution associated with the classical oscillator. Since the displacement $x(t)$ of the classical particle is as given by (4.12-4), it follows that the kinetic and potential energy of the classical oscillator are given by

$$\varepsilon_k = \tfrac{1}{2}m(dx/dt)^2 = \tfrac{1}{2}mA^2\omega_0^2 \cos^2(\omega_0 t - \delta)$$

$$\varepsilon_p = \tfrac{1}{2}m\omega_0^2 x^2 = \tfrac{1}{2}mA^2\omega_0^2 \sin^2(\omega_0 t - \delta),$$

respectively, whence the total energy must be

$$\varepsilon = \varepsilon_k + \varepsilon_p = \tfrac{1}{2}mA^2\omega_0^2. \tag{4.12-32}$$

This equation may be solved for A in terms of ε, giving $A = \sqrt{2\varepsilon/m\omega_0^2}$, and if the energy is assumed to have the value $(n + \tfrac{1}{2})\hbar\omega_0$ (as it must for the quantum oscillator to which we wish to make a direct comparison), we obtain for the amplitude of the *classical* oscillator having this energy

$$A = \sqrt{\frac{(2n + 1)\hbar}{m\omega_0}} = \sqrt{\frac{2n + 1}{\alpha}}, \tag{4.12-33}$$

where α is defined by (4.12-8).

The probability amplitude is the probability of finding the particle within a given region dx about the point x. For the classical oscillator, this is simply the ratio of the time dt which the particle spends in this region during the course of one vibration to the period of oscillation $T\ (= 2\pi/\omega_0)$. Since $dx = v(x)\,dt$, the time spent in this interval during one period is

$$2\,dt = \frac{2dx}{v(x)}, \tag{4.12-34}$$

the particle passing through the region *twice* during a full period of the motion. Since, from (4.12-4), $v(t) = dx/dt = A\omega_0 \cos(\omega_0 t - \delta)$, and since $\omega_0 t - \delta = \sin^{-1}(x/A) = \cos^{-1}(1 - (x/A)^2)^{1/2}$, v can be expressed as a function of x by

$$v(x) = A\omega_0 \sqrt{1 - \frac{x^2}{A^2}}. \tag{4.12-35}$$

The fraction of time spent in dx is $P(x)\,dx$, where $P(x)$ is the classical probability amplitude, corresponding to the quantum mechanical quantity $\psi^*\psi$. Since this quantity is equal to $2dt/T$, we must have, in view of (4.12-34) and of the relation $T = 2\pi/\omega_0$,

$$P(x)\,dx = \frac{2\,dt}{T} = \frac{\omega_0\,dx}{\pi v(x)} = \frac{dx}{\pi\sqrt{A^2 - x^2}}. \tag{4.12-36}$$

Using (4.12-33), the probability amplitude of a classical oscillator having energy $(n + \frac{1}{2})\hbar\omega_0$ becomes

$$P(x) = \frac{1}{\sqrt{\dfrac{2n + 1}{\alpha} - x^2}}. \qquad (4.12\text{-}37)$$

We shall compare this quantity with the quantum probability amplitudes $\psi_n^* \psi_n$ obtained from (4.12-30).

Figure 4.18 shows the general behavior of the wave functions for the harmonic oscillator. It will be noted that here again the wave function penetrates into the region beyond the limits of amplitude of the classical motion. In addition, as we might

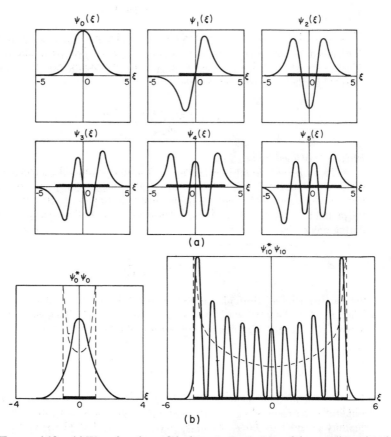

(a)

(b)

FIGURE 4.18. (a) Wave functions of the lowest energy states of the one-dimensional quantum harmonic oscillator. (b) Quantum probability amplitudes (solid curves) and classical probability distributions (dashed curves) for the one-dimensional harmonic oscillator at two energy values, $\frac{1}{2}\hbar\omega_0$ and $\frac{21}{2}\hbar\omega_0$. [After L. Pauling and E. B. Wilson, *Introduction to Quantum Mechanics*, McGraw-Hill, New York (1935).]

expect from the fact that $V(x)$ is an even function of x, the wave functions are alternately even and odd functions of x for ascending values of n. For the lowest energy state the classical probability amplitude as computed from (4.12-37), shown as the dotted curve in the figure, does not resemble the quantum mechanical probability amplitude at all. The classical probability amplitude predicts that the particle is most likely to be found near the ends of the region of allowed motion, while the quantum result predicts that it will most likely be found at the center. For a state of larger amplitude and correspondingly higher energy, however, there is good agreement between the classical and quantum probability distributions, apart from the fact that, as usual, the quantum probability amplitude oscillates rapidly about a mean value close to the classical result. As the amplitude and energy associated with the motion become detectable on a macroscopic scale, the quantum and classical results are not appreciably different. There are, as usual, points along the classical track of the particle where the probability of finding the particle, as predicted by the quantum probability amplitude, becomes vanishingly small.

4·13 ORTHOGONALITY OF EIGENFUNCTIONS AND SUPERPOSITION OF STATES

It can be shown that eigenfunctions of the one-dimensional Schrödinger equation satisfying the physical requirements for wave functions, always exhibit the property of *orthogonality*, that is, they always obey the equation

$$\int_{-\infty}^{\infty} \psi_m^* \psi_n \, dx = 0 \qquad (m \neq n)$$
$$= 1 \qquad (m = n). \qquad (4.13\text{-}1)$$

These relations are a characteristic property of the Schrödinger equation and the boundary conditions, and are found to hold *no matter what the potential function $V(x)$ is* nor what mathematical form the wave functions are found to have. That this must be true can be seen by writing the one-dimensional Schrödinger equation for ψ_n,

$$\frac{d^2\psi_n}{dx^2} - \frac{2m}{\hbar^2} [\varepsilon_n - V(x)]\psi_n(x) = 0, \qquad (4.13\text{-}2)$$

and the wave equation for ψ_m^*, which, from (4.7-12) and (4.8-6) must be

$$\frac{d^2\psi_m^*}{dx^2} - \frac{2m}{\hbar^2} [\varepsilon_m - V(x)]\psi_m^*(x) = 0. \qquad (4.13\text{-}3)$$

Multiplying the first equation by ψ_m^*, the second by ψ_n, subtracting the two equations, and integrating from ∞ to $-\infty$, we obtain

$$\int_{-\infty}^{\infty} \left[\psi_m^* \frac{d^2\psi_n}{dx^2} - \psi_n \frac{d^2\psi_m^*}{dx^2} \right] dx = \frac{2m}{\hbar^2} (\varepsilon_n - \varepsilon_m) \int_{-\infty}^{\infty} \psi_m^* \psi_n \, dx. \qquad (4.13\text{-}4)$$

The integrand of the term on the left-hand side of this equation can be expressed as the derivative of $\psi_m^*(d\psi_n/dx) - \psi_n(d\psi_m^*/dx)$ with respect to x, whence (4.13-4) becomes

$$\left[\psi_m^* \frac{d\psi_n}{dx} - \psi_n \frac{d\psi_m^*}{dx}\right]_{-\infty}^{\infty} = \frac{2m}{\hbar^2}(\varepsilon_n - \varepsilon_m) \int_{-\infty}^{\infty} \psi_m^* \psi_n \, dx. \qquad (4.13\text{-}5)$$

For physically well-behaved wave functions ψ and $d\psi/dx$ approach zero as $x \to \pm\infty$; therefore, the quantity on the left vanishes at both endpoints and

$$\frac{2m}{\hbar^2}(\varepsilon_n - \varepsilon_m) \int_{-\infty}^{\infty} \psi_m^* \psi_n \, dx = 0. \qquad (4.13\text{-}6)$$

If the wave functions ψ_m^* and ψ_n are assumed to belong to two distinct energy levels, so that $\varepsilon_n \neq \varepsilon_m$, the only way in which (4.13-6) can be satisfied is if

$$\int_{-\infty}^{\infty} \psi_m^* \psi_n \, dx = 0 \qquad (m \neq n). \qquad (4.13\text{-}7)$$

For $m = n$, the factor $\varepsilon_n - \varepsilon_n$ in (4.13-6) vanishes; in this case from the way in which the wave functions are defined (4.8-10) we must have

$$\int_{-\infty}^{\infty} \psi_n^* \psi_n \, dx = 1. \qquad (4.13\text{-}8)$$

The wave functions ψ_m, ψ_n thus exhibit the property of orthogonality. The property of orthogonality is not restricted to solutions of Schrödinger's equation, but is also associated with the characteristic solutions of a large class of differential equations, including the heat equation, the equation of the vibrating string, and the electromagnetic wave equations. It is easily seen that the wave functions (4.10-10) associated with the infinitely deep potential well will satisfy the orthogonality conditions, because it is well known (and easily verified) that

$$\int_0^L \sin \frac{m\pi x}{L} \sin \frac{n\pi x}{L} \, dx = 0. \qquad (m \neq n) \qquad (4.13\text{-}9)$$

Although it is more difficult to prove for the finite well and for the harmonic oscillator, it is, of course, true that the wave functions (4.11-5,6,7) and (4.12-30) pertaining to these systems are also orthogonal.[10]

Suppose now that we have an infinite set of eigenfunctions $\psi_n(x)$ which are solutions of Schrödinger's equation for some particular potential $V(x)$. We know that these eigenfunctions must be orthogonal and normalized in accord with (4.8-10), so that they obey (4.13-1). It is possible to combine these eigenfunctions as a linear superposition in such a way that the sum will represent an *arbitrary* function of x, subject to certain mathematical restrictions which are not very severe. To show how this may be done, let us assume that an arbitrary function $f(x)$ may be represented as a

---

[10] *Ibid.*, p. 219. See also R. V. Churchill, *Fourier Series and Boundary Value Problems*, Mc-Graw-Hill, New York (1941), pp. 46-52.

linear combination of the eigenfunctions, such that

$$f(x) = \sum_{n=0}^{\infty} a_n \psi_n(x). \tag{4.13-10}$$

Multiplying this equation by $\psi_m^*(x)$ and integrating, we find

$$\int_{-\infty}^{\infty} f(x)\psi_m^*(x)\,dx = \sum_{n=0}^{\infty} \int_{-\infty}^{\infty} a_n \psi_m^*(x)\psi_n(x)\,dx. \tag{4.13-11}$$

In the summation all the integrals are zero by virtue of (4.13-1) except that for which $n = m$, which yields the value a_m. The final result is

$$a_m = \int_{-\infty}^{\infty} f(x)\psi_m^*(x)\,dx. \tag{4.13-12}$$

Substituting this value for a_m into (4.13-10) it is clear that $f(x)$ can be represented as

$$f(x) = \sum_{n} \left[\int_{-\infty}^{\infty} f(x')\psi_n^*(x')\,dx' \right] \psi_n(x), \tag{4.13-13}$$

where the dummy variable of integration x' has been substituted for x to avoid confusion.

For the sine eigenfunctions of Section 4.10, this expansion turns out to be nothing more than a Fourier series expansion. Similar expansions may be made in terms of the Hermite functions associated with the harmonic oscillator and in terms of other wave functions related to other potential functions. We have already seen in Section 4.9 an example of a situation where the eigenvalues and eigenfunctions are continuously distributed rather than discrete. In that case we found that an arbitrary function could be represented as an *integral* combination of the eigenfunctions in the form of a Fourier integral. This result can be generalized to eigenfunctions representing *any* system whose eigenvalues are continuously distributed, the result being expressable as Equation (4.13-13) in the limit where the eigenvalues crowd closely together, the sum in that equation then becoming an integral. Although this integral representation is important and must be mentioned for completeness, since we have no immediate use for it we shall not discuss it in detail. Its application to the case of free particle eigenfunctions should be clear from the discussion of Section 4.9.

A linear combination of eigenfunctions such as (4.13-10) is no longer a solution of the time independent Schrödinger equation associated with some particular value of energy ε_n, since it no longer has the form (4.8-7), but a linear combination of time-dependent eigenfunctions of the form

$$f(x,t) = \sum_{n=0}^{\infty} a_n \Psi_n(x,t) = \sum_{n=0}^{\infty} a_n \psi_n(x)e^{-i\varepsilon_n t/\hbar}, \tag{4.13-14}$$

which reduces to (4.13-10) for $t = 0$ is indeed a solution to the *time dependent* Schrödinger equation (4.7-3). This can easily be verified directly, with the help of the fact that the time-independent eigenfunctions ψ_n always, according to (4.8-8) obey the relation $\mathscr{H}\psi_n = \varepsilon_n\psi_n$. It is clear from this that although a superposition of wave functions which obeys the time-dependent Schrödinger equation can be constructed so as to represent any function or any probability amplitude *at a given time* ($t = 0$ in the above example), the superposition of wave functions will not, in general, *continue* to represent that same function or probability amplitude for later times, on account of the different time dependences of the individual components of the superposition. A superposition wave function thus does not generally represent a stationary state of the system. This state of affairs has been touched upon previously in Section 4.8.

If a superposition of eigenfunctions such as (4.13-14) is to serve as any sort of wave function for a physical system, it must be normalized so as to satisfy (4.7-6). Accordingly, for wave functions of the form (4.13-14) we must have

$$\int_{-\infty}^{\infty} f^*f\,dx = \sum_m \sum_n a_m^* a_n e^{i(\varepsilon_n - \varepsilon_m)t/\hbar} \int_{-\infty}^{\infty} \psi_m^* \psi_n\,dx = 1. \qquad (4.13\text{-}15)$$

In view of (4.13-1), all the terms of the double summation are zero except those for which $m = n$, where the value of the integral is unity. It is found thus that the normalization requirement will be met provided that the coefficients are chosen such that

$$\sum_n a_n^* a_n = 1. \qquad (4.13\text{-}16)$$

4·14 EXPECTATION VALUES AND QUANTUM NUMBERS

Suppose that a large number of measurements are made of a physical quantity f which is a property of some specific dynamical system. These measurements are always made upon the same system, or upon identical systems which are always in the same state. If the property f is conserved, that is, if it is constant with respect to the motion of the system, then (within experimental error, which we shall assume to be negligible) the same value for f will be obtained from every measurement, and a plot of the frequency $P(f)df$ with which a measured value in a range df about f is obtained will simply be a "spike" located at some value f_0 which is then the expectation value $\langle f \rangle$. This situation is shown in Figure 4.19(a). If the property f is not conserved in the motion, then we may expect to obtain different results from each measurement and a plot of the frequency with which values in a range df about f are observed as a function of f will exhibit a statistical "spread" about some average or expectation value $\langle f \rangle$, as shown in Figure 4.19(b). The extent of the statistical spread of the measurements can be expressed in terms of the standard deviation σ, defined as

$$\sigma^2 = \langle [f - \langle f \rangle]^2 \rangle. \qquad (4.14\text{-}1)$$

From this it is clear that if f is always the same as $\langle f \rangle$, as for the "spike" distribution of Figure 4.19(a), then $\sigma = 0$, while if f and $\langle f \rangle$ are not always the same, then σ is the root-mean-square deviation of the values of f from the average. The expression (4.19-1) can be written in a different way by noting that

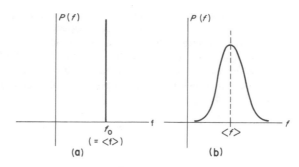

FIGURE 4.19. (a) Statistical distribution of many determinations of a quantity which is a constant of the motion. (b)
Corresponding distribution for a quantity which is not a
constant of the motion.

$$[f - \langle f \rangle]^2 = f^2 - 2f\langle f \rangle + \langle f \rangle^2$$

whence
$$\sigma^2 = \langle f^2 \rangle - 2\langle f \rangle\langle f \rangle + \langle f \rangle^2$$

$$= \langle f^2 \rangle - \langle f \rangle^2. \tag{4.14-2}$$

One of the fundamental postulates is that the expectation value of a dynamical
quantity is given by

$$\langle f \rangle = \int_{-\infty}^{\infty} \Psi^* f_{op} \Psi \, dx, \tag{4.14-3}$$

where f_{op} is the operator corresponding to the dynamical variable f and where Ψ
represents the wave function of the system for which f is being measured. Let us now
investigate certain expectation values connected with the energy operator $(-\hbar/i)(\partial/\partial t)$,
first for a system in one of its eigenstates and then for the same system in a state which
is represented as a superposition of two eigenstates.

First of all, it is clear that for any system the expectation value of the energy, $\langle \varepsilon \rangle$,
can be calculated from the time-dependent wave function $\Psi(x,t)$ by (4.14-3), using
also (4.7-4). The result is

$$\langle \varepsilon \rangle = \int_{-\infty}^{\infty} \Psi^* \left(-\frac{\hbar}{i} \frac{\partial \Psi}{\partial t} \right) dx, = \int_{-\infty}^{\infty} \Psi^* \mathcal{H} \Psi \, dx = \langle \mathcal{H} \rangle. \tag{4.14-4}$$

The expectation value of the energy is thus the same as the expectation value of the
Hamiltonian operator, evaluated over the time dependent wave functions. If $\Psi(x,t)$
is a wave function which represents a stationary eigenstate of the system, and which
is thus of the form (4.8-7), we may write (4.14-4) as

$$\langle \varepsilon \rangle = \langle \mathcal{H} \rangle = \int_{-\infty}^{\infty} \psi^*(x) e^{i\varepsilon t/\hbar} \mathcal{H} \psi(x) e^{-i\varepsilon t/\hbar} \, dx = \int_{-\infty}^{\infty} \psi^*(x) \mathcal{H} \psi(x) \, dx. \tag{4.14-5}$$

The expectation value of the Hamiltonian operator, and thus of the energy, in this case is the same when taken over the time-independent wave functions $\psi(x)$ as over the time-dependent functions $\Psi(x,t)$. Since $\mathcal{H}$, according to (4.7-5) is a time-independent operator, it does not operate on the time factors in (4.14-5); these factors may thus be moved past the $\mathcal{H}$ operator and cancelled one against the other.

If the wave function $\Psi(x,t)$ is a superposition of stationary state eigenfunctions of the form (4.13-14),

$$\Psi(x,t) = \sum_n a_n \psi_n(x) e^{-i\varepsilon_n t/\hbar}, \qquad (4.14\text{-}6)$$

then, from (4.14-4), again noting that $\mathcal{H}$ does not operate on the time factors,

$$\langle \varepsilon \rangle = \int_{-\infty}^{\infty} \Psi^* \mathcal{H} \Psi \, dx = \sum_m \sum_n e^{-i(\varepsilon_n - \varepsilon_m)t/\hbar} \int_{-\infty}^{\infty} a_m^* \psi_m^*(x) \mathcal{H}[a_n \psi_n(x)] \, dx. \qquad (4.14\text{-}7)$$

Using (4.8-8), which tells us that $\mathcal{H}\psi_n = \varepsilon_n \psi_n$, we find

$$\langle \varepsilon \rangle = \sum_m \sum_n \varepsilon_n e^{-i(\varepsilon_n - \varepsilon_m)t/\hbar} \int_{-\infty}^{\infty} a_m^* a_n \psi_m^* \psi_n \, dx, \qquad (4.14\text{-}8)$$

which, in view of the orthogonality property of the eigenfunctions, reduces to

$$\langle \varepsilon \rangle = \sum_n a_n^* a_n \varepsilon_n, \qquad (4.14\text{-}9)$$

all the integrals in (4.14-8) vanishing except those for which $m = n$. This is, however, the same value which is obtained by taking the expectation value of the Hamiltonian operator $\mathcal{H}$ over the *time-independent* wave function

$$\psi(x) = \Psi(x,0) = \sum_n a_n \psi_n(x), \qquad (4.14\text{-}10)$$

because then, utilizing (4.8-8) and the orthogonality properties of the eigenfunctions

$$\langle \mathcal{H} \rangle = \int_{-\infty}^{\infty} \psi^* \mathcal{H} \psi \, dx = \sum_m \sum_n a_m^* a_n \int_{-\infty}^{\infty} \psi_m^* \mathcal{H} \psi_n \, dx$$

$$= \sum_m \sum_n a_m^* a_n \varepsilon_n \int_{-\infty}^{\infty} \psi_m^* \psi_n \, dx = \sum_n a_n^* a_n \varepsilon_n = \langle \varepsilon \rangle. \qquad (4.14\text{-}11)$$

The operator $\mathcal{H}$ operating on the time-independent function (4.14-10) is thus seen to be fully equivalent to the operator $(-\hbar/i)(\partial/\partial t)$ operating on the time-dependent function (4.14-6). We shall utilize this result, which is due to the time-independence of the $\mathcal{H}$ operator and the orthogonality of the eigenfunctions, when we calculate expectation values for superposition wave functions.

If the system is in one of its stationary eigenstates, its wave function is simply $\psi_n(x)$, and from (4.14-5),

$$\langle \varepsilon \rangle = \langle \mathscr{H} \rangle = \int_{-\infty}^{\infty} \psi_n^* \mathscr{H} \psi_n \, dx = \int_{-\infty}^{\infty} \psi_n^* (\varepsilon_n \psi_n) \, dx = \varepsilon_n \int_{-\infty}^{\infty} \psi_n^* \psi_n \, dx = \varepsilon_n.$$

(4.14-12)

In like fashion, the expectation value of ε^2 may be found by evaluating the corresponding time-independent operator $\mathscr{H}^2$ over the time independent wave function,[11] whereby

$$\langle \varepsilon^2 \rangle = \langle \mathscr{H}^2 \rangle = \int_{-\infty}^{\infty} \psi_n^* \mathscr{H} (\mathscr{H} \psi_n) \, dx = \varepsilon_n \int_{-\infty}^{\infty} \psi_n^* \mathscr{H} \psi_n \, dx = \varepsilon_n^2. \quad (4.14\text{-}13)$$

The standard deviation associated with a series of determinations of energy upon a system in one of its stationary eigenstates will be, according to (4.14-2), (4.14-12), and (4.14-13),

$$\sigma^2 = \langle \varepsilon^2 \rangle - \langle \varepsilon \rangle^2 = \varepsilon_n^2 - \varepsilon_n^2 = 0. \quad (4.14\text{-}14)$$

A "spike" distribution will thus result, wherein every measurement will yield the same value of energy ε_n.

Now let us assume that the wave function of the system is a superposition of *two* eigenfunctions, of the form

$$\Psi(x,t) = a_m \psi_m(x) e^{-i\varepsilon_m t/\hbar} + a_n \psi_n(x) e^{-i\varepsilon_n t/\hbar}, \quad (4.14\text{-}15)$$

so that
$$\psi(x) = \Psi(x,0) = a_m \psi_m(x) + a_n \psi_n(x), \quad (4.14\text{-}16)$$

where, in accord with (4.13-16),

$$a_m^* a_m + a_n^* a_n = 1. \quad (4.14\text{-}17)$$

The expectation value of the energy may now be evaluated by computing the expectation value of the $\mathscr{H}$ operator over the wave function (4.14-16). The results, of course, can be obtained as a special case of (4.4-11), whereby

$$\langle \varepsilon \rangle = a_m^* a_m \varepsilon_m + a_n^* a_n \varepsilon_n = \alpha \varepsilon_m + (1 - \alpha) \varepsilon_n \quad (4.14\text{-}18)$$

where
$$\alpha = a_m^* a_m \quad \text{and} \quad 1 - \alpha = a_n^* a_n. \quad (4.14\text{-}19)$$

Likewise, the expectation value of ε^2 can be found by evaluating $\langle \mathscr{H}^2 \rangle$ using the wave function (4.14-16), giving

$$\langle \mathscr{H}^2 \rangle = \int_{-\infty}^{\infty} (a_m^* \psi_m^* + a_n^* \psi_n^*) \mathscr{H} \mathscr{H} (a_m \psi_m + a_n \psi_n) \, dx$$

$$= \int_{-\infty}^{\infty} a_m^* a_m \psi_m^* \mathscr{H} (\mathscr{H} \psi_m) \, dx + \int_{-\infty}^{\infty} a_n^* a_n \psi_n^* \mathscr{H} (\mathscr{H} \psi_n) \, dx$$

$$+ \int_{-\infty}^{\infty} a_m^* a_n \psi_m^* \mathscr{H} (\mathscr{H} \psi_n) \, dx + \int_{-\infty}^{\infty} a_n^* a_m \psi_n^* \mathscr{H} (\mathscr{H} \psi_m) \, dx. \quad (4.14\text{-}20)$$

---

[11] Strictly speaking, this assertion should be (and can be) proved by the methods used in deriving (4.14-9) and (4.14-11).

But $\mathscr{H}\psi_m = \varepsilon_m\psi_m$ and $H\psi_n = \varepsilon_n\psi_n$; applying these formulas twice to each of the four integrals in (4.14-20), we obtain

$$\langle\mathscr{H}^2\rangle = a_m^*a_m\varepsilon_m^2 \int_{-\infty}^{\infty} \psi_m^*\psi_m\,dx + a_n^*a_n\varepsilon_n^2 \int_{-\infty}^{\infty} \psi_n^*\psi_n\,dx$$

$$+ a_m^*a_n\varepsilon_n^2 \int_{-\infty}^{\infty} \psi_m^*\psi_n\,dx + a_n^*a_m\varepsilon_m^2 \int_{-\infty}^{\infty} \psi_n^*\psi_m\,dx. \qquad (4.14\text{-}21)$$

The first two integrals in (4.14-21) are unity, the second two zero, due to the orthogonality of the eigenfunctions ψ_n. The equation then reduces to

$$\langle\varepsilon^2\rangle = \langle\mathscr{H}^2\rangle = a_m^*a_m\varepsilon_m^2 + a_n^*a_n\varepsilon_n^2 = \alpha\varepsilon_m^2 + (1-\alpha)\varepsilon_n^2. \qquad (4.14\text{-}22)$$

The standard deviation associated with measurements of energy on a system whose wave function is the superposition (4.14-15) is

$$\sigma^2 = \langle\varepsilon^2\rangle - \langle\varepsilon\rangle^2 = \alpha\varepsilon_m^2 + (1-\alpha)\varepsilon_n^2 - [\alpha\varepsilon_m + (1-\alpha)\varepsilon_n]^2$$

$$= \alpha(1-\alpha)(\varepsilon_m - \varepsilon_n)^2. \qquad (4.14\text{-}23)$$

The standard deviation is zero only when $\alpha = 0$, $\alpha = 1$, or $\varepsilon_m = \varepsilon_n$; in each of these instances the wave function (4.14-16) reduces to a single eigenfunction of the system. In general, however, there will be some sort of "spread" associated with energy measurements on a system of this sort, the determinations not yielding the same value of

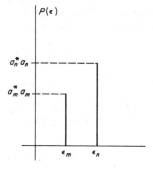

FIGURE 4.20. Statistical distribution of energies for systems whose wave functions are superpositions of the mth and nth eigenfunctions.

energy every time. As a matter of fact, a bit of reflection will serve to convince one that the "two-spike" frequency distribution of Figure 4.20 is consistent with the values of $\langle\varepsilon\rangle$ and $\langle\varepsilon^2\rangle$ given by (4.14-18) and (4.14-22). A consideration of expectation values of higher powers of ε leads to the conclusion that this distribution is the *only* one which is consistent with all these expectation values.[12] The currently prevailing interpretation of this result is that when measurement of energy is made on a system, the system is forced by the act of measurement into one of its eigenstates, if it is not

---

[12] C. W. Sherwin, *op. cit.* pp. 118–122.

already in one of them, and that the probability that the measured value of energy will be that associated with one or the other eigenstate is proportional to the quantity $a_m^* a_m$ associated with that eigenstate in the superposition wave function.

If the quantity σ^2 associated with the expectation value of some dynamical quantity should be zero, then that quantity is said to be *conserved* in the motion, or to be a constant of the motion. From (4.14-14) it is clear that the energy of a system in one of its eigenstates is conserved. If we can find an operator A with the property that

$$A\psi_n = \lambda_n \psi_n, \tag{4.14-24}$$

where λ_n is a constant, it is said that λ_n is an eigenvalue of the operator A. For such an operator

$$\langle A \rangle = \int_{-\infty}^{\infty} \psi_n^* A \psi_n \, dx = \lambda_n \int_{-\infty}^{\infty} \psi_n^* \psi_n \, dx = \lambda_n \tag{4.14-25}$$

and

$$\langle A^2 \rangle = \int_{-\infty}^{\infty} \psi_n^* A A \psi_n \, dx = \lambda_n \int_{-\infty}^{\infty} \psi_n^* A \psi_n \, dx = \lambda_n^2, \tag{4.14-26}$$

whereby

$$\sigma^2 = \langle A^2 \rangle - \langle A \rangle^2 = \lambda_n^2 - \lambda_n^2 = 0, \tag{4.14-27}$$

and the dynamical quantity represented by A is conserved, the expectation value of the operator A being given by the eigenvalue λ_n. It is obvious that the Hamiltonian operator $\mathcal{H}$ is just such an operator, whose eigenvalues are the energies ε_n. Any other operator A which has the property (4.14-24) with respect to the eigenfunctions of the system will always *commute* with the Hamiltonian operator, that is, it will always be found that

$$A\mathcal{H} = \mathcal{H}A, \tag{4.14-28}$$

since then

$$\mathcal{H} A\psi_n = \mathcal{H}(\lambda_n \psi_n) = \lambda_n \mathcal{H} \psi_n = \lambda_n \varepsilon_n \psi_n \tag{4.14-29}$$

and

$$A\mathcal{H}\psi_n = A(\varepsilon_n \psi_n) = \varepsilon_n A\psi_n = \varepsilon_n \lambda_n \psi_n = \mathcal{H} A\psi_n. \tag{4.14-30}$$

The converse of this statement is also true; if an operator A commutes with the Hamiltonian operator, as in (4.14-28), then that operator has the property (4.14-24) with respect to the eigenfunctions ψ_n, and the dynamical variable which it represents is a constant of the motion, because then (4.14-27) will be satisfied.[13]

In dealing with more complex systems, such as the hydrogen atom, which we shall treat in the next section, the relations (4.14-24) and (4.14-28) are very useful in exhibiting the fact that certain dynamical quantities are constants of the motion. We shall see, by using (4.14-24) that the total orbital angular momentum of the electron in the hydrogen atom, and its z-component are constants of the motion.

---

[13] For a proof of this converse statement, see E. Merzbacher, *op. cit.*, p. 160.

4·15 THE HYDROGEN ATOM

The hydrogen atom is a system consisting of a proton and a single electron, which interact through their mutual electrostatic attraction. The classical potential for this system is simply

$$V(r) = -e^2/r, \qquad (4.15\text{-}1)$$

where e is the electronic charge and r is the distance between the electron and the proton. In our treatment we shall regard the mass of the proton as infinite compared to the electron mass. In this approximation, any motion of the proton is neglected, the motion being assumed to be that of the electron with respect to a fixed nucleus. Since the proton mass is some 1850 times the electron mass, the approximation is a good one. The motion of the proton can be taken into account through the use of a reduced mass, as discussed in Section 4.5 in connection with the Bohr atom.

The spherical symmetry of the potential (4.15-1) suggests the use of the familiar spherical coordinate system, with

$$x = r \sin \theta \cos \phi$$

$$y = r \sin \theta \sin \phi$$

$$z = r \cos \theta,$$

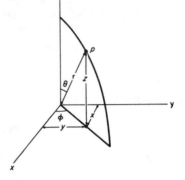

FIGURE 4.21. Spherical coordinate system used in discussing the hydrogen atom.

as shown in Figure 4.21, in which we may write the time-independent Schrödinger equation as

$$\nabla^2 \psi + \frac{2m_0}{\hbar^2}\left[\varepsilon + \frac{e^2}{r}\right]\psi(r,\theta,\phi) = 0, \qquad (4.15\text{-}2)$$

with $$\nabla^2 \psi = \frac{1}{r^2}\frac{\partial}{\partial r}\left(r^2\frac{\partial\psi}{\partial r}\right) + \frac{1}{r^2 \sin\theta}\frac{\partial}{\partial\theta}\left(\sin\theta\,\frac{\partial\psi}{\partial\theta}\right) + \frac{1}{r^2 \sin^2\theta}\frac{\partial^2\psi}{\partial\phi^2}. \qquad (4.15\text{-}3)$$

In (4.15-2), m_0 refers to the electron mass.

We may proceed by the familiar separation of variables technique, in which it is assumed that the solutions have the form

$$\psi(r,\theta,\phi) = R(r)\Theta(\theta)\Phi(\phi), \tag{4.15-4}$$

where $R(r)$ is a function of r *alone*, independent of θ and ϕ, $\Theta(\theta)$ is a function of θ alone and $\Phi(\phi)$ is a function of ϕ alone. Substituting this form for the solution into (4.15-2), noting that $\partial\psi/\partial r = \Theta\Phi\, dR/dr$, $\partial\psi/\partial\theta = R\Phi\, d\Theta/d\theta$ and $\partial\psi/\partial\phi = R\Theta\, d\Phi/d\phi$, multiplying the resulting equation by $r^2 \sin^2\theta/(R\Theta\Phi)$, and transposing, we find

$$\frac{\sin^2\theta}{R}\frac{d}{dr}\left(r^2\frac{dR}{dr}\right) + \frac{\sin\theta}{\Theta}\frac{d}{d\theta}\left(\sin\theta\frac{d\Theta}{d\theta}\right) + \frac{2m_0}{\hbar^2}\sin^2\theta\left(\varepsilon + \frac{e^2}{r}\right) = -\frac{1}{\Phi}\frac{d^2\Phi}{d\phi^2} = m^2. \tag{4.15-5}$$

Since the expression on the right, which is a function of r and θ only is equal to another expression which is a function of ϕ only, the two expressions must separately be equal to a constant, which for reasons which will presently become clear, will be denoted by m^2.

Now let us confine our attention for the moment to the latter equality shown in (4.15-5). This may be rearranged slightly to read

$$\frac{d^2\Phi}{d\phi^2} = -m^2\Phi(\phi), \tag{4.15-6}$$

which is a simple differential equation whose solution may be written as

$$\Phi_m(\phi) = e^{\pm im\phi}. \tag{4.15-7}$$

Since the total wave function ψ must be single-valued, it must be required that $\Phi_m(\phi)$ be single-valued also. In other words, one must demand that

$$\Phi_m(\phi + 2\pi) = \Phi_m(\phi), \tag{4.15-8}$$

whereby, from (4.15-7),

$$e^{\pm im\phi}e^{\pm 2\pi mi} = e^{\pm im\phi}$$

or
$$e^{\pm 2\pi mi} = 1. \tag{4.15-9}$$

This condition can be fulfilled only if m is a positive or negative *integer* or *zero*. The allowed values for m are thus restricted to

$$m = 0, \pm 1, \pm 2, \pm 3, \cdots,$$

otherwise the wave function will be multiple-valued. It can now be seen why the separation constant was chosen to be m^2. If m is a real number, then m^2 must be positive. Had the separation constant in (4.15-5) been negative, then the solutions (4.15-7) would have been exponential rather than sinusoidal in character, and the single-valuedness condition (4.15-8) could never have been fulfilled. The separation constant was thus chosen in such a way as to insure that it would always have a positive value, so as to exclude at once such unwanted solutions.

In (4.15-5), the expression on the left may be set equal to m^2, and the resulting equation divided through by $\sin^2 \theta$ and rearranged so as to read

$$\frac{1}{R}\frac{d}{dr}\left(r^2\frac{dR}{dr}\right) + \frac{2m_0r^2}{\hbar^2}\left(\varepsilon + \frac{e^2}{r}\right) = -\frac{1}{\Theta \sin \theta}\frac{d}{d\theta}\left(\sin \theta \frac{d\Theta}{d\theta}\right) + \frac{m^2}{\sin^2 \theta} = \beta,$$

$$(4.15\text{-}10)$$

Again, since the first expression is a function of r alone and the second a function of θ alone, the two must separately be equal to a constant β. The second equality of (4.15-10) may be written in the form

$$\frac{1}{\sin \theta}\frac{d}{d\theta}\left(\sin \theta \frac{d\Theta}{d\theta}\right) - \frac{m^2\Theta}{\sin^2 \theta} + \beta\Theta(\theta) = 0. \qquad (4.15\text{-}11)$$

This equation is a familiar one in mathematical physics, known as *Legendre's equation*. Without going into the mathematical details,[14] we shall note that the only solutions of this equation which do not violate the requirements of finiteness for the wave functions are those for which β has the values $0,2,6,12,20, \cdots$, thus for which

$$\beta = l(l + 1) \quad \text{with} \quad (l = 0,1,2,3, \cdots). \qquad (4.15\text{-}12)$$

In these instances, the solution of (4.15-11) reduces to a polynomial in $\cos \theta$ and $\sin \theta$ in much the same way as the solutions of Hermite's equation reduce to polynomials under certain conditions. These polynomial solutions may be represented as the *associated Legendre functions* $P_l^m(\theta)$, defined by

$$P_l^m(\theta) = \sin^{|m|} \theta \frac{d^{|m|}}{d(\cos \theta)^{|m|}} P_l(\cos \theta) \qquad (4.15\text{-}13)$$

where

$$P_l(\cos \theta) = \frac{1}{2^l l!}\frac{d^l(\cos^2 \theta - 1)^l}{d(\cos \theta)^l}. \qquad (4.15\text{-}14)$$

The functions $P_l (\cos \theta)$ are the so-called *Legendre polynomials*, the first few of which are

$$P_0(\cos \theta) = 1$$

$$P_1(\cos \theta) = \cos \theta$$

$$P_2(\cos \theta) = \tfrac{1}{2}(3 \cos^2 \theta - 1) \qquad (4.15\text{-}15)$$

$$P_3(\cos \theta) = \tfrac{1}{2}(5 \cos^3 \theta - \cos \theta)$$

etc.

---

[14] See, for example, R. B. Leighton, *Principles of Modern Physics*, McGraw-Hill Book Co., Inc., New York (1959), p. 168.

When $m = 0$, of course, the associated functions $P_l^0(\theta)$ reduce to the simple Legendre polynomials $P_l(\cos \theta)$. The function $\Theta(\theta)$ of (4.15-11) is usually written in the form

$$\Theta_{lm}(\theta) = \sqrt{\frac{(2l + 1)(l - m)!}{2(l + m)!}}\, P_l^m(\theta), \qquad (4.15\text{-}16)$$

the factor multiplying $P_l^m(\theta)$ being a normalization constant. The *angular part* of the wave function, represented as a product of $\Theta_{lm}(\theta)$ and $\Phi_m(\phi)$ is often written in the form

$$Y_{lm}(\theta,\phi) = \frac{(-1)^m}{\sqrt{2\pi}}\, \Theta_{lm}(\theta)\Phi_m(\phi)$$

$$= (-1)^m \sqrt{\frac{(2l + 1)(l - m)!}{4\pi(l + m)!}}\, P_l^m(\theta)e^{im\phi}. \qquad (4.15\text{-}17)$$

The functions $Y_{lm}(\theta,\phi)$ are referred to as *spherical harmonics*. For a central force, where the potential is a function of r alone, the angular part of the wave function is *always* given by (4.15-17), no matter what particular form $V(r)$ may take.

The radial part of Equation (4.15-10), taking $\beta = l(l + 1)$ as required by (4.15-12) now becomes

$$\frac{1}{r^2}\frac{d}{dr}\left(r^2 \frac{dR}{dr}\right) + \frac{2m_0}{\hbar^2}\left(\varepsilon + \frac{e^2}{r}\right)R(r) - \frac{l(l + 1)R(r)}{r^2} = 0. \qquad (4.15\text{-}18)$$

This equation is called the Laguerre equation. The only solutions whose behavior at $r = \infty$ is such as to qualify them as wave functions are those which may be expressed in terms of *Laguerre polynomials*. These functions, with which are associated *a discrete set of eigenvalues of the energy* ε, may be written

$$R_{nl}(r) = -\left[\left(\frac{2m_0 e^2}{n\hbar^2}\right)^3 \cdot \frac{(n - l - 1)!}{(2n[n + l]!)^3}\right]^{1/2} e^{-\rho/2}\rho^l L_{n+l}^{2l+1}(\rho), \qquad (4.15\text{-}19)$$

where

$$n = 1,2,3, \cdots . \qquad (4.15\text{-}20)$$

and where

$$\rho = \frac{2m_0 e^2}{n\hbar^2} r. \qquad (4.15\text{-}21)$$

The functions $L_{n+l}^{2l+1}(\rho)$ are *associated Laguerre functions*, defined by

$$L_r^s(\rho) = \frac{d^s}{d\rho^s} L_r(\rho), \qquad (4.15\text{-}22)$$

where $L_r(\rho)$ are the *Laguerre polynomials*, in turn defined as

$$L_r(\rho) = e^\rho \frac{d^r}{d\rho^r}(\rho^r e^{-\rho}). \qquad (4.15\text{-}23)$$

The energies ε_n for which acceptable wave functions exist are related to the values of n in (4.15-19) by[15]

$$\varepsilon_n = -\frac{m_0 e^4}{2n^2\hbar^2}.$$
(4.15-24)

Equation (4.15-24) for the energy levels of the hydrogen atom is in agreement with the result (4.5-8) of the Bohr theory; the energy levels as given by the Bohr theory and illustrated in Figure 4.5 are therefore the same as those predicted by wave mechanics.

There are three quantum numbers, n, l, and m in the wave mechanical treatment of the hydrogen atom. The possible values for n are given by (4.15-20). The Laguerre polynomials $L_r(\rho)$ of (4.15-23) are easily seen to be polynomials of degree r. Such polynomials may be differentiated just r times before the result of the repeated differentiation yields zero. Therefore, for nonzero wave functions of the form L_r^s to result from the polynomial L_r, the number of differentiations s required to generate the associated function L_r^s according to (4.15-22) must be less than or equal to r. For physically meaningful wave functions, then, in (4.15-22) we must have

$$s \leqslant r,$$
(4.15-25)

or in (4.15-19) $$2l + 1 \leqslant n + l.$$
(4.15-26)

Subtracting $l + 1$ from both sides of this inequality, (4.15-26) reduces to

$$l \leqslant n - 1.$$
(4.15-27)

In like fashion, the associated Legendre function $P_l^m(\theta)$ results from the Legendre polynomial $P_l(\cos\theta)$ by m-fold differentiation. For a nonzero wave function of the form $P_l^m(\theta)$, the number of differentiations m must be less than or equal to the degree l of the polynomial. We must then have

$$m \leqslant l$$
(4.15-28)

for wave functions of physical interest. Conditions (4.15-20), (4.15-27), and (4.15-28) together serve to define the well-known expressions for the possible range of values for n, l, and m,

$$n = 1,2,3,4, \cdots$$

$$l = 0,1,2,3, \cdots n - 1$$
(4.15-29)

$$m = 0,\pm 1, \pm 2, \pm 3, \cdots \pm l.$$

It is customary to refer to the states for which $l = 0,1,2,3$ as s,p,d, and f-states, respectively. Thus a state for which $n = 2$ and $l = 0$ is referred to as a $2s$-state, while one for which $n = 3$ and $l = 2$ is called a $3d$-state.

---

[15] The mathematical details of this development are given by R. B. Leighton, *op. cit.*, pp. 171–175.

The complete wave functions for the hydrogen atom, from (4.15-4), must be

$$\psi_{nlm}(r,\theta,\phi) = N_{nlm}R_{nl}(r)\Theta_{lm}(\theta)\Phi_m(\phi), \tag{4.15-30}$$

where $R_{nl}(r)$, $\Theta_{lm}(\theta)$, and $\Phi_m(\phi)$ are given by (4.15-7), (4.15-16), and (4.15-19), respectively, and where N_{nlm} is a normalization constant chosen so that

$$\int \psi_{nlm}^*\psi_{nlm} \cdot r^2 \sin\theta \, dr \, d\theta \, d\phi = 1. \tag{4.15-31}$$

The wave functions (4.15-30) will be found to be orthogonal, so that

$$\int \psi_{n'l'm'}^*\psi_{nlm} \cdot r^2 \sin\theta \, dr \, d\theta \, d\phi = 0, \tag{4.15-32}$$

unless $n = n'$, $l = l'$ and $m = m'$ *simultaneously*. The time-dependent wave functions are, as usual, obtained from the time-independent functions by multiplying by $e^{-i\varepsilon_n t/\hbar}$.

TABLE 4.1.

Wave Functions for the Hydrogen Atom

$n = 1$	$l = 0$	$m = 0$	$\psi_{100} = N_{100}e^{-\frac{1}{2}\rho}$	$1s$
$n = 2$	$l = 0$	$m = 0$	$\psi_{200} = N_{200}(2 - \rho)e^{-\frac{1}{2}\rho}$	$2s$
	$l = 1$	$m = 0$	$\psi_{210} = N_{210}\rho e^{-\frac{1}{2}\rho}\cos\theta$	$2p^3$
(4 states)		$m = \pm 1$	$\psi_{21\pm1} = N_{211}\rho e^{-\frac{1}{2}\rho}\sin\theta\, e^{\pm i\phi}$	
$n = 3$	$l = 0$	$m = 0$	$\psi_{300} = N_{300}(6 - 6\rho + \rho^2)e^{-\frac{1}{2}\rho}$	$3s$
	$l = 1$	$m = 0$	$\psi_{310} = N_{310}\rho(4 - \rho)e^{-\frac{1}{2}\rho}\cos\theta$	$3p^3$
		$m = \pm 1$	$\psi_{31\pm1} = N_{311}\rho(4 - \rho)e^{-\frac{1}{2}\rho}\sin\theta\, e^{\pm i\phi}$	
	$l = 2$	$m = 0$	$\psi_{320} = N_{320}\rho^2 e^{-\frac{1}{2}\rho}(3\cos^2\theta - 1)$	$3d^5$
		$m = \pm 1$	$\psi_{32\pm1} = N_{321}\rho^2 e^{-\frac{1}{2}\rho}\sin\theta\cos\theta\, e^{\pm i\phi}$	
(9 states)		$m = \pm 2$	$\psi_{32\pm2} = N_{322}\rho^2 e^{-\frac{1}{2}\rho}\sin^2\theta e^{\pm 2i\phi}$	
$n = 4$	$l = 0$	$m = 0$	$\psi_{400} = N_{400}(24 - 36\rho + 12\rho^2 - \rho^3)e^{-\frac{1}{2}\rho}$	$4s$
	$l = 1$	$m = 0$	$\psi_{410} = N_{410}\rho e^{-\frac{1}{2}\rho}(20 - 10\rho + \rho^2)\cos\theta$	$4p^3$
		$m = \pm 1$	$\psi_{41\pm1} = N_{411}\rho e^{-\frac{1}{2}\rho}(20 - 10\rho + \rho^2)\sin\theta\, e^{\pm i\phi}$	
	$l = 2$	$m = 0$	$\psi_{420} = N_{420}\rho^2(6 - \rho)e^{-\frac{1}{2}\rho}(3\cos^2\theta - 1)$	
		$m = \pm 1$	$\psi_{42\pm1} = N_{421}\rho^2(6 - \rho)e^{-\frac{1}{2}\rho}\sin\theta\cos\theta\, e^{\pm i\phi}$	$4d^5$
		$m = \pm 2$	$\psi_{42\pm2} = N_{422}\rho^2(6 - \rho)e^{-\frac{1}{2}\rho}\sin^2\theta\, e^{\pm 2i\phi}$	
	$l = 3$	$m = 0$	$\psi_{430} = N_{430}\rho^3 e^{-\frac{1}{2}\rho}(\frac{5}{3}\cos^3\theta - \cos\theta)$	
		$m = \pm 1$	$\psi_{43\pm1} = N_{431}\rho^3 e^{-\frac{1}{2}\rho}(5\cos^2\theta - 1)\sin\theta\, e^{\pm i\phi}$	$4f^7$
(16 states)		$m = \pm 2$	$\psi_{43\pm2} = N_{432}\rho^3 e^{-\frac{1}{2}\rho}\sin^2\theta\cos\theta\, e^{\pm 2i\phi}$	
		$m = \pm 3$	$\psi_{43\pm3} = N_{433}\rho^3 e^{-\frac{1}{2}\rho}\sin^3\theta\, e^{\pm 3i\phi}$	

$$N_{nlm} = -\left[\left(\frac{2m_0 e^2}{n\hbar^2}\right)^3 \frac{(n - l - 1)!(l - m)!(2l + 1)}{4\pi(2n[n + l]!)^3(l + m)!}\right]^{\frac{1}{2}}$$

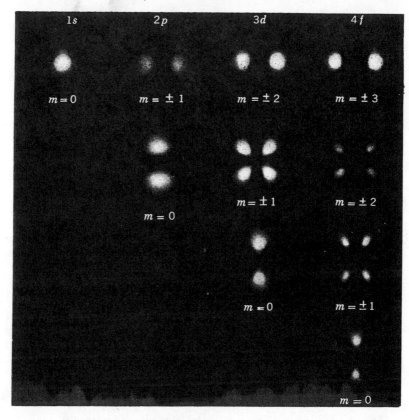

FIGURE 4.22. Photographic representations of the electron density associated with several eigenstates of the hydrogen atom. The polar axis lies in the plane of the page and is oriented vertically.

From (4.15-24) it is clear that the energy depends only upon the principal quantum number n in this particular system. There are in general many independent quantum states [corresponding to the various permitted values for l and m, according to (4.15-29)] which have the same principal quantum number n, and which thus belong to the *same* energy level. This phenomenon is called *degeneracy* and an energy level which contains more than one independent quantum state is referred to as a *degenerate* level. It can be seen from (4.15-29), simply by counting states, that the degeneracy or multiplicity associated with the level of energy ε_n, whose principal quantum number is n, is just n^2.

The algebraic expressions for some of the wave functions of the hydrogen atom are given in Table 4.1, and some representations of the probability density for some of these states are shown in Figure 4.22. It will be noted that the s-states ($l = 0$) are always spherically symmetric, while the p-, d-, f-states, etc., have angular dependences which may be quite complex. The $2p$-state for which $m = 0$ has two lobes which point along the $\pm z$-directions, while the $2p$-states for which $m = \pm 1$ are "doughnut-shaped" and symmetric about the z-axis. All the $2p$-states thus result in probability

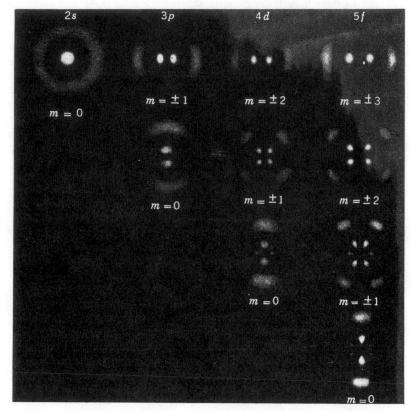

FIGURE 4.22 (*Cont'd*). The scale is not uniform among all the diagrams. [After R. B. Leighton, *Principles of Modern Physics*, McGraw-Hill, New York (1959).]

amplitudes which are symmetric about the z-axis. The question of why the z-axis is thus seemingly favored or preferred may well arise. After all, the potential $V(r)$ is spherically symmetric, which permitted us in the first place to choose any polar axis at all for our spherical coordinate system, and we chose the one we did purely arbitrarily. Why, then does it appear in this seemingly favored aspect—or, more to the point, is it really a preferred axis in a physical sense?

We have noted in connection with Section 4.13 that linear superpositions of stationary state wave functions belonging to different energy levels no longer represent probability amplitudes which are independent of time. Since the different independent stationary state eigenfunctions belonging *to the same degenerate energy level* all have the same time dependence factor $e^{-i\varepsilon_n t/\hbar}$, however, it follows that *states of this nature can be superposed* and the superposition probability amplitude will *still* be independent of time. We have already run across one example of this in Section 4.10 (in connection with equation (4.10-13)). It is clear, for example, from Table 4.1 and from (4.15-29) that there are *three* independent 2p states for the hydrogen atom. It makes no difference, however, whether we regard those three states as being those given in Table 4.1, or as three linearly independent superpositions of those states. By properly super-posing the three states given in Table 4.1, we could produce three independent super-position wave functions corresponding to probability amplitudes with lobes pointing along the x-, y- and z-axes, instead of the polar lobe and doughnut configuration of the functions in the table. By further proper linear superposition of these wave functions, these three lobes could be made to point in *any* three given orthogonal directions. It is clear from this that the wave functions of Table 4.1, in which an arbitrarily chosen polar axis seems to have a unique or preferred status, form only one *representation* of the wave functions of the system, and that by proper superposition of *degenerate* quantum states, one might form a representation in which the polar axis was in some other direction. In addition, of course, if we so desired, we could superpose the $2s$ state along with the three $2p$ states. We could *not*, however, add in a $3s$ or $3p$ state and still have a time-independent probability amplitude.

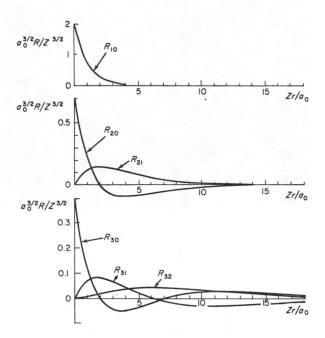

FIGURE 4.23. Plots of the radial wave function $R_{nl}(r)$ for several of the states of the hydrogen atom. [After R. B. Leighton, *Principles of Modern Physics*, McGraw-Hill, New York (1959).]

Figure 4.23 shows a plot of the radial wave function $R_{nl}(r)$ for several of the lower energy states of the hydrogen atom, and Figure 4.24 shows a plot of the radial probability density, which is the probability of finding the electron within a spherical shell of thickness dr about radius r. Since the volume of such a shell is $4\pi r^2 dr$, the radial probability density must be proportional to $r^2 R_{nl}^2(r)$. The proof of this statement is assigned as an exercise. The maxima in the radial probability density for the

s-wave functions as shown in Figure 4.24 are in close (but not exact) agreement with the Bohr radii r_n as given by (4.5-4).

We have already noted from Equation (4.15-24) that the principal quantum number n is associated with the total energy of the system. We shall find that the quantum numbers l and m are related to the angular momentum of the system. In classical

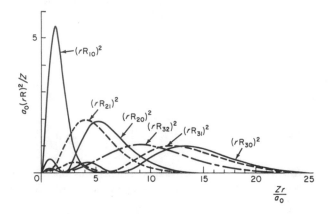

FIGURE 4.24. Plots of the radial probability distribution $r^2 R_{nl}^2(r)$ for several of the states of the hydrogen atom. [After R. B. Leighton, *Principles of Modern Physics*, McGraw-Hill, New York (1959).]

planetary motion, the orbital angular momentum is conserved, and this, as we shall soon see, is also true in the quantum mechanical treatment. The classical angular momentum **L** of a particle is defined by

$$\mathbf{L} = \mathbf{r} \times \mathbf{p}, \tag{4.15-33}$$

where **p** is the linear momentum and **r** is the radius vector from some arbitrary origin to the particle. From this definition and from the definition of the linear momentum operators (4.7-2), a set of operators representing the Cartesian components of the angular momentum can be constructed in a very direct manner. The result is

$$L_x = y p_z - z p_y = \frac{\hbar}{i} \left[y \frac{\partial}{\partial z} - z \frac{\partial}{\partial y} \right]$$

$$L_y = z p_x - x p_z = \frac{\hbar}{i} \left[z \frac{\partial}{\partial x} - x \frac{\partial}{\partial z} \right] \tag{4.15-34}$$

$$L_z = x p_y - y p_x = \frac{\hbar}{i} \left[x \frac{\partial}{\partial y} - y \frac{\partial}{\partial x} \right].$$

These operators may be expressed in terms of the spherical coordinates (r, θ, ϕ) by a tedious but straightforward coordinate transformation. The operators then become

$$L_x = \frac{\hbar}{i}\left[-\sin\phi\,\frac{\partial}{\partial\theta} - \operatorname{ctn}\theta\cos\phi\,\frac{\partial}{\partial\phi}\right]$$

$$L_y = \frac{\hbar}{i}\left[\cos\phi\frac{\partial}{\partial\theta} - \operatorname{ctn}\theta\sin\phi\,\frac{\partial}{\partial\phi}\right] \qquad (4.15\text{-}35)$$

$$L_z = \frac{\hbar}{i}\frac{\partial}{\partial\phi}.$$

The operator representing the square of the total angular momentum can be obtained from (4.15-35) as

$$L^2 = L_x^2 + L_y^2 + L_z^2 = -\hbar^2\left[\frac{1}{\sin\theta}\frac{\partial}{\partial\theta}\left(\sin\theta\,\frac{\partial}{\partial\theta}\right) + \frac{1}{\sin^2\theta}\frac{\partial^2}{\partial\phi^2}\right]. \qquad (4.15\text{-}36)$$

Now let us examine what happens when this operator operates upon the wave functions ψ_{nlm} of the hydrogen atom, as given by (4.15-30). We find then that

$$L^2\psi_{nlm} = -\hbar^2\left[\frac{1}{\sin\theta}\frac{\partial}{\partial\theta}\left(\sin\theta\,\frac{\partial}{\partial\theta}\right) + \frac{1}{\sin^2\theta}\frac{\partial^2}{\partial\phi^2}\right]R_{nl}(r)\Theta_{lm}(\theta)\Phi_m(\phi). \qquad (4.15\text{-}37)$$

The operator for L^2, however, is independent of r, and operates on the product wave function (4.15-30) in such a manner that (4.15-37) can be written

$$L^2\psi_{nlm} = -\hbar^2 R_{nl}(r)\left[\frac{\Phi_m(\phi)}{\sin\theta}\frac{\partial}{\partial\theta}\left(\sin\theta\,\frac{\partial\Theta_{lm}}{\partial\theta}\right) + \frac{\Theta_{lm}(\theta)}{\sin^2\theta}\frac{\partial^2\Phi_m}{\partial\phi^2}\right]. \qquad (4.15\text{-}38)$$

Applying, successively, the relations (4.15-6), (4.15-11), and (4.15-12) to this expression, it is evident that

$$L^2\psi_{nlm} = -\hbar^2 R_{nl}(r)\Phi_m(\phi)\left[\frac{1}{\sin\theta}\frac{\partial}{\partial\theta}\left(\sin\theta\,\frac{\partial\Theta_{lm}}{\partial\theta}\right) - m^2\frac{\Theta_{lm}(\theta)}{\sin^2\theta}\right]$$

$$= \hbar^2 R_{nl}(r)\Phi_m(\phi)\cdot\beta\Theta_{lm}(\theta) = \hbar^2 l(l+1)\psi_{nlm}. \qquad (4.15\text{-}39)$$

This equation shows that the operator L^2 is an operator which has the special property given by (4.14-24) with respect to the wave functions ψ_{nlm}. Therefore, according to (4.14-27) and (4.14-25), the dynamical variable associated with the operator must be a constant of the motion, and its expectation value must be $\hbar^2 l(l+1)$. The square of the angular momentum is thus conserved, whereby the angular momentum itself must be conserved, and the expectation value of the angular momentum must be

$$\langle L\rangle = \sqrt{\langle L^2\rangle} = \hbar\sqrt{l(l+1)}. \qquad (4.15\text{-}40)$$

In a similar manner, one can show that the z-component of the angular momentum is conserved. In this case, from (4.15-35) and (4.15-30) we have

$$L_z \psi_{nlm} = \frac{\hbar}{i} \frac{\partial}{\partial \phi} R_{nl}(r) \Theta_{lm}(\theta) \Phi_m(\phi) = \frac{\hbar}{i} R_{nl}(r) \Theta_{lm}(\theta) \frac{\partial \Phi_m}{\partial \phi}. \qquad (4.15\text{-}41)$$

But $\partial \Phi_m / \partial \phi = im e^{im\phi} = im \Phi_m(\phi)$, whence (4.15-41) becomes

$$L_z \psi_{nlm} = \frac{\hbar}{i} \cdot im \, R_{nl}(r) \Theta_{lm}(\theta) \Phi_m(\phi) = m\hbar \psi_{nlm}. \qquad (4.15\text{-}42)$$

The operator L_z thus also has the special property (4.14-24), and L_z is therefore a constant of the motion whose expectation value must be

$$L_z = m\hbar. \qquad (4.15\text{-}43)$$

The total angular momentum and its component along a polar axis (whose direction, as we have already seen, may be chosen arbitrarily) are simultaneously conserved. The quantum number l, which specifies the total orbital angular momentum is often referred to as the *orbital* quantum number. The quantum number m, which specifies the z-component of angular momentum, is sometimes called the *magnetic* quantum number because the degeneracy of the states having the same values for n and l but different values of m can be removed when a magnetic field is applied.

The possible orientations of the angular momentum vector **L** under the restrictions imposed by the quantum conditions (4.15-40) and (4.15-43) are illustrated by Figure 4.25 for states for which $l = 0, 1,$ and 2. The **L** vector must at the same time

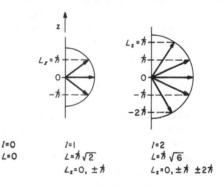

FIGURE 4.25. Possible orientations for the angular momentum vector under the restrictions imposed by (4.15-40) and (4.15-43) for $l = 0, 1,$ and 2.

have a total length given by (4.15-40) and a projection along the z-axis given by (4.15-43). This results in a discrete set of possible orientations for the vector, as shown in the diagrams. The extension of these diagrams to systems with higher values for l is obvious. It should also be clear that if a constant magnetic field is applied along the z-axis, the magnetic moment of the electron in its orbit (which is proportional to the total orbital angular momentum) will interact with the applied field in such a way that the atom will experience a force tending to align its magnetic moment vector (hence its orbital angular momentum vector) with the applied field. The states whose value of m permits partial or complete alignment then must lie *lower in energy* than those corresponding to nearly antiparallel orientations. The quantum states having different *m*-values, which are degenerate in the absence of a magnetic field, will thus be split into *m* closely spaced but separate levels when the field is applied. This phenomenon is the well-known Zeeman effect.

These results serve to explain quite well the observed electric, magnetic, and spectroscopic properties of the hydrogen atom. More complex atoms can, in principle, be treated by the same methods, although since such atoms represent systems of three or more bodies, no closed-form analytic solutions can be obtained, and a perturbation approach must be adopted. Such an approach is based upon the technique of separating the Hamiltonian of the system into a major part whose solution is known or can be obtained analytically, and a smaller part, hopefully containing the terms which cause the analytical difficulties. The solution is then assumed to be of the form of a linear combination of eigenfunctions of the first part of the Hamiltonian, and the constants involved in the linear combination are determined in such a way that the wave function is an approximate eigenfunction of the complete Hamiltonian. These methods, while very useful in solving a variety of practical problems will not be discussed further here.[16]

4·16 ELECTRON SPIN, THE PAULI EXCLUSION PRINCIPLE AND THE PERIODIC SYSTEM

In the quantum mechanical calculation of the wave functions and energy levels of many-electron atoms (He, Be, Li, etc.), it is generally assumed that each electron moves in a spherically symmetric potential due to the nucleus and the average spatial distribution of all the other electrons. In this approximation the total wave function of the system may be expressed as a product of one-electron wave functions. The potential function for each electron is spherically symmetric, and the angular part of the one-electron wave functions is still expressable in terms of the spherical harmonics (4.15-i *₁*). The radial part of such wave functions, however, is more complex than that for the hydrogen atom. Due to the separability of the one-electron wave functions into
        ̆ular parts, each electron can be regarded as being represented by three quantum numbers, *n*, *l*, and *m*, just as in hydrogen. Since the radial wave function does not now have the form (4.15-19) corresponding to a simple Coulomb potential, however, states having the same value of *n* but a different value of *l are no longer degenerate* and the total energy of the electron depends upon the orbital quantum number *l* as well as upon *n*.

From these considerations the energy levels of complex atoms may be calculated and predictions regarding the spectra of those atoms made. These predictions, however, are *not* in complete agreement with experiment unless two additional corrections are made.

The first of these is the hypothesis of electron spin, which was introduced by Goudsmit and Uhlenbeck in 1925, in which every electron, in addition to whatever orbital angular momentum it may have, is assumed to have an *intrinsic* angular

---

[16] For a complete treatment of this subject see for example C. W. Sherwin, *op. cit.*, Chapters 7–10, or E. Merzbacher, *op. cit.*, Chapters 16–20.

momentum or *spin* angular momentum of magnitude $L_s = \frac{1}{2}\hbar\sqrt{3}$, whose component along any given field axis may have *two* possible values, $m_s = \pm\hbar/2$. The electron can thus exist in two "spin states" corresponding to the two possible values of m_s. These two spin states are often referred to as the "spin up" and "spin down" states. The electron can be regarded as having a total spin angular momentum quantum number s which can have only one value, $\frac{1}{2}$, in which case, in analogy with (4.15-40) the spin angular momentum is

$$\langle L_s \rangle = \hbar\sqrt{s(s+1)} = \hbar\sqrt{3}/2, \qquad (4.16\text{-}1)$$

and a quantum number m_s representing the z-component of spin angular momentum which may have the values $\pm\frac{1}{2}$, so that in analogy with (4.15-43), the z-component of spin angular momentum is

$$\langle L_{sz} \rangle = m_s\hbar = \pm\hbar/2. \qquad (4.16\text{-}2)$$

The degeneracy of all the electronic states in hydrogen and in more complex atoms is thus doubled by the existence of the two spin states of the electron. The intrinsic angular momentum of the electron is a consequence of relativity and is obtained from first principles when the wave equation is formulated in a relativistically covariant manner, as shown by Dirac in 1928. The incorporation of electron spin into the formalism of wave mechanics thus really involves no new *ad hoc* physical assumption at all.

The other factor which must be introduced into the wave-mechanical explanation of the behavior of many-particle systems is the *Pauli Exclusion Principle*, which states that in a given *system* (which may be an atom, a molecule, or a whole crystal made up of many interacting atoms), no two electrons may occupy the same quantum state, counting states of opposite spin as different states. Applied to a complex atom, this means that no two electrons in the atom can have quantum numbers n, l, m, and m_s, *all* of which are the same. For other systems, it means that no more than g electrons may occupy an energy level which is g-fold degenerate. Although electrons obey the Pauli principle, certain other particles, such as photons, do not. In general, particles such as electrons whose intrinsic spin is half-integral ($s = \frac{1}{2}, \frac{3}{2}, \frac{5}{2}, \cdots$) obey the Pauli principle, while those, such as photons, whose intrinsic spin is integral ($s = 0,1,2, \cdots$) do not. This fact introduces very important differences into the statistical behavior of assemblies of electrons and photons, as we shall see in the next chapter.

The addition of electron spin and the Pauli principle to the wave-mechanical treatment of many-electron atoms made it possible to explain the periodic system of the elements on the basis of the successive occupation of electronic states of the atoms. This scheme is shown in Table 4.2, which lists the electronic configuration of each atom.

In this table, the simplest atom, H, is represented as having a single electron, which in the lowest energy configuration of the system must be in a $1s$ state. The next atom, He, with two electrons, in the lowest energy configuration (or *ground state*), may have two $1s$ electrons with opposite spins. This configuration is written $1s^2$. Since there are only two states associated with the level $n = 1$, $l = 0$, the next most complex atom, Li, must in accord with the Pauli principle have the ground state configuration $1s^2)2s$, the third electron going into the next lowest energy state, which is the $2s$ state. The fourth element, Be, adds another $2s$ electron, but since there are only two

$2s$ states, the Pauli principle requires that the fifth element, B, have an electron in the $2p$ level, which save for the $1s$ and $2s$ is the lowest in energy; B thus has the ground state configuration $1s^2)2s^22p$. There are 6 possible $2p$ states, corresponding to quantum numbers $n = 2$, $l = 1$, $m = 0$, ± 1 and $m_s = \pm\frac{1}{2}$. The next five elements, C, N, O F, and Ne, each add one additional $2p$ electron to fill the $2p$ level. The filled $2p$ shell

TABLE 4.2.

Electronic Configurations and the Periodic System

$1s$	H	$1s$
	He	$1s^2)$
$2s$	Li	$1s^2)2s$
	Be	$1s^2)2s^2$
	B	$1s^2)2s^22p$
	C	$1s^2)2s^22p^2$
	N	$1s^2)2s^22p^3$
	O	$1s^2)2s^22p^4$
	F	$1s^2)2s^22p^5$
	Ne	$1s^2)2s^22p^6)$
$3s$	Na	$1s^2)2s^22p^6)3s$
	Mg	$1s^2)2s^22p^6)3s^2$
$3p$	Al	$1s^2)2s^22p^6)3s^23p$
	Si	$1s^2)2s^22p^6)3s^23p^2$
	P	$1s^2)2s^22p^6)3s^23p^3$
	S	$1s^2)2s^22p^6)3s^23p^4$
	Cl	$1s^2)2s^22p^6)3s^23p^5$
	A	$1s^2)2s^22p^6)3s^23p^6)$
$4s$	K	$1s^2)2s^22p^6)3s^23p^6)4s$
	Ca	$1s^2)2s^22p^6)3s^23p^6)4s^2$
$3d$	Sc	$1s^2)2s^22p^6)3s^23p^63d)4s^2$
	Ti	$1s^2)2s^22p^6)3s^23p^63d^2)4s^2$
	V	$1s^2)2s^22p^6)3s^23p^63d^3)4s^2$
	Cr	$1s^2)2s^22p^6)3s^23p^63d^5)4s$
	Mn	$1s^2)2s^22p^6)3s^23p^63d^5)4s^2$
	Fe	$1s^2)2s^22p^6)3s^23p^63d^6)4s^2$
	Co	$1s^2)2s^22p^6)3s^23p^63d^7)4s^2$
	Ni	$1s^2)2s^22p^6)3s^23p^63d^8)4s^2$
	Cu	$1s^2)2s^22p^6)3s^23p^63d^{10})4s$
	Zn	$1s^2)2s^22p^6)3s^23p^63d^{10})4s^2$
$4p$	Ga	$1s^2)2s^22p^6)3s^23p^63d^{10})4s^24p$
	Ge	$1s^2)2s^22p^6)3s^23p^63d^{10})4s^24p^2$
	As	$1s^2)2s^22p^6)3s^23p^63d^{10})4s^24p^3$
	Se	$1s^2)2s^22p^6)3s^23p^63d^{10})4s^24p^4$
	Br	$1s^2)2s^22p^6)3s^23p^63d^{10})4s^24p^5$
	Kr	$1s^2)2s^22p^6)3s^23p^63d^{10})4s^24p^6)$ ... etc.

is spherically symmetric and very stable, as is the filled $1s$ shell, and the elements Ne and He which have these configurations are inert gases. The succeeding elements then fill up the $3s$ and $3p$ shells as shown in the table, ending up with another inert gas, A, which has a full $3p$ shell containing 6 electrons. The elements which have similar arrangements of outer or valence electrons in the $n = 2$ and $n = 3$ groups have similar chemical properties.

When the $3p$ level is filled, the next two elements add electrons in the $4s$ states rather than in the $3d$ levels, as might have been expected. This is simply because the $4s$ state happens to fall lower in energy than the $3d$ state. There is then a series of eight "transition elements" which have an incompletely filled inner $3d$-shell. When this $3d$ level is at length completely filled, the filling of the $4p$ shell resumes, ending with another inert gas, Kr. The rest of the table may be constructed along generally similar lines. In every case the outer shell structure of chemically related elements, such as Li, Na, K, Rb, and Cs is *identical*. The quantum theory is thus found to account satisfactorily not only for the spectroscopically observed energy level structure of complex atoms, but also for the arrangement of the atoms in the periodic system and the regularity of chemical properties associated with the periodic table of the elements.

EXERCISES

1. Show that the expectation value of the energy is equal to the separation constant ε for a system with time-dependent wave functions of the form (4.8-8). Show that the expectation value of the Hamiltonian operator taken over the time-independent wave function is equal to ε for this type of system.

2. Show that if z_1 and z_2 are any two complex numbers, then $(z_1 z_2)^* = z_1^* z_2^*$; show that the complex conjugate of a complex quantity expressed in polar form as $f e^{i\theta}$ (f, θ real) is $f^* e^{-i\theta}$.

3. For the free particle of Section 4.9, the phase velocity, from (4.9-5) is $\omega/k = \hbar k/2m = p/2m$. Classically, however, the velocity of a free particle is p/m. Explain the discrepancy of a factor of 2.

4. Verify directly that the wave function (4.9-11) for a free particle satisfies Schrödinger's equation (4.7-3) with $V = 0$.

5. By expressing the function of (4.9-17) as a Fourier integral with the help of (4.9-16), verify Equation (4.9-18).

6. Find $\langle x \rangle_n$, $\langle px \rangle_n$, $\langle p_x^2 \rangle_n$ for the nth eigenstate of the infinitely deep potential well of Section 4.10, and show explicitly that $\varepsilon_n = \langle p_x^2 \rangle_n / 2m$, as for the free particle.

7. What is the ratio of the amplitudes of the *continuum* eigenfunctions of the finite potential well of Section 4.11 in the region $(-a < x < a)$ to the amplitude in the regions $(x < -a; x > a)$?

8. From the definition (4.12-29) prove that $dH_n(\xi)/d\xi = 2nH_{n-1}(\xi)$. *Hint:* Note that $d^n(xf(x))/dx^n = x\, d^n f/dx^n + nd^{n-1}f/dx^{n-1}$.

9. From (4.12-29), prove that $H_{n+1}(\xi) = 2\xi H_n(\xi) - 2nH_{n-1}(\xi)$.

10. From the results of Problems 8 and 9, show that the functions $H_n(\xi)$ satisfy Hermite's equation (4.12-16).

11. Find the expectation value of the potential energy $\langle V \rangle_n$ for the nth eigenstate of a 1-dimensional quantum harmonic oscillator. *Hint:* Use the orthogonality properties of the wave functions.

12. Find the expectation value of the kinetic energy $\langle K \rangle_n$ for the nth eigenstate of a

1-dimensional quantum harmonic oscillator, and using the result of Problem 11, show that $\langle V \rangle_n + \langle K \rangle_n = \hbar\omega_0(n + \tfrac{1}{2})$, in agreement with intuition.

13. Show by evaluating the standard deviation that the momentum of a free particle is a constant of the motion.

14. Show for the eigenstates of the hydrogen atom that the probability of finding the electron in a spherical shell of thickness dr and radius r is $r^2 R_{nl}^2(r)$. Assume that the spherical harmonics $Y_{lm}(\theta, \phi)$ are normalized.

15. Show directly that the standard deviation formula yields $\sigma = 0$ for the operators L^2 and L_z for the hydrogen atom.

16. Obtain the expression (4.15-36) for L^2 from the expressions (4.15-35) for L_x, L_y, and L_z.

17. Using the methods of Section 4.15, show that if $\langle \varepsilon^n \rangle$ can be evaluated as $\langle \mathscr{H}^n \rangle$ using the time-independent wave functions, then $\langle \varepsilon^{n+1} \rangle$ can be represented as $\langle \mathscr{H}^{n+1} \rangle$ using these same functions. Indicate how, by induction, this means that $\langle f(\varepsilon) \rangle$ can be obtained from $\langle f(\mathscr{H}) \rangle$ using time-independent wave functions, where $f(\varepsilon)$ is any function of ε which can be represented as a convergent power series.

GENERAL REFERENCES

D. Bohm, *Quantum Mechanics*, Prentice-Hall, Inc., Englewood Cliffs, N.J. (1951).

E. Ikenberry, *Quantum Mechanics for Mathematicians and Physicists*, Oxford University Press, New York (1962).

R. B. Leighton, *Principles of Modern Physics*, McGraw-Hill Book Co., Inc., New York (1959).

E. Merzbacher, *Quantum Mechanics*, John Wiley & Sons, Inc., New York (1961).

L. Pauling and E. B. Wilson, *Introduction to Quantum Mechanics*, McGraw-Hill Book Co., Inc., New York (1935).

J. L. Powell and B. Craseman, *Quantum Mechanics*, Addison-Wesley Publishing Company, Inc., Reading, Mass. (1961).

C. W. Sherwin, *Introduction to Quantum Mechanics*, Holt, Rinehart and Winston, Inc., New York (1959).

J. C. Slater, *Quantum Theory of Matter*, McGraw-Hill Book Co., Inc., New York (1951).

CHAPTER 5

OUTLINE OF STATISTICAL MECHANICS

5·1 INTRODUCTION

The objective of statistical mechanics is to treat the behavior of a very large assembly of identical particles or systems in a statistical or probabilistic fashion, deriving the most probable values of the properties of the *ensemble* without inquiring in detail what the values of these properties are for any particular particle at any given time.

We shall restrict the present treatment to an assembly of identical systems, which we assume to be independent of one another except that they may interact with each other, or with their external surroundings, only through instantaneous processes which conserve energy and momentum. Such systems might, for example, represent the particles of a free-particle ideal monatomic gas, which could interact with one another or with their surroundings by instantaneous collisions, although systems of other types might equally well be envisioned. External force fields (electric, magnetic, or gravitational) might also exert their influences upon the systems. Under circumstances such as these, we may treat each system independently of the others, and we may describe the behavior of any system in terms of its coordinates in a six-dimensional phase space whose coordinates are (x,y,z,p_x,p_y,p_z). These coordinates are assumed to be independent and orthogonal, and a complete specification of the state of motion of any particle belonging to the ensemble is given by assigning values to these six-phase space coordinates; its subsequent behavior can then, in principle, be described on the basis of this knowledge.

The basic postulate of statistical mechanics is that the *a priori* probability for a system to be in any given quantum state is the same for all quantum states of the system. This means, for example, that the *a priori* probability that a free particle will be in a region $(\Delta x, \Delta y, \Delta z)$ about (x,y,z) is the same for all values of x, y, z and the probability that its momentum will be in a region $(\Delta p_x, \Delta p_y, \Delta p_z)$ about (p_x, p_y, p_z) is the same for all values of p_x, p_y, and p_z. It should be emphasized that this postulate refers only to the *a priori* probabilities, that is, to the probabilities which prevail in the absence of any dynamical restrictions. The *a priori* probabilities in any given ensemble of systems will generally be modified by external constraints imposed upon the system, such as requirements that the total number and total energy of all the systems belonging to the ensemble remain constant.

We shall first treat the particles belonging to the ensemble as if they were classical "billiard balls," regarding them as distinguishable one from another and allowing any number to occupy a single quantum state. Then, subsequently, we shall investigate what effect the Pauli exclusion principle has, and what the effect of the *indistinguishability* of the elementary particles which we are discussing is. We shall draw freely upon the results of the preceding chapter, and even in the case of the statistical treatment of classical particles, in the interest of consistency and uniformity, we shall use the quantum framework of energy levels and quantum states, although it is not absolutely necessary to do so.

5·2 THE DISTRIBUTION FUNCTION AND THE DENSITY OF STATES

In order to compute the average properties of an ensemble of particles, it is generally necessary to know how those particles are, on the average, distributed in energy. If we denote by $f(\varepsilon)$ the average number of particles of the system that occupy a single quantum state of energy ε, and if we let $g(\varepsilon)d\varepsilon$ be the number of quantum states of the system whose energy is in a range $d\varepsilon$ about ε, then the number of particles of the system whose energy is in the range $d\varepsilon$ about ε is given by $N(\varepsilon)d\varepsilon$, where

$$N(\varepsilon)\,d\varepsilon = f(\varepsilon)g(\varepsilon)\,d\varepsilon. \tag{5.2-1}$$

The quantity $f(\varepsilon)$ as defined above is referred to as the *distribution function* of the system, and depends upon the probabilities associated with the distribution of particles of the system among the available quantum states. The quantity $g(\varepsilon)$, which depends only upon how the quantum states themselves are situated in energy, is called the *density of states*.

If both these quantities are known, then the average value of any quantity α which can be expressed as a function of the total energy ε can be evaluated as

$$\langle\alpha\rangle = \frac{\displaystyle\int \alpha(\varepsilon)N(\varepsilon)\,d\varepsilon}{\displaystyle\int N(\varepsilon)\,d\varepsilon}$$

$$= \frac{1}{N}\int \alpha(\varepsilon)f(\varepsilon)g(\varepsilon)\,d\varepsilon, \tag{5.2-2}$$

the integral of $N(\varepsilon)d\varepsilon$ taken over all possible values of ε giving simply N, the total number of particles in the system. If the quantity α is a function of the system coordinates q_i ($= x,y,z$ for $i = 1,2,3$, respectively) and the momenta p_i ($= p_x,p_y,p_z$, in like fashion) which cannot be expressed as a function of the total energy ε alone, then ε may be expressed as a function of the coordinates and momenta as the sum of kinetic and potential energies,

$$\varepsilon = \frac{1}{2m}\sum_i p_i^2 + V(q_i), \tag{5.2-3}$$

and the average value of α evaluated as an integral over the phase space coordinates as

$$\langle \alpha \rangle = \frac{\iint \alpha(p_i,q_i) f(p_i,q_i)\, d\mathbf{p}\, d\mathbf{q}}{\iint f(p_i,q_i)\, d\mathbf{p}\, d\mathbf{q}}. \tag{5.2-4}$$

Here $f(p_i,q_i)$ represents the distribution function $f(\varepsilon)$ with ε expressed in terms of p_i and q_i by (5.2-3) and the quantities $d\mathbf{p}$ and $d\mathbf{q}$ represent volume elements in momentum and coordinate space expressed in any form (Cartesian, cylindrical, spherical, etc.,) which may be convenient for integration. Equation (5.2-4) contains no density of states factor $g(p_i,q_i)$ because, as we shall soon see, for systems of the type we are considering, the density of quantum states in phase space is constant and will cancel in the numerator and denominator of (5.2-4). We shall often refer to (5.2-2) or (5.2-4) when computing averages.

Since the density of states factor $g(\varepsilon)$ as defined in connection with (5.2-1) depends only upon the distribution of quantum states of the system in energy, it should be possible to calculate $g(\varepsilon)$ from the solution of Schrödinger's equation. We may do this for free particles as defined in Section 5.1 by assuming that our system is confined within a rigid container whose dimensions are x_0, y_0, z_0 in the x-, y-, and z-directions, respectively. This container is, in effect, an infinitely deep potential well for the particles inside, the potential being zero for each particle within the container, provided that interactions between particles are confined to instantaneous collisions. In this instance, inside the container, Schrödinger's equation, according to (4.8-5) can be written

$$\nabla^2 \psi + k^2 \psi(x,y,z) = 0, \tag{5.2-5}$$

where
$$k^2 = 2m\varepsilon/\hbar^2. \tag{5.2-6}$$

Outside the container, from the results of Sections 4.10 and 4.11, we may conclude that $\psi = 0$. From (4.9-3) we may expect that a valid solution to (5.2-5) would be a plane wave of the form

$$\psi(x,y,z) = Ae^{ik_x x}e^{ik_y y}e^{ik_z z} = Ae^{i(k_x x + k_y y + k_z z)} = Ae^{i(\mathbf{k}\cdot\mathbf{r})}. \tag{5.2-7}$$

It is easy to verify that this solution does actually satisfy (5.2-5). Equation (5.2-7) may indeed be obtained as a solution of (5.2-5) by the separation of variables technique used in connection with the hydrogen atom, assuming initially a product solution of the form $\psi = X(x)Y(y)Z(z)$. In order that (5.2-7) be a solution of (5.2-5), however, the quantities k_x, k_y, and k_z in (5.2-7) must be related in such a way that

$$k_x^2 + k_y^2 + k_z^2 = k^2 = \text{const.} \tag{5.2-8}$$

and these quantities may thus be considered to behave like the components of a vector $\mathbf{k}$, which is referred to as the propagation vector. From a consideration of the physical character of the time-dependent wave function which is related to (5.2-7) by (4.8-7),

$$\Psi(x,y,z,t) = Ae^{i(\mathbf{k}\cdot\mathbf{r}-\omega t)}$$

$$(\omega = \varepsilon/\hbar),$$

$$\tag{5.2-9}$$

it should be clear that the direction of **k** is the same as the direction in which the wave front advances. The actual physical explanation of why this is so is assigned as an exercise for the reader. It can be shown by the methods developed in the last chapter that the expectation value of the vector momentum **p** is given by

$$\langle \mathbf{p} \rangle = \hbar \mathbf{k}, \tag{5.2-10}$$

and that **p**, (hence **k**), is a constant of the motion.

For boundary conditions, we shall demand that the wave function on any one face of the container be equal to the wave function on the opposite face. These *periodic* or *cyclic* boundary conditions have been discussed previously in connection with lattice vibrations (Section 3.3). In a one-dimensional geometry these boundary conditions represent a system which is topologically equivalent to a ring. In three dimensions, the periodic boundary conditions divide all space up into exactly similar regions of dimensions (x_0, y_0, z_0) in each of which the wave function is the same, and any one of these regions may be used to represent the interior of the potential well corresponding to the container. Proceeding in this manner and assuming that edges of the container extend along the x-axis from 0 to x_0, along the y-direction from 0 to y_0, and along the z-direction from 0 to z_0, the boundary conditions become

$$\psi(0,y,z) = \psi(x_0,y,z)$$

$$\psi(x,0,z) = \psi(x,y_0,z) \tag{5.2-11}$$

$$\psi(x,y,0) = \psi(x,y,z_0),$$

which require that in (5.2-7) we take

$$k_x = 2\pi n_x/x_0 \qquad (n_x = 0,\pm 1,\pm 2,\cdots)$$

$$k_y = 2\pi n_y/y_0 \qquad (n_y = 0,\pm 1,\pm 2,\cdots) \tag{5.2-12}$$

$$k_z = 2\pi n_z/z_0 \qquad (n_z = 0,\pm 1,\pm 2,\cdots),$$

(5.2-7) then reducing to

$$\psi(x,y,z) = Ae^{2\pi i \left(\frac{n_x x}{x_0} + \frac{n_y y}{y_0} + \frac{n_z z}{z_0} \right)}. \tag{5.2-13}$$

According to (5.2-8) and (5.2-6) this means that only a certain discrete set of energy values, given by

$$\varepsilon_{n_x n_y n_z} = \frac{2\pi^2 \hbar^2}{m} \left[\frac{n_x^2}{x_0^2} + \frac{n_y^2}{y_0^2} + \frac{n_z^2}{z_0^2} \right] \tag{5.2-14}$$

are allowed. From (5.2-10), the allowed values of (k_x, k_y, k_z) can be expressed as allowed values of (p_x, p_y, p_z) in the form

$$p_x = hn_x/x_0$$

$$p_y = hn_y/y_0 \qquad\qquad (5.2\text{-}15)$$

$$p_z = hn_z/z_0.$$

If we were to plot all the allowed values of the momentum, corresponding to all possible integer values for (n_x, n_y, n_z) in (5.2-15), as points in an orthogonal momentum space with coordinates (p_x, p_y, p_z), we would obtain a simple orthogonal lattice of points representing allowed momentum values, with unit cell dimensions $(h/x_0, h/y_0, h/z_0)$ corresponding to unit changes in (n_x, n_y, n_z) in (5.2-15). The volume of momentum space corresponding to a single quantum state of the system is simply the unit cell volume

$$V_p = \frac{h^3}{x_0 y_0 z_0} = h^3/V, \qquad\qquad (5.2\text{-}16)$$

where V is the physical volume of the container. From this (and from the plot of allowed momentum points) we see that the density with which the allowed quantum states are distributed in momentum space is uniform over all momentum space (which was mentioned previously in connection with Equation (5.2-4)). In (5.2-16) we have taken no account of spin; for particles of spin $\frac{1}{2}$ like electrons, there will be *two* allowed momentum states for each lattice point in momentum space, corresponding to the two possible spin orientations of the particle. Taking this into account, (5.2-16) should be rewritten in the form

$$V_p = \tfrac{1}{2} h^3/V. \qquad\qquad (5.2\text{-}17)$$

For large values of x_0, y_0, and z_0, such as would be associated with a container of macroscopic dimensions, the volume V_p of (5.2-17) becomes very small, and the momentum states (5.12-15) and energy states (5.2-14) become crowded together very closely compared to energy and momentum intervals which are at all appreciable macroscopically. For a system of reasonable dimensions, then, we shall be justified in regarding the allowed energy and momentum values as substantially *continuously* distributed in momentum space.

Now let us consider a surface in the momentum space (p_x, p_y, p_z), all points of which are at constant energy ε. From (5.2-6), (5.2-8), and (5.2-10), the equation of this surface must be that of a sphere of radius $p = \sqrt{2m\varepsilon}$, or

$$p^2 = p_x^2 + p_y^2 + p_z^2 = 2m\varepsilon. \qquad\qquad (5.2\text{-}18)$$

This surface is illustrated in Figure 5.1. A similar sphere whose surface represents all points of energy $\varepsilon + d\varepsilon$ is shown in that figure. The spherical shell between these two spheres represents that region of momentum space corresponding to energies in the range ε to $\varepsilon + d\varepsilon$. The volume of momentum space contained within this shell is just

$$dV_p = 4\pi p^2 \, dp. \qquad\qquad (5.2\text{-}19)$$

However, from (5.2-18)

$$p \, dp = m \, d\varepsilon,$$ (5.2-20)

whereby $$dV_p = 4\pi p \cdot m \, d\varepsilon = 4\pi m \sqrt{2m\varepsilon} \, d\varepsilon.$$ (5.2-21)

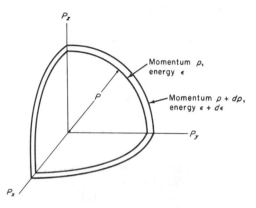

FIGURE 5.1. Spherical surfaces corresponding to constant energies ε and $\varepsilon + d\varepsilon$ plotted in the momentum space (p_x, p_y, p_z) of a particle.

The number of quantum states to be found in this volume of momentum space is determined simply by dividing this result by the volume of momentum space associated with a single quantum state as given by (5.2-17). The result,

$$g(\varepsilon) \, d\varepsilon = \frac{8\sqrt{2\pi} V}{h^3} \, m^{3/2} \sqrt{\varepsilon} \, d\varepsilon,$$ (5.2-22)

is by definition the density of states factor referred to in (5.2-1). This result, of course, refers only to free particles subject to instantaneous collision interactions only. For other systems, particularly where long range mutual interactions between particles are involved, the density of states factor would be much more complex. Also, the result (5.2-22) pertains, strictly speaking, only to a container in the shape of a rectangular solid. However, since only the volume of the container appears in the final result, it is intuitively clear that the same density of states would be obtained for a container of like volume independent of its shape. This can indeed be shown to be true.

5·3   THE MAXWELL-BOLTZMANN DISTRIBUTION

If there are no restrictions on the amount of energy or momentum a particle of the system may possess, then the probability associated with all quantum states is the same, and the average number of particles per quantum state will be independent of energy. This means that the distribution function $f(\varepsilon)$ will be constant. This rather simple situation is not very important physically, since a system which is thermally isolated from its surroundings must obey the restriction that the sum of the energies of all the particles of the system be constant, and it is systems of this type with which we shall have to deal. In this more realistic case the proportion of particles occupying states of extremely high energy is reduced and the distribution function is no longer constant with respect to energy.

To find out just what the distribution function is under these conditions, we shall proceed initially along classical lines, imagining the particles of the system to be *identifiable* objects such as numbered billiard balls. We shall retain the framework of quantum states and energy levels, although, ignoring the Pauli exclusion principle, we shall permit any number of particles to occupy a given quantum state of the system. We shall suppose that we are dealing with an isolated system of N distinguishable particles, with constant total energy U, which may be distributed among n energy levels $\varepsilon_1, \varepsilon_2, \varepsilon_2, \cdots \varepsilon_i, \cdots \varepsilon_n$. Statistically, the ensemble will have that energy distribution which corresponds to some chance distribution of particles among levels, the number in each level being given by $N_1, N_2, N_3, \cdots N_i, \cdots N_n$. Of all the chance distributions of N particles among n energy levels, some will occur with relatively high probability on a purely statistical basis, and others will be quite unlikely. This situation is analogous to the simultaneous tossing of two coins, in which the distribution (1 head, 1 tail) is more probable than the distribution (2 heads, 0 tails) or the distribution (0 heads, 2 tails). The distribution which has the maximum probability of occurrence is that distribution of particles among levels *which can be realized in a maximum number of statistically independent ways.* The equilibrium state of the system will then be assumed to correspond closely to this statistical distribution of particles among levels which is of maximum probability under the conditions of the problem.

The calculation of the number of ways in which indentifiable particles can be distributed among energy levels of a system is equivalent to the calculation of the number of ways in which numbered objects can be distributed among a set of numbered containers. We must thus consider the problem of determining the number of ways of distributing N identifiable objects among n containers such that there are N_1 in the first, N_2 in the second, $\cdots$, N_i in the ith, etc. This number will be proportional to the probability with which a distribution where $N_1, N_2, \cdots N_n$ objects are in containers 1, 2, 3, $\cdots n$, will occur.

To begin, let us assume that we have only two containers in which the objects may be placed, as shown in Figure 5.2. In general there will be N_1 objects in container 1 and N_2 in container 2, with

$$N_1 + N_2 = N = \text{const.} \tag{5.3-1}$$

Let us denote by $Q(N_1, N_2)$ the number of statistically independent ways of arriving at the distribution (N_1, N_2) objects in containers (1,2). Now $Q(0, N_2)$ is certainly equal to one, because the only way of achieving this distribution is to put all N objects into the second box, as shown in the figure. Likewise, $Q(1, N_2) = N$, because to realize this distribution, we may put object 1 in the first container and the rest in the second, object 2 in the first container, and the rest in the second, and so on through the N objects. There are N ways of choosing the object to be put in the first container and thus N ways of arriving at the distribution in question. If there are two objects in the first box, then $Q(2, N_2) = N(N-1)/2!$. This result follows because there are N ways of choosing the first object which goes into the first container, but only $N-1$ ways of choosing the second from the remaining objects; also, the two distributions where the object numbered α was chosen as the first object and the object numbered β as the second object to be put into the first container and where β was chosen as the first and α as the second are really identical, the same two objects ending up finally in the first box; the factor $N(N-1)$ must thus be divided by two. For the case where there are three objects in the first box, there are N ways of choosing the first,

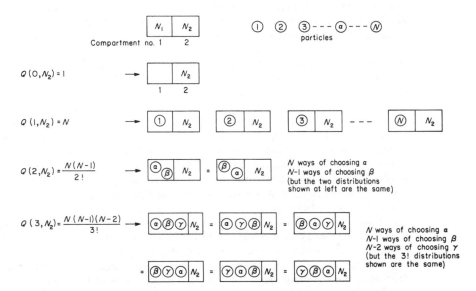

FIGURE 5.2. The possible ways of distributing N identifiable particles among two containers.

$N-1$ ways of choosing the second and $N-2$ ways of choosing the third, while there are 3! ways of permuting numbered objects α, β, and γ among themselves as having been chosen first, second or third, whereby $Q(3,N_2) = N(N-1)(N-2)/3!$. By an obvious extension of this process, and recalling (5.3–1), it is readily established that

$$Q(N_1,N_2) = \frac{N(N-1)(N-2) \cdots (N-N_1+1)}{N_1!} = \frac{N!}{N_1!(N-N_1)!} = \frac{N!}{N_1!N_2!}.$$

$$(5.3\text{-}2)$$

Suppose now that the second container is divided into two subcompartments containing ν_1 and ν_2 objects, where, of course,

$$\nu_1 + \nu_2 = N_2 = N - N_1, \qquad (5.3\text{-}3)$$

as shown in Figure 5.3. The number of independent ways of realizing the distribution (ν_1,ν_2) among the sub-compartments of the second box is, from (5.3-3)

$$Q(\nu_1,\nu_2) = \frac{N_2!}{\nu_1!\nu_2!}. \qquad (5.3\text{-}4)$$

But now we may, if we wish, regard the system as having *three* distinct containers, and the total number of ways of arranging the objects so that there are N_1 in the first container, v_1 in the second and v_2 in the third [which we shall call $Q(N_1,v_1,v_2)$] will be just the product of $Q(N_1,N_2)$ and $Q(v_1,v_2)$, since for *each* arrangement (N_1,N_2)

FIGURE 5.3. The subdivision of the second container into two subcompartments, leading to a system which can be thought of as actually having been divided into *three* distinct containers.

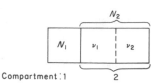

Compartment : 1   2

there will be $Q(v_1,v_2)$ ways of arranging particles among the subcompartments, and there are $Q(N_1,N_2)$ basic ways of arriving at a distribution (N_1,N_2) for the two original containers. We shall find thus, that

$$Q(N_1,v_1,v_2) = Q(N_1,N_2)Q(v_1,v_2) = \frac{N!}{N_1!\,N_2!}\frac{N_2!}{v_1!\,v_2!} = \frac{N!}{N_1!\,v_1!\,v_2!}. \qquad (5.3\text{-}5)$$

We may now simply relabel the three containers as boxes 1, 2, and 3, and call the number of objects in each N_1, N_2, and N_3, rather than N_1, v_1, and v_2, in which case (5.3-5) may be written

$$Q(N_1,N_2,N_3) = \frac{N!}{N_1!\,N_2!\,N_3!}. \qquad (5.3\text{-}6)$$

Again, one may imagine the *third* compartment to be subdivided as before, and obtain in the same manner an expression for the number of ways of arranging N objects among four boxes so as to obtain a distribution (N_1,N_2,N_3,N_4). By repeated application of this process, the result may be extended indefinitely by induction to cover the case where there may be n boxes. It is clear from the form of (5.3-3) and (5.3-6) that one will find

$$Q(N_1,N_2,N_3,\cdots N_n) = \frac{N!}{N_1!\,N_2!\,N_3!\cdots N_n!} = \frac{N!}{\prod\limits_{i=1}^{n} N_i!}, \qquad (5.3\text{-}7)$$

where the $\prod$ notation is used to indicate an extended product in the same way that the more familiar $\sum$ notation is used to express a summation.

The actual *probability* associated with a given distribution of N objects among n boxes is the number of independent ways of arranging objects among boxes $Q(N_1, \cdots N_n)$ which lead to that particular distribution, divided by the total number of ways of arranging N objects among n boxes *without regard for what distribution of objects among boxes results*. It is easily shown that this latter factor is simply n^N. The probability associated with a given distribution will then be $Q(N_1, \cdots N_n)/n^N$. We shall find it more convenient, however, to deal exclusively with the quantities $Q(N_1, \cdots N_n)$, which are *proportional* to the actual probabilities, since we shall only be interested in finding the values $N_1, \cdots N_n$ which render the probability a maximum, and if $Q(N_1, \cdots N_n)$ is a maximum, then so is the associated probability.

Now let us identify the ith container with the ith energy level of the system and the number of objects N_i in that container with the number of particles belonging to that level. We must, however, allow for the fact that the energy levels may be degenerate, and we shall assume there are g_i independent quantum states associated with the ith energy level, each of which has the same *a priori* probability of being occupied, according to the fundamental postulate of Section 5.1. Each energy level must then be thought of not as a single container, but as a *group* of g_i containers, as illustrated in Figure 5.4. For a given energy level, say the ith, containing N_i particles and g_i

FIGURE 5.4. The subdivision of energy levels into separate quantum states according to their degeneracy. The separate quantum states play the role of the containers in the statistical development, each having equal statistical weight.

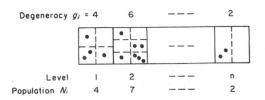

independent quantum states, there are $g_i^{N_i}$ ways of arranging the particles among the states or containers pertaining to that level, because there are g_i independent choices of where the first particle may be placed, and for *each* of these there are g_i choices of where the second may be placed, and so on to N_i factors. Each of these arrangements would constitute a separate statistically independent way of distributing particles among the quantum states of the system, since it is the quantum states themselves which are of equal probability, not the energy levels. If the first energy level, ε_1, were the *only* degenerate level, then there would be $g_1^{N_1}$ times as many independent ways of forming a given distribution of particles among energy levels as is given by (5.3-7), because there are that many ways of permuting particles among separate degenerate states belonging to the first level, and for each of these there still remain the same number of ways of assigning particles to the other levels of the system in such a way that the distribution $(N_1, N_2, \cdots N_n)$ results. If, in addition, the second level were degenerate, there would be $g_2^{N_2}$ possibilities for each one that existed before, the total number of independent ways of achieving a given distribution of particles among levels being now $g_1^{N_1} g_2^{N_2}$ times that given by (5.3-7). It is clear that when the degeneracy of all the levels is included, the result (5.3-7) will have to be multiplied by $g_1^{N_1} g_2^{N_2} g_3^{N_3} \cdots g_n^{N_n}$, giving

$$Q(N_1, N_2, \cdots N_n) = \frac{N!}{\displaystyle\prod_{i=1}^{n} N_i!} \prod_{i=1}^{n} g_i^{N_i} \tag{5.3-8}$$

when degeneracy is included.

We shall now assume that the actual distribution $(N_1, N_2, \cdots N_n)$ of particles among energy states which is observed in equilibrium is essentially that which can be realized in the *maximum* number of statistically independent ways, in other words, that for which the quantity Q of (5.3-8) is a maximum. Q must then be maximized with respect to the parameters $N_1, N_2, \cdots N_n$, subject to the restrictions

$$\sum_{i=1}^{n} N_i = N = \text{const.} \tag{5.3-9}$$

and
$$\sum_{i=1}^{n} \varepsilon_i N_i = U = \text{const.} \qquad (5.3\text{-}10)$$

This maximization is most easily carried out by the use of a mathematical technique called the method of *Lagrangean multipliers*, and we shall digress briefly to describe this subject. Suppose we are given a function $f(x_1, x_2, \cdots x_n)$ of n variables, and we are asked to find the values of x_1, x_2, $\cdots x_n$ which make f a maximum *under the restriction that some other given function* $\phi(x_1, x_2, \cdots x_n)$ *remain constant.* For a maximum (or minimum) value of f, $df = 0$, and if ϕ remains constant, $d\phi = 0$, so that if f is a maximum or minimum under the stated conditions, then

$$df + \alpha \, d\phi = 0 \qquad (5.3\text{-}11)$$

no matter what value the arbitrary undetermined multiplier α may have. But (5.3-11) may be written

$$\left(\frac{\partial f}{\partial x_1} + \alpha \frac{\partial \phi}{\partial x_1}\right) dx_1 + \left(\frac{\partial f}{\partial x_2} + \alpha \frac{\partial \phi}{\partial x_2}\right) dx_2 + \cdots + \left(\frac{\partial f}{\partial x_n} + \alpha \frac{\partial \phi}{\partial x_n}\right) dx_n = 0. \quad (5.3\text{-}12)$$

According to this equation, condition (5.3-11) will certainly be fulfilled if for every i

$$\frac{\partial f}{\partial x_i} + \alpha \frac{\partial \phi}{\partial x_i} = 0 \qquad (i = 1, 2, \cdots n), \qquad (5.3\text{-}13)$$

while, in addition, we were given to begin with the fact that

$$\phi(x_1, x_2, \cdots x_n) = \text{const.} \qquad (5.3\text{-}14)$$

Equations (5.3-13) and (5.3-14), taken together, represent a set of $n + 1$ simultaneous equations which can be solved for the n quantities x_1, x_2, $\cdots x_n$ which maximize (or minimize) f *and* for the undetermined multiplier α which was introduced in (5.3-11).

If there are *two* auxiliary functions $\phi(x_1, x_2, \cdots x_n)$ and $\psi(x_1, x_2, \cdots x_n)$ which are to remain constant as the function f is maximized, then *two* arbitrary multipliers α and β are introduced and it is required that

$$df + \alpha \, d\phi + \beta \, d\psi = 0, \qquad (5.3\text{-}15)$$

which leads, in the same manner, to a set of $n + 2$ equations

$$\frac{\partial f}{\partial x_i} + \alpha \frac{\partial \phi}{\partial x_i} + \beta \frac{\partial \psi}{\partial x_i} = 0 \qquad (i = 1, 2, \cdots n) \qquad (5.3\text{-}16)$$

$$\phi(x_1, x_2, \cdots x_n) = \phi_0 = \text{const.} \qquad (5.3\text{-}17)$$

$$\psi(x_1, x_2, \cdots x_n) = \psi_0 = \text{const.} \qquad (5.3\text{-}18)$$

for the $n + 2$ unknowns x_1, x_2, $\cdots x_n$, α and β.

We shall now proceed to apply this method to find the distribution $N_1, N_2, \cdots N_n$ for which (5.3-8) is a maximum. As a matter of mathematical convenience, we shall actually maximize $\ln Q$ rather than Q itself, but since the logarithm is a single-valued monotonic function of all the variables involved, when $\ln Q$ is a maximum, so also is Q. If we take the logarithm of both sides of (5.3-8), we find

$$\ln Q(N_1, N_2, \cdots N_n) = \ln N! + \sum_{i=1}^{n} N_i \ln g_i - \sum_{i=1}^{n} \ln N_i!. \tag{5.3-19}$$

We shall assume that our system is so large that for each level $N_i!$ may be approximated by Stirling's approximation, which states that for $x \gg 1$,

$$\ln x! \cong x \ln x - x. \tag{5.3-20}$$

Making use of this approximation, (5.3-19) becomes

$$\ln Q = \ln N! + \sum_i N_i \ln g_i - \sum_i N_i \ln N_i + \sum_i N_i, \tag{5.3-21}$$

which must be maximized under the restrictions

$$\phi(N_1, N_2, \cdots N_n) = \sum_i N_i = N \tag{5.3-22}$$

$$\psi(N_1, N_2, \cdots N_n) = \sum_i \varepsilon_i N_i = U. \tag{5.3-23}$$

From (5.3-16), this requires that

$$\frac{\partial(\ln Q)}{\partial N_j} + \alpha \frac{\partial \phi}{\partial N_j} + \beta \frac{\partial \psi}{\partial N_j} = \frac{\partial}{\partial N_j}\left[\sum_i N_i \ln g_i - \sum_i N_i \ln N_i + \sum_i N_i\right]$$

$$+ \alpha \frac{\partial}{\partial N_j}\left(\sum_i N_i\right) + \beta \frac{\partial}{\partial N_j}\left(\sum_i \varepsilon_i N_i\right) = 0 \qquad (j = 1, 2, \cdots n). \tag{5.3-24}$$

Working out the derivatives in (5.3-24), noting that the only terms of the summations whose derivatives with respect to N_j are other than zero are those for which $i = j$, (5.3-24) can be reduced to

$$\ln g_j - \ln N_j + \alpha + \beta \varepsilon_j = 0 \qquad (j = 1, 2, \cdots n). \tag{5.3-25}$$

Solving for $\ln (N_j/g_j)$ and exponentializing, this can be rewritten in the form

$$\frac{N_j}{g_j} = e^\alpha e^{\beta \varepsilon_j} = f(\varepsilon_j). \tag{5.3-26}$$

Equation (5.3-26) gives the average number of particles per quantum state of the system and thus, by definition, represents the energy distribution function $f(\varepsilon)$. This particular energy distribution function obtained as (5.3-26) under the classical assumption of identifiable particles and without the use of the Pauli exclusion principle, is called the *Maxwell-Boltzmann* distribution function.

We must now discuss how the constants α and β are related to the physical properties of the system. To begin with, we shall assign to the constant β the value

$$\beta = -1/kT,\tag{5.3-27}$$

where T is the absolute temperature of the system and k is a constant called Boltzmann's constant, and we shall regard this equation, along with (5.3-26) as *defining* what is meant by temperature. We shall see, in due course, that this definition leads to all the familiar characteristics of temperature as related to an ideal gas, and it will then be clear that the definition of temperature *could* have been postponed until the properties of an ideal gas had been worked out from (5.3-26). An identification of β with the value given above could then have been assigned on the basis of a comparison of the results so obtained and the familiar thermodynamic equation of state for an ideal gas (from which the temperature is more commonly defined). It will be shown that the value of k can be related to the measured ideal gas constant R and Avogadro's number.

Using the value given by (5.3-27) for β, (5.3-26) may now be written

$$N_j = g_j e^{\alpha} e^{-\varepsilon_j/kT}.\tag{5.3-28}$$

The value of the constant α may now be expressed in terms of the total number of particles N, since from (5.3-9) and (5.3-28)

$$N = \sum_j N_j = e^{\alpha} \sum_j g_j e^{-\varepsilon_j/kT},\tag{5.3-29}$$

whereby

$$e^{\alpha} = \frac{N}{\sum\limits_j g_j e^{-\varepsilon_j/kT}},\tag{5.3-30}$$

and

$$N_j = g_j e^{\alpha} e^{-\varepsilon_j/kT} = \frac{N g_j e^{-\varepsilon_j/kT}}{\sum\limits_j g_j e^{-\varepsilon_j/kT}}.\tag{5.3-31}$$

If the energy levels of the system are crowded very closely together, as are the levels of the gas of free particles which was discussed in the previous section, then the quantity g_j in (5.3-28) may be regarded as $g(\varepsilon)d\varepsilon$ and the quantity N_j in that equation may be regarded as $N(\varepsilon)d\varepsilon$, as discussed in connection with Equation (5.2-1). The number of particles in an energy range $d\varepsilon$ about ε in this limit will, according to (5.3-28), be given by

$$N(\varepsilon)\,d\varepsilon = e^{\alpha} e^{-\varepsilon/kT} g(\varepsilon)\,d\varepsilon = f(\varepsilon)g(\varepsilon)\,d\varepsilon.\tag{5.3-32}$$

For an ideal gas of free particles, the density of states factor $g(\varepsilon)$ will be given by (5.2-22). Note that the form of (5.3-32) is the same as that of (5.2-1) with $f(\varepsilon) = e^{\alpha} e^{-\varepsilon/kT}$. Again, as discussed in connection with (5.3-31), we may evaluate the constant α by the condition that the total number of particles in the system shall be a constant N, whereby

$$N = \int N(\varepsilon)\, d\varepsilon = e^{\alpha} \int g(\varepsilon) e^{-\varepsilon/kT}\, d\varepsilon, \qquad (5.3\text{-}33)$$

and

$$e^{\alpha} = \frac{N}{\displaystyle\int g(\varepsilon) e^{-\varepsilon/kT}\, d\varepsilon}, \qquad (5.3\text{-}34)$$

the integrals being taken over all energies available to the particles of the system.

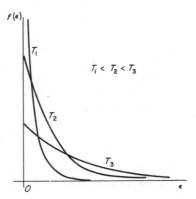

FIGURE 5.5. A schematic representation of the Maxwell-Boltzmann energy distribution function for three different temperatures.

We should note that the quantity Q of (5.3-8) is closely connected with the thermo-dynamic *entropy* of the system. At equilibrium, the state of the system is such that Q, the number of statistically independent ways of distributing particles among quantum states, is a maximum, as is the entropy of the system. We shall not go into details at this point, but it can indeed be demonstrated[1] that the relation between these two quantities is

$$S = k \ln Q, \qquad (5.3\text{-}35)$$

where S is the entropy.

5·4 MAXWELL-BOLTZMANN STATISTICS OF AN IDEAL GAS

We shall now discuss the properties of an ideal gas of free particles of the type discussed in Section 4.2. For this system, the density of states factor $g(\varepsilon)\, d\varepsilon$ is given by (5.2-22), whereby, according to (5.3-34),

---

[1] J. E. Mayer and M. Göppert-Mayer, *Statistical Mechanics*, John Wiley and Sons, New York (1940), Chapter 4.

$$e^\alpha = \frac{N}{\dfrac{8\sqrt{2\pi}Vm^{3/2}}{h^3}\displaystyle\int_0^\infty \sqrt{\varepsilon}\,e^{-\varepsilon/kT}\,d\varepsilon}.$$ (5.4-1)

The integration is taken between the limits zero and infinity, since the energy of the particles is entirely kinetic, hence positive. The integral in (5.4-1) may be expressed in a simpler form by the substitution

$$x = \varepsilon/kT,$$ (5.4-2)

whence (5.4-1) becomes

$$e^\alpha = \frac{N}{8\sqrt{2\pi}V\left(\dfrac{mkT}{h^2}\right)^{3/2}\displaystyle\int_0^\infty x^{1/2}e^{-x}\,dx}.$$ (5.4-3)

The integral can now be expressed as a Γ-function,[2] since $\Gamma(n)$ is defined as

$$\Gamma(n) = \int_0^\infty x^{n-1}e^{-x}\,dx,$$ (5.4-4)

the integral in (5.4-3) thus being equal to $\Gamma(3/2)$, which in turn equals $\sqrt{\pi}/2$. Inserting this value into (5.4-3), we find

$$e^\alpha = \frac{N}{2V}\left(\frac{h^2}{2\pi mkT}\right)^{3/2},$$ (5.4-5)

which makes it possible to write the Maxwell-Boltzmann distribution function for an ideal gas as

$$f(\varepsilon) = e^\alpha e^{-\varepsilon/kT} = \frac{N}{2V}\left(\frac{h^2}{2\pi mkT}\right)^{3/2}e^{-\varepsilon/kT}.$$ (5.4-6)

It will be noted that the value of e^α given by (5.4-5) is *temperature dependent*. It should also be emphasized that this particular value for e^α pertains *only* to the free particle density of states function (5.2-22) and that for systems having other density of states functions associated with them the value for e^α will be different from that given by (5.4-5). A plot of the Boltzmann distribution (5.4-6) for several temperatures is shown in Figure 5.5. From (5.4-6) and (5.2-22) it is clear that the actual distribution of particle density with energy is given by

$$N(\varepsilon)\,d\varepsilon = f(\varepsilon)g(\varepsilon)\,d\varepsilon = \frac{2\pi N}{(\pi kT)^{3/2}}\sqrt{\varepsilon}\,e^{-\varepsilon/kT}\,d\varepsilon.$$ (5.4-7)

---

[2] See, for instance, I. S. and E. S. Sokolnikoff, *Higher Mathematics for Engineers and Physicists*, McGraw-Hill Book Co., Inc., New York (1941), pp. 273–276.

The total internal energy and specific heat of the gas can be obtained very easily from these results. Since the energy of the particles in the range $d\varepsilon$ about ε is simply $\varepsilon N(\varepsilon)\, d\varepsilon$, the total internal energy of the gas is

$$U = \int_0^\infty \varepsilon N(\varepsilon)\, d\varepsilon, \tag{5.4-8}$$

or, using (5.4-7),

$$U = \frac{2\pi N}{(\pi kT)^{3/2}} \int_0^\infty \varepsilon^{3/2} e^{-\varepsilon/kT}\, d\varepsilon. \tag{5.4-9}$$

This integral can be evaluated in terms of Γ-functions by making the substitution (5.4-2). Working out the integral in this way, noting that $\Gamma(5/2) = (3/2)\Gamma(3/2) = 3\sqrt{\pi}/4$, we find

$$U = \tfrac{3}{2}NkT, \tag{5.4-10}$$

whence the average internal energy per particle, U/N, is $\tfrac{3}{2}kT$, an important and familiar result. The heat capacity of the gas at constant volume is the rate of increase of internal energy with respect to temperature, whereby from (5.4-10),

$$C_v = \left(\frac{\partial U}{\partial T}\right)_v = \tfrac{3}{2}Nk, \tag{5.4-11}$$

independent of temperature. The specific heat c_v is simply the heat capacity per unit volume.

Our objective, eventually, is to derive the equation of state for an ideal Boltzmann gas from the dynamical properties of the particles and from the distribution function. Before doing this, however, we must convert the *energy* distributions (5.4-6) and (5.4-7) into appropriate velocity distributions, and investigate briefly how to use these velocity distribution functions. Table 5.1 gives the value of some of the definite integrals which are frequently encountered when working with the Boltzmann velocity distributions.

In a gas of free particles possessing no internal degrees of freedom, all the energy resides in the kinetic energy of the particles. We may relate this to the velocity by writing

$$\varepsilon = \tfrac{1}{2}mv^2 = \tfrac{1}{2}m(v_x^2 + v_y^2 + v_z^2) \tag{5.4-12}$$

whence

$$d\varepsilon = mv\, dv, \tag{5.4.13}$$

and the energy distribution (5.4-7) may be written directly as a velocity distribution of the form

$$N(v)\, dv = 4\pi N\left(\frac{m}{2\pi kT}\right)^{3/2} v^2 e^{-mv^2/2kT}\, dv. \tag{5.4-14}$$

This function expresses the number of particles of the system whose *speeds* lie in a range dv about v, or the number of particles which lie within a spherical shell of thickness dv and radius v in *velocity* space.

TABLE 5.1.

Maxwell-Boltzmann Integrals*

$$\int_0^\infty e^{-\alpha x^2}\, dx = \frac{1}{2\sqrt{}}\sqrt{\frac{\pi}{\alpha}} = \frac{1}{2\sqrt{}}\sqrt{\frac{2\pi kT}{m}} = \frac{\pi \bar{c}}{4}$$

$$\int_0^\infty x e^{-\alpha x^2}\, dx = \frac{1}{2\alpha} = \frac{kT}{m} = \frac{\pi \bar{c}^2}{8}$$

$$\int_0^\infty x^2 e^{-\alpha x^2}\, dx = \frac{1}{4\alpha\sqrt{}}\sqrt{\frac{\pi}{\alpha}} = \frac{\sqrt{\pi}}{4}\left(\frac{2kT}{m}\right)^{3/2} = \frac{\pi^2 \bar{c}^3}{32}$$

$$\int_0^\infty x^3 e^{-\alpha x^2}\, dx = \frac{1}{2\alpha^2} = \frac{1}{2}\left(\frac{2kT}{m}\right)^2 = \frac{\pi^2 \bar{c}^4}{32}$$

$$\int_0^\infty x^4 e^{-\alpha x^2}\, dx = \frac{3}{8\alpha^2\sqrt{}}\sqrt{\frac{\pi}{\alpha}} = \frac{3\sqrt{\pi}}{8}\left(\frac{2kT}{m}\right)^{5/2} = \frac{3\pi^3 \bar{c}^5}{256}$$

$$\int_0^\infty x^5 e^{-\alpha x^2}\, dx = \frac{1}{\alpha^3} = \left(\frac{2kT}{m}\right)^3 = \frac{\pi^3 \bar{c}^6}{64}$$

$$\int_0^\infty x^6 e^{-\alpha x^2}\, dx = \frac{15}{16\alpha^3\sqrt{}}\sqrt{\frac{\pi}{\alpha}} = \frac{15\sqrt{\pi}}{16}\left(\frac{2kT}{m}\right)^{7/2} = \frac{15\pi^4 \bar{c}^7}{2^{11}}$$

$$\int_0^\infty x^7 e^{-\alpha x^2}\, dx = \frac{3}{\alpha^4} = 3\left(\frac{2kT}{m}\right)^4 = \frac{3\pi^4 \bar{c}^8}{256}$$

$$\int_0^\infty x^8 e^{-\alpha x^2}\, dx = \frac{105}{32\alpha^4\sqrt{}}\sqrt{\frac{\pi}{\alpha}} = \frac{105\sqrt{\pi}}{32}\left(\frac{2kT}{m}\right)^{9/2} = \frac{105\pi^5 \bar{c}^9}{2^{14}}$$

* The second column in the table gives the value of the definite integral shown in the first column. The third column gives the value of the definite integral when α is set equal to $m/2kT$, which is usually the case when working with Boltzmann velocity distributions. The fourth column expresses the result shown in the third column as a multiple of the mean thermal speed $\bar{c}$, where

$$\bar{c} = \sqrt{8kT/\pi m}.$$

Suppose now that we wish to know how many particles of the system have velocities such that the x-component of velocity is in a range dv_x about v_x, the y-component is in a range dv_y about v_y and the z-component is in a range dv_z about v_z, thus the number of particles in a rectangular volume element $(dv_x dv_y dv_z)$ in velocity space centered on the value (v_x, v_y, v_z). We shall call this number $N(v_x, v_y, v_z) dv_x\, dv_y\, dv_z$. We must now proceed more carefully, starting with (5.4-6), which according to (5.4-12) we may write as

$$f(v_x, v_y, v_z) = \frac{N}{2V}\left(\frac{h^2}{2\pi m kT}\right)^{3/2} e^{-m(v^2_x + v^2_y + v^2_z)/2kT}. \tag{5.4-15}$$

Then, as required by (5.2-1),

$$N(v_x, v_y, v_z)\, dv_x\, dv_y\, dv_z = f(v_x, v_y, v_z) g(v_x, v_y, v_z)\, dv_x\, dv_y\, dv_z. \tag{5.4-16}$$

where $g(v_x,v_y,v_z)dv_x\,dv_y\,dv_z$ is the number of quantum states within the velocity space element $dv_x dv_y dv_z$. From (5.2-17) the number of states in a volume element $dp_x\,dp_y\,dp_z$ of momentum space is $(2V/h^3)(dp_x dp_y dp_z)$, whereby

$$\frac{2V}{h^3}\,dp_x\,dp_y\,dp_z = \frac{2m^3V}{h^3}\,dv_x\,dv_y\,dv_z = g(v_x,v_y,v_z)\,dv_x\,dv_y\,dv_z. \qquad (5.4\text{-}17)$$

Since the density of states in momentum space is constant, so also is the density of states in velocity space. Substituting (5.4-15) and (5.4-17) into (5.4-16) and simplifying, we find

$$N(v_x,v_y,v_z)\,dv_x\,dv_y\,dv_z = N\left(\frac{m}{2\pi kT}\right)^{3/2}e^{-m(v^2_x+v^2_y+v^2_z)/2kT}\,dv_x\,dv_y\,dv_z.$$

$$(5.4\text{-}18)$$

A third velocity distribution function, $N(v_x)dv_x$, representing the number of particles whose x-component of velocity lies in a range dv_x about v_x, regardless of what values the y- and z-components of velocity for those particles may have, is closely related to the distribution function (5.4-18). The distribution function $N(v_x)dv_x$ can be calculated from (5.4-18) by integrating over all possible values of v_y and v_z, the result being

$$N(v_x)dv_x = \int_{-\infty}^{\infty}\int_{-\infty}^{\infty}[N(v_x,v_y,v_z)dv_x]\,dv_y\,dv_z = N\sqrt{\frac{m}{2\pi kT}}\,e^{-mv^2_x/2kT}dv_x.$$

$$(5.4\text{-}19)$$

The values of the definite integrals required in integrating (5.4-18) have been taken from Table 5.1. A plot of the distributions $N(v)$ and $N(v_x)$ as functions of the appropriate velocity coordinate is shown in Figure 5.6.

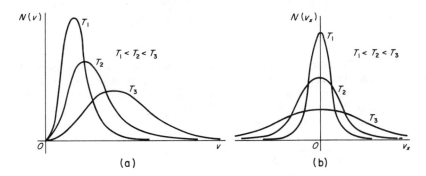

FIGURE 5.6. (a) A schematic representation of the Maxwell-Boltzmann distribution of particle speeds, $N(v)$, for three different temperatures. (b) The corresponding distribution of x-components of velocity, $N(v_x)$.

Knowing these three velocity distribution functions it is an easy matter to evaluate averages over the velocity distributions. For example, the average thermal speed $\bar{c}$

for a particle in a Boltzmann distribution is obtained quite simply from (5.4-14) by writing

$$\langle v \rangle = \frac{\int_0^\infty vN(v)dv}{\int_0^\infty N(v)dv} = \frac{1}{N} \cdot 4\pi N \left(\frac{m}{2\pi kT}\right)^{3/2} \int_0^\infty v^3 e^{-mv^2/2kT} \, dv$$

$$= \sqrt{8kT/\pi m}, \qquad (5.4\text{-}20)$$

the integral having been evaluated with the help of Table 5.1.

We are now in a position to discuss the equation of state of an ideal gas. Consider the particles striking unit area of the wall of a container filled with such a gas, as shown in Figure 5.7. If the wall is a plane oriented normal to the x-axis, the momentum components p_y and p_z of the particles striking the wall are conserved if the collisions of the

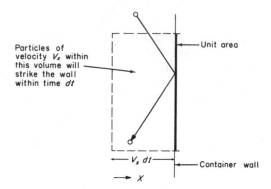

Particles of velocity V_x within this volume will strike the wall within time dt

—Unit area

— $V_x \, dt$ — ——Container wall

— x

FIGURE 5.7.　The elastic collision of a particle whose x-component of velocity is v_x with the wall of a rigid, fixed container.

particles with the walls are elastic, which we shall assume to be the case. In each collision the velocity component v_x directed toward the wall is changed to $-v_x$, directed away from the wall. The transfer of momentum from the particle to the wall, per collision, is thus $2mv_x$. The transfer of momentum to the wall in time dt is this quantity times the number of particles which collide with the wall during that time; for particles whose x-component of velocity is v_x, this is simply the number of such particles in a volume extending a distance $d = v_x dt$ behind the wall, or the number per unit volume times $v_x dt$, thus $v_x dt \cdot N(v_x)dv_x/V$. The momentum transfer in time dt for particles of velocity in the range dv_x about v_x is then

$$(dp_x)_{v_x} = 2mv_x \frac{v_x \, dt \cdot N(v_x)dv_x}{V}$$

or,

$$\left(\frac{dp_x}{dt}\right)_{v_x} = \frac{2m}{V} v_x^2 N(v_x)dv_x. \qquad (5.4\text{-}21)$$

The total rate of momentum transfer to the walls by collisions involving particles with all possible values of v_x can be obtained by integrating over v_x, giving, with the help of (5.4-19) and Table 5.1,

$$\frac{dp_x}{dt} = \frac{2m}{V} N \sqrt{\frac{m}{2\pi kT}} \int_0^\infty v_x^2 e^{-mv^2_x/2kT} \, dv_x = \frac{NkT}{V}. \tag{5.4-22}$$

The time rate of transfer of momentum to unit area of the container wall, however, is according to Newton's law equal to the force experienced by a unit area of the wall, which by definition is the pressure P. Equation (5.4-22) then reduces to

$$PV = NkT, \tag{5.4-23}$$

which is the equation of state for a Boltzmann gas of independent particles.

This equation has the form of the familiar ideal gas law, which is more commonly written in the form

$$PV = nRT, \tag{5.4-24}$$

where n is the number of moles of gas in the system and R is an experimentally measured "molar gas constant" which is the same for all "ideal" gases. If there are n moles of gas present, then the number of particles must be given by $N = nN_A$, where N_A is Avogadro's number, equal to 6.026×10^{23} molecules per mole. Under these conditions (5.4-23) takes the form

$$PV = nN_A kT, \tag{5.4-25}$$

whence, comparing (5.4-24) with (5.4-25), it is clear that Boltzmann's constant k must be given by

$$k = R/N_A. \tag{5.4-26}$$

Boltzmann's constant is thus simply the gas constant *per particle* of the system. Its value can be derived from N_A and R, and is equal to 1.380×10^{-16} ergs/°K, or 8.615×10^{-5} eV/°K.

It is clear now that the value for the constant β in the distribution function which was assumed in (5.3-27) was chosen correctly. Had we carried through all our calculations up to this point without ever having assumed any value for β, we should have found for equation (5.4-23), instead,

$$PV = -N/\beta, \tag{5.4-27}$$

and in order that our results be in agreement with the experimentally established gas law (5.4-24) we should have been *forced* to choose

$$-N/\beta = nRT = \frac{N}{N_A} RT,$$

or,
$$\beta = -\frac{N_A}{RT} = -1/kT. \tag{5.4-28}$$

5·5 FERMI-DIRAC STATISTICS

In the development of the Maxwell-Boltzmann distribution function, the particles are considered to be distinguishable, while in actual fact it is quite impossible to distinguish one electron or other elementary particle from another. Furthermore, we permitted any number of particles to occupy the same quantum state of the system, in spite of the fact that many particles, *electrons* in particular, obey the Pauli exclusion principle, which allows each quantum state to accept no more than one particle. If these additional conditions are imposed upon the system, the calculations of Section 5.4 must be modified, resulting in another distribution function, which is called the Fermi-Dirac distribution function. This energy distribution is of the greatest importance, since it describes the statistical behavior of free electrons in metals and semiconductors, and since many of the electrical and thermal properties of solids which could not be understood at all on the basis of classical statistics follow as a direct consequence of the Fermi-Dirac statistics.

If the particles of the system are indistinguishable, they cannot be identified by number in the manner which was adopted when discussing the various possibilities illustrated in Figure 5.2. As a matter of fact, all the various distributions in each row of Figure 5.2 leading to $Q(N_1, N_2)$ for a particular value of N_1 are *the same* if the numbered labels are removed from the particles, in which case the factor (5.3-7) reduces to unity. It is still possible, however, to permute the N_i particles in the ith energy level among the g_i quantum states belonging to that level in many ways, each of which constitutes a statistically independent way of achieving an arrangement wherein N_i particles are in the ith energy level of the system, as shown in Figure 5.8. The product

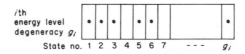

FIGURE 5.8. A possible distribution of particles among quantum states in the ith energy level of a system wherein the Pauli exclusion principle is applicable.

of the possible numbers of permutations of particles among quantum states over all energy levels of the system then gives the number of independent ways of realizing a given distribution of particles among energy levels.

For the Fermi-Dirac case, in which the Pauli exclusion principle is assumed to apply, a maximum of one particle per quantum state is allowed. Referring to Figure 5.8, which depicts the ith energy level of the system, there are g_i ways of choosing where to insert a first particle, $g_i - 1$ ways of choosing where to insert a second, since the second cannot occupy the same quantum state as the first, $g_i - 2$ ways of choosing where to insert a third, and so forth. The total number of ways of arranging N_i particles in the ith level obtained on this basis is

$$g_i(g_i - 1)(g_i - 2) \cdots (g_i - N_i + 1) = \frac{g_i!}{(g_i - N_i)!}. \qquad (5.5\text{-}1)$$

However, since the particles are indistinguishable, the $N_i!$ ways of permuting the particles among themselves in any given arrangement of particles among states do not count as separate arrangements. The actual number of independent ways of realizing a distribution of N_i particles in the ith level is thus arrived at by dividing (5.5-1) by $N_i!$, giving

$$\frac{g_i!}{N_i!(g_i - N_i)!}. \tag{5.5-2}$$

The total number of independent ways of realizing a distribution of $(N_1, N_2, \cdots N_n)$ indistinguishable particles among n energy levels, no more than one per quantum state, is just the product of individual factors of the form (5.5-2) over all the levels, that is,

$$Q_f(N_1, N_2, \cdots N_n) = \prod_{i=1}^{n} \frac{g_i!}{N_i!(g_i - N_i)!}. \tag{5.5-3}$$

We now proceed to maximize the logarithm of this quantity with respect to the variables $N_1, N_2, \cdots N_n$ by the method of Lagrangean multipliers used in Section 5.3. We find from (5.5-3) that

$$\ln Q_f = \sum_i \ln g_i! - \sum_i \ln N_i! - \sum_i \ln(g_i - N_i)!, \tag{5.5-4}$$

which by using Stirling's approximation (5.3-20) can be written

$$\ln Q_f = \sum_i \left[g_i \ln g_i - N_i \ln N_i - (g_i - N_i) \ln(g_i - N_i) \right]. \tag{5.5-5}$$

Again, it is required that the total number of particles in the system and the total energy of the system be constant, which means that Equations (5.3-22) and (5.3-23) must hold here also, in which case one may write, as before,

$$\frac{\partial(\ln Q_f)}{\partial N_j} + \alpha \frac{\partial \phi}{\partial N_j} + \beta \frac{\partial \psi}{\partial N_j} = 0 \qquad (j = 1, 2, \ldots n), \tag{5.5-6}$$

where ϕ and ψ are given by (5.3-22) and (5.3-23). Substituting (5.5-5), (5.3-22), and (5.3-23) into this equation, one obtains

$$-\frac{\partial}{\partial N_j} \left[\sum_i N_i \ln N_i + \sum_i (g_i - N_i) \ln(g_i - N_i) \right] + \alpha \frac{\partial}{\partial N_j} \left(\sum_i N_i \right) + \beta \frac{\partial}{\partial N_j} \left(\sum_i \varepsilon_i N_i \right) = 0. \tag{5.5-7}$$

Working out the derivatives as in Section 5.3, we find

$$\ln(g_j - N_j) - \ln N_j = -\alpha - \beta \varepsilon_j, \tag{5.5-8}$$

or, rearranging, exponentializing and solving for $N_j/g_j = f(\varepsilon_j)$,

$$f(\varepsilon_j) = N_j/g_j = \frac{1}{1 + e^{-\alpha - \beta \varepsilon_j}}. \tag{5.5-9}$$

This is the Fermi-Dirac distribution function.

As in Section 5.4, we shall take the value of β to be

$$\beta = -1/kT, \tag{5.5-10}$$

postponing until later the justification for this step. It is customary to write α in the form

$$\alpha = \varepsilon_f/kT, \tag{5.5-11}$$

where ε_f is a parameter with the dimensions of energy, which is called the *Fermi energy*, or the Fermi level, of the system. Equation (5.5-9) then becomes

$$N_j = \frac{g_j}{1 + e^{(\varepsilon_j - \varepsilon_f)/kT}}, \tag{5.5-12}$$

or, if the levels are assumed to crowd together into a continuum, so that $g_j(\varepsilon_j) \rightarrow g(\varepsilon)d\varepsilon$, then

$$N(\varepsilon)d\varepsilon = g(\varepsilon)f(\varepsilon)\,d\varepsilon = \frac{g(\varepsilon)d\varepsilon}{1 + e^{(\varepsilon - \varepsilon_f)/kT}}. \tag{5.5-13}$$

For a gas of independent particles such as free electrons, $g(\varepsilon)d\varepsilon$ is represented by (5.2-22), just as for a Maxwell-Boltzmann gas.

The Fermi energy ε_f is in general a function of the temperature, whose form and temperature dependence are critically dependent upon the density of states function for the system, just as for the corresponding parameter e^{α} of the Maxwell-Boltzmann distribution. Its value is determined by the condition (5.3-22), or for a continuum of levels, by

$$N = \text{const.} = \int \frac{g(\varepsilon)d\varepsilon}{1 + e^{(\varepsilon - \varepsilon_f)/kT}} = \int g(\varepsilon)f(\varepsilon)d\varepsilon. \tag{5.5-14}$$

The integral is taken over all energies available to the particles of the system. For a Fermi gas of independent particles, $g(\varepsilon)$ is given by (5.2-22) and ε_f is determined by

$$N = \frac{8\sqrt{2}\pi V m^{3/2}}{h^3} \int_0^\infty \frac{\sqrt{\varepsilon}\,d\varepsilon}{1 + e^{(\varepsilon - \varepsilon_f)/kT}}. \tag{5.5-15}$$

Unfortunately, this integral cannot be evaluated in closed analytic form, so that ε_f cannot be determined as a simple function of the temperature. For a *two-dimensional* independent particle gas, however, the density of states function can be shown by the methods used to derive (5.2-22) to be

$$g(\varepsilon)d\varepsilon = \frac{4\pi mA}{h^2}\,d\varepsilon, \tag{5.5-16}$$

independent of energy (Exercise 3, Chapter 5). In this formula, A represents the area of the two-dimensional "container." For this system (5.5-14) becomes

$$N = \frac{4\pi mA}{h^2} \int_0^\infty \frac{d\varepsilon}{1 + e^{(\varepsilon - \varepsilon_f)/kT}},$$ (5.5-17)

which can be evaluated in closed form, allowing one to solve for ε_f and obtain

$$\varepsilon_f(T) = kT \ln(e^{\varepsilon_f(0)/kT} - 1),$$ (5.5-18)

where $$\varepsilon_f(0) = \frac{Nh^2}{4\pi mA}$$ (5.5-19)

is the value which $\varepsilon_f(T)$ as given by (5.5-18) assumes as T approaches zero. The details of calculating this result are very instructive, and are assigned as an exercise for the reader. The variation of the Fermi energy of the two-dimensional independent particle gas with temperature is shown in Figure 5.9. It will be noted for this case that the Fermi energy is a monotonic decreasing function of temperature. The Fermi energy for the three-dimensional Fermi gas, with the density of states function (5.2-22) will be found to exhibit the same general behavior, except that in this case the variation of the Fermi energy with temperature is *linear* with temperature at reasonably low temperatures, while for the two-dimensional example the variation is much more complex at low temperatures. For many systems, including these two, the variation of the Fermi level with temperature is quite small over the range of physically realizable temperatures; in Figure 5.9 the temperature for which $\varepsilon_f = 0$ would be of the order of

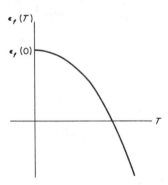

FIGURE 5.9. Schematic representation of the variation of Fermi energy with temperature in a two-dimensional free electron gas, according to (5.5-18).

75 000°K for an electron gas with the free-electron density of metallic copper. For this reason, in many applications the temperature dependence of the Fermi energy may either be neglected or approximated by a linear or other appropriate function of temperature.

The Fermi distribution function itself,

$$f(\varepsilon) = \frac{1}{1 + e^{(\varepsilon - \varepsilon_f)/kT}},$$ (5.5-20)

is plotted in Figure 5.10 for several values of the temperature. Since only one particle may occupy a given quantum state, the value of $f(\varepsilon)$ for a Fermi distribution at a particular energy is just equal to the probability that a quantum state of that energy will be occupied. At absolute zero, it is easily seen from Figure 5.10 and from (5.5-20) that the Fermi distribution function becomes simply the step function

$$f(\varepsilon) = 1 \qquad (\varepsilon < \varepsilon_f)$$

$$= 0 \qquad (\varepsilon > \varepsilon_f). \qquad (5.5\text{-}21)$$

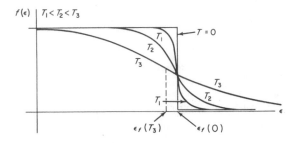

FIGURE 5.10. Schematic representation of the Fermi distribution function for four different temperatures. Note the variation of the Fermi energy with temperature. The temperature dependence of the Fermi energy depicted here is typical of a three-dimensional free-electron gas, but the actual variation in any particular system will depend critically upon the density of states function (or level degeneracies) for that system.

As the temperature increases, the edges of the step are rounded off, and the distribution function varies rapidly from nearly unity to nearly zero over an energy range of a few times kT around the value $\varepsilon = \varepsilon_f$. At the same time, the value of ε_f itself changes, the variation illustrated in Figure 5.10 being approximately that associated with the three-dimensional electron gas whose density of states is given by (5.2-22). At very high temperatures, the distribution function loses its step-like character and varies much more slowly with energy. From (5.5-20), it is clear that the value of $f(\varepsilon)$ at $\varepsilon = \varepsilon_f$ is just $\frac{1}{2}$, that is,

$$f(\varepsilon_f) = \tfrac{1}{2}, \qquad (5.5\text{-}22)$$

hence a quantum state at the Fermi level has a probability of occupation of $\frac{1}{2}$.

Figure 5.11 shows the actual distribution of particle density $N(\varepsilon)$ as a function of energy for a Fermi gas of independent particles, as given by (5.5-13) with the density of states function (5.2-22). Again, at $T = 0$ the curve has a step-like character, the portion for which $(\varepsilon < \varepsilon_f)$ being the density of states parabola (5.2-22) and that for which $(\varepsilon > \varepsilon_f)$ being zero. As the temperature increases this step-like aspect becomes less and less pronounced, as shown in the drawing. At low temperatures, when the Fermi distribution function is step-like, the distribution is said to be highly *degenerate*.

At low temperatures, the Fermi-Dirac distribution may be represented as a sphere in momentum space in which all or most of the quantum states of energy less than ε_f

are filled, while all or most of the states of energy greater than ε_f are empty. From (5.2-18), the equation of the surface of this "Fermi sphere" must be

$$p_x^2 + p_y^2 + p_z^2 = 2m\varepsilon_f, \qquad (5.5-23)$$

the radius therefore being $\sqrt{2m\varepsilon_f}$, as shown in Figure 5.12. At very high temperatures the surface of the Fermi sphere becomes poorly defined, due to the disappearance of the step-like aspect of $f(\varepsilon)$, and the concept loses some of its usefulness.

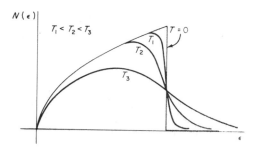

FIGURE 5.11. Schematic representation of electron density as a function of energy for a three-dimensional free-electron gas.

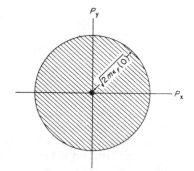

FIGURE 5.12. The representation of the absolute zero Fermi distribution for a free-electron gas as a sphere of electrons in momentum space. This representation is a useful one so long as $T \ll T_F$, where $T_F = \varepsilon_f(0)/k$.

(z - axis ⊥ to paper)

Although, as we have already seen, the fact that the integral (5.5-15) cannot be evaluated in closed form prevents us from finding a simple expression for the Fermi energy of a three-dimensional Fermi gas at all temperatures, the simple character (5.5-21) which the function assumes at $T = 0$ permits one to evaluate ε_f very simply at absolute zero. Using (5.5-21) to represent $f(\varepsilon)$ at $T = 0$, we may rewrite (5.5-14), using the density of states function (5.2-22), as

$$\frac{N}{V} = \frac{8\sqrt{2}\pi m^{3/2}}{h^3} \int_0^{\varepsilon_f(0)} \sqrt{\varepsilon}\, d\varepsilon = \frac{16\sqrt{2}\pi m^{3/2}[\varepsilon_f(0)]^{3/2}}{3h^3}. \qquad (5.5-24)$$

Solving this equation for $\varepsilon_f(0)$, we may obtain

$$\varepsilon_f(0) = \frac{h^2}{8m}\left(\frac{3N}{\pi V}\right)^{2/3} \qquad (5.5-25)$$

In a somewhat similar manner, it is possible to obtain the internal energy of a Fermi gas at absolute zero, the result being

$$\frac{U_0}{V} = \frac{\pi h^2}{40m} \left(\frac{3N}{\pi V}\right)^{5/3} = \frac{3}{5} \frac{N}{V} \varepsilon_f(0). \tag{5.5-26}$$

The details of this calculation are assigned as an exercise.

For energies which are much greater than ε_f, $e^{(\varepsilon - \varepsilon_f)/kT}$ is much larger than unity and for such energies the Fermi-Dirac distribution function (5.5-20) may be written approximately as

$$f(\varepsilon) \cong e^{\varepsilon_f/kT} e^{-\varepsilon/kT}. \tag{5.5-27}$$

If all the energies available to the system satisfy the condition

$$\varepsilon - \varepsilon_f \gg kT, \tag{5.5-28}$$

that is, if ε_f is many kT units smaller than any energy a particle belonging to the system may have, then (5.5-27) will be a good approximation to (5.5-20) for all particles of the system. The approximate distribution function (5.5-27), however, is simply the Maxwell-Boltzmann distribution function of (5.4-6) with $\alpha = \varepsilon_f/kT$. If the condition (5.5-28) holds, then, for all particles of the system, the Fermi-Dirac distribution and the Maxwell-Boltzmann distribution are very nearly the same.

For the two-dimensional Fermi gas, where the Fermi energy is given by (5.5-18), if T is so large that $kT \gg \varepsilon_f(0)$, the exponent $\varepsilon_f(0)/kT$ will be small, so that the exponential can be approximated by $1 + [\varepsilon_f(0)/kT]$, giving

$$\varepsilon_f(T) = kT \ln \frac{\varepsilon_f(0)}{kT}$$

$$= \varepsilon_f(0) \left[\frac{kT}{\varepsilon_f(0)} \ln \frac{\varepsilon_f(0)}{kT}\right]$$

$$= -\varepsilon_f(0) \left[\frac{kT}{\varepsilon_f(0)} \ln \frac{kT}{\varepsilon_f(0)}\right]. \tag{5.5-29}$$

From this, we see that as T becomes large, $\varepsilon_f(T) \to -\infty$. Since the lowest energy any particle of the system may have is zero, it is clear that condition (5.5-28) will be fulfilled for sufficiently high temperatures, and the distribution function will therefore be approximately the same as the Maxwell-Boltzmann distribution at very high temperatures. The same result can be shown to hold for the three-dimensional case. These results are to be expected on physical grounds, since at high temperatures the particles are distributed over a very wide range of energy states, the number of particles in every range of available energies being so small that there are always many more available quantum states than there are particles to occupy them. Under these circumstances, the probability of *two or more* particles occupying the same quantum state becomes vanishingly small in any case, so that there is not much difference between the distribution function for which the Pauli principle is obeyed (the Fermi-Dirac distribution) and that for which it is not (the Maxwell-Boltzmann distribution).[3] Since the quantity $\varepsilon_f(0)$ which is required to be much less than kT for the two distributions to coincide is proportional to the density of particles N/A by (5.5-19), the reduction of the Fermi distribution to a Maxwell-Boltzmann distribution will take place at lower temperatures in less dense gases. By the same token, according to (5.5-19) it will take place at lower temperatures in gases where the particle mass m is large. It is for these reasons that ordinary gaseous substances obey the Maxwell-Boltzmann statistics at normal temperatures rather than the Fermi-Dirac (or Bose-Einstein) statistics. For a dense gas of very light particles, such as the free electrons in a metal, however, the Fermi energy at absolute zero is quite large, and the condition (5.5-28) can be satisfied for all particles of the system only at temperatures so high as to be unrealizable physically. A dense free electron gas must therefore be treated using Fermi-Dirac statistics. In semiconductors, however, the peculiar form of the density of states function is such that the Maxwell-Boltzmann distribution is virtually *always* a good approximation to the Fermi-Dirac distribution. We shall examine this situation in considerable detail in a later chapter.

Had we made a choice for the undetermined multiplier β other than that given by (5.5-10) we should *not* in general have found the correspondence between the Fermi-Dirac and Maxwell-Boltzmann systems in the high-temperature limit, which, as we have seen, we have every right to expect on physical grounds. We must therefore conclude that the value for β as given by (5.5-10) is physically justified.

5·6   THE BOSE-EINSTEIN DISTRIBUTION

In the previous section we showed that the statistical distribution which characterizes the behavior of an ensemble of indistinguishable particles which obey the Pauli exclusion principle is the Fermi-Dirac distribution (5.5-20). Since not all elementary particles obey the Pauli principle (photons being the most conspicuous exception) it is necessary to consider the statistical behavior of particles which, though indistinguishable, do not obey the Pauli exclusion principle. In this case again, since the particles are not numbered, the factor (5.3-7) reduces to unity and we need only consider the possible permutations of N_i identical particles among the g_i quantum states of the ith energy level, but now with no restrictions with regard to the number of particles which may occupy any given quantum state.

Consider a linear array of N_i particles and $g_i - 1$ *partitions* which would be necessary to divide these particles into g_i groups, as shown in Figure 5.13. It is not difficult to see that the number of ways of permuting the N_i particles among g_i levels is equal to the number of independent permutations of objects and partitions in Figure 5.13.

---

[3] It might be contended that the indistinguishability of particles renders the Fermi-Dirac distribution distinct from the Maxwell-Boltzmann distribution even in this limit. It can, however, be shown, as we shall see in the next section, that the distribution function for *indistinguishable* particles which do not obey the Pauli exclusion principle (the Bose-Einstein distribution) also approaches the Maxwell-Boltzmann distribution function in this limit. A more accurate way of explaining the situation would be to say that in the high-temperature limit the Fermi-Dirac distribution approaches the Bose-Einstein distribution, which in turn approaches the Maxwell-Boltzmann distribution.

Since there are a total of $N_i + g_i - 1$ particles plus partitions, these can be arranged linearly in $(N_i + g_i - 1)!$ ways, but since permutations of particles among themselves or of partitions among themselves do not count as independent arrangements, we must divide by the number of ways of permuting particles among themselves $(N_i!)$

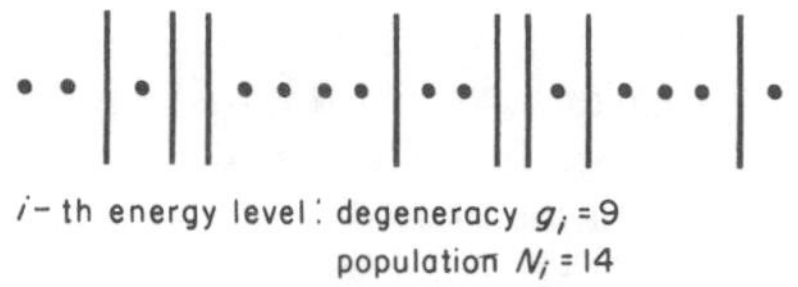

FIGURE 5.13. A possible distribution of particles among quantum states in the ith energy level of a system to which the Pauli exclusion principle does not apply.

$i-$ th energy level: degeneracy $g_i = 9$
population $N_i = 14$

and again by the number of ways of permuting partitions among themselves $((g_i - 1)!)$, giving

$$\frac{(N_i + g_i - 1)!}{N_i!(g_i - 1)!} \tag{5.6-1}$$

ways of realizing a distribution of $N_i!$ indistinguishable particles among g_i states which may accommodate any number of particles. The number Q_b of statistically independent ways of achieving a distribution $(N_1, N_2, \cdots N_n)$ particles among the energy levels of the system according to these rules is then just the product of factors of the form (5.6-1) over all levels of the system, whence

$$Q_b(N_1, N_2, \cdots N_n) = \prod_{i=1}^{n} \frac{(N_i + g_i - 1)!}{N_i!(g_i - 1)!}. \tag{5.6-2}$$

It is now possible to maximize this quantity with respect to the variables N_1, N_2, $\cdots N_n$, using the method of Lagrangean multipliers, under the restrictions (5.3-22) and (5.3-23). The actual calculations will not be set forth here, but will be left as an exercise for the reader. The result is

$$f(\varepsilon_j) = N_j/g_j = \frac{1}{e^{-\alpha}e^{-\beta \varepsilon_j} - 1}. \tag{5.6-3}$$

This formula is referred to as the Bose-Einstein distribution function. It is again possible to identify β as

$$\beta = -1/kT, \tag{5.6-4}$$

while α may be determined in terms of the number of particles in the system, just as in the Maxwell-Boltzmann and Fermi-Dirac cases. For the case of a continuum of closely spaced levels, $g_j \to g(\varepsilon)d\varepsilon$ and $N_j \to N(\varepsilon)d\varepsilon$, giving

$$f(\varepsilon) = \frac{1}{e^{-\alpha}e^{+\varepsilon/kT} - 1}. \tag{5.6-5}$$

For the independent particle density of states function (5.5-16) for a two-dimensional free particle gas, the parameter α may be explicitly calculated in the same way as ε_f for this system under Fermi-Dirac statistics. It may then be shown that as T becomes large, $\alpha \to -\infty$, in which case the exponential factor in the denominator of (5.6-5) becomes much larger than unity, and the Bose-Einstein distribution (5.6-5) approaches

$$f(\varepsilon) \cong e^{\alpha} e^{-\varepsilon/kT}, \qquad (5.6\text{-}6)$$

which is a distribution function of the Maxwell-Boltzmann type. The same general behavior is obtained for a three-dimensional independent-particle Bose-Einstein gas, although in this case it is not possible to obtain an expression for α in closed form.

In the limit of very high temperatures, the particles of the system will be distributed over a very wide range of energies, and the number of particles in every available energy range will become much smaller than the number of quantum states in that range, whence for all states $g_i \gg N_i$. In this situation, we may write, approximately,

$$\frac{(N_i + g_i - 1)!}{(g_i - 1)!} \cong g_i^{N_i}, \qquad (5.6\text{-}7)$$

whereby (5.6-2) becomes

$$Q_b(N_1, N_2, \cdots N_n) \cong \prod_{i=1}^{n} \frac{g_i^{N_i}}{N_i!}. \qquad (5.6\text{-}8)$$

But this, apart from a constant factor $N!$, is just equal to $Q(N_1, N_2, \cdots N_n)$ as given by (5.3-8) for a Maxwell-Boltzmann system! We should then expect the Bose-Einstein and Maxwell-Boltzmann distributions to coincide in the high-temperature limit on purely physical grounds. The choice of the value given by (5.6-4) for β is thus justified, for this choice, as we have seen above, leads directly to the correspondence between the Bose-Einstein and Maxwell-Boltzmann Statistics shown by (5.6-6).

In the case of the two-dimensional and three-dimensional independent particle Bose-Einstein gases, as the temperature approaches zero, the value of α tends toward zero, the result being that all the particles of the system tend to condense into the lowest energy state of the system at absolute zero. This phenomenon, called the *Bose condensation*, is characteristic of systems obeying Bose-Einstein statistics.

In some applications it is of interest to obtain the Bose-Einstein distribution function without making the restriction that the number of particles in the system be constant. It can be seen from (5.3-15) and (5.3-16) that this result can be obtained from (5.6-5) if α is taken to be identically zero. In this case (5.6-5) reduces to

$$f(\varepsilon) = \frac{1}{e^{\varepsilon/kT} - 1}. \qquad (5.6\text{-}9)$$

EXERCISES

1. Show that the wave front associated with the plane wave $\Psi = e^{i(\mathbf{k}\cdot\mathbf{r} - \omega t)}$ advances along the $\mathbf{k}$ direction.

2. Show for the free particle in three dimensions, whose wave function is given by

(5.2-7) that the expectation value of the vector momentum $\mathbf{p}$ is equal to $\hbar\mathbf{k}$, and that $\mathbf{p}$ (hence $\mathbf{k}$) is a constant of the motion.

3. Calculate the density of states factor $g(\varepsilon)d\varepsilon$ for a two-dimensional system of free particles, with instantaneous collision interactions only, contained within a rigid container of area A and dimensions x_0 and y_0. Start with Schrödinger's equation.

4. Suppose that 4 coins are tossed simultaneously; what are the probabilities associated with the distributions (0 heads, 4 tails), (1 head, 3 tails) $\cdots$ (4 heads, 0 tails)?

5. Find the dimensions and area of the rectangle of maximum area with sides parallel to the coordinate axes which can be inscribed within an ellipse whose major axis is $2a$ and whose minor axis is $2b$. The axes of the ellipse may be taken to be parallel with the coordinate axes. Use the method of Lagrangean multipliers.

6. Find the root-mean-square speed and the most probable speed of a particle in an ideal Boltzmann gas.

7. Show that the *flux* of particles in an ideal Boltzmann gas, whose x-components of velocity are positive, per unit area across a plane normal to the x-axis is $\frac{1}{4}N\bar{c}/V$. *Hint:* The flux or current density is defined as the number of particles per unit volume times their velocity component along the normal to the plane across which the flux is observed.

8. Show that the Fermi energy of a two-dimensional Fermi gas of free particles, whose density of states function is given by the result of Exercise 3, is $\varepsilon_f(T) = kT \ln(e^{\varepsilon_f(0)/kT} - 1)$, where $\varepsilon_f(0) = Nh^2/(4\pi mA)$.

9. Calculate the Fermi energy, in electron volts, for the free electrons in copper at absolute zero, assuming one free electron per copper atom. For what temperature would kT be equal to $\varepsilon_f(0)$?

10. Show that the internal energy per unit volume of a three-dimensional Fermi gas of free particles at absolute zero is $\frac{3}{5}\frac{N}{V}\varepsilon_f(0)$; show that the corresponding result for a two-dimensional Fermi gas is $\frac{1}{2}\frac{N}{V}\varepsilon_f(0)$.

11. Show by the method of Lagrangean multipliers that if $Q(N_1, N_2, \cdots N_n)$ is given by Equation (5.6-2), the distribution function corresponding to the most probable values of $N_1, N_2, \cdots N_n$ is given by (5.6-3).

12. Using the density of states function calculated in Exercise 3, show that the parameter α of a two-dimensional Bose-Einstein gas of free particles is given by
$$\alpha = \ln(1 - e^{-h^2N/(2\pi AmkT)}).$$
Discuss the properties of the resulting distribution function for $T \to \infty$ and for $T \to 0$.

GENERAL REFERENCES

W. Band, *An Introduction to Quantum Statistics*, D. Van Nostrand, Princeton, N.J. (1955).

R. W. Gurney, *Introduction to Statistical Mechanics*, McGraw-Hill Book Co., Inc., New York (1949).

D. ter Haar, *Elements of Statistical Mechanics*, Holt, Rinehart and Winston, Inc., New York (1954).

R. B. Lindsay, *Introduction to Physical Statistics*, John Wiley and Sons, New York (1941).

J. E. Mayer and Maria Göppert-Mayer, *Statistical Mechanics*, John Wiley and Sons, New York (1940).

R. C. Tolman, *The Principles of Statistical Mechanics*, Oxford University Press, London (1938).

CHAPTER 6

LATTICE VIBRATIONS AND THE THERMAL PROPERTIES OF CRYSTALS

6·1   CLASSICAL CALCULATION OF LATTICE SPECIFIC HEAT

In Section 4.4, we have already outlined some of the problems which arise when the specific heat of a crystal is calculated on the assumption that the atoms of the crystal behave as independent classical harmonic oscillators. We must now examine these problems in detail, and study how they may be resolved, using the results obtained in Chapters 3–5 relating to lattice dynamics, quantum mechanics, and statistical mechanics. We shall throughout these investigations deal with a crystal composed of N atoms which are held together in a periodic array. These atoms are assumed to be free to vibrate about their equilibrium positions subject to constraining forces which, to a first approximation, obey Hooke's law. It is also assumed that there are no free electrons, such as one might find in a metal, and that the entire heat capacity of the crystal is due to the excitation of thermal vibrations of the lattice. When free electrons are present, their motion may also be excited by an external heat source, and so they also will contribute to the observed specific heat, but this electronic contribution will be discussed in a later chapter.

The classical calculation assumes that each atom is a classical three-dimensional harmonic oscillator which vibrates independently of all other atoms in the crystal. Under these circumstances, one may calculate the total internal thermal energy of the crystal by finding the average energy of a single oscillator and multiplying the result by N. For a single three-dimensional isotropic harmonic oscillator, the total energy is

$$\varepsilon = \frac{p^2}{2m} + V(\mathbf{r}) \tag{6.1-1}$$

with

$$V(\mathbf{r}) = \tfrac{1}{2}m\omega_0^2(x^2 + y^2 + z^2) = \tfrac{1}{2}m\omega_0^2 r^2, \tag{6.1-2}$$

where m is the mass and ω_0 the natural frequency of the oscillator. If we assume that the distribution of the oscillators in energy obeys the Maxwell-Boltzmann distribution law, then the distribution function which expresses the probability for an oscillator to have a certain energy ε is

$$f(\varepsilon) = Ae^{-\varepsilon/kT} = Ae^{-\frac{p^2}{2mkT}}e^{-\frac{m\omega_0^2 r^2}{2kT}} = f(\mathbf{p},\mathbf{q}) \qquad (6.1\text{-}3)$$

where A is a constant. According to (5.2-4), the average energy will be given by

$$\langle\varepsilon\rangle = \frac{\displaystyle\int_p\int_r \left(\frac{p^2}{2m} + \frac{1}{2}m\omega_0^2 r^2\right)e^{-\frac{p^2}{2mkT}}e^{-\frac{m\omega_0^2 r^2}{2kT}} \cdot p^2 \sin\theta_p\, dp\, d\theta_p\, d\phi_p \cdot r^2 \sin\theta\, dr\, d\theta\, d\phi}{\displaystyle\int_p\int_r e^{-\frac{p^2}{2mkT}}e^{-\frac{m\omega_0^2 r^2}{2kT}} \cdot p^2 \sin\theta_p\, dp\, d\theta_p\, d\phi_p \cdot r^2 \sin\theta\, dr\, d\theta\, d\phi}.$$

$$(6.1\text{-}4)$$

Here we have chosen to use spherical coordinates (r,θ,ϕ) in coordinate space, and (p,θ_p,ϕ_p) in momentum space for the purposes of integration, which, of course, must be taken over all phase space. Since the integrand has no dependence upon θ, ϕ, θ_p, or ϕ_p the integration over these variables may be performed very simply, giving a factor $(4\pi)^2$ in both numerator and denominator, and leaving

$$\langle\varepsilon\rangle = \frac{\dfrac{1}{2m}\displaystyle\int_0^\infty p^4 e^{-\frac{p^2}{2mkT}}\, dp}{\displaystyle\int_0^\infty p^2 e^{-\frac{p^2}{2mkT}}\, dp} + \frac{\dfrac{m\omega_0^2}{2}\displaystyle\int_0^\infty r^4 e^{-\frac{m\omega_0^2 r^2}{2kT}}\, dr}{\displaystyle\int_0^\infty r^2 e^{-\frac{m\omega_0^2 r^2}{2kT}}\, dr} = \langle\varepsilon_k\rangle + \langle\varepsilon_p\rangle. \quad (6.1\text{-}5)$$

In the first term of (6.1-5), each double integral in (6.1-4) has been expressed as a product of a p-integral and an r-integral; the r-integrals in the numerator and denominator are the same, and cancel, giving the first term of the expression shown above. The second term is treated in the same way, but here the p-integrals in numerator and denominator cancel. It is clear that the first term above represents the average kinetic energy and the second the average potential energy. Evaluating the integrals with the help of Table 5.1, one finds

$$\langle\varepsilon_k\rangle = \langle\varepsilon_p\rangle = \tfrac{3}{2}\, kT \qquad \text{whence} \quad \langle\varepsilon\rangle = 3kT. \qquad (6.1\text{-}6)$$

For an assembly of N *independent* oscillators, the total internal energy U is simply

$$U = N\langle\varepsilon\rangle = 3\,NkT, \qquad (6.1\text{-}7)$$

and the heat capacity C_v is, by definition,

$$C_v = (\partial U/\partial T)_v = 3Nk. \qquad (6.1\text{-}8)$$

The specific heat is simply the heat capacity per gram (or per unit volume). The heat capacity is found simply to be a constant, independent of temperature. For a mole of any substance N is equal to Avogadro's number N_a and the *molar* heat capacity C_{vm} is given by

$$C_{vm} = 3N_a k = 5.96 \text{ cal/mole-}^\circ K. \qquad (6.1\text{-}9)$$

This result is known as the law of Dulong and Petit and is familiar to students of elementary chemistry as the basis of a rudimentary way of estimating the atomic weight of an unknown element. The law of Dulong and Petit agrees fairly well with experimental results for most substances at and above room temperature, but breaks

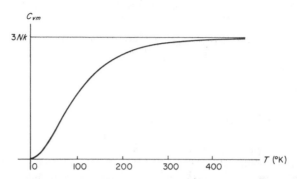

FIGURE 6.1. The heat capacity of a typical solid substance plotted as a function of temperature.

down seriously at low temperatures. In the low-temperature range, the specific heat of all substances is found to approach zero, being proportional to T^3 as T approaches zero, as shown in Figure 6.1.

6·2 THE EINSTEIN THEORY OF SPECIFIC HEAT

The discrepancies in the classical theory of specific heat were investigated by Einstein, who in 1911 succeeded in explaining, qualitatively, at least, the experimentally observed form of the specific heat as a function of temperature. In the Einstein theory, the atoms are again regarded as identical independent harmonic oscillators with a single natural vibration frequency ω_0, but they are regarded as *quantum* harmonic oscillators which may have only discrete energy values[1] such that

$$\varepsilon_n = (n + \tfrac{1}{2})\hbar\omega_0 \qquad (n = 0,1,2, \cdots), \qquad (6.2\text{-}1)$$

as given by (4.12-27). The oscillators may still be regarded as having the Maxwell-Boltzmann distribution of energies, since they form an assembly of systems which are *distinguishable* or identifiable by virtue of their location at separate and distinct lattice sites, and since *any number* of oscillators may be in the same quantum state of

---

[1] Originally, of course, Einstein used the Planck result $\varepsilon_n = n\hbar\omega$ rather than the wave-mechanical result as given by (6.2-1) for the energy levels of a harmonic oscillator. If the original Planck result is used, the final expressions for internal energy and specific heat which are obtained differ from the results derived here only in the absence of a temperature-independent zero-point energy contribution to the internal energy.

the system. Thus, although the atoms are quantized oscillators, the classical Maxwell-Boltzmann distribution is still the appropriate one with which to describe their statistical behavior. The actual quanta of vibrational energy which are absorbed or emitted, by the oscillators, on the other hand, are not distinguishable entities, and therefore Bose-Einstein statistics must be used to describe their statistical behavior. As we shall see later, the same results can be obtained by discussing the properties of the system of oscillators using Maxwell-Boltzmann statistics, or by treating the dynamics of the quanta of vibrational energy themselves using Bose-Einstein statistics.

For the sake of simplicity, we shall assume that the crystal contains $3N$ one-dimensional harmonic oscillators, whose energy levels are given by (6.2-1), rather than N three-dimensional isotropic oscillators. This assumption is justified by the fact that there are indeed three independent vibrational degrees of freedom associated with each atom of the crystal, and that according to the principle of equipartition, the available vibrational energy is distributed equally, on the average, among these three degrees of freedom. If the oscillators are distributed in energy according to the Boltzmann law, then the number of oscillators at energy ε is proportional to $e^{-\varepsilon/kT}$, and the average energy is

$$\langle \varepsilon \rangle = \frac{\sum_{n=0}^{\infty} \varepsilon_n e^{-\varepsilon_n/kT}}{\sum_{n=0}^{\infty} e^{-\varepsilon_n/kT}} = \hbar\omega_0 \frac{\sum_n (n + \frac{1}{2})e^{(n+\frac{1}{2})x}}{\sum_n e^{(n+\frac{1}{2})x}} \tag{6.2-2}$$

where
$$x = -\hbar\omega_0/kT. \tag{6.2-3}$$

Equation (6.2-2) can, however, be written as

$$\langle \varepsilon \rangle = \hbar\omega_0 \frac{\frac{1}{2}e^{\frac{1}{2}x} + \frac{3}{2}e^{\frac{3}{2}x} + \frac{5}{2}e^{\frac{5}{2}x} + \cdots}{e^{\frac{1}{2}x} + e^{\frac{3}{2}x} + e^{\frac{5}{2}x} + \cdots} = \hbar\omega_0 \frac{d}{dx} \ln[e^{\frac{1}{2}x}(1 + e^x + e^{2x} + \cdots)]$$

$$= \hbar\omega_0 \frac{d}{dx} [\tfrac{1}{2}x - \ln(1 - e^x)] = \hbar\omega_0 \left[\tfrac{1}{2} + \frac{1}{e^{\hbar\omega_0/kT} - 1} \right]. \tag{6.2-4}$$

The internal energy is obtained by multiplying the average energy per oscillator by the number of oscillators, as before, whereby

$$U = 3N\langle \varepsilon \rangle = \frac{3N\hbar\omega_0}{2} + \frac{3N\hbar\omega_0}{e^{\hbar\omega_0/kT} - 1} \tag{6.2-5}$$

and
$$C_v = \left(\frac{\partial U}{\partial T} \right)_v = 3Nk\left(\frac{\hbar\omega_0}{kT} \right)^2 \frac{e^{\hbar\omega_0/kT}}{(e^{\hbar\omega_0/kT} - 1)^2}. \tag{6.2-6}$$

This result can be expressed in a somewhat simpler way by defining an "Einstein Temperature" Θ_E such that

$$\hbar\omega_0 = k\Theta_E. \tag{6.2-7}$$

Substituting $k\Theta_E$ for $\hbar\omega_0$ in (6.2-6), one finds

$$C_v = 3Nk\left(\frac{\Theta_E}{T}\right)^2 \frac{e^{\Theta_E/T}}{(e^{\Theta_E/T} - 1)^2}. \tag{6.2-8}$$

For temperatures high enough that $T \gg \Theta_E$, $e^{\Theta_E/T} \cong 1$; by using the power series expansion for the exponential and retaining only first-order terms it is easily seen that in this limit C_v approaches the classical result $3Nk$. For low temperatures ($T \ll \Theta_E$), $e^{\Theta_E/T} \gg 1$ and (6.2-8) can be approximated by

$$C_v = 3Nk\left(\frac{\Theta_E}{T}\right)^2 e^{-\Theta_E/T} \tag{6.2-9}$$

which approaches zero as T approaches zero, although it does not agree with the observed T^3 behavior at low temperatures. The first term in (6.2-5) represents the contribution of the zero-point energy of the oscillators to the internal energy of the system. Since it is independent of temperature, it makes no contribution to the specific heat.

Physically, the reason that the specific heat becomes quite small at low temperatures can be understood by assuming that the crystal is placed in contact with an external heat bath consisting of an ideal monatomic gas at some given temperature, and allowed to absorb energy from the atoms of the ideal gas. The average energy of the gas atoms is $\frac{3}{2}kT$, and if the temperature of the system is high enough so that kT is of the order of, or larger than, the energy $\hbar\omega_0$ required to excite one of the vibrating atoms of the crystal to a higher energy state, then such excitations will occur frequently when gas atoms from the heat bath collide with the crystal. These collisions will then be *inelastic* in the mechanical sense, and energy initially belonging to the gas atoms will readily be transferred to the crystal lattice as vibrational energy. On the other hand, at sufficiently low temperatures, kT will be much smaller than the excitation energy $\hbar\omega_0$ and only an occasional gas atom having much higher energy than the average will be capable of effecting such an excitation and thus transferring heat to the crystal lattice. The crystal is then relatively incapable of absorbing heat from its surroundings, and a unit temperature change in the surroundings will transfer but little heat to the crystal, compared to that which may be transferred under similar conditions at a higher temperature. As the temperature approaches absolute zero, the fraction of atoms in the heat bath possessing the minimum excitation energy $\hbar\omega_0$ approaches zero, and consequently so also does the specific heat.

The Einstein temperature, $\Theta_E = \hbar\omega_0/k$ can easily be computed if ω_0 is known. This natural vibrational frequency can be calculated from the atomic mass and the observed elastic constants of the crystal.[2] For many metallic elements the value thus calculated is of the order of 100–200°K, and thus the transition from the low temperature to high temperature behavior should occur in this temperature range. Experimentally, the Einstein theory is found to fit the observations fairly well at all but very low temperatures, where the observed T^3 behavior is not obtained. The calculated values of Θ_E and those obtained by fitting the theoretical expression (6.2-8) to observed data also generally agree moderately well. Obviously, the Einstein theory provided a

---

[2] See Chapter 3, Exercise 6.

much better explanation of the experimental data than the classical theory, despite the fact that the agreement with experiment is by no means perfect. The discrepancy between theory and the low temperature experimental results was removed, and the general fit to the experimental data improved by the Debye theory of 1912. In the Debye theory, interactions between atoms which result in there being a range of possible values of vibrational frequency rather than just a single value ω_0, as described in Chapter 3, are taken into account.

6·3   THE DEBYE THEORY OF SPECIFIC HEAT

The Debye theory of specific heat, first proposed in 1912,[3] treated the atoms of the crystals as oscillators which were coupled together and which were capable of propagating elastic waves whose frequency might vary over a wide range of values, as predicted by the calculations of Chapter 3. Although still involving certain approximations, it was found to be superior to the Einstein model in predicting the specific heat of substances in the low-temperature region, and has formed the basis for a number of subsequent and more detailed investigations.

Since the atoms of the crystal no longer vibrate independently of one another, it is more convenient to work with the normal modes of vibration of the system than with the vibrational motion of a single atom. We have already seen (in Section 3.3 and Exercise 7, Chapter 3) that for a one-dimensional linear chain of N atoms there are essentially N normal modes of vibration. For a three-dimensional crystal of N atoms, each atom can vibrate independently along three coordinate directions, so that for this system there are $3N$ possible normal modes. Since any vibrational motion of the system can be thought of as a superposition of independent normal-mode vibrations, the normal modes can be regarded as independent harmonic oscillations whose allowed energy levels are given by (6.2-1) and whose average energy is given by (6.2-4), provided that the number of excited normal modes are distributed in energy according to the Maxwell-Boltzmann distribution law.

Suppose now that in a frequency range $d\omega$ about some frequency ω there are $g(\omega)\, d\omega$ normal modes of vibration of the crystal. The quantity $g(\omega)$ must thus represent the number of normal modes per unit frequency at frequency ω. The contribution to the internal vibrational energy of the crystal from these modes of vibration is then given by

$$dU = \langle \varepsilon(\omega) \rangle g(\omega)\, d\omega \qquad (6.3\text{-}1)$$

where $\langle \varepsilon(\omega) \rangle$ is the average energy of a vibrational mode of frequency ω, as given by the formula (6.2-4). It is somewhat more convenient to calculate the number of normal modes $g(k)\, dk$ in an interval dk about wave number k, and since $dk = (dk/d\omega)\, d\omega$, we may write (6.3-1) as

$$dU = \langle \varepsilon(\omega(k)) \rangle g(k)\, dk = \langle \varepsilon(\omega) \rangle g(k) \frac{dk}{d\omega}\, d\omega. \qquad (6.3\text{-}2)$$

---

[3] P. Debye, *Ann. Physik* **39**, 789 (1912).

In order to calculate $g(k)$, we may assume a rectangular crystal of dimensions (L_x, L_y, L_z) and assume that the origin is at a corner of the crystal, the crystal edges coinciding with the $+x$, $+y$, and $+z$ axes. The equations of motion for the mechanical vibrational amplitude $u(x,y,z,t)$ then lead to vibrational solutions of the form

$$u(x,y,z,t) = Ae^{i(\mathbf{k}\cdot\mathbf{r}-\omega t)} = Ae^{i(k_x x + k_y y + k_z z - \omega t)}. \tag{6.3-3}$$

If periodic boundary conditions of the form

$$u(L_x,y,z,t) = u(0,y,z,t)$$

$$u(x,L_y,z,t) = u(x,0,z,t) \tag{6.3-4}$$

$$u(x,y,L_z,t) = u(x,y,0,t)$$

are applied to (6.3-3), then it is clear that $k_x L_x = 2\pi n_x$, $k_y L_y = 2\pi n_y$, $k_z L_z = 2\pi n_z$, or

$$k_x = 2\pi n_x / L_x$$

$$k_y = 2\pi n_y / L_y \tag{6.3-5}$$

$$k_z = 2\pi n_z / L_z$$

where n_x, n_y, n_z are positive or negative *integers*. The allowed values (k_x, k_y, k_z), to each of which corresponds a single normal mode of vibration satisfying periodic boundary conditions, can be plotted as points in an orthogonal k-space, such as that shown in Figure 6.2. The allowed normal modes form a simple orthorhombic lattice in such a space, the dimensions of the unit cell being $(2\pi/L_x, 2\pi/L_y, 2\pi/L_z)$.

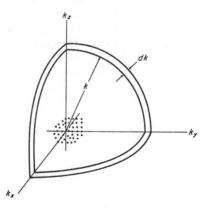

FIGURE 6.2. Spherical surfaces corresponding to constant values of propagation constant k and $k + dk$ plotted in a (k_x, k_y, k_z) coordinate system.

If the crystal dimensions L_x, L_y, L_z are sufficiently large, the unit cell will become small, the points representing allowed normal modes crowding together very closely in a diagram such as Figure 6.2. In this case one may calculate $g(k)\, dk$ by evaluating the volume of k-space between the spherical surface of radius k

$$k_x^2 + k_y^2 + k_z^2 = k^2 \tag{6.3-6}$$

which is the locus of all points corresponding to wave number k, and another spherical surface of radius of $k + dk$ which is the locus of all points whose wave number is $k + dk$. The required volume is simply the volume between the two spheres of Figure 6.2 or $4\pi k^2\ dk$. The number of allowed normal modes is then obtained by dividing this quantity by the volume of the unit cell corresponding to a single normal mode of the system. This volume is simply $8\pi^3/(L_x L_y L_z) = 8\pi^3/V$, where V is the volume of the crystal. This procedure leads to the value

$$g(k)\ dk = \frac{k^2 V}{2\pi^2}\ dk.$$

Actually, however, there are *three* independent normal modes for each allowed point in k-space, one corresponding to a longitudinal vibration (and discussed above), and two others corresponding to two mutually orthogonal *transverse* vibrations, which we have not included in our reckoning so far. To take these possibilities into account, the right-hand side of the above equation should be increased by a factor of 3, giving finally

$$g(k)\ dk = \frac{3k^2 V}{2\pi^2}\ dk. \tag{6.3-7}$$

At this point, what one would wish to do would be to use some relation such as (3.3-7) which expresses the variation of k as a function of ω for a monatomic lattice, find $dk/d\omega$, and substitute these results along with (6.3-7) and (6.2-4) into (6.3-2), integrating finally over ω. This procedure was first outlined by Born and von Kármán.[4] Unfortunately, the mathematical form of the equation is such that the resulting expression is very cumbersome and cannot be integrated in closed form or handled numerically in any convenient way. To get around this difficulty, Debye *approximated* the sinusoidal function (3.3-7) as a linear function, writing

$$\omega(k) = v_0 k \tag{6.3-8}$$

where v_0 is a (constant) phase velocity equal, in principle, to the velocity of long-wave sound in the crystal. This is simply what is given by (3.3-7) for wavelengths long enough to satisfy the condition $ka \ll \pi$, although Debye used it to describe the behavior of *all* the vibrations which can be propagated in the crystal.

The integration over ω indicated in (6.3-2) is performed between the limits zero and a *maximum frequency* ω_m which is chosen in such a way that the *total number* of normal modes of frequency less than ω_m shall be just $3N$, since it is known that that is the actual number of normal modes which exist for a crystal of N atoms.[5] Corresponding to ω_m there is a wave number k_m, and all the normal modes for which $k < k_m$ have frequencies $\omega < \omega_m$, according to (3.3-7) or (6.3-8). There must, therefore, be just $3N$ normal modes of oscillation within a sphere of radius k_m in k-space. Since the volume

---

[4] M. Born and T. von Kármán, *Physikalische Zeitschrift* **13**, 297 (1912).

[5] Had it been possible to use an exact expression for relating ω and k, the integration would have been carried out within the limits of the first Brillouin zone. This would have guaranteed that the condition that the total number of normal modes equal $3N$ would have been satisfied *automatically*.

of k-space occupied by a single normal mode is $8\pi^3/3V$, we may thus write

$$\frac{4}{3}\pi k_m^3 = 3N \cdot \frac{8\pi^3}{3V}$$

or

$$k_m = (6N\pi^2/V)^{1/3} \tag{6.3-9}$$

whence, using the Debye approximation (6.3-8)

$$\omega_m = v_0 k_m = v_0(6N\pi^2/V)^{1/3}. \tag{6.3-10}$$

The variation of ω with k as assumed by the Debye approximation is shown in comparison with the value given by (3.3-7) in Figure 6.3. We are assuming in (6.3-10) that

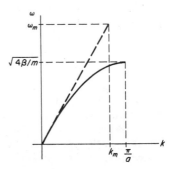

FIGURE 6.3. The actual dispersion relation (solid curve) and the dispersion relation (dotted curve) according to the Debye approximation.

the velocity v_0 is the same for longitudinal and transverse elastic waves. This is not generally true, but the difference is not usually great enough to affect the final result significantly.

If we now substitute (6.2-4) and (6.3-7) into (6.3-2), using (6.3-8) wherever necessary to express k in terms of ω, and integrate from zero to ω_m, we obtain

$$U = \frac{3\hbar V}{2\pi^2 v_0^3} \int_0^{\omega_m} \left[\frac{\omega^3}{2} + \frac{\omega^3}{e^{\hbar\omega/kT} - 1} \right] d\omega = \frac{3\hbar\omega_m^4 V}{16\pi^2 v_0^3} + \frac{3\hbar V}{2\pi^2 v_0^3} \int_0^{\omega_m} \frac{\omega^3 \, d\omega}{e^{\hbar\omega/kT} - 1}. \tag{6.3-11}$$

The integral of the first term above is easily evaluated and gives rise to a zero-point contribution to the internal energy which, however, is independent of temperature and contributes nothing to the specific heat. The integral of the second term cannot be evaluated in closed form but must be worked out numerically.

It is customary to express these results in a slightly different form. If one lets

$$x = \hbar\omega/kT, \tag{6.3-12}$$

$$x_m = \hbar\omega_m/kT = \frac{\hbar v_0}{kT}\left(\frac{6N\pi^2}{V}\right)^{1/3} \tag{6.3-13}$$

and
$$\Theta = \frac{\hbar\omega_m}{k} = \frac{\hbar v_0}{k}\left(\frac{6N\pi^2}{V}\right)^{1/3} \tag{6.3-14}$$

then (6.3-12) can be written in the form

$$U = \frac{9}{8}Nk\Theta + 9NkT\left(\frac{T}{\Theta}\right)^3 \int_0^{\Theta/T} \frac{x^3\,dx}{e^x - 1}. \tag{6.3-15}$$

The parameter Θ is seen to have the dimensions of temperature and to play the role of a characteristic temperature in (6.3-15) in much the same way as does the Einstein Temperature in (6.2-8). Θ is usually referred to as the *Debye temperature*. It is, of course, independent of temperature, except for the slight temperature variation introduced by the variation of V and v_0 with temperature.

By differentiating (6.3-11) under the integral with respect to temperature, the heat capacity can be found; the result is easily shown to be

$$C_v = \frac{3\hbar^2 V}{2\pi^2 v_0^3 kT^2} \int_0^{\omega_m} \frac{\omega^4 e^{\hbar\omega/kT}\,d\omega}{(e^{\hbar\omega/kT} - 1)^2} = 9Nk\left(\frac{T}{\Theta}\right)^3 \int_0^{\Theta/T} \frac{x^4 e^x\,dx}{(e^x - 1)^2}. \tag{6.3-16}$$

The integral in (6.3-16), which is a function of Θ/T alone, can be evaluated numerically, yielding a function a plot of which is shown in Figure 6.4. At high temperatures,

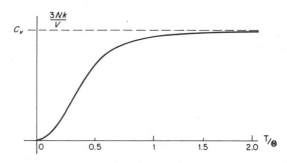

FIGURE 6.4. The heat capacity of a monatomic insulating solid according to the Debye theory plotted against the reduced temperature variable T/Θ.

$T \gg \Theta$ and the exponential term in the center expression of (6.3-16) can be expanded just as was done in connection with (6.2-8), giving an expression which is easily integrated; subsequently, using (6.3-10) it is easy to show that the classical result $C_v = 3Nk$ is obtained. At low temperatures, in the right-hand expression of (6.3-15), the limit of the integral approaches infinity, and the resulting definite integral can be shown by methods of contour integration to equal $\pi^4/15$, giving

$$\left(\frac{\partial U}{\partial T}\right)_v = C_v = \frac{12\pi^4}{5} Nk\left(\frac{T}{\Theta}\right)^3 \qquad (T \ll \Theta). \tag{6.3-17}$$

This is the well-known "Debye T^3 Law"; its predictions for specific heats in the low temperature region are in good agreement with experimental data for many substances.

The value of the Debye temperature may be calculated from (6.3-14), or obtained experimentally by choosing that value for Θ which leads to the best fit between the experimental data and the theoretical expression (6.3-16). The calculated and experimentally determined values are generally in fairly good agreement for substances which obey the assumptions which have been made in the Debye theory. For the more common metallic elements the Debye temperature generally lies in the range 150–450°K. A table of Debye temperatures and a detailed discussion of the agreement between experimentally determined and calculated values of the Debye temperature is given by Seitz.[6]

In the preceding development, a crystal containing N *identical* atoms was assumed. In a crystal where the atoms are not all identical, such as NaCl, elastic vibrations belonging to either optical or acoustical mode can be excited and the results calculated by this procedure are no longer directly applicable. In such a crystal, however, it is possible to treat the acoustical modes and optical modes separately. The acoustical modes may be treated by the above procedure, while the variation of ω with k for the optical modes as shown, for example by Figure 3.11, is so small that it is usually a fairly good approximation to regard the optical modes as vibrations of a single frequency given by (3.4-10). The optical modes may then be treated by the Einstein method developed in the preceding section, and the contributions from optical and acoustical modes combined to give the total heat capacity of the crystal. These and other corrections to the simple Debye theory have been considered in some detail by Blackman,[7,8] who concluded that in certain specific cases the Debye procedure might lead to significant errors, and that under these circumstances the more exact Born-von Kármán procedure would have to be used instead.

It is customary to think of a metallic substance as possessing a large number of free electrons which may serve to produce a flow of current under an applied electric field, thus accounting for the very high electrical conductivity of metals. On the basis of this picture, one might expect that these free electrons would acquire additional kinetic energy when the crystal is exposed to an external heat source, and in this way contribute significantly to the specific heat. Actually, and somewhat surprisingly, this contribution to the specific heat is so small as to be negligible at temperatures greater than about 5°K. Below this temperature in metallic crystals, it is possible to distinguish a *linear* variation of specific heat with temperature, which is indeed due to the free electrons. The reasons for this rather peculiar behavior of the electronic specific heat will be discussed in detail in the next chapter.

6·4   THE PHONON

So far, we have discussed the specific heat of solid materials in terms of quantum harmonic oscillators, and calculated the internal energy by ascertaining the number of oscillators, on the average, in each allowed energy state. If the vibrations of the lattice can be treated in this way, then there must exist quanta of vibrational energy

[6] F. Seitz, *Modern Theory of Solids*, McGraw-Hill Book Co., Inc., New York (1940), pp. 110–111.

[7] M. Blackman, Z. Physik **86**, 421 (1933).

[8] M. Blackman, Proc. Roy. Soc. **148**, 384 (1935); **159**, 416 (1937).

which are emitted or absorbed when transitions between one quantum state and another occur. It is a result of the quantum mechanics of the harmonic oscillator that to a first order of approximation transitions occur only between *adjacent* states of the system;[9] in other words, in any such transition the quantum number of the oscillator may change only by ± 1. Accordingly the energy change $\Delta\varepsilon$ of the oscillator, corresponding to the energy of the quantum which is absorbed or emitted must be

$$\Delta\varepsilon = \Delta n \cdot \hbar\omega = \pm\hbar\omega \qquad (6.4\text{-}1)$$

where ω is the frequency of the oscillator. The plus sign in (6.4-1) refers to absorption, the minus sign to emission.

Such a quantum of acoustical energy is commonly referred to as a *phonon*, in analogy with the terminology *photon* for a quantum of electromagnetic energy. A phonon must be regarded as having characteristics both of a wave and a particle, just as does the photon; in fact it is quite possible to treat the interaction between two phonons or between a phonon and an electron as a scattering "collision" between two particles. The Debye result for the specific heat can indeed be obtained by regarding the lattice vibrations of the crystal as a *gas of phonons*. Since the phonons are indistinguishable particles, the Bose-Einstein distribution function must be used to describe the distribution of particles among the energy states of the system, and since the number of phonons in the system is not constant with respect to temperature the form which the distribution function takes is (5.6-9), in which $\alpha = 0$.

The number of phonons in the frequency range $d\omega$ about ω (corresponding to an energy range $d\varepsilon$ about $\varepsilon = \hbar\omega$) is then

$$N(\omega)\,d\omega = \frac{g(\omega)\,d\omega}{e^{\hbar\omega/kT} - 1} \qquad (6.4\text{-}2)$$

where $g(\omega)\,d\omega$ is the number of possible normal vibrations of the system in the range $d\omega$ about ω. But in (6.3-2) and (6.3-7) we have already shown that

$$g(\omega)\,d\omega = g(k)\frac{dk}{d\omega}\,d\omega = \frac{3k^2 V}{2\pi^2}\frac{dk}{d\omega}\,d\omega. \qquad (6.4\text{-}3)$$

Using the Debye approximation (6.3-8) to express k and $dk/d\omega$ in terms of ω, (6.4-2) becomes

$$N(\omega)\,d\omega = \frac{3V}{2\pi^2 v_0^3}\frac{\omega^2\,d\omega}{e^{\hbar\omega/kT} - 1}. \qquad (6.4\text{-}4)$$

The contribution of the internal energy from the frequency range $d\omega$ about ω is just the quantum energy $\hbar\omega$ times the number of quanta in that frequency range. Thus

$$dU = \varepsilon(\omega)N(\omega)\,d\omega = \hbar\omega N(\omega)\,d\omega$$

$$= \frac{3\hbar V}{2\pi^2 v_0^3}\frac{\omega^3\,d\omega}{e^{\hbar\omega/kT} - 1}. \qquad (6.4\text{-}5)$$

---

[9] For a derivation of this *selection rule* see R. B. Leighton, *Principles of Modern Physics*, McGraw-Hill Book Co., Inc., New York (1959), pp. 211–222.

Integrating over all allowed frequencies from zero to the Debye maximum frequency ω_m, one obtains immediately the result (6.3-11). The zero-point energy term, of course, is not obtained when the internal energy is calculated in this way, since the zero-point energy is not connected with the distribution of phonons in any way. This term, however, does not contribute to the specific heat and the two methods lead to the same value (6.3-16) for the specific heat, the result which is of physical interest and accessible to measurement. One may thus treat the system as an ensemble of distinguishable harmonic oscillators, obeying Maxwell-Boltzmann statistics, or as a gas of phonons or vibrational energy quanta, which are indistinguishable, and which obey Bose-Einstein statistics.

6·5   THERMAL EXPANSION OF SOLIDS

The picture of a solid as a system of atoms which are free to vibrate about a periodic array of equilibrium positions, which has been developed in the preceding sections to account for the observed behavior of the specific heat, also suffices to explain the thermal expansion of solids. In this case, however, it is found that if the atoms of the crystal are ideal simple harmonic oscillators, there is no thermal expansion at all, either by classical or quantum mechanical reckoning, and the entire thermal expansion effect in a real crystal depends upon the *anharmonicity* of the lattice vibrations.

This result can be understood qualitatively by considering the potential energy of interaction between neighboring atoms, which must have the form shown in Figure 6.5 as a function of interatomic distance. The minimum in the potential energy curve, at B, is the classical equilibrium position of the atom in question if it is at rest. If the interatomic forces were such that the atom if set in motion thermally would vibrate about its equilibrium position as an ideal classical harmonic oscillator, then the potential curve would be a perfect parabola whose vertex would be located at B. In such a potential "well," the atom would execute harmonic vibrations about the equilibrium point B, the maximum excursions from the equilibrium point in either direction being equal, and the average interatomic distance $\langle x \rangle$ would be equal to the zero-temperature lattice constant a. There would thus be no thermal expansion whatever.

In reality, however, since the crystal can be torn apart by the expenditure of a finite amount of energy, the potential well in which the atoms vibrate, must have approximately the appearance shown in Figure 6.5. In this case, although *nearly* parabolic about the minimum point B, the actual well deviates from the parabolic form more and more as the distance from the minimum point increases. If the atom has energy ε_0, it should, according to the classical picture vibrate between the extreme amplitude limits A and C, the vibrations being somewhat anharmonic in character. But the distance DC between the equilibrium position and the maximum extension position is now greater than the distance AD between the equilibrium position and the maximum compression position. The average interatomic distance $\langle x \rangle$ is thus *greater* than the zero-temperature lattice constant, and thermal expansion is observed. This classical view of the effect can be shown to lead to the conclusion that the thermal expansion $\langle x \rangle - a$ is directly proportional to the temperature, or that the coefficient of thermal expansion $d(\langle x \rangle - a)/dT$ is independent of temperature. This is in fairly good agreement for most substances in the room temperature range or above.

At low temperatures, however, the observed thermal expansion coefficient for most materials becomes much smaller than the classical value, and indeed approaches

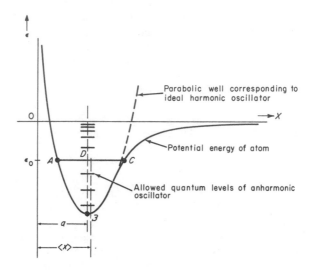

FIGURE 6.5. A representation of the potential energy of a single atom of the crystal as a function of interatomic spacing, illustrating the role of the anharmonic component in the thermal expansion.

zero as the absolute temperature approaches zero. Here again the classical picture breaks down, and the atoms of the crystal must be regarded as quantum oscillators rather than classical ones. In order to calculate the thermal expansion one must first solve Schrödinger's equation using an anharmonic potential (it is usually sufficient to modify the harmonic potential (4.12-6) by the addition of a cubic term in x) to obtain the energy levels ε_n and wave functions corresponding to this modified potential, and then find the expectation value $\langle x \rangle_n$ for each allowed energy state by (4.8-11). The interatomic distance, averaged over a Maxwell-Boltzmann distribution may then be written (in analogy with (6.2-1)) as

$$\langle x \rangle = \frac{\sum_n \langle x \rangle_n e^{-\varepsilon_n/kT}}{\sum_n e^{-\varepsilon_n/kT}}, \tag{6.5-1}$$

the sum being taken over all the levels of the anharmonic oscillator. The details of this procedure are rather involved and will not be presented here. The coefficient of thermal expansion calculated by this method approaches zero as the absolute temperature approaches zero, and is in reasonably good agreement with experiment over a wide range of temperature.

6·6   LATTICE THERMAL CONDUCTIVITY OF SOLIDS

The conduction of heat in solids is a process in which the lattice vibrations also play an important role. In metallic crystals, where large concentrations of free electrons are present, the free electrons contribute significantly to the thermal conductivity, and indeed provide the dominant mechanism for the transport of thermal energy through the crystal. The discussion of electronic thermal conductivity in metals is best presented in connection with the free-electron theory of metals, and hence will be postponed to a later chapter. In nonmetallic substances, however, the lattice vibrations are solely responsible for the transport of heat through the crystal.

The thermal conductivity is customarily defined in terms of a one-dimensional heat flow situation in which a temperature difference ΔT exists over a distance Δx, causing a flow of heat ΔQ over an area A in a time Δt. The thermal conductivity is the coefficient of proportionality between the heat flow per unit area and the temperature gradient. In accord with this definition, one may write

$$J_{qx} = \frac{1}{A}\frac{\Delta Q}{\Delta t} = -K\frac{\Delta T}{\Delta x}, \qquad (6.6\text{-}1)$$

where K is the thermal conductivity coefficient and J_{qx} the heat flow per unit area or thermal current density. The minus sign is required since K is regarded as a positive quantity and since a positive temperature gradient leads to a negative thermal current. If thermal energy is conserved, then the thermal current must everywhere obey a continuity equation of the form

$$\nabla \cdot \mathbf{J}_q = \frac{\partial J_{qx}}{\partial x} = -\frac{\partial Q}{\partial t}. \qquad (6.6\text{-}2)$$

By combining (6.6-2) and (6.6-1) it is easily seen that the temperature (hence the thermal energy per atom) must obey a differential equation of the form

$$\frac{K}{c_v}\frac{\partial^2 T}{\partial x^2} = \frac{\partial T}{\partial t}. \qquad (6.6\text{-}3)$$

This is a partial differential equation which is essentially the same as the equation which describes the *diffusive* transport of mobile particles in a medium in which a concentration gradient exists. It differs from the equation for radiative propagation (3.1-12) in that the right-hand side is a first derivative with respect to time rather than a second derivative. Its characteristic solutions have an exponential rather than harmonic behavior with respect to time, as we shall see in a later chapter.

In a system made up of *ideal* harmonic oscillators, the various normal modes of oscillation of the system are completely independent of one another, there being no coupling whatever between different normal modes of vibration. If a given normal mode of such a system is excited, the system will vibrate in a stable fashion in that same normal mode for an indefinite period of time. In such a system the energy of vibration is propagated directly by acoustical *radiation* rather than by the much slower process which is associated with thermal conduction. Energy thus flows through the system with the velocity of sound, there is no mechanism by which this energy is randomized and by which thermal resistance is created, and the thermal conductivity of the specimen is essentially infinite. These observations are true for quantum oscillators as well as classical oscillators. This state of affairs can be described in terms of phonons by the statement that phonons corresponding to different normal-mode harmonic vibrations do not interact with one another at all—or that their mutual interaction cross-section is zero.

If a small amount of anharmonicity is introduced into such a system, then a *coupling* between the various normal modes arises. In this case, if a single normal mode is excited at the outset, the energy initially belonging to this mode will gradually be transferred to the other possible normal modes of the system and thus randomized. In the language of phonons, one may say that phonons representing the various normal-mode vibrations of the system interact (or *collide*) with one another. In other words, a group of phonons corresponding to a single normal mode of the harmonic system will now interact (or be *scattered* by) one another in such a way that after a time their individual momenta and propagation directions will be altered in an essentially random fashion. The phonon distribution is thus converted to one which represents a more or less random selection of normal modes of the system, within the restrictions imposed by conservation of energy and momentum in the individual interactions.

It is clear from this that the thermal conductivity would be infinite were it not for the anharmonicity of the lattice vibrations. The strength of the anharmonic component of the interatomic potential is thus a major factor in determining the thermal conductivity. Since the magnitude of the anharmonicity increases with vibration amplitude, one would expect that the thermal conductivity would decrease as this amplitude increases, leading to a decrease of lattice thermal conductivity with increasing temperature. This effect is quite generally observed in experiment at sufficiently high temperatures. An inverse measure of the strength of the anharmonic interaction between atoms of the crystal is given by the average distance which a phonon can travel between randomizing collisions or scattering events involving other phonons. This distance is referred to as the *mean free path* for phonon-phonon interactions, and the thermal conductivity of solids can be understood quite easily on a qualitative basis in terms of the mean free path. We shall have occasion to deal with the mean free path in a rather more precise and quantitative way in later sections.

The thermal conductivity of solids is most easily understood by regarding the crystal as a container enclosing a "gas" of phonons. We shall first develop a formula for the thermal conductivity of an ideal monatomic gas and then examine how it may be applied to the lattice of a crystalline solid. To begin with, we must find the number of particles of an ideal gas having velocity in the range dv about v, which cross an area element dS per unit time from a direction lying within an angular range $d\theta$ about a polar angle θ. This number is just half the number within that velocity range inside a column of cross section $dS \cos \theta$ (which is the projection of the area dS onto a plane normal to the θ-direction) which lie with a distance of v centimeters from the area element itself. This situation is illustrated in Figure 6.6. The factor one-half is included because on the average only half the particles within the column going along the θ-direction are headed *toward* the area element, the rest going in the opposite direction. If $n(v)\, dv$ is the number of particles per unit volume having velocity in the range dv about v, then the number crossing the area element is

$$dn = \tfrac{1}{2}v \cdot n(v)\, dv \cdot \frac{4\pi \sin \theta\, d\theta}{4\pi} \cdot dS \cos \theta = \tfrac{1}{2}vn(v) \sin \theta \cos \theta\, dv\, d\theta\, dS. \qquad (6.6\text{-}4)$$

The factor involving $\sin \theta$ above represents the ratio of the solid angle subtended by the element $d\theta$ about θ to the solid angle 4π subtended by all space; it is clear that this factor also must appear in the expression. Integrating over v, using the Maxwell-Boltzmann distribution (5.4-14) (and remembering that the velocity distribution is independent of θ), one may obtain

$$n(\theta) \, d\theta = \tfrac{1}{2} n \bar{c} \sin \theta \cos \theta \, d\theta \, dS \qquad (6.6\text{-}5)$$

with $\bar{c}$ given by (5.4-20), as the desired result.

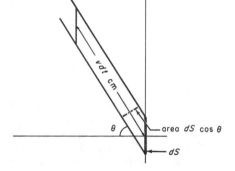

FIGURE 6.6. Geometry of particles streaming across a plane used in the calculations of Section 6.6.

It should be noted at this point that since the problem of thermal conduction always involves a flow of energy and a temperature gradient, one always has to deal with a system which is *not* in a state of thermal equilibrium. For this reason it is not strictly correct to use the statistical distribution functions which apply to systems at equilibrium. If the temperature gradient in the system is so small that the fractional variation of absolute temperature over a mean free path is small, however, the use of the equilibrium distribution as an approximation is justified. We shall in all cases limit ourselves to a discussion of systems of this sort. The question of the use of statistical distributions in systems which are not in equilibrium will be considered further in the next chapter.

Consider now the situation illustrated in Figure 6.7. A temperature gradient is assumed to exist along the x-axis, so that particles at the origin have average energy ε and those at a short distance to the left have average energy $\varepsilon + \Delta\varepsilon$. Suppose that a particle starts out a distance l away from the origin and gives up its energy to the distribution at the origin in a collision which takes place there. By Taylor's expansion the energy of this particle is

$$\varepsilon + \Delta\varepsilon = \varepsilon + \frac{\partial \varepsilon}{\partial x} \Delta x = \varepsilon + (l \cos \theta) \frac{\partial \varepsilon}{\partial x}. \qquad (6.6\text{-}6)$$

For each such particle, of course, there will be a particle going in the opposite direction which transports energy ε from the origin to the starting point of the first particle. The *net* transport of energy is just $\Delta\varepsilon = (l \cos \theta)(\partial \varepsilon / \partial x)$, and since the average value of l is simply the *mean free path* between randomizing collisions, which we shall call λ, the net energy transport per particle, averaged over the distribution of path lengths l, is

$$\Delta\varepsilon = (\lambda \cos \theta) \frac{\partial \varepsilon}{\partial x}. \qquad (6.6\text{-}7)$$

For particles of polar angle θ, according to (6.6-5) and (6.6-7), a net energy flux ΔF per unit area arises, where

$$\Delta F = \frac{\Delta\varepsilon \cdot n(\theta)\, d\theta}{dS} = \frac{1}{2}\, n\bar{c}\lambda\, \frac{\partial \varepsilon}{\partial x} \cos^2 \theta \sin \theta\, d\theta. \qquad (6.6\text{-}8)$$

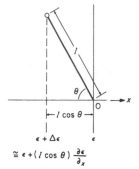

FIGURE 6.7. The path of a particle in a system containing a temperature gradient along the x-direction.

Integrating over the polar angle θ between limits of 0 and π, it is easily seen that the total energy flux must be $\frac{1}{3} n\bar{c}\lambda\, (\partial\varepsilon/\partial x)$. But from the definition of the thermal conductivity K (6.6-1) this must also equal $K\, (\partial T/\partial x)$. Accordingly,

$$K \frac{\partial T}{\partial x} = \frac{1}{3}\, n\bar{c}\lambda\, \frac{\partial \varepsilon}{\partial x} = \frac{1}{3}\, n\bar{c}\lambda\, \frac{\partial \varepsilon}{\partial T} \frac{\partial T}{\partial x}, \qquad (6.6\text{-}9)$$

whereby

$$K = \frac{1}{3}\, n\bar{c}\lambda\, \frac{\partial \varepsilon}{\partial T} = \frac{1}{3}\, c_v \lambda \bar{c}, \qquad (6.6\text{-}10)$$

since, by definition, the specific heat c_v is just $n(\partial\varepsilon/\partial T)$.

This result, derived above for an ideal Maxwell-Boltzmann gas, is also good for a distribution of phonons, provided that the velocity $\bar{c}$ is understood to refer to an average phonon velocity, that is, to an average *sound* velocity in the crystal. [It will be recalled that the Boltzmann distribution was used only to average velocities in obtaining (6.6-5).] According to (6.6-10), then, the problem of determining the lattice thermal conductivity may be reduced essentially to determining the mean free path for phonons in the crystal, since the sound velocity and the specific heat are easily measured or calculated. This problem, unfortunately, is not an easy one, and to reproduce the work involved in its calculation is quite beyond the scope of the present work. The problem was first solved by Peierls,[10] who found λ to be proportional to

---

[10] R. Peierls, Ann. Physik **3**, 1055 (1929).

$1/T$ at high temperatures and to $e^{\Theta/2T}$ at low temperatures. According to this, the thermal conductivity would become infinitely large as the temperature approaches absolute zero, since the product of c_v and λ would diverge to infinity in this limit. Actually, however, at very low temperatures, the phonon mean free path ceases to be limited by phonon-phonon interactions, and is instead limited by scattering of phonons by impurities and imperfections in the crystal, or, in very pure crystals, by the *surfaces* of the crystal sample itself. In the latter case, it is found that the thermal conductivity of the crystal is a function of the size of the sample! In either instance, the mean free path tends to some finite limit as T approaches zero, and the thermal conductivity approaches zero as T approaches zero, due to the fact that c_v vanishes in the zero-temperature limit. With increasing temperature, then, the thermal conductivity rises from zero to some maximum value, then decreases as the temperature is further increased. This qualitative behavior is indeed observed experimentally in most instances, the magnitudes of the values calculated by (6.6-10) from calculated values of λ agreeing with experimental data in order of magnitude. The effect of scattering by crystal boundaries was first considered by Casimir,[11] and the effect of certain types of imperfections has been treated in detail by Klemens.[12]

EXERCISES

1. Show explicitly, using Maxwell-Boltzmann statistics, that the average energy of a three-dimensional classical harmonic oscillator is just three times that of a one-dimensional classical harmonic oscillator at the same temperature, and that it is thus immaterial whether a crystal of N atoms is regarded as an assembly of N three-dimensional oscillators or $3N$ one-dimensional oscillators.

2. Calculate by the Debye method the heat capacity of a two-dimensional periodic crystal lattice containing N identical atoms, and show that at low temperatures the heat capacity varies with temperature as T^2. What is the high-temperature limit for the heat capacity?

3. Calculate the Debye temperature Θ for (a) diamond, (b) germanium, (c) copper, and (d) lead, and compare your results with published values derived from experimental heat capacity data.

4. Using the Planck relation to express the energy of a phonon in terms of its frequency and the de Broglie relation to express the momentum in terms of the wave number, find an expression for the equivalent "mass" of a phonon of average thermal energy at 300°K, and compare this value with the electronic mass. You may assume that k is small enough so that dispersion effects may be neglected.

5. Prove that the average value $\langle x \rangle_n$ of the amplitude of a quantum harmonic oscillator in the nth energy state is zero.

6. Show, using the laws of thermodynamics, that the difference between the specific heat of a substance at constant pressure and at constant volume is given by

$$c_p - c_v = \frac{9\alpha^2 VT}{K_{iso}}$$

[11] H. B. G. Casimir, Physica **5**, 495 (1938).
[12] P. G. Klemens, Proc. Roy. Soc. **A208**, 108 (1951).

where α is the linear coefficient of expansion and K_{iso} is the isothermal compressibility $-V^{-1}(\partial V/\partial P)_T$. *Hint:* note that, by Maxwell's relations, $(\partial S/\partial P)_T = -(\partial V/\partial T)_P$.

7. Consider an assembly of N identical systems having just two energy levels, at energies zero and ε_0. Assuming that the systems are distinguishable, find (a) the internal energy of the assembly, (b) the heat capacity, (c) approximate expressions for the heat capacity in the high- and low-temperature limits. Plot the heat capacity as a function of temperature.

8. For the assembly of systems discussed in connection with Exercise 7, prove that the heat capacity reaches a maximum value at a temperature T_0 for which

$$x \tanh x = 1$$

where $x = \varepsilon_0/2kT$. Show also that the heat capacity at this temperature is given by

$$c_v(\max) = Nk(x_0^2 - 1)$$

where $x_0 = \varepsilon_0/2kT_0$. Contributions to the heat capacity from systems of this sort are often encountered at low temperatures in paramagnetic substances, and are referred to as *Schottky anomalies*.

9. From the quantum-mechanical expression $\varepsilon = (\frac{3}{2} + n_x + n_y + n_z)\hbar\omega_0$ (where n_x, n_y, and n_z may be zero or positive integers) for the energy levels of a three-dimensional harmonic oscillator, show, in the limit of large energies where $\varepsilon \gg \hbar\omega_0$ and $d\varepsilon \gg \hbar\omega_0$, that the number of quantum states in a range $d\varepsilon$ about energy ε is given by $g(\varepsilon)\,d\varepsilon = \varepsilon^2 d\varepsilon/2(\hbar\omega_0)^3$.

10. Using the result obtained in Problem 9 above, show that the expression $\langle\varepsilon\rangle = 3kT$ for the average energy of a *classical* three-dimensional harmonic oscillator in a system whose members are distributed in energy according to the Boltzmann law may be obtained simply by evaluating the expression

$$\langle\varepsilon\rangle = \frac{\displaystyle\int \varepsilon f(\varepsilon)g(\varepsilon)\,d\varepsilon}{\displaystyle\int f(\varepsilon)g(\varepsilon)\,d\varepsilon}.$$

Compare critically the physical aspects of this method and the one used to obtain the same result in Section 6.1.

GENERAL REFERENCES

P. G. Klemens, *Thermal Conductivity and Lattice Vibrational Modes*, in *Solid State Physics, Advances in Research and Applications*, Vol. 7, Academic Press, New York (1959), pp. 1–98.

J. de Lannay, *The Theory of Specific Heats and Lattice Vibrations*, in *Solid State Physics, Advances in Research and Applications*, Vol. 2, Academic Press, New York (1956), pp. 219–303.

R. B. Leighton, *Principles of Modern Physics*, McGraw-Hill Book Co., Inc., New York (1959), Chapter 6.

R. B. Peierls, *Quantum Theory of Solids*, Clarendon Press, Oxford (1955), Chapter II.

F. Seitz, *Modern Theory of Solids*, McGraw-Hill Book Co., Inc., New York (1940), Chapter III.

CHAPTER 7

THE FREE-ELECTRON THEORY OF METALS

7·1   INTRODUCTION

The idea that the large electrical and thermal conductivity of metallic substances might be explained by the presence of large concentrations of mobile *free electrons* in these materials was first proposed by Drude[1] in 1900. The implications of this hypothesis were exhaustively investigated subsequently by Lorentz.[2] Drude and Lorentz assumed that the free electrons in a metal could be treated as an ideal gas of free particles which, when in thermal equilibrium, would obey Maxwell-Boltzmann statistics. In order to examine what might happen when electric or thermal currents were allowed to flow, thus establishing a nonequilibrium state, it was necessary to investigate how the equilibrium distribution would be modified by a (small) electrical or thermal current. It was also necessary to consider the kinetic behavior of the electrons as being that of free particles subject to instantaneous collisions which serve to return the distribution to the equilibrium condition, and to express the final result for electrical and thermal conductivity in terms of a mean free path or mean free time between these randomizing collisions. The Drude-Lorentz theory accounted satisfactorily for the well known experimental law of Wiedemann and Franz,[3] according to which the ratio of electrical to thermal conductivity for most metals is nearly the same. In addition, the magnitudes of the electrical and thermal conductivity could be obtained using values for the mean free path which were quite reasonable.

The free-electron theory in its simplest form, however, led to a prediction of the electronic component of the specific heat which (assuming that the Debye theory for the lattice component is right) was in serious disagreement with experimental results. This difficulty (along with certain others) was resolved by Sommerfeld[4] by the use of Fermi-Dirac statistics rather than the classical Boltzmann statistics. The Fermi-Dirac free-electron picture serves as a very simple and conceptually quite direct way of discussing and visualizing transport effects in metals. Since it starts with a preconceived notion of what a metal is like, it is of no value in explaining just *why* some substances are metals and have free electrons in abundance and others are insulators with few, if any, free electrons. For an explanation of these underlying questions, one must begin by examining the quantum-mechanical behavior of electrons in periodic potential fields, thus discussing essentially the same subject on a much more fundamental level. The latter approach will be postponed until the next chapter, where it will be shown that

---

[1] P. Drude, Ann. Physik **1**, 566 (1900).
[2] H. A. Lorentz, *The Theory of Electrons*, Teubner Verlag, Leipzig (1909).
[3] Wiedemann and Franz, Ann. Physik **89**, 497 (1853).
[4] A. Sommerfeld, Z. Physik, **47**, 1 (1928).

the properties of metals and insulators can be understood in terms of quantum theory, and that the foundations upon which the free-electron theory rests, are, for the most part, compatible with the requirements of quantum mechanics.

7·2   THE BOLTZMANN EQUATION AND THE MEAN FREE PATH

In any problem of statistical mechanics it is of central importance to know the form of the distribution function. In a system in thermal equilibrium, of course, the distribution function will be one of those which have been discussed in Chapter 5, but if the state of the system is not one of equilibrium (for example, if a current is flowing or if a temperature gradient is present), then the distribution function will be somewhat different. In the limit where the perturbation from the equilibrium condition vanishes, naturally, the nonequilibrium distribution function must approach the appropriate equilibrium distribution. As a preliminary to any statistical discussion of nonequilibrium processes, then, one must find a way of evaluating the distribution function when the system is in a nonequilibrium state. This is accomplished by means of the Boltzmann equation.

To derive this equation, consider a region of phase space about the point (x,y,z, p_x,p_y,p_z). The number of particles entering this region in time dt is equal to the number which were in the region of phase space at $(x - v_x dt, y - v_y dt, z - v_z dt, p_x - F_x dt, p_y - F_y dt, p_z - F_z dt)$ at a time dt earlier. Here $\mathbf{F}$ represents the force acting on the particles of the distribution at the point (x,y,z) and time t, and it follows from Newton's Law that $\mathbf{F} = d\mathbf{p}/dt$. If $f(x,y,z,p_x,p_y,p_z)$ is the distribution function, which expresses the number of particles per quantum state in the region, then the change df which occurs during time dt due to the motion of the particles in coordinate space and due to the fact that force fields acting on the particles tend to move them from one region to another in momentum space is

$$df = f(x - v_x dt, y - v_y dt, z - v_z dt, p_x - F_x dt, p_y - F_y dt, p_z - F_z dt)$$
$$- f(x,y,z,p_x,p_y,p_z). \quad (7.2\text{-}1)$$

This, using Taylor's expansion and retaining only first-order terms in the limit $dt \to 0$ may be written

$$df = \left(-v_x \frac{\partial f}{\partial x} - v_y \frac{\partial f}{\partial y} - v_z \frac{\partial f}{\partial z} - F_x \frac{\partial f}{\partial p_x} - F_y \frac{\partial f}{\partial p_y} - F_z \frac{\partial f}{\partial p_z} \right) dt$$

or

$$\frac{df}{dt} = -\mathbf{v} \cdot \nabla f - \mathbf{F} \cdot \nabla_p f, \quad (7.2\text{-}2)$$

where the symbol ∇_p refers to the gradient operator in the momentum space (p_x,p_y,p_z). So far, only the change in the distribution function due to the motion of the particles in coordinate space and due to the momentum changes arising from force fields acting on the particles have been accounted for. Particles may also be transferred into or out of a given region of phase space by *collisions* or *scattering interactions* involving other particles of the distribution or scattering centers external to the assembly of particles under consideration. If the rate of change of the distribution function due to collisions or scattering is denoted by $(\partial f/\partial t)_{\text{coll}}$, then the total rate of change of f may be obtained by simply adding this quantity to the right-hand side of (7.2-2), giving

$$\frac{df}{dt} = -\mathbf{v} \cdot \nabla f - \mathbf{F} \cdot \nabla_p f + \left(\frac{\partial f}{\partial t}\right)_{\text{coll}}. \tag{7.2-3}$$

This is the Boltzmann equation. If the force field $\mathbf{F}$ and the rate of change $(\partial f/\partial t)_{\text{coll}}$ are known, then the differential equation (7.2-3) can, in principle, be solved for the distribution function.

To actually do this, of course, one must adopt some model for the scattering interaction which can be made to yield the rate at which particles may enter or leave any region in phase space due to collisions. For example, one may assume steady state conditions, whereby $df/dt = 0$, and in addition specify that there are no force fields acting on the particles, and that the particles of the distribution are hard, perfectly elastic spheres of given radius, which undergo collisions that may be treated mechanically by the classical laws of motion to obtain an average value for $(\partial f/\partial t)_{\text{coll}}$. This mechanical calculation yields essentially a *cross-section* for scattering by a particle of the distribution. In this case it can be shown that the Boltzmann equation leads directly to the Maxwell-Boltzmann distribution of Chapter 5.

The Boltzmann equation is more commonly used to discuss the properties of a system which is displaced from equilibrium by a small perturbing force, such as an electric field or a temperature gradient, when the distribution function f_0 which is the equilibrium distribution function in the absence of the perturbation is known. In such cases, it is frequently assumed that the collision term has the form

$$\left(\frac{\partial f}{\partial t}\right)_{\text{coll}} = -\frac{f - f_0}{\tau} \tag{7.2-4}$$

where τ is a parameter called the relaxation time or mean free time. Equation (7.2-4) represents the rate at which the distribution function approaches the equilibrium condition (the perturbing force having been suddenly removed) as being proportional at any time to the deviation from the equilibrium condition.

For example, suppose that an electric field E_0 sets up a current in a circuit, which is composed of a uniform metallic substance, so that f is the same at all points and hence $\nabla f = 0$. As long as the field is on, the force $\mathbf{F}$ in (7.2-3) must be represented as $-eE_0$, but if the field is suddenly removed, say at $t = 0$, then for positive values of t, $F = 0$ and Boltzmann's equation, using the relaxation time approximation (7.2-4) becomes

$$\frac{df}{dt} = \left(\frac{\partial f}{\partial t}\right)_{\text{coll}} = -\frac{f - f_0}{\tau}, \qquad (t > 0) \tag{7.2-5}$$

which can easily be solved to obtain

$$f - f_0 = (\text{const.})e^{-t/\tau} \qquad (t > 0). \tag{7.2-6}$$

The distribution function thus approaches the equilibrium distribution function exponentially with time constant τ, and the current in the circuit will be found to decrease from its initial value in essentially the same way, approaching zero exponentially with time constant τ.

This result may seem at first to contradict experimental evidence, but it really does not, since the decay constant τ under ordinary circumstances is very short—typically of the order of 10^{-13} sec. More careful consideration will serve to verify the plausibility of (7.2-6), since the action of the electric field is to give the electrons in the circuit a drift velocity opposite the direction of the current, which, by the law of inertia, *must* persist for a short time after the removal of the force, until the electrons in the wire undergo processes which randomize their velocity, reducing the average drift velocity to zero. It is clear that the collisions exert an influence upon a distribution which is perturbed in one way or another, which tends to *restore* it to the equilibrium state, and that if the perturbation is removed, this restoration is accomplished in a time of the order of the relaxation time τ. The details of the scattering process determine the relaxation time which is in general a function of energy due to the fact that scattering cross-section may depend upon energy.

This restraining effect of collisions can be seen to be essential in determining the law relating current density and electric field in an ordinary conductor. If a system containing perfectly free electrons is acted upon by a constant electric field, in the absence of collisions, the electrons, subject to a constant force, acquire a constant acceleration, thus a linearly *increasing* velocity. With a given field, the current density in a long conductor would then be greater than in a short one, since the average velocity of electrons starting from rest at one end and flowing to the other would be larger over a long path than over a short one. This, of course, is not what happens at all. In reality an electron starting from rest is accelerated by the field and acquires a linearly increasing drift velocity for a short time, but then undergoes a collision which (on the average) reduces its drift velocity to zero, whereupon the process is repeated. The electron, over a period of time, can be assigned a *constant average drift* velocity, proportional to the field and inversely proportional to the relaxation time. This leads at once to Ohm's law, wherein the current density is proportional to electric field strength. A crude estimate of the conductivity can be made on the basis of this picture. The force on an electron is $-eE_0$ where E_0 is the applied field, whereby the acceleration is $-eE_0/m$, with m the electron mass. If the force acts for time t, it produces a drift velocity $-eE_0t/m$, and if the average time between randomizing collisions is τ, the average drift velocity is given by

$$\bar{v} = -\frac{eE_0\tau}{2m},$$
(7.2-7)

assuming, of course, that the drift velocity is on the average reduced to zero by a collision. If the free electron density is n_0, the current density I will be, by definition,

$$I = -n_0 e\bar{v} = \frac{n_0 e^2 \tau}{2m} E_0 = \sigma E_0$$
(7.2-8)

with the conductivity σ expressed by

$$\sigma = \frac{n_0 e^2 \tau}{2m}.$$
(7.2-9)

Equation (7.2-8) is simply Ohm's law; this simple-minded approach, which neglects the niceties of averaging yields a value for σ which is too low by a factor of 2, as we shall soon see.

The justification for regarding the relaxation time τ as the mean free time between randomizing collisions can be obtained by examining a group which consists initially of n_0 particles at time $t = 0$, each of which may be expected to undergo a collision which completely randomizes its velocity with probability dt/τ in any time interval dt. Suppose that at some time t there remain $n(t)$ particles which have not yet experienced collision. The number colliding in a time dt about t is then, on the average, $n(t)dt/\tau$, so that

$$\frac{dn(t)}{dt} = -\frac{n(t)}{\tau},$$ (7.2-10)

which upon integration (demanding that $n = n_0$ at $t = 0$) gives

$$n(t) = n_0 e^{-t/\tau}.$$ (7.2-11)

The number of particles in the original group thus decays exponentially, returning to the equilibrium distribution according to (7.2-11). This is exactly the behavior predicted by the Boltzmann equation in the relaxation time approximation (7.2-6). Furthermore, the distribution of free times, according to (7.2-11) is also exponential, such that the number of free times $dn(t)$ in an interval dt about t is simply

$$dn(t) = \frac{n_0}{\tau} e^{-t/\tau} \, dt.$$ (7.2-12)

The *average* free time may be computed in the usual way, giving

$$\bar{t} = \frac{\displaystyle\int_0^\infty t \, dn(t)}{\displaystyle\int_0^\infty dn(t)} = \frac{\displaystyle\int_0^\infty te^{-t/\tau} \, dt}{\displaystyle\int_0^\infty e^{-t/\tau} \, dt} = \tau.$$ (7.2-13)

We may conclude, then, that the assumption of a collision probability which for a given particle is *constant* with time leads to the same result as the relaxation time approximation in the Boltzmann equation, and that the relaxation time of the Boltzmann equation is identical with the mean free time between collisions. In the above development we have assumed that a single collision is sufficient to return a particle to the equilibrium distribution. If this is not so (for example, if several collisions are required to completely randomize the velocity of a particle) then the relaxation time must be interpreted as referring to the time required, on the average, for a particle to acquire a completely random velocity, even though this may be several times larger than the actual mean time between collisions.

In the above discussion it is assumed that the *initial* group of particles all have the same energy. If the cross-section for collision varies with incident particle energy, as it often does, then the relaxation time will be different for each value of particle energy, and τ will be a function of energy. It is also possible for the relaxation time to be nonisotropic, varying with the direction of the particle momentum vector. We shall, however, limit ourselves to discussing in detail only the isotropic case where τ may be a function of energy but not of direction.

The mean free path between collisions may be defined in the same manner as the relaxation time. If the probability for a single particle to undergo collision in traversing a distance dx is the constant value dx/λ, and if there are $n(x)$ particles of an initial group numbering n_0 remaining after having travelled a distance x, the number colliding in a path interval dx about x will be $n(x)dx/\lambda$, whence

$$\frac{dn(x)}{dx} = -\frac{n(x)}{\lambda}.$$ (7.2-14)

As before, this leads to

$$n(x) = n_0 e^{-x/\lambda}.$$ (7.2-15)

Again, the distribution of path lengths $dn(x)$ in an interval dx about x is simply

$$dn(x) = \frac{n_0}{\lambda} e^{-x/\lambda}\, dx$$ (7.2-16)

and the average free path is given by

$$\bar{x} = \frac{\displaystyle\int_0^\infty x\, dn(x)}{\displaystyle\int_0^\infty dn(x)} = \frac{\displaystyle\int_0^\infty xe^{-x/\lambda}\, dx}{\displaystyle\int_0^\infty e^{-x/\lambda}\, dx} = \lambda.$$ (7.2-17)

The quantity λ, which from the discussion above is seen to be the reciprocal of the probability of collision per unit distance, is called the mean free path.

The mean free path is also defined with reference to a group of particles all of which have the same initial energy, and must in general be regarded as a function of energy. For the case of elastic mutual interactions between hard spheres or elastic collisions between rigid spherical particles and fixed hard sphere scattering centers it is easily seen that the mean free path is simply a geometric property of the system and is therefore *independent* of energy. We shall see that scattering of electrons by acoustical mode phonons in metals and semiconductors can be approximated by this model, and hence may be described in terms of a velocity-independent mean free path. Any individual free path l and the corresponding free time t are, of course, related by

$$l = vt,$$ (7.2-18)

but the relation between the *mean* free path λ and the *mean* free time τ must be obtained by averaging over the distribution of velocities and will depend in detail upon the way in which the mean free path and relaxation time depend upon energy.

7.3 ELECTRICAL CONDUCTIVITY OF A FREE-ELECTRON GAS

The Boltzmann equation will now be used to predict the electrical and thermal conductivity of a free-electron gas. We shall consider first a uniform isotropic substance at constant temperature in a steady-state condition under the influence of a constant applied electric field $\mathbf{E}_0$. Since the system is in a steady state, $df/dt = 0$, and since the material is uniform and at constant temperature, the distribution function must be the same in every part of the sample, whereby $\nabla f = 0$. Under these conditions, the force upon an electron is $-e\mathbf{E}_0$ and the Boltzmann equation in the relaxation time approximation may, using (7.2-3) and (7.2-4), be written as

$$e\mathbf{E}_0 \cdot \nabla_p f = \frac{f - f_0}{\tau}. \tag{7.3-1}$$

If the coordinate system is chosen so that the z-axis is in the direction of $\mathbf{E}_0$, then $\mathbf{E}_0$ has only a z-component, and the above equation may be written

$$\frac{eE_0\tau}{m} \frac{\partial f}{\partial v_z} = f - f_0. \tag{7.3-2}$$

In this equation the difference between f and f_0 is assumed to be quite small, so that $\partial f/\partial v_z$ can be approximated by $\partial f_0/\partial v_z$. If the equilibrium state of the system is a Maxwell-Boltzmann distribution of the form (5.4-15), then
Boltzmann distribution of the form (5.4-15), then

$$\frac{\partial f_0}{\partial v_z} = -\frac{mv_z}{kT} f_0. \tag{7.3-3}$$

Under these conditions, the Boltzmann equation becomes

$$f = f_0\left(1 - \frac{eE_0\tau v_z}{kT}\right) = f_0\left(1 - \frac{eE_0\tau}{kT} v \cos\theta\right), \tag{7.3-4}$$

since $v_z = v \cos\theta$, where θ is the angle between the polar z-axis and the direction $\mathbf{v}$.
The electrical current density I_z is by definition

$$I_z = -ne\bar{v}_z \tag{7.3-5}$$

and $\bar{v}_z$ may be expressed as

$$\bar{v}_z = \frac{\displaystyle\int v_z f(v)g(v)d^3v}{\displaystyle\int f(v)g(v)d^3v}, \tag{7.3-6}$$

where $g(v)$ is the density of states per unit volume of velocity space and d^3v is an appropriate volume element in velocity space. The notation we shall adopt, and use consistently hereafter, is that a barred quantity, $\bar{\alpha}$, refers to an average over the *actual* nonequilibrium distribution, while a bracketed quantity, $\langle \alpha \rangle$, refers to the average of that quantity taken over the *equilibrium* distribution function. The single exception is the thermal velocity, which according to this convention should really be written $\langle v \rangle$, but for which, in deference to convention, we shall retain the notation $\bar{c}$. Since we shall find it most convenient to integrate over spherical coordinates (v, θ, ϕ) in velocity space, we must use the volume element $d^3v = v^2 \sin \theta \, dv d\theta d\phi$. Using this volume element in conjunction with the density of states in velocity space as given by (5.4-17) and the distribution function (7.3-4), noting that $v_z = v \cos \theta$, we may write (7.3-6) as

$$\bar{v}_z = \frac{\int_0^\infty \int_0^\pi f_0(v) \left(1 - \frac{eE_0\tau}{kT} v \cos \theta \right) v^3 \cos \theta \sin \theta \, dv \, d\theta}{\int_0^\infty \int_0^\pi f_0(v) \left(1 - \frac{eE_0\tau}{kT} v \cos \theta \right) v^2 \sin \theta \, dv \, d\theta}. \tag{7.3-7}$$

In (7.3-7) the integration over ϕ from 0 to 2π has already been performed. Since the integrands are both independent of ϕ the factors 2π obtained in numerator and denominator cancel. Also, since $\langle v_z \rangle$ is zero (for the equilibrium condition) the integral involving only f_0 in the numerator of (7.3-7) must be zero; furthermore, in the field dependent term in the denominator an angular integral of $\cos \theta \sin \theta = \frac{1}{2} \sin 2\theta$ over the range $0 < \theta < \pi$ appears, and hence this term also integrates to zero. Equation (7.3-7) can thus be reduced to

$$\bar{v}_z = - \frac{\frac{eE_0}{kT} \int_0^\infty \int_0^\pi v^2 \tau(v) \cdot f_0(v) \, v^2 \cos^2 \theta \sin \theta \, dv \, d\theta}{\int_0^\infty \int_0^\pi f_0(v) \cdot v^2 \sin \theta \, dv \, d\theta}. \tag{7.3-8}$$

The angular integrals can easily be evaluated, giving

$$\bar{v}_z = - \frac{eE_0}{3kT} \frac{\int_0^\infty v^2 \tau(v) \cdot v^2 f_0(v) \, dv}{\int_0^\infty v^2 f_0(v) \, dv} = - \frac{eE_0}{3kT} \langle v^2 \tau \rangle, \tag{7.3-9}$$

the remaining integrals representing simply the average $\langle v^2 \tau(v) \rangle$ over the equilibrium distribution function. But, in a Boltzmann gas of free particles, the average kinetic energy per particle is $3kT/2$, so that

$$\langle \tfrac{1}{2} mv^2 \rangle = \tfrac{1}{2} m \langle v^2 \rangle = \tfrac{3}{2} kT,$$

whence

$$kT = \tfrac{1}{3} m \langle v^2 \rangle. \tag{7.3-10}$$

Substituting this into (7.3-9), we may write, finally

$$\bar{v}_z = -\frac{eE_0\bar{\tau}}{m} \tag{7.3-11}$$

where

$$\bar{\tau} = \frac{\langle v^2\tau \rangle}{\langle v^2 \rangle}. \tag{7.3-12}$$

The result of this more accurate treatment is in agreement with (7.2-7), which was obtained in a very crude fashion, except for a factor of 2, and except for the fact that the quantity $\bar{\tau}$ as given by (7.3-12) and (7.3-9) is not precisely equal to τ except when τ is independent of v.

It is quite common in this connection to define the *mobility* of a particle as the magnitude of the average drift velocity per unit field. From (7.3-11), the mobility μ must be given by

$$\mu = \left| \frac{\bar{v}_z}{E_0} \right| = \frac{e\bar{\tau}}{m}. \tag{7.3-13}$$

The unit of mobility is velocity per unit field, which in the cgs-gaussian system is cm^2 volt^{-1} sec^{-1}. The current density I_z is, by (7.3-5) and (7.3-11)

$$I_z = -ne\bar{v}_z = \frac{ne^2\bar{\tau}}{m}E_0 = \sigma E_0 \tag{7.3-14}$$

whereby

$$\sigma = \frac{ne^2\bar{\tau}}{m}. \tag{7.3-15}$$

It should be noted that these results can be also stated in terms of the mobility as

$$I_z = \sigma E_0 = ne\mu E_0 \tag{7.3-16}$$

whence

$$\sigma = ne\mu. \tag{7.3-17}$$

To evaluate σ explicitly in terms of the relaxation time τ, it is necessary to specify the dependence of τ upon v. If τ is independent of v, then, from (7.3-9) and (7.3-12) $\bar{\tau} = \tau$. For the case where the mean free path is independent of velocity, $\tau = \lambda/v$ and (7.3-12), (7.3-10), and (5.4-20) lead to

$$\bar{\tau} = \frac{\lambda\langle v \rangle}{\langle v^2 \rangle} = \frac{\lambda\bar{c}}{3kT/m} = \frac{8}{3\pi}\frac{\lambda}{\bar{c}}. \tag{7.3-18}$$

In the preceding discussion we assumed, as did Drude and Lorentz, that the electron gas could be described adequately by a Maxwell-Boltzmann distribution of energies. Actually, of course, since the electrons obey the Pauli exclusion principle, we should really use the Fermi-Dirac distribution function if the results we obtain are to be perfectly correct. If the equilibrium distribution is the Fermi-Dirac distribution (5.5-20) we should write, in place of (7.3-3),

$$\frac{\partial f_0}{\partial v_z} = \frac{\partial f_0}{\partial \varepsilon}\frac{\partial \varepsilon}{\partial v_z} = -\frac{mv_z}{kT}\frac{e^{\varepsilon - \varepsilon_f/kT}}{(1 + e^{\varepsilon - \varepsilon_f/kT})^2} = \frac{-mv_z}{kT}f_0(1 - f_0). \qquad (7.3\text{-}19)$$

Substituting this result into (7.3-2), again assuming that $\partial f/\partial v_z$ can be replaced by $\partial f_0/\partial v_z$, one finds

$$f = f_0 - \frac{eE_0\tau}{kT}f_0(1 - f_0)v\cos\theta. \qquad (7.3\text{-}20)$$

Using this as the distribution function in (7.3-6) and noting that the integral involving the first term of (7.3-20) in the numerator and the integral involving the second term of (7.3-20) in the denominator of the resulting expression vanish exactly as before, we are led after integrating over the angular coordinates to

$$\bar{v}_z = -\frac{eE_0}{3kT}\frac{\displaystyle\int_0^\infty v^4\tau(v)f_0(1 - f_0)dv}{\displaystyle\int_0^\infty v^2 f_0 dv}. \qquad (7.3\text{-}21)$$

Although it is impossible to evaluate these integrals over the Fermi-Dirac distribution function analytically in the general case, it is quite easy to obtain an approximate result which is very nearly correct for $T \ll T_F$, which, of course, is usually the case for most metallic conductors under conditions which are encountered in practice.

To accomplish this, it is most convenient to express the integrals in (7.3-21) as integrals over energy rather than velocity, whereby we obtain

$$\bar{v}_z = -\frac{2eE_0}{3mkT}\frac{\displaystyle\int_0^\infty \varepsilon^{3/2}\tau(\varepsilon)f_0(\varepsilon)(1 - f_0(\varepsilon))\,d\varepsilon}{\displaystyle\int_0^\infty \varepsilon^{1/2}f_0(\varepsilon)\,d\varepsilon}. \qquad (7.3\text{-}22)$$

It must now be noted that for $T \ll T_F$ (i.e., $kT \ll \varepsilon_f$) the Fermi function $f_0(\varepsilon)$ is very small for ε significantly in excess of ε_f and the complementary function $1 - f_0$ is very small for values of ε which are substantially smaller than ε_f. The product function $f_0(1 - f_0)$ is therefore very small *except* in the neighborhood of $\varepsilon = \varepsilon_f$, as shown in Figure 7.1. It is easy to show analytically, from (7.3-19), that $f_0(1 - f_0)$ is a maximum at $\varepsilon = \varepsilon_f$, at which point $f_0(1 - f_0) = \frac{1}{4}$, and that the width of the region over which $f_0(1 - f_0)$ is not small compared to this maximum value is of the order of kT. There will thus be no significant contribution to the integral in the numerator of (7.3-22) except where ε takes on values within a few kT units of ε_f. If $\varepsilon^{3/2}\tau(\varepsilon)$ is a function which does not change much over this energy range, we may then regard $f_0(1 - f_0)$

as essentially a Dirac δ-function of $\varepsilon - \varepsilon_f$, and write[5]

$$f_0(\varepsilon)(1 - f_0(\varepsilon)) \cong A\delta(\varepsilon - \varepsilon_f) \qquad (7.3\text{-}23)$$

where A is a normalizing factor which must be chosen in such a way that when (7.3-23) is integrated over energy the correct value for the integral on the left-hand side is obtained.

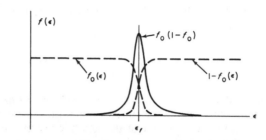

FIGURE 7.1. The functions $f_0(\varepsilon)$, $1 - f_0$ and $f_0(1 - f_0)$ for a Fermi system, plotted as a function of energy. The vertical scale factor for the $f_0(1 - f_0)$ curve is not the same as for the other curves.

Since by direct integration (using the substitution $u = (\varepsilon - \varepsilon_f)/kT$) it is possible to show that

$$\int_0^\infty f_0(1 - f_0) \, d\varepsilon = kT \int_{-\varepsilon_f/kT}^\infty \frac{e^u \, du}{(1 + e^u)^2} = \frac{kT}{1 + e^{-\varepsilon_f/kT}} \cong kT \qquad (7.3\text{-}24)$$

and since the integral of the δ-function is unity, it is clear that we must take $A = kT$ in (7.3-23), whereby

$$f_0(\varepsilon)(1 - f_0(\varepsilon)) \cong kT\delta(\varepsilon - \varepsilon_f) \qquad (kT \ll \varepsilon_f). \qquad (7.3\text{-}25)$$

Using this in (7.3-22), and evaluating the integral in the denominator of that equation by regarding $f_0(\varepsilon)$ to be unity for $0 < \varepsilon < \varepsilon_f$ and zero for $\varepsilon > \varepsilon_f$, which is a good approximation for $kT \ll \varepsilon_f$, we obtain

$$\bar{v}_z = -\frac{2eE_0}{3m} \frac{\displaystyle\int_0^\infty \varepsilon^{3/2}\tau(\varepsilon)\delta(\varepsilon - \varepsilon_f) \, d\varepsilon}{\displaystyle\int_0^{\varepsilon_f} \varepsilon^{1/2} d\varepsilon} = -\frac{eE_0\tau(\varepsilon_f)}{m}. \qquad (7.3\text{-}26)$$

The current density $I_z = -ne\bar{v}_z$ and the conductivity σ can be evaluated as before, in

---

[5] See Appendix A for a definition and discussion of the Dirac δ-function.

connection with (7.3-15), giving

$$\sigma = \frac{ne^2\tau(\varepsilon_f)}{m} = ne\mu \quad \text{with} \quad \mu = \frac{e\tau(\varepsilon_f)}{m}. \tag{7.3-27}$$

It is apparent that these results are the same as those which were found for the Maxwell-Boltzmann case (7.3-11) etc. except that the average mean free time τ is represented by the value of $\tau(v)$ at the "Fermi velocity" v_f, where

$$\tfrac{1}{2}mv_f^2 = \varepsilon_f. \tag{7.3-28}$$

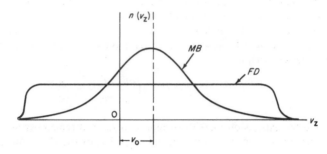

FIGURE 7.2. The Maxwell-Boltzmann (MB) and Fermi-Dirac (FD) z-velocity component distributions in the presence of an electric field.

The statistics in this instance makes a difference only in the details of averaging, and does not affect the conductivity by orders of magnitude. This is understandable on physical grounds, since the effect of the field in either case is simply to superimpose a drift velocity of the order of $eE_0\tau/m$ upon all the electrons of the distribution as shown in Figure 7.2, irrespective of the details of their distribution in energy. In any case the introduction of Fermi-Dirac statistics, while altering some of the details, preserves the essential features of the Drude-Lorentz-Boltzmann free electron picture of electrical conduction, which are basically correct. As we shall see in due time, however, the *thermal capacities* predicted by Maxwell and Fermi-Dirac statistics for the free electrons in metals are very different indeed, and in this case the use of Fermi-Dirac statistics is essential in explaining the experimental findings.

7·4 THERMAL CONDUCTIVITY AND THERMOELECTRIC EFFECTS IN FREE-ELECTRON SYSTEMS

In discussing the thermal conductivity of free electrons, it is necessary to consider a sample wherein there is no electric current, but in which there exists a temperature gradient. The distribution function at any point is then characterized by the local temperature, and must be regarded as a function of position as well as energy. As before, we shall assume that we are dealing with a homogeneous, isotropic sample, and for simplicity we assume that the temperature varies only along the z-direction. We shall work with Fermi-Dirac statistics from the outset; the case of Maxwell-Boltzmann statistics may be obtained as a special case of the Fermi-Dirac result when the Fermi energy is allowed to become large and negative.

Under these conditions in the steady state, using the usual relaxation time approximation with $\tau = \tau(v)$, the Boltzmann equation can be written as

$$\frac{df}{dt} = 0 = -v_z \frac{\partial f_0}{\partial z} + \frac{eE_0}{m} \frac{\partial f_0}{\partial v_z} - \frac{f - f_0}{\tau(v)}. \tag{7.4-1}$$

In this equation the derivatives of the actual distribution function with respect to z and v_z have been approximated by the values of the corresponding equilibrium distribution function, in analogy with the reasoning used in connection with Equation (7.3-2). The field $\mathbf{E}_0$ is assumed to have only a z-component. The above equation may be rewritten as

$$f = f_0 + \tau\left(\frac{eE_0}{m} \frac{\partial f_0}{\partial v_z} - v_z \frac{\partial f_0}{\partial z}\right). \tag{7.4-2}$$

Using the expression (5.5-20) to represent f_0, and remembering that ε_f is a function of temperature, we may with the aid of (7.3-19) write the quantity $\partial f_0/\partial z$ as

$$\frac{\partial f_0}{\partial z} = \frac{\partial f_0}{\partial T} \frac{\partial T}{\partial z} = \frac{\partial}{\partial T}\left(\frac{1}{1 + e^{\frac{\varepsilon - \varepsilon_f}{kT}}}\right)\frac{\partial T}{\partial z} = \frac{1}{kT}\left[\frac{\varepsilon - \varepsilon_f}{T} + \frac{\partial \varepsilon_f}{\partial T}\right]\frac{e^{\frac{\varepsilon - \varepsilon_f}{kT}}}{\left(1 + e^{\frac{\varepsilon - \varepsilon_f}{kT}}\right)^2}\frac{\partial T}{\partial z}$$

$$= \frac{f_0(1 - f_0)}{kT}\left[\frac{\varepsilon - \varepsilon_f}{T} + \frac{\partial \varepsilon_f}{\partial T}\right]\frac{\partial T}{\partial z} = \left[T\frac{\partial}{\partial T}\left(\frac{\varepsilon_f}{T}\right) + \frac{\varepsilon}{T}\right]\frac{f_0(1 - f_0)}{kT}\frac{\partial T}{\partial z}. \tag{7.4-3}$$

The Boltzmann Equation can then be written in the form

$$f = f_0 - \frac{v_z \tau}{kT} f_0(1 - f_0)\left[eE_0 + \left(T\frac{\partial(\varepsilon_f/T)}{\partial T} + \frac{\varepsilon}{T}\right)\frac{\partial T}{\partial z}\right]. \tag{7.4-4}$$

As before, the electrical current density I_z will be

$$I_z = -en\bar{v}_z = -en\frac{\int v_z f(v,z)g(v)\, d^3v}{\int f(v,z)g(v)\, d^3v}, \tag{7.4-5}$$

while the thermal current density Q_z will be represented by the energy carried per particle times the z-component of its velocity times the particle density, thus,

$$Q_z = n\overline{(\varepsilon v_z)} = n\frac{\int \varepsilon v_z f(v,z)g(v)\, d^3v}{\int f(v,z)g(v)\, d^3v}. \tag{7.4-6}$$

Upon substituting the distribution function (7.4-4) into (7.4-5) and (7.4-6), one obtains expressions analogous to (7.3-7). For exactly the same reasons discussed in connection with that equation the integral of the first term in the numerator of these expressions (involving f_0) must vanish, and the integral of the second term in the denominator (involving $f_0(1 - f_0)$) must also vanish. We have then

$$I_z = \frac{en \int \frac{v_z^2 \tau}{kT} f_0(1 - f_0) \left[eE_0 + \left(T \frac{\partial(\varepsilon_f/T)}{\partial T} + \frac{\varepsilon}{T} \right) \frac{\partial T}{\partial z} \right] d^3v}{\int f_0(v) \, d^3v} \tag{7.4-7}$$

and

$$Q_z = \frac{-n \int \frac{v_z^2 \tau}{kT} f_0(1 - f_0) \left[eE_0\varepsilon + \left(\varepsilon T \frac{\partial(\varepsilon_f/T)}{\partial T} + \frac{\varepsilon^2}{T} \right) \frac{\partial T}{\partial z} \right] d^3v}{\int f_0(v) \, d^3v} . \tag{7.4-8}$$

Without going into the details involved in evaluating these integrals, it is clear from the results of the preceding section that these equations may be expressed as

$$I_z = \frac{en}{m} \left[\left(eE_0 + T \frac{\partial(\varepsilon_f/T)}{\partial T} \frac{\partial T}{\partial z} \right) \bar{\tau} + \frac{1}{T} \frac{\partial T}{\partial z} \overline{(\varepsilon\tau)} \right] \tag{7.4-9}$$

and

$$Q_z = \frac{-n}{m} \left[\left(eE_0 + T \frac{\partial(\varepsilon_f/T)}{\partial T} \frac{\partial T}{\partial z} \right) \overline{(\varepsilon\tau)} + \frac{1}{T} \frac{\partial T}{\partial z} \overline{(\varepsilon^2\tau)} \right] \tag{7.4-10}$$

where the quantities over which a bar has been written are averages defined by

$$\bar{\alpha} = \frac{\int \frac{mv_z^2}{kT} \alpha(v) f_0(1 - f_0) \, d^3v}{\int f_0(v) \, d^3v} . \tag{7.4-11}$$

If we note that $mv_z^2 = mv^2 \cos^2 \theta$, and if we assume that α is independent of angular variables, we may integrate over angles in (7.4-11) to obtain

$$\bar{\alpha} = \frac{m}{3kT} \frac{\int v^2 \alpha(v) f_0(1 - f_0) \cdot v^2 \, dv}{\int f_0 \cdot v^2 dv} = \frac{2}{3kT} \frac{\int \varepsilon^{3/2} \alpha(\varepsilon) f_0(1 - f_0) \, d\varepsilon}{\int \varepsilon^{1/2} f_0 \, d\varepsilon} . \tag{7.4-12}$$

It is easy to see from (7.4-12) and (7.3-10) that when the Fermi distribution reduces to a Maxwellian one (whereupon $1 - f_0 \cong 1$ and $f_0 = Ae^{-\varepsilon/kT}$) the average $\bar{\alpha}$ reduces to

$$\bar{\alpha} = \frac{\langle v^2\alpha \rangle}{\langle v^2 \rangle} = \frac{\langle \varepsilon\alpha \rangle}{\langle \varepsilon \rangle} \tag{7.4-13}$$

in analogy to (7.3-12), the averages being taken over the Maxwell-Boltzmann distribution. Likewise, if the distribution is a Fermi distribution for which $kT \ll \varepsilon_f$, we may use (7.3-25) to represent $f_0(1 - f_0)$, whereupon, by the same methods used to arrive at (7.3-26), it is a simple matter to show that

$$\bar{\alpha} = \alpha(\varepsilon_f). \tag{7.4-14}$$

If no electric current flows, then $I_z = 0$, whereby, from (7.4-9) we must have

$$eE_0 + T \frac{\partial(\varepsilon_f/T)}{\partial T} \frac{\partial T}{\partial z} = -\frac{1}{T} \frac{\partial T}{\partial z} \frac{\overline{(\varepsilon\tau)}}{\bar{\tau}}, \tag{7.4-15}$$

whence

$$E_0 = -\frac{1}{e} \left[\frac{1}{T} \frac{\overline{(\varepsilon\tau)}}{\bar{\tau}} + T \frac{\partial(\varepsilon_f/T)}{\partial T} \right] \frac{\partial T}{\partial z} . \tag{7.4-16}$$

If the temperature dependences of the averages $\bar{\tau}$ and $\overline{(\varepsilon\tau)}$ may be neglected (and this is *not* always possible), equation (7.4-16) may be written in the form

$$E_0 = T \frac{\partial}{\partial T} \left[\frac{\overline{(\varepsilon\tau)} - \varepsilon_f \bar{\tau}}{e\bar{\tau}T} \right] \frac{\partial T}{\partial z} . \tag{7.4-17}$$

From this it appears that even in the absence of current, there must be an electric field if a temperature gradient is present. This is indeed the case, and the above equation is the fundamental description of the *thermoelectric effect*. Phenomenologically one may define the *Thomson coefficient* as the coefficient of proportionality relating the electric field and the temperature gradient from which it arises, as follows;

$$E_0 = -\mathscr{T} \frac{\partial T}{\partial z} \tag{7.4-18}$$

whereby the Thomson coefficient $\mathscr{T}$ must be given by

$$\mathscr{T} = -T \frac{\partial}{\partial T} \left[\frac{\overline{(\varepsilon\tau)} - \varepsilon_f \bar{\tau}}{e\bar{\tau}T} \right]. \tag{7.4-19}$$

From (7.4-9) and (7.4-10) it is obvious also that if there is an applied electric field which causes an electric current to flow, then even when $\partial T/\partial z = 0$ the heat current Q_z does not vanish. There is instead a heat current which is proportional to the electric field E_0 and hence to the electric current. This effect is known as *Thomson heating*, and is quite distinct from the ordinary irreversible Joule heating associated with the passage of current through a resistive medium. The voltage developed per unit temperature difference by a thermocouple is expressed as the difference in the *thermoelectric powers* associated with the two constituents of the couple. It can be shown by a simple thermodynamic argument[6] that the absolute thermoelectric power

---

[6] M. W. Zemansky, *Heat and Thermodynamics*, McGraw-Hill Book Co., Inc., New York (1951) p. 301.

$\mathscr{P}$ of a substance is related to the Thomson coefficient by

$$\mathscr{T} = T \frac{\partial \mathscr{P}}{\partial T}, \tag{7.4-20}$$

whence, from (7.4-19), we must have

$$\mathscr{P} = \frac{\varepsilon_f \bar{\tau} - \overline{(\varepsilon\tau)}}{e\bar{\tau}T}.$$

If $I_z = 0$, then by using (7.4-15) the term involving the electric field plus a contribution involving the temperature gradient and the temperature derivative of the Fermi energy can be eliminated from (7.4-10) to give

$$Q_z = -\sigma_t \frac{\partial T}{\partial z} = -\frac{n}{mT} \left[\frac{\overline{\tau(\varepsilon^2\tau)} - \overline{(\varepsilon\tau)}^2}{\bar{\tau}} \right] \frac{\partial T}{\partial z} \tag{7.4-21}$$

whereby the thermal conductivity σ_t is simply

$$\sigma_t = \frac{n}{mT} \left[\frac{\overline{\tau(\varepsilon^2\tau)} - \overline{(\varepsilon\tau)}^2}{\bar{\tau}} \right]. \tag{7.4-22}$$

It is clear from this that the Fermi free electron theory and the Boltzmann free electron picture differ only in details of averaging, and that both approaches lead to a thermal conductivity of the same order of magnitude, just as both predict about the same electrical conductivity.

According to the experimental investigations of Wiedemann and Franz, the ratio $\sigma_t/(T\sigma_e)$, where σ_e is the electrical conductivity, is about the same for nearly all metallic conductors, independent of temperature. Since $\sigma_e = ne^2\bar{\tau}/m$, the Wiedemann-Franz ratio can be expressed as

$$L = \frac{\sigma_t}{T\sigma_e} = \frac{1}{e^2 T^2} \frac{\overline{\tau(\varepsilon^2\tau)} - \overline{(\varepsilon\tau)}^2}{\bar{\tau}^2}. \tag{7.4-23}$$

From this it is apparent that if $\tau(\varepsilon)$ is the same function of energy for all metallic substances, the Wiedemann-Franz ratio will be the same for all materials. Since each factor of ε contributes kT to the averages, the Wiedemann-Franz ratio will be independent of temperature. As a matter of fact, if $\tau(\varepsilon)$ is slowly-varying with respect to ε, the Wiedemann-Franz ratio will be nearly the same even if $\tau(\varepsilon)$ is not the same function of energy for all substances. In particular, if the mean free path λ is independent of energy, then, by (7.3-18), using the classical Boltzmann free-electron picture of Drude and Lorentz, the thermal conductivity is just given by

$$\sigma_t = \frac{nm\lambda}{4T} \frac{\langle v \rangle \langle v^5 \rangle - \langle v^3 \rangle^2}{\langle v \rangle \langle v^2 \rangle} = \frac{4}{3} n\lambda k \sqrt{\frac{2kT}{\pi m}} \tag{7.4-24}$$

while the Wiedemann-Franz ratio becomes

$$L = \frac{m^2}{4e^2T^2} \frac{\langle v \rangle \langle v^5 \rangle - \langle v^3 \rangle^2}{\langle v \rangle^2} = \frac{2k^2}{e^2}, \tag{7.4-25}$$

independent of the mean free path or the temperature. The magnitude of L as given by (7.4-25) is in fair agreement with experimentally determined values for most pure metals.

If we attempt to evaluate the thermal conductivity and the Wiedemann-Franz ratio using Fermi statistics and the average values as calculated according to (7.4-14), we obtain the result $\sigma_t = L = 0$! This rather surprising state of affairs arises simply because the representation of $f_0(1 - f_0)$ as a δ-function and the rather crude treatment of the Fermi integral in the denominator (as in (7.3-26)) of the expressions for the averages, which leads to (7.4-14), is not quite accurate enough for these calculations. If ε_f is expressed as kT_F in (7.4-23), it becomes apparent that what is involved here is really a small difference between two very large quantities of the order of (k^2/e^2) $(T_F/T)^2$, and the two quantities themselves must be expressed quite accurately in order that their difference be even approximately correct. A more precise treatment, such as that given by Smith,[7] leads to the result

$$\sigma_t = \frac{\pi^2 n k^2 T \tau(\varepsilon_f)}{3m} \tag{7.4-26}$$

and

$$L = \frac{\sigma_t}{T\sigma_e} = \frac{\pi^2}{3} \frac{k^2}{e^2}. \tag{7.4-27}$$

7·5   SCATTERING PROCESSES

In all the foregoing discussions the idea of free electrons undergoing instantaneous collisions which tend to return them to the equilibrium distribution has been of central importance, but up to this point we have studiously avoided answering the question, "collisions with what?" It is the purpose of this section to discuss in a qualitative way the answer to this question. As we shall see in detail in the next chapter, free electrons moving independently in a *perfectly periodic* crystal lattice potential are subject to *no scattering interactions* at all with the atoms of the lattice. Furthermore, since in elastic collisions between electrons themselves energy and each component of momentum is conserved, there is no resulting net change either in electrical or heat current. Therefore, the collision mechanisms which do result in randomizing the electron distribution must be associated with impurities, imperfections or aperiodicities of one sort or another in the crystal.

---

[7] R. A. Smith, *Wave Mechanics of Crystalline Solids*, Chapman and Hall, London (1961) p. 328.

Consider a crystal which contains no impurity atoms or structural imperfections whatsoever. If there were no thermal motion of the atoms about their equilibrium positions, the electric potential experienced by an electron within the crystal would be perfectly periodic and there would be no mechanism at all to return electrons acted on by external electric or magnetic forces to the thermal equilibrium state. Due to the thermal vibrations of the atoms, however, there exists at any given time a slight *aperiodicity* of the potential within the crystal which serves to scatter the conduction electrons, dissipating whatever drift velocity they might have acquired from externally applied fields and returning them to the thermal equilibrium state. Obviously the higher the temperature, the stronger the lattice vibrations and the higher the probability of scattering per unit time. This mechanism of scattering leads therefore to a mean free time [and thus, according to (7.3-15) to a conductivity] which *decreases* with rising temperature. This is in accord with experimental observations on pure metals.

Since the lattice vibrations can be thought of as particle-like quanta of vibrational energy (phonons), the scattering interaction between electrons and phonons can be described as a quasi-mechanical collision process involving the free electrons and phonons, the latter behaving like neutral particles whose mass is much greater than the electron mass.[8] Because of the fact that the phonons are neutral, current is no longer conserved in these collisions. Because the phonon "mass" is much greater than the electronic mass, the fractional energy loss suffered by the electron in an electron-phonon interaction is quite small, although, of course, the electron momentum change can be quite large. Electrons can interact with either acoustical-mode or optical-mode phonons, although since much more thermal energy is required to excite optical-mode phonons, the chief interaction at moderate temperatures is likely to be with acoustical phonons. The two interactions are similar in their qualitative aspects, although they lead to certain quantitative differences in the transport coefficients, particularly the temperature dependence of the relaxation time (thus the temperature dependence of the mobility). At any given temperature, the number of available phonons is constant. Since the phonons behave like massive quasi-stationary particles, the probability of scattering per unit distance along the path of an electron is a purely geometrically determined quantity, dependent only on the effective geometrical cross-section associated with the phonon, and *independent* of the velocity of the electron. The mean free path is therefore independent of particle velocity under these circumstances, as noted in Section 7.2. The mean free time is then given by

$$\tau(v) = \frac{\lambda}{v} \qquad (7.5\text{-}1)$$

where λ is independent of v, leading to the relation (7.3-18) between $\bar{\tau}$ and λ. Of course, since the number of available phonons is a function of temperature, λ will be a function of *temperature* despite the fact that it is independent of velocity. This *lattice scattering* or *phonon scattering* mechanism is the dominant scattering process in relatively pure and structurally perfect crystals, especially in the higher temperature ranges.

The presence of an impurity atom in a crystal will usually alter the electrostatic potential in the neighborhood and create an aperiodicity in the potential field within the crystal which can act to scatter conduction electrons. The details of the scattering

---

[8] See, for example, Problem 4 at the end of Chapter 6.

process will naturally depend upon the nature of the impurity atom, its ionic size, its valence, and the way it is bonded into the crystal lattice. The presence of impurity atoms in an otherwise pure crystal will thus decrease the mean free time between scattering events and lead to a decrease in electrical conductivity which will be proportional to impurity content. This effect is indeed observed experimentally in a large number of substances.[9] The impurity scattering mechanism is usually dominant in crystals which are relatively impure, or even in very pure samples at very low temperatures, when the phonon mechanism is quite weak. Structural imperfections in the crystal lattice, such as lattice vacancies, interstitial atoms, dislocations, and grain boundaries also lead to aperiodicities in the crystal potential and hence to scattering centers with which the conduction electrons may interact. The transport properties of the crystal will thus be influenced to a greater or lesser extent by the structural perfection of the lattice. In some circumstances the scattering of electrons by the surfaces of the sample may also be significant. We have already seen (in Section 6.6) how surface scattering of phonons is sometimes an important factor in the lattice component of thermal conductivity at low temperatures.

If two independent scattering mechnisms (for example, lattice scattering and impurity scattering) operate simultaneously to thermalize the electron energy distribution function, then there will be two relaxation times, τ_1 and τ_2, associated with the respective mechanisms. In the above example τ_1 may represent the mean free time between lattice scattering events and τ_2 the mean free time between impurity scatterings. Under these conditions the collision term of the Boltzmann equation can be written

$$\left(\frac{\partial f}{\partial t}\right)_{coll} = -\frac{f-f_0}{\tau_1} - \frac{f-f_0}{\tau_2} = -\frac{f-f_0}{\tau} \tag{7.5-2}$$

where

$$\frac{1}{\tau} = \frac{1}{\tau_1} + \frac{1}{\tau_2}. \tag{7.5-3}$$

From this it is clear that the combined effect of the two scattering mechanisms can be represented by a *single* relaxation time τ related to τ_1 and τ_2 by the reciprocal addition formula (7.5-3). Likewise, if there are more than two mechanisms, their effect can be represented by a single relaxation time τ given by

$$\frac{1}{\tau} = \sum_i \frac{1}{\tau_i}. \tag{7.5-4}$$

Since mobilities, according to (7.3-13) are directly proportional to relaxation times, the mobilities arising from separate independent scattering mechanisms may be added reciprocally in the same way to give a combined mobility of the form

$$\frac{1}{\mu} = \sum_i \frac{1}{\mu_i}. \tag{7.5-5}$$

---

[9] This is the reason why, for example, copper of very high purity is required for electrical conductors in commercial power transmission systems.

Mean free paths arising from separate scattering processes can also be added reciprocally to arrive at an overall mean free path. It should be noted, however, that the addition formulas (7.5-4) and (7.5-5) should be applied *before*, not after, averaging over velocity.

7·6 THE HALL EFFECT AND OTHER GALVANOMAGNETIC EFFECTS

So far, we have considered only cases where an electric field acts upon the particles of a free-electron system. If in addition there is a magnetic field, the force upon a particle of charge q is given by the Lorentz force

$$\mathbf{F} = q(\mathbf{E} + \frac{1}{c}\mathbf{v} \times \mathbf{B}) \tag{7.6-1}$$

and the Boltzmann equation in the steady state, using the relaxation time approximation becomes

$$\mathbf{v} \cdot \nabla f + q(\mathbf{E} + \frac{1}{c}\mathbf{v} \times \mathbf{B}) \cdot \nabla_p f + \frac{f - f_0}{\tau} = 0. \tag{7.6-2}$$

The solution of this equation exhibits a number of *galvanomagnetic effects* arising out of the interaction of electric and magnetic fields and thermal gradients with the particles of the distribution. We shall not go through the details of the solution of the Boltzmann equation at this point, but shall simply restrict ourselves to a phenomenological description of these effects. A more rigorous treatment of galvanomagnetic phenomena as applied to semiconductors, where it is easier to arrive at exact analytic solutions, will be given in a later chapter.

The most important of the galvanomagnetic effects is the *Hall Effect*, in which an electric field in the y-direction is produced as a result of a current flowing in the x-

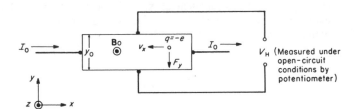

FIGURE 7.3.   The geometry of the Hall effect experiment.

direction and a magnetic field along the z-direction, as shown in Figure 7.3. In that figure an electron moving with velocity v_x is subject to a downward thrust from the Lorentz force. In the steady state there can be no net force upon the electron, so what must happen is that an excess concentration of electrons builds up at the lower edge of the sample until an electrostatic field E_y is generated which exactly balances the Lorentz force. Then

$$F_y = \frac{q}{c} (\mathbf{v} \times \mathbf{B_0})_y - eE_y = \frac{ev_x B_0}{c} - eE_y = 0, \qquad (7.6\text{-}3)$$

whence

$$E_y = \frac{v_x B_0}{c}. \qquad (7.6\text{-}4)$$

But the current density I_0 is

$$I_0 = -nev_x. \qquad (7.6\text{-}5)$$

Expressing v_x in (7.6-4) in terms of I_0 by (7.6-5) we may write the Hall field E_y as

$$E_y = \frac{-I_0 B_0}{nec} = R I_0 B_0 \qquad (7.6\text{-}6)$$

where the *Hall coefficient* R is simply given by

$$R = -\frac{1}{nec}. \qquad (7.6\text{-}7)$$

The Hall field is directly proportional to both I_0 and B_0; the Hall coefficient R is merely the proportionality constant between the Hall field and the product $I_0 B_0$. The Hall field can be obtained by measuring the Hall voltage V_H of Figure 7.3 potentiometrically. The Hall field and Hall voltage are then related by

$$V_H = E_y y_0 \qquad (7.6\text{-}8)$$

where y_0 is the width of the sample. The Hall coefficient R can be determined if V_H, I_0, and B_0 are known.

The Hall coefficient is inversely proportional to the density n of charge carriers in the sample, and a measurement of R affords a simple way of determining n. By combining measurements of the Hall coefficient and conductivity the mobility μ may be obtained, since from (7.6-7) and (7.3-17)

$$R\sigma = -\frac{1}{nec} (ne\mu) = -\frac{\mu}{c}$$

or

$$\mu = -R\sigma c. \qquad (7.6\text{-}9)$$

The above analysis of the Hall effect is oversimplified in that all charge carriers do not have the same velocity component v_x, and hence one should average over the distribution of v_x. The effect of this is generally to change the value (7.6-7) for R by a small numerical factor.

According to (7.6-7), if the charge carriers are electrons R should always be negative. Experimentally it has been found that R is *positive* for many metallic and semiconducting substances. The occurrence of a positive Hall coefficient can be explained on the assumption that the charge carriers are positive rather than negative, but the free-electron theory of metals provides no reasonable way of accounting for the presence of positive charge carriers. The occurrence of positive Hall coefficients and the existence of positive charge carriers can be explained by the quantum theory, as we shall see in the next chapter.

In the experimental arrangement of Figure 7.3, a temperature gradient is set up along the y-direction along with the Hall field E_y. The reason for this is that since the downward force on the electrons is proportional to v_x, according to the Lorentz force law, the faster more energetic electrons experience a greater $\mathbf{v} \times \mathbf{B}_0$ force and hence tend to accumulate in the lower part of the sample, raising its temperature with respect to the top part where the "cooler" electrons remain. The resulting temperature gradient gives rise to a thermoelectric field in the y-direction, as given by (7.4-17) which results in an additional voltage, distinct from the Hall voltage, across the y-terminals of the sample. This effect is called the *Ettingshausen effect*, and in some instances it can interfere seriously with measurements of the Hall voltage. Its effect on Hall measurements can be eliminated by using an alternating sample current whose period is short compared to the thermal time constant of the sample, in conjunction with an ac detector.

When the electric current I_0 in the Hall effect experiment is replaced by a thermal current, one may still measure an electric field along the y-direction. In this case there is on the average, a net transport of faster electrons along the x-direction from the hot end to the cold end of the sample, and the $\mathbf{v} \times \mathbf{B}_0$ force acts upon them to produce an electric field in much the same way as the Hall field is produced when an electric current flows. This phenomenon is known as the *Nernst effect*. The Nernst field is accompanied by a temperature gradient along the y-direction just as is the Hall field; this causes an additional thermoelectric voltage analogous to the Ettingshausen voltage. The thermoelectric effect arising from the Nernst effect is called the *Righi-Leduc effect*.

Due to the fact that in the presence of a magnetic field the electrons travel in curved rather than straight paths, the conductivity is in general found to depend upon the magnetic field. This phenomenon is called *magnetoresistance* and will be dealt with in more detail in a later chapter.

7.7 THE THERMAL CAPACITY OF FREE-ELECTRON SYSTEMS

The classical Drude-Lorentz free-electron theory, in which the free electrons are assumed to obey Maxwell-Boltzmann statistics leads one to the conclusion that (assuming one free electron per atom) the heat capacity of the free electrons should be, as given by (5.4-11), $\frac{3}{2}nk$. The total heat capacity would then be the sum of the lattice contribution, given by the Debye theory, and the electronic contribution $\frac{3}{2}nk$, and at temperatures large compared to the Debye temperature the limiting value of the heat capacity would be $\frac{3}{2}nk$. This conclusion is contradicted by experiment, however, the observations showing that the asymptotic value of the heat capacity for high temperatures differs little from the value $3nk$ predicted by the Debye theory for the lattice contribution alone. It would appear from this that the heat capacity of the free electrons is much smaller than the value predicted by the Drude-Lorentz-Boltzmann model. This was one of the major shortcomings of the original Drude-Lorentz model as first proposed.

It turns out, as we shall see, that when Fermi-Dirac statistics are used to describe the free-electron energy distribution, the calculated electronic heat capacity is much smaller than the value $\frac{3}{2}nk$ given by Boltzmann statistics, and is in fact negligible compared to the lattice contribution at moderate and high temperatures. The Fermi free-electron model is thus in much better accord with the experimental observations than the original Drude-Lorentz theory. The explanation of the electronic specific heat of metallic substances was one of the original triumphs of quantum statistics.

The reason for the much smaller heat capacity of the Fermi-Dirac free-electron system is that only the electrons within a few kT of the surface of the Fermi sphere can receive energy from an external heat source. Ordinarily a particle of an external heat bath at temperature T has an energy of only a few times kT to give to an electron of the Fermi free-electron system. Electrons deep inside the Fermi sphere are incapable of interacting with such external exciting bodies, because there are *no unoccupied states within a few times kT in energy into which they can be excited*. Only electrons near the Fermi surface where there are unoccupied states available can participate in interactions with an external heat source.

In order to make an accurate calculation of the heat capacity, it is necessary to take into account the first-order variation of the Fermi energy with temperature. In order to do this we shall have to evaluate the integral

$$I = \int_0^\infty f_0(\varepsilon) \frac{\partial \phi(\varepsilon)}{\partial \varepsilon} \, d\varepsilon. \qquad (7.7\text{-}1)$$

where f_0 is the Fermi function and $\phi(\varepsilon)$ is a function of ε which has the property that

$$\phi(0) = 0. \qquad (7.7\text{-}2)$$

If (7.7-1) is integrated by parts, it is easy to see that I can be expressed in the form

$$I = \left[f_0(\varepsilon)\phi(\varepsilon) \right]_0^\infty - \int_0^\infty \phi(\varepsilon) \frac{\partial f_0}{\partial \varepsilon} \, d\varepsilon = -\int_0^\infty \phi(\varepsilon) \frac{\partial f_0}{\partial \varepsilon} \, d\varepsilon, \qquad (7.7\text{-}3)$$

since the product $f_0(\varepsilon)\phi(\varepsilon)$ vanishes at both limits. Expanding $\phi(\varepsilon)$ as a Taylor's series about the point $\varepsilon = \varepsilon_f$, we may write

$$\phi(\varepsilon) = \phi(\varepsilon_f) + (\varepsilon - \varepsilon_f) \left(\frac{\partial \phi}{\partial \varepsilon} \right)_{\varepsilon_f} + \frac{1}{2}(\varepsilon - \varepsilon_f)^2 \left(\frac{\partial^2 \phi}{\partial \varepsilon^2} \right)_{\varepsilon_f} + \cdots, \qquad (7.7\text{-}4)$$

whereby (7.7-3) can be expressed as

$$I = a_0 \phi(\varepsilon_f) + a_1 \left(\frac{\partial \phi}{\partial \varepsilon} \right)_{\varepsilon_f} + a_2 \left(\frac{\partial^2 \phi}{\partial \varepsilon^2} \right)_{\varepsilon_f} + \cdots, \qquad (7.7\text{-}5)$$

where

$$a_n = -\frac{1}{n!} \int_0^\infty (\varepsilon - \varepsilon_f)^n \frac{\partial f_0}{\partial \varepsilon} d\varepsilon. \tag{7.7-6}$$

Now, if $kT \ll \varepsilon_f$, according to (7.3-19), $\partial f_0/\partial \varepsilon$ is negligible for negative values of ε, and the limits of the integral (7.7-6) may be extended to ∞ and $-\infty$. If this is done, then we shall find

$$a_0 = -\int_{-\infty}^\infty \frac{\partial f_0}{\partial \varepsilon} d\varepsilon = \left[f_0 \right]_{-\infty}^\infty = 1 \tag{7.7-7}$$

$$a_1 = -\int_{-\infty}^\infty (\varepsilon - \varepsilon_f) \frac{\partial f_0}{\partial \varepsilon} d\varepsilon = 0 \tag{7.7-8}$$

$$a_2 = -\frac{1}{2} \int_{-\infty}^\infty (\varepsilon - \varepsilon_f)^2 \frac{\partial f_0}{\partial \varepsilon} d\varepsilon = \frac{(kT)^2}{2} \int_{-\infty}^\infty \frac{x^2 e^x \, dx}{(1 + e^x)^2} = \frac{\pi^2}{6} (kT)^2. \tag{7.7-9}$$

The integral in (7.7-8) vanishes because $\partial f_0/\partial \varepsilon$ is an even function of $\varepsilon - \varepsilon_f$ and the integrand is thus an odd function of that argument, which when integrated from $-\infty$ to ∞ must yield zero. The integral of (7.7-9) is put into the form shown above by the substitution $x = (\varepsilon - \varepsilon_f)/kT$; this form is listed in most standard tables of definite integrals. Substituting these values into (7.7-5), the result is

$$I = \int_p^\infty f_0(\varepsilon) \frac{\partial \phi}{\partial \varepsilon} d\varepsilon = \phi(\varepsilon_f) + \frac{\pi^2}{6} (kT)^2 \left(\frac{\partial^2 \phi}{\partial \varepsilon^2} \right)_{\varepsilon_f} + \cdots. \tag{7.7-10}$$

This formula is convenient for working out approximate values of Fermi integrals where the δ-function approximation for $\partial f_0/\partial \varepsilon$ is not good enough. It is restricted by the condition (7.7-2) and by the condition that $kT \ll \varepsilon_f$. This approach can be used to evaluate the Wiedemann-Franz ratio for a Fermi gas; the details are left as an exercise.

Suppose now that we choose

$$\phi(\varepsilon) = \int_0^\varepsilon g(\varepsilon) \, d\varepsilon \tag{7.7-11}$$

whence

$$\partial \phi/\partial \varepsilon = g(\varepsilon) \qquad \partial^2 \phi/\partial \varepsilon^2 = \partial g(\varepsilon)/\partial \varepsilon, \tag{7.7-12}$$

with $g(\varepsilon)$ as given by (5.2-22) for a free-electron system. Then, from (7.7-10),

$$n = \int_0^\infty f_0(\varepsilon) g(\varepsilon) \, d\varepsilon = \int_0^{\varepsilon_f(T)} g(\varepsilon) \, d\varepsilon + \frac{\pi^2}{6} (kT)^2 \left(\frac{\partial g}{\partial \varepsilon} \right)_{\varepsilon_f}. \tag{7.7-13}$$

But also, from (5.5-24),

$$n = \int_0^{\varepsilon_f(0)} g(\varepsilon) \, d\varepsilon. \tag{7.7-14}$$

Subtracting (7.7-14) from (7.7-13) we have

$$0 = \int_{\varepsilon_f(0)}^{\varepsilon_f(T)} g(\varepsilon)\, d\varepsilon + \frac{\pi^2}{6}(kT)^2\left(\frac{\partial g}{\partial \varepsilon}\right)_{\varepsilon_f} \cong g(\varepsilon_f)\left[\varepsilon_f(T) - \varepsilon_f(0)\right] + \frac{\pi^2}{6}(kT)^2\left(\frac{\partial g}{\partial \varepsilon}\right)_{\varepsilon_f}.$$

$$(7.7\text{-}15)$$

In the above equation we have assumed that $g(\varepsilon)$ does not vary much in the interval from $\varepsilon_f(0)$ to $\varepsilon_f(T)$, which for $kT \ll \varepsilon_f$ will be only a small fraction of $\varepsilon_f(0)$. Since $g(\varepsilon)$ is of the form $c\varepsilon^{1/2}$ with c a constant, (7.7-15) can be written

$$\varepsilon_f(T) = \varepsilon_f(0) - \frac{\pi^2}{12}\frac{(kT)^2}{\varepsilon_f(T)}.$$

$$(7.7\text{-}16)$$

For $kT \ll \varepsilon_f$, the second term will be a small correction to be subtracted from the relatively large quantity $\varepsilon_f(0)$ and the difference between $\varepsilon_f(T)$ and $\varepsilon_f(0)$ will be small compared to $\varepsilon_f(0)$. Not much error will be made under these circumstances if $\varepsilon_f(T)$ in the correction term of (7.7-16) is replaced with $\varepsilon_f(0)$, giving finally

$$\varepsilon_f(T) = \varepsilon_f(0)\left[1 - \frac{\pi^2}{12}\left(\frac{kT}{\varepsilon_f(0)}\right)^2\right].$$

$$(7.7\text{-}17)$$

To evaluate the electronic specific heat, we shall choose

$$\phi(\varepsilon) = \int_0^\varepsilon \varepsilon\, g(\varepsilon)\, d\varepsilon$$

$$(7.7\text{-}18)$$

whence $\partial\phi/\partial\varepsilon = \varepsilon\, g(\varepsilon)$ and $\partial^2\phi/\partial\varepsilon^2 = \partial(\varepsilon\, g(\varepsilon))/\partial\varepsilon.$ $(7.7\text{-}19)$

Then, substituting in (7.7-10), we find

$$\int_0^\infty f_0(\varepsilon)\frac{\partial\phi}{\partial\varepsilon}\, d\varepsilon = \int_0^\infty \varepsilon\, g(\varepsilon) f_0(\varepsilon)\, d\varepsilon = U$$

$$(7.7\text{-}20)$$

and

$$U = \int_0^{\varepsilon_f(T)} \varepsilon\, g(\varepsilon)\, d\varepsilon + \frac{\pi^2}{6}(kT)^2\left[\frac{\partial}{\partial\varepsilon}(\varepsilon\, g(\varepsilon))\right]_{\varepsilon_f}.$$

Again using the fact that $g(\varepsilon) = c\varepsilon^{1/2}$, this can be written as

$$U = \int_0^{\varepsilon_f(0)} \varepsilon\, g(\varepsilon)\, d\varepsilon + \int_{\varepsilon_f(0)}^{\varepsilon_f(T)} \varepsilon\, g(\varepsilon)\, d\varepsilon + \frac{\pi^2}{6}(kT)^2 \cdot \frac{3}{2}g(\varepsilon_f)$$

$$\cong U_0 + \left[\varepsilon_f(T) - \varepsilon_f(0)\right]\varepsilon_f(0)g(\varepsilon_f(0)) + \frac{\pi^2}{4}(kT)^2 g(\varepsilon_f(0)). \quad (7.7\text{-}21)$$

Here U_0 is the absolute zero value of the internal energy as represented by the integral of $\varepsilon\, g(\varepsilon)$ from 0 to $\varepsilon_f(0)$, and the integral of this quantity over the small range $\varepsilon_f(0)$

to $\varepsilon_f(T)$ has been approximated by the same procedure which was used in connection with (7.7-15). In the last term of (7.7-21) $g(\varepsilon)$ is evaluated at $\varepsilon_f(0)$ rather than $\varepsilon_f(T)$, which introduces only a small error. Substituting the value given by (7.7-17) for $\varepsilon_f(T) - \varepsilon_f(0)$ into (7.7-21), one may write

$$U = U_0 + \frac{\pi^2}{6}(kT)^2 g(\varepsilon_f(0)). \tag{7.7-22}$$

From (5.2-22) and (5.5-25), $g(\varepsilon_f(0))$ may be expressed as

$$g(\varepsilon_f(0)) = \frac{8\sqrt{2\pi}}{h^3} m^{3/2}[\varepsilon_f(0)]^{1/2} = \frac{4\pi m}{h^2}\left(\frac{3n}{\pi}\right)^{1/3},$$

which, since from (5.5-25) the ratio m/h^2 equals $(3n/\pi)^{2/3}/(8\varepsilon_f(0))$, can be written

$$g(\varepsilon_f(0)) = \frac{3n}{2\varepsilon_f(0)} = \frac{3n}{2kT_F}, \tag{7.7-23}$$

where T_F is a "Fermi temperature" defined by the relation $\varepsilon_f(0) = kT_F$. Substituting this value into (7.7-22) gives finally

$$U = U_0 + \frac{\pi^2}{4}\frac{nkT^2}{T_F}. \tag{7.7-24}$$

The heat capacity is obtained in the usual way, the result being

$$C_v = \frac{\partial U}{\partial T} = \frac{\pi^2}{2} nk\left(\frac{T}{T_F}\right). \tag{7.7-25}$$

This expression differs from the classical result $\frac{3}{2}nk$ by a factor $(\pi^2/3)(T/T_F)$, which amounts to 0.03 for $T = 300°K$ and $T_F = 30\,000°K$ [a typical value for simple metals calculated from (5.5-25)]. The electronic component of specific heat as calculated using Fermi statistics thus amounts to only a small fraction of the Drude-Lorentz value $\frac{3}{2}nk$, and except at very low temperatures is small compared to the lattice contribution. At temperatures small compared to the Debye temperature, the lattice contribution becomes quite small, approaching zero like T^3 as $T \to 0$. In this range the electronic contribution is often a significant factor, and in certain temperature ranges may even be the dominant effect. It is easily recognized in the experimental data due to its *linear* temperature dependence, as exhibited in (7.7-25).

EXERCISES

1. Show that if τ is independent of v, the distribution function

$$f(v) = Ae^{\frac{-m(\mathbf{v}-\mathbf{v}_0)^2}{2kT}} \qquad \left(\text{where } \mathbf{v}_0 = \frac{e\tau}{m}\mathbf{E}_0\right)$$

is an approximate solution of the Boltzmann equation in the case where the Maxwell-Boltzmann distribution represents the equilibrium distribution function and where $v_0 \ll (kT/m)^{1/2}$.

2. Assuming Maxwell-Boltzmann statistics, evaluate $\bar{\tau}$ from (7.3-12), assuming that the dependence of τ on energy is given by $\tau(\varepsilon) = A\varepsilon^{-s}$, where A and s are constants.

3. In a metal such as copper (resistivity 1.7×10^{-6} ohm-cm), what (approximately) is the maximum electric field for which the Drude or Fermi free-electron treatment of Section 7.3 might be expected to hold? What is the current density for this value of field?

4. Starting with (7.7-10) and (7.7-17) derive formulas (7.4-26) and (7.4-27) for the thermal conductivity and Wiedemann-Franz ratio of a Fermi free-electron gas, using the approximation $kT \ll \varepsilon_f(0)$.

5. Adopting a Fermi free-electron picture of a metal surface as shown in the accompanying diagram, show that there is a thermonic-emission current which may flow out across the surface potential barrier, and that the thermonic current density is given by

$$I = \frac{4\pi em(kT)^2}{h^3} e^{-\phi_0/kT} i_x.$$

Assume ϕ_0, the work function, to be $\gg kT$.

6. Show that if the mean free path is independent of the velocity, the electrical conductivity of a Maxwell-Boltzmann free-electron gas may be expressed as

$$\sigma = \tfrac{4}{3}ne^2\lambda/\sqrt{2\pi mkT}.$$

7. Describe physically the origin of the thermoelectric effect. (Hint: Consider a long bar, half of which is maintained at temperature T_1, the other half at temperature T_2 ($>T_1$) and calculate the electron fluxes in either direction at the interface between the two regions. Use Maxwell-Boltzmann statistics.)

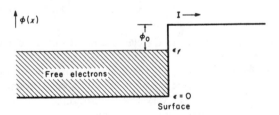

8. Assuming that the mean free path for electron-phonon interaction is independent of velocity, and that $T \gg \Theta$, show that the electrical conductivity of a Fermi-Dirac free electron gas for which $T \ll T_F$ is proportional to T^{-1}. Show that the electrical conductivity of a Maxwell-Boltzmann free electron gas under similar circumstances is proportional to $T^{-3/2}$. Note that the former result is in agreement with experimental data for pure metals, wherein Fermi statistics are applicable, and the latter agrees with experimental results in slightly impure semiconductors wherein the number of free electrons is essentially independent of temperature, and where Maxwell-Boltzmann statistics may be used. *Hint:* What is the total number of phonons, and how may the mean free path be expected to vary with this number, ignoring the dependence of scattering cross-section with temperature?

9. Show that for $kT \ll \varepsilon_f(0)$ (i.e., $T \ll T_F$), the heat capacity of a two-dimensional Fermi free electron gas can be expressed approximately as $C_v = (\pi^2/3)nk(T/T_F)$.

GENERAL REFERENCES

F. J. Blatt, *Theory of Mobility of Electrons in Solids*, Solid State Physics, Advances in Research and Applications, Vol. 4, Academic Press, Inc., New York (1957), pp. 199–366.

J. R. Drabble and H. J. Goldsmid, *Thermal Conduction in Semiconductors*, Pergamon Press, Oxford (1961).

J.-P. Jan, *Galvanomagnetic and Thermomagnetic Effects in Metals*, Solid State Physics, Advances in Research and Applications, Vol. 5, Academic Press, Inc., New York (1957), pp. 1–96.

E. H. Putley, *The Hall Effect and Related Phenomena*, Butterworth & Co., Ltd., London (1960).

F. Seitz, *Modern Theory of Solids*, McGraw-Hill Book Co., Inc., New York (1940), Chapter 4.

R. A. Smith, *Wave Mechanics of Crystalline Solids*, Chapman & Hall, Ltd., London (1961), Chapter 10.

J. Tauc, *Photo- and Thermoelectric Effects in Semiconductors*, Pergamon Press, Oxford (1962).

CHAPTER 8

QUANTUM THEORY OF ELECTRONS IN PERIODIC LATTICES

8·1 INTRODUCTION

The free-electron theory of metals developed in the previous chapter is based upon the notion that the conduction electrons within a metallic substance act like the classical free particles of a gas, subject only to the limitations of the Fermi-Dirac statistics. It is not immediately obvious, from the quantum mechanical point of view, why this should be the case. It is also not clear why some substances have large numbers of free electrons and are thus very good conductors, while others have hardly any and behave as insulators.

The simplest quantum mechanical view of an electron in a crystal is that of a single electron in a perfectly periodic potential which has the periodicity of the lattice. In this *one-electron* model of a solid, the periodic potential may be thought of as arising from the periodic charge distribution associated with the ion cores situated on the lattice sites, *plus* the (constant) average "smeared out" potential contribution due to all the other free electrons belonging to the crystal, so that the interaction of the single electron with all the others is accounted for in an average sense. The solution of Schrödinger's equation for the single electron in this potential then provides a set of "one-electron" states which the single electron may occupy, and in fact, which may be occupied (subject to the limitations of the Pauli principle) by all the electrons of the crystal, since the single electron which is considered initially may be regarded as typical of all electrons of the system. A one-dimensional representation of a periodic crystal potential, such as might be obtained by following a path along one of the $\langle 100 \rangle$ directions of a cubic crystal of lattice constant a is shown in Figure 8.1. So far as the one-electron quantum-mechanical picture is concerned, the crystal periodicity will usually be assumed to extend to infinity in all directions, but, of course, at a surface of any actual crystal the periodicity will be interrupted, the potential function then behaving somewhat as shown at the left-hand edge of Figure 8.1. The lattice spacing will not be *quite* uniform near such a surface, but as a practical matter the lattice periodicity will usually be found to be almost perfect after a few atomic spacings within the crystal.

We shall find that the one-electron wave functions (called Bloch functions) calculated by the above prescription always have certain properties closely related to the lattice periodicity, and that the allowed electronic energies occur in bands of permitted states separated by forbidden energy regions. Within the allowed energy bands, the dynamical behavior of electrons will be found to be in many ways similar to that of free particles, and the question of whether a crystal is a conductor or an insulator will depend upon whether the electronic states within a given allowed band or set of bands is completely filled or partially empty. In addition, we shall see that the collective behavior of electrons in an *almost* filled band of allowed states is very

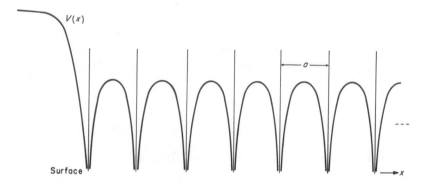

FIGURE 8.1. Schematic representation of the potential within a perfectly periodic crystal lattice. The surface potential barrier is shown at the left.

much like that of a few *positive* charge carriers in an almost empty band! The one-electron picture will thus serve to justify the free-electron model from a fundamental point of view, and also to answer certain rather puzzling questions which the free-electron theory, of itself, could not even attempt to explain. It should be remembered, however, that the one-electron treatment is an approximation in which the *details* of of electron-electron interactions are completely overlooked.

8·2   THE BLOCH THEOREM

The Bloch theorem[1] is a mathematical statement regarding the form of the one-electron wave functions for a perfectly periodic potential. Consider a differential equation of the form

$$\frac{d^2\psi}{dx^2} + f(x)\psi(x) = 0 \tag{8.2-1}$$

where $f(x)$ is a given function which is *periodic* with period a so that

$$f(x + a) = f(x). \tag{8.2-2}$$

---

[1] F. Bloch, *Z. Physik* **52**, 555 (1928); the same result had been known to mathematicians previously as *Floquet's Theorem*. See, for example, E. I. Whittaker and G. N. Watson, *Modern Analysis*, Cambridge University Press (1948), p. 412.

If the potential function $V(x)$ for the one-dimensional Schrödinger equation, of the form (4.8-5), is periodic, then Schrödinger's equation will be a special case of equation (8.2-1). Since (8.2-1) is a linear second-order differential equation, there are (for any given value of the energy ε in 4.8-5) two independent solutions $g(x)$ and $h(x)$ such that

$$\psi(x) = Ag(x) + Bh(x) \tag{8.2-3}$$

represents the most general solution of (8.2-1). Since $f(x + a) = f(x)$, not only $g(x)$ and $h(x)$ but *also* $g(x + a)$ and $h(x + a)$ satisfy (8.2-1). But *any* solution of (8.2-1) must be expressable as a linear combination of $g(x)$ and $h(x)$ of the form (8.2-3). In particular, then,

$$g(x + a) = \alpha_1 g(x) + \alpha_2 h(x)$$

$$h(x + a) = \beta_1 g(x) + \beta_2 h(x) \tag{8.2-4}$$

where α_1, α_2, β_1, and β_2 are constants. Then

$$\psi(x + a) = Ag(x + a) + Bh(x + a)$$

$$= (\alpha_1 A + \beta_1 B)g(x) + (\alpha_2 A + \beta_2 B)h(x). \tag{8.2-5}$$

Now $\psi(x + a)$ can always be expressed in the form

$$\psi(x + a) = \lambda\psi(x), \tag{8.2-6}$$

where λ is a constant, provided that the proper value for λ is chosen. Comparing (8.2-5) and (8.2-3) we see that if (8.2-6) is to be satisfied, then

$$(\alpha_1 - \lambda)A + \beta_1 B = 0$$

and

$$\alpha_2 A + (\beta_2 - \lambda)B = 0. \tag{8.2-7}$$

This system of homogeneous equations in A and B has solutions other than $A = B = 0$ only if

$$\begin{vmatrix} \alpha_1 - \lambda & \beta_1 \\ \alpha_2 & \beta_2 - \lambda \end{vmatrix} = \lambda^2 - (\alpha_1 + \beta_2)\lambda + (\alpha_1\beta_2 - \alpha_2\beta_1) = 0. \tag{8.2-8}$$

The solution of this quadratic equation in λ serves to determine the two possible values of λ for which (8.2-6) is true. Designating them λ_1 and λ_2, then,

$$\psi(x + a) = \lambda_1\psi(x)$$

$$\psi(x + a) = \lambda_2\psi(x). \tag{8.2-9}$$

If we now define k_1 and k_2 such that

$$\lambda_1 = e^{ik_1 a}$$

$$\lambda_2 = e^{ik_2 a}, \tag{8.2-10}$$

and define $u_{k_1}(x)$ and $u_{k_2}(x)$ as

$$u_{k_1}(x) = e^{-ik_1 x}\psi(x)$$

$$u_{k_2}(x) = e^{-ik_2 x}\psi(x),$$

(8.2-11)

then, using (8.2-11), (8.2-9), and (8.2-10) it is clear that

$$u_{k_1}(x + a) = e^{-ik_1(x+a)}\psi(x + a) = e^{-ik_1(x+a)}\lambda_1\psi(x)$$

$$= e^{-ik_1(x+a)}e^{ik_1 a}\psi(x) = e^{-ik_1 x}\psi(x) = u_{k_1}(x).$$

(8.2-12)

The function $u_{k_1}(x)$ is thus *periodic* with period a; the same is found to be true for $u_{k_2}(x)$ in a similar way. According to (8.2-11), then, $\psi(x)$ can then always be written in the form

$$\psi_k(x) = e^{ikx}u_k(x)$$

(8.2-13)

where $u_k(x)$ is a *periodic* function of period a, and where k represents either k_1 or k_2 as determined above. This is Bloch's theorem; all one-electron wave functions for periodic potentials can be written in this way. In three dimensions Bloch's theorem becomes

$$\psi_{\mathbf{k}}(\mathbf{r}) = e^{i\mathbf{k}\cdot\mathbf{r}}u_{\mathbf{k}}(\mathbf{r}).$$

(8.2-14)

The wave functions (8.2-13) and (8.2-14) clearly have the form of plane waves with propagation vector $\mathbf{k}$ modulated by a function whose periodicity is that of the crystal lattice.

The Bloch function (8.2-13) can be thought of as the most general way of writing a solution of Schrödinger's equation which leads to the same probability density $\psi^*\psi$ in each unit cell of the crystal. From (8.2-13) it is apparent that in the nth unit cell, due to the periodicity of u_k,

$$\psi(x + na) = e^{ik(x+na)}u_k(x + na) = e^{ik(x+na)}u_k(x)$$

$$= e^{ikna}\psi(x).$$

(8.2-15)

Likewise, $$\psi^*(x + na) = e^{-ikna}\psi^*(x)$$

(8.2-16)

and $$\psi^*(x + na)\psi(x + na) = \psi^*(x)\psi(x).$$

(8.2-17)

The 3-dimensional wave function (8.2-14) exhibits the same sort of behavior.

In the one-dimensional case, for an *infinite* crystal, one can find a formal solution to Schrödinger's equation for every value of the energy ε within certain ranges, and corresponding to each such value of ε, by (8.2-8) and (8.2-10), are two values of k. The allowed energy eigenvalues for such a system within these ranges are continuous. If the crystal is of finite extent, however, in which case appropriate physical boundary conditions must be satisfied at the surfaces, then solutions which satisfy Schrödinger's equation *and* the boundary conditions can be found only for certain discrete energy eigenvalues. Since for each energy eigenvalue there are only two related values of k, the allowed values of k also form a discrete set. The situation here is very similar to that which was encountered in connection with mechanical vibrations of a periodic lattice and which was discussed in Section 3.3. Of course, when the number of atoms in the crystal becomes very large, the allowed values of ε and k, although still discrete, crowd very closely together and in most cases can be regarded as quasi-continuous bands of allowed values. The same qualitative behavior is found for three-dimensional crystals, except that in this case there are more than two allowed values of **k** for each energy eigenvalue.

For example, if periodic boundary conditions are imposed in a one-dimensional crystal of N atoms (or if the lattice is assumed to have the form of a closed ring of N atoms), then if the wave function ψ is to be single-valued, we must have (using (8.2-15))

$$\psi(x + Na) = e^{ikNa}\psi(x) = \psi(x) \qquad (8.2\text{-}18)$$

whereby

$$e^{ikNa} = 1$$

or

$$e^{ika} = (1)^{1/N} = e^{\frac{2\pi i n}{N}} \qquad (n = 0,1,2, \cdots N - 1), \qquad (8.2\text{-}19)$$

recalling the form of the Nth roots of unity. Taking the logarithm of both sides of this equation and solving for k, we find that the possible values for k under these boundary conditions are

$$k_n = \frac{2\pi n}{Na} \qquad (n = 0,1,2, \cdots N). \qquad (8.2\text{-}20)$$

If N is large, there will be very many allowed values of k as given by (8.2-20), and in this case they may be regarded as forming a quasi-continuous range of values. Of course the N distinct values of k allowed by (8.2-20) do not correspond to the same value of energy. For each constant value of ε, there are only the *two* allowed values of k given by (8.2-10).

8·3 THE KRONIG-PENNEY MODEL OF AN INFINITE ONE-DIMENSIONAL CRYSTAL

For the case of the infinite periodic one-dimensional square-well potential shown in Figure 8.2, it is possible to arrive at an exact solution of the Schrödinger equation. This solution was first investigated by Kronig and Penney,[2] and although it is related to a somewhat idealized periodic potential which is only a crude approximation to that found in the actual crystal, it is nevertheless very useful because it serves to illustrate in a most explicit way many of the important characteristic features of the quan-

---

[2] R. de L. Kronig and W. G. Penney, *Proc. Roy. Soc.* **A130**, 499 (1931).

tum behavior of electrons in periodic lattices. The wave functions associated with this model may be calculated in the one-electron approximation by solving Schrödinger's equation

$$\frac{d^2\psi}{dx^2} + \frac{2m}{\hbar^2}(\varepsilon - V(x))\,\psi(x) = 0 \tag{8.3-1}$$

for a single electron in the periodic potential $V(x)$ as given by Figure 8.2. Since, from

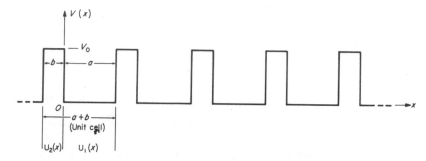

FIGURE 8.2.   Ideal periodic square well potential used by Kronig and Penney to illustrate the general characteristics of the quantum behavior of electrons in periodic lattices.

the previous section, the wave functions must have the Bloch form, we may expect that

$$\psi(x) = e^{ikx}u(x). \tag{8.3-2}$$

Substituting (8.3-2) into (8.3-1), it is found that the function $u(x)$ must satisfy

$$\frac{d^2u}{dx^2} + 2ik\frac{du}{dx} - \left(k^2 - \alpha^2 + \frac{2mV(x)}{\hbar^2}\right)u(x) = 0 \tag{8.3-3}$$

where

$$\alpha = (2m\varepsilon/\hbar^2)^{1/2}. \tag{8.3-4}$$

For the potential of Figure 8.2, then, we shall find that

$$\frac{d^2u_1}{dx^2} + 2ik\frac{du_1}{dx} - (k^2 - \alpha^2)u_1(x) = 0 \qquad (0 < x < a) \tag{8.3-5}$$

and

$$\frac{d^2u_2}{dx^2} + 2ik\frac{du_2}{dx} - (k^2 - \beta^2)u_2(x) = 0 \qquad (-b < x < 0), \tag{8.3-6}$$

where $u_1(x)$ represents the value of $u(x)$ in the interval $(0 < x < a)$ and $u_2(x)$ represents the value of $u(x)$ in $(-b < x < 0)$, and where

$$\beta = (2m(\varepsilon - V_0)/\hbar^2)^{1/2}. \tag{8.3-7}$$

The differential equations (8.3-5) and (8.3-6) are easily solved by standard procedures to give

$$u_1(x) = Ae^{i(\alpha-k)x} + Be^{-i(\alpha+k)x} \qquad (0 < x < a) \qquad (8.3\text{-}8)$$

$$u_2(x) = Ce^{i(\beta-k)x} + De^{-i(\beta+k)x} \qquad (-b < x < 0), \qquad (8.3\text{-}9)$$

where A, B, C, and D are arbitrary constants. Note that the quantity β is a purely imaginary one for $0 < \varepsilon < V_0$.

The requirement of continuity for the wave function ψ and its derivative at $x = a$ and $x = -b$ demands that the functions $u(x)$ satisfy these same conditions, since in (8.3-2) e^{ikx} is a well-behaved function. Applying these boundary conditions (and recalling that since $u(x)$ has the periodicity of the lattice, $u_1(a) = u_2(-b)$), we find that

$$A + B = C + D$$

$$i(\alpha - k)A - i(\alpha + k)B = i(\beta - k)C - i(\beta + k)D$$

$$Ae^{i(\alpha-k)a} + Be^{-i(\alpha+k)a} = Ce^{-i(\beta-k)b} + De^{i(\beta+k)b} \qquad (8.3\text{-}10)$$

$$i(\alpha - k)Ae^{i(\alpha-k)a} - i(\alpha + k)Be^{-i(\alpha+k)a} = i(\beta - k)Ce^{-i(\beta-k)b} - i(\beta + k)De^{i(\beta+k)b}.$$

The coefficients A, B, C, and D can thus be determined as the solution of a set of four simultaneous linear homogeneous equations in those quantities. There is no solution other than $A = B = C = D = 0$, of course, unless the determinant of the coefficients vanishes. This requires that

$$\begin{vmatrix} 1 & 1 & 1 & 1 \\ \alpha - k & -(\alpha + k) & \beta - k & -(\beta + k) \\ e^{i(\alpha-k)a} & e^{-i(\alpha+k)a} & e^{-i(\beta-k)b} & e^{i(\beta+k)b} \\ (\alpha - k)e^{i(\alpha-k)a} & -(\alpha + k)e^{-i(\alpha+k)a} & (\beta - k)e^{-i(\beta-k)b} & -(\beta + k)e^{i(\beta+k)b} \end{vmatrix} = 0.$$

$$(8.3\text{-}11)$$

Expanding the determinant, one can show after quite a lot of tedious but straightforward algebra that (8.3-11) can be expressed as

$$-\frac{\alpha^2 + \beta^2}{2\alpha\beta} \sin \alpha a \sin \beta b + \cos \alpha a \cos \beta b = \cos k(a + b). \qquad (8.3\text{-}12)$$

Since in the range $(0 < \varepsilon < V_0)$, β as defined by (8.3-7) is imaginary, for these values of energy it is most convenient to express (8.3-12) in a slightly different form. Letting

$$\beta = i\gamma \qquad (8.3\text{-}13)$$

in this region, and noting that $\cos ix = \cosh x$ and $\sin ix = i \sinh x$, Equation (8.3-12) can be written as

$$\frac{\gamma^2 - \alpha^2}{2\alpha\gamma} \sinh \gamma b \sin \alpha a + \cosh \gamma b \cos \alpha a = \cos k(a + b), \qquad (8.3\text{-}14)$$

where γ is a real positive quantity in the interval $(0 < \varepsilon < V_0)$, just as β is in the interval $(V_0 < \varepsilon < \infty)$. We may thus use (8.3-12) most conveniently when $(V_0 < \varepsilon < \infty)$ and (8.3-14) when $(0 < \varepsilon < V_0)$.

The wave functions (8.3-2) must, like all wave functions, be well-behaved functions as x approaches $\pm\infty$. Since $u(x)$ is a periodic function whose values are the same in each unit cell, no difficulties arise on its account, provided that the factor e^{ikx} in (8.3-2) is well behaved under these conditions. But e^{ikx} is well behaved at *both* $+\infty$ and $-\infty$ only if k is real, whereby e^{ikx} is oscillatory. If k were imaginary, e^{ikx} would diverge to infinity either at $+\infty$ or $-\infty$, and the resulting expression for $\psi(x)$ would not behave properly as a wave function. We must therefore accept only functions of the form (8.3-2) with *real* values for k. The above expressions (8.3-12) and (8.3-14) have, on the left-hand side, a function of the form $K_1 \sin \alpha a + K_2 \cos \alpha a$ which must equal $\cos k(a + b)$. If for a given value of energy the function on the left-hand side of these equations should have a value in the range between $+1$ and -1, then the required value for $\cos k(a + b)$ is obtained with a *real* value for the argument $k(a + b)$. On the other hand, if the value of the function on the left-hand side of (8.3-12) or (8.3-14) should lie outside that range, it would mean that $\cos k(a + b)$ would have to be greater than $+1$ or less than -1, which would, in turn, require that the argument $k(a + b)$ be a *complex* number with an imaginary part other than zero. Under these circumstances the solutions (8.3-2) would not behave properly at infinity and would not satisfy the physical requirement for wave functions of the system. The energies associated with such values of k would simply be forbidden to the electron.

It is possible to write the left-hand sides of (8.3-12) and (8.3-14) in the form $K_3 \cos(\alpha a - \delta)$, where $K_3 = (K_1^2 + K_2^2)^{1/2}$ and $\tan \delta = K_1/K_2$. In this way (8.3-12) and (8.3-14) can be written as

$$\left[1 + \frac{(\alpha^2 - \beta^2)^2}{4\alpha^2\beta^2} \sin^2 \beta b\right]^{1/2} \cos(\alpha a - \delta) = \cos k(a + b), \tag{8.3-15}$$

where
$$\tan \delta = -\frac{\alpha^2 + \beta^2}{2\alpha\beta} \tan \beta b \qquad (V_0 < \varepsilon < \infty)$$

and
$$\left[1 + \frac{(\alpha^2 + \gamma^2)^2}{4\alpha^2\gamma^2} \sinh^2 \gamma b\right]^{1/2} \cos(\alpha a - \delta') = \cos k(a + b), \tag{8.3-16}$$

where
$$\tan \delta' = \frac{\alpha^2 + \gamma^2}{2\alpha\gamma} \tanh \gamma b \qquad (0 < \varepsilon < V_0).$$

From these expressions it is clear that in both cases the left-hand side has the form of a cosine function times a modulating factor whose amplitude is *invariably* greater than unity. The value of this modulating factor is actually a maximum for $\alpha = 0$ (hence for $\varepsilon = 0$) and approaches unity in the limit of large energies, where $\alpha \cong \beta$.

When the left-hand sides of (8.3-15) and (8.3-16) are plotted as a function of energy, in which connection one should note from (8.3-4) and (8.3-7) that

$$\alpha^2 - \beta^2 = \alpha^2 + \gamma^2 = 2mV_0/\hbar^2 = \text{const.}, \tag{8.3-17}$$

the results are as illustrated in Figure 8.3. In this figure the left-hand side of (8.3-15)

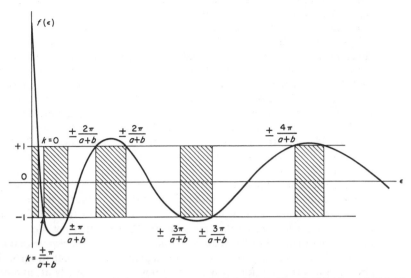

FIGURE 8.3. A plot of the functions on the left-hand side of (8.3-15) and (8.3-16) *versus* energy. The shaded regions show forbidden energy bands where the value of k is complex, the unshaded regions allowed energy bands corresponding to real values of k.

or (8.3-16) is plotted as a function of energy. When the ordinate of the curve lies between $+1$ and -1, there exists a real value for k corresponding to physically possible wave functions. Outside these limits, however, k must be complex with a nonzero imaginary part. Such values for k can never lead to physically well-behaved wave functions; the corresponding ranges of energy are forbidden and are shown in Figure 8.3 as shaded regions. There are thus formed alternate regions of allowed energy eigenvalues and forbidden regions. These regions are usually referred to as allowed and forbidden *energy bands*, and the grouping of the permitted energy values into these bands is one of the most important and characteristic features of the behavior of electrons in periodic lattices. It can be shown that energy bands having the same qualitative aspects as those shown in Figure 8.3 are formed no matter what the detailed form of the potential is, so long as it is periodic.

Using (8.3-15), (8.3-16), (8.3-4), and (8.3-7) it is possible to plot a curve showing the energy ε as a function of k. The result is illustrated schematically in Figure 8.4. The forbidden energy bands and the relationship of ε *vs* k within the various allowed bands have been assigned in accord with the scheme shown in Figure 8.3. Since $\cos^{-1} k(a + b)$ is not a single valued function this assignment is of necessity somewhat arbitrary. For large energies, it is apparent that the function $\varepsilon(k)$ approaches the free-electron relation $\varepsilon = \hbar^2 k^2/2m$ (shown as a dotted curve in Figure 8.4) quite closely within the allowed bands. Also, for large energies it is found that the allowed bands become very broad and the forbidden regions quite narrow. Nevertheless, the curve $\varepsilon(k)$ always has *zero* slope at the edges of the allowed bands, that is, at $k = \pm n\pi/(a + b)$, where n is an integer. This result can be proved by direct differentiation of (8.3-15) and (8.3-16), but we shall not go through the details of the cal-

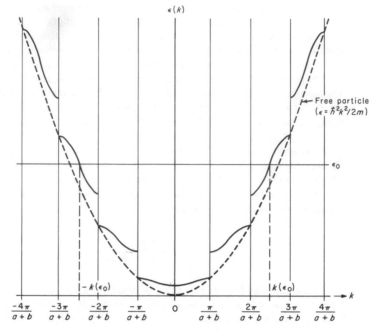

FIGURE 8.4. The energy ε plotted as a function of k according to (8.3-15) and (8.3-16).

culation. This feature of the $\varepsilon(k)$ curve is quite general, and can be shown to occur even in the three-dimensional case independent of the precise mathematical form of the potential function.

8·4   CRYSTAL MOMENTUM AND EFFECTIVE MASS

If the function $u_k(x)$ in (8.2-13) is a constant, then the wave function has the form $e^{\pm ikx}$, corresponding to a perfectly free electron of momentum $p = \pm \hbar k$ whose energy, according to the results of Section 4.9, would be

$$\varepsilon = \frac{\hbar^2 k^2}{2m}. \qquad (8.4\text{-}1)$$

This relation is shown as the dashed curve in Figure 8.4. For large values of ε the actual ε *vs* k relation conforms quite closely to this relation. It is clear from (8.2-13) and from the results of the previous section that k is a constant of the motion, that $\hbar k$ will have the dimensions of a momentum, and that as the energy of the electron increases (the particle becoming thereby more nearly "free") the values of k in general approximate those of the free-particle momentum divided by $\hbar$. It can easily be seen that these conclusions must hold no matter what particular form the periodic potential takes. It is therefore customary to refer to $\hbar k$ as the *crystal momentum*, and we shall soon see that in many cases the dynamical behavior of the electron in the crystal lattice with respect to crystal momentum is very similar to that of a free particle with respect to the real momentum. To make the distinction between the actual momentum and the crystal momentum perfectly clear, we must note that due to the presence of the lattice potential, the true instantaneous momentum of an electron is not a constant of the motion at all and is not directly calculable by the methods of quantum mechanics, except as an average value, while, for a state of given energy, the crystal momentum $\hbar k$ is a perfectly well-defined constant value, just as is the true momentum of a free particle of given energy.

Let us now consider the motion of an electron in the crystal under the influence of an applied electric field. For convenience in visualizing the motion of an electron in such a situation we shall have to localize the wave function by superposing solutions having different values of k, as in Section (4.9). If this is done, then the group velocity associated with the electron "wave packet" is

$$v_g = \frac{d\omega}{dk} = \frac{1}{\hbar}\frac{d\varepsilon}{dk}, \qquad (8.4\text{-}2)$$

where, of course, ε and ω are connected by the Planck relation $\varepsilon = \hbar\omega$. Suppose that the electron is acted upon by an external electric field E, acquiring an increase in velocity dv_g over a distance dx in a time dt. Then, using (8.4-2) we see that

$$d\varepsilon = \frac{d\varepsilon}{dk}dk = -eE\,dx = -eEv_g\,dt = -\frac{eE}{\hbar}\frac{d\varepsilon}{dk}dt, \qquad (8.4\text{-}3)$$

whereby

$$dk = -\frac{eE}{\hbar}dt, \qquad (8.4\text{-}4)$$

or

$$\hbar\frac{dk}{dt} = \frac{dp}{dt} = -eE = F, \qquad (8.4\text{-}5)$$

where we now use the symbol p to denote the *crystal* momentum. Equation (8.4-5) tells us simply that the time rate of change of crystal momentum equals the force $-eE$. It is thus the analogue of Newton's law, showing that the crystal momentum of the electron in a periodic lattice changes under the influence of an applied field in the same way as does the true momentum of a free electron *in vacuo*.

If one differentiates (8.4-2) with respect to time, the result is

$$\frac{dv_g}{dt} = \frac{1}{\hbar}\frac{d}{dt}\left(\frac{d\varepsilon}{dk}\right) = \frac{1}{\hbar}\frac{d^2\varepsilon}{dk^2}\frac{dk}{dt}, \qquad (8.4\text{-}6)$$

which, using (8.4-5), can be written as

$$\frac{dv_g}{dt} = \frac{-eE}{\hbar^2}\frac{d^2\varepsilon}{dk^2} = \frac{F}{m^*} \qquad (8.4\text{-}7)$$

where the *effective mass* m^* is given by

$$m^* = \frac{\hbar^2}{(d^2\varepsilon/dk^2)}.$$ (8.4-8)

Equation (8.4-7) is essentially Newton's force equation. The proportionality factor relating the force eE and the acceleration dv_g/dt may be regarded as the *effective mass* of the electron. The actual gravitational mass of the electron, however, has nothing directly to do with the effective mass; it is the quantity $d^2\varepsilon/dk^2$ which is the important factor. If the electron is really a free electron, then ε and k are related by (8.4-1) and (8.4-8) reduces to $m^* = m$. If the energy is a *parabolic* function of k, having the form

$$\varepsilon = C(k - k_0)^2,$$ (8.4-9)

then the effective mass has the *constant* value $m^* = \hbar^2/2C$ and the dynamical behavior of the electron will be the same as that of a free particle with this effective mass. If the energy is not a parabolic function of k, then the effective mass will not be constant with energy, the dynamical behavior will be that of a particle of variable mass, and the situation will be much more involved. In any case, to a first approximation, the entire effect of the periodic crystal potential is to replace the free electron mass with an effective mass.

Fortunately, the relation connecting ε and k is almost always parabolic or nearly parabolic over the range of energies which is accessible to an electron in the crystal. From Figure 8.4 it is apparent that the ε *vs* k curve is always parabolic in form at the bottom and top of the allowed energy bands. Even when the electrons of the system are in other regions, however, what is important, usually, is that the effective mass be essentially constant over an energy interval of the order of the average amount of energy which a particle may gain or lose in a single scattering event, or between scattering events in a time of the order of the relaxation time. This amount of energy is of the order of kT. But at 300°K, kT amounts to only 0.025 eV, and the total width in energy of a band which might play an important role in transport processes in typical crystalline substances is much greater than this—ordinarily of the order of 1 eV. In single scattering events, or in the interval between successive scattering events, then, an electron will usually be confined to a short segment of the ε *vs* k curve, which can ordinarily be regarded as approximately parabolic.

It is clear from this that the free-electron picture of a metal is to a very great extent justified, the only important correction being that the free electron mass should be in all cases replaced with an appropriate effective mass related to the form of the ε *versus* k curve by (8.4-8). We may thus conclude that all the results derived in Chapter 7 on the basis of the free-electron theory are correct, provided that the electron mass is replaced with the proper effective mass. The effective mass can be deduced by comparison of experimental electronic specific heat data with the free-electron result given as Equation (7.7-25); it may also be calculated quantum mechanically starting with self-consistent atomic wave functions for the atoms of the crystal, which are then combined to give a crystal potential appropriate for the substance in question. The values so calculated are usually in reasonable agreement with the experimentally determined ones. For most metals the effective mass ranges between one-half and twice the free-electron mass, although for many of the transition metals it is much higher, and for certain semiconducting compounds of the III-V type it may be much lower.

8·5 REDUCED ZONE REPRESENTATION; ELECTRONS AND HOLES

The assignment of ranges of values for k within each allowed energy band, leading to the curve of Figure 8.4, is shown in Figure 8.3. Since the only requirement which must be fulfilled within each band is that $-1 \leqslant \cos k(a + b) \leqslant +1$, it is clear that this assignment of values for k for each band is not a unique one. The advantage of the assignment shown in Figure 8.3 is that it leads to the so-called *expanded representation* of Figure 8.4 in which the relationship with the free particle ε *versus* k curve is clearly apparent. We could, however, translate any of the curve segments shown in Figure 8.4 to the right or left parallel to the k-axis at a distance $2n\pi/(a + b)$, where n is an integer, and still satisfy the relations (8.3-15) and (8.3-16), since any such transformation leaves the value of $\cos k(a + b)$ unchanged.

It is sometimes very useful to make a particular transformation of this type, by translating the various segments of the ε *versus* k curve to the right or to the left, parallel to the k-axis, through distances which are integral multiples of $2\pi/(a + b)$,

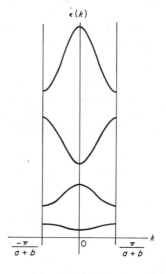

FIGURE 8.5. Schematic representation of the ε *versus* k plot of Figure 8.4 transformed to the reduced-zone representation.

so that they all fit within the interval $-\pi/(a + b) < k < \pi/(a + b)$. The resulting representation of the energy versus propagation constant relationship is called the *reduced zone representation*. The reduced zone representation of the curve of Figure 8.4 is shown in Figure 8.5. It will in either case be noted that the Bloch wave functions (8.2-13) satisfy the Bragg reflection condition (2.1-1) at the points $k = n\pi/(a + b)$, as discussed in connection with mechanical vibrations of the lattice in Section 3.3. At these points the group velocity $\hbar^{-1}\, d\varepsilon/dk$ is zero, corresponding to a standing wave which represents an electron at rest. The electron may be regarded as undergoing an internal diffraction by the lattice potential under these circumstances. We shall have more to say about this subject in a later section.

In the infinitely large crystal of Section 8.3, the permitted energies form a continuum of values within the allowed bands. If the crystal is of finite extent and contains a total of N atoms, however, we have already seen from (8.2-20) that there will be instead just N distinct allowed eigenstates of the crystal momentum k within each allowed energy band. From Figure 8.4 it is easy to see that for a given energy ε_0 the two allowed crystal momentum values are $k(\varepsilon_0)$ and $-k(\varepsilon_0)$. Physically, the former represents a state in which an electron is moving to the right with positive crystal momentum $k(\varepsilon_0)$, while the latter represents a state in which an electron moves to the left with equal and opposite crystal momentum; the energy is clearly the same for either case. The effect of electron spin here, as in isolated atomic systems, is simply to *double* the degeneracy factor associated with all the energy levels of the system, since each quantum state must be regarded as having split into two, one accommodating an electron with "spin up" and the other an electron with "spin down." When the effect of electron spin is included, then, we must conclude that within each allowed energy band of this system there are just $2N$ quantum states.

At a temperature of absolute zero the electrons of the system will occupy these states, one per state, as required by the Pauli exclusion principle, from the lowest state up to a given energy determined by the number of available states, their distribution in energy and the number of electrons in the crystal. This energy is, of course, the Fermi energy of the crystal at zero temperature. In a one-dimensional crystal some of the allowed bands will be entirely filled, some will be entirely empty, and one may be partially filled.[3] If an electric field is applied, it is obvious that no current can be contributed from the unfilled bands; it is equally true, however, that no current can arise from the filled bands. This can be understood by noting that the current density arising from a given band will be

$$I = -n_0 e \bar{v} \qquad (8.5\text{-}1)$$

where $\bar{v}$ is the average velocity, and n_0 is the number of electrons per unit volume belonging to that band. But $\bar{v}$ can be expressed as

$$\bar{v} = \frac{1}{n_0 V} \sum_i v_i, \qquad (8.5\text{-}2)$$

where the summation is taken over all the velocities associated with individual electrons within the volume V of material. Using (8.5-2), (8.5-1) can then be written as

$$I = -\frac{e}{V} \sum_i v_i. \qquad (8.5\text{-}3)$$

---

[3] The situation in two- and three-dimensional systems is more complex. In those instances different allowed bands may *overlap* in energy, and more than one band may be partially filled. These systems will be discussed in detail later.

This, however, must yield zero when summed over a full band, because due to the symmetry of the curves of Figure 8.4 or Figure 8.5 about the $\varepsilon(k)$ axis, for every state of positive velocity $\hbar^{-1}\,\partial\varepsilon/\partial k$ corresponding to a point of positive slope, there is a state corresponding to a negative velocity of equal magnitude (with negative slope) at $k' = -k$. We must conclude that only bands which are *partially* filled can contribute to a current flow. This can be understood on physical grounds by observing that in a partially filled band there are always electrons which can be excited gradually to unoccupied states of higher energy and momentum, whereas in a filled band, on account of the Pauli exclusion principle, this gradual field excitation can never occur, since all the states are already occupied.

A single electron at rest in an otherwise unoccupied band will occupy a state of lowest energy at the bottom of that band in the absence of thermal excitation, as shown in Figure 8.6. If an electric field E_0 is now impressed upon the crystal. the

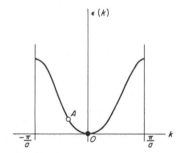

FIGURE 8.6. An electron near the bottom of an allowed band, subject to forces which tend to accelerate it along the $-x$ direction.

electron will experience a force $-eE_0$ and will gradually be excited through states of ever-increasing energy and (negative) momentum, acquiring a higher and higher momentum in the $-x$ direction, according to (8.4-5). According to the predictions of the theory of Sections 8.2–8.4, this process would continue to excite the electron through successively higher energies until it reached the top of the band at $x = \pi/a$, where a is the lattice spacing. Since the points $x = \pi/a$ and $x = -\pi/a$ are equivalent, the electron may then be regarded as "reappearing" at $x = \pi/a$, decreasing in energy along the right-hand portion of the curve as k goes from π/a to zero, arriving finally at the starting point, whereby the whole process is repeated. The motion of the electron under these circumstances would be oscillatory (Zener's oscillation). It is important to note that there is no mechanism inherent in the theory of Sections 8.2–8.4 which serves to limit this motion; it will continue indefinitely according to the theory outlined there. If the potential were perfectly parabolic instead of having the form shown in Figure 8.6, the electron would acquire unlimited velocity and energy. We must conclude in either case that the motion of a free electron in a *perfectly periodic* potential is *unimpeded* by the lattice, in the sense that there is no scattering process associated with the presence of the lattice itself which stops or randomizes the velocity of an electron.

If there is a *deviation* from the perfect periodicity of the lattice, however, this result no longer holds true. In this case there will always be some probability of a sudden random transition to a state associated with another value of k within the band. On the *average* these random transitions bring the electron back to $k = 0$, and we conclude that there is a *scattering process* associated with the departure of the lattice potential from perfect periodicity. The deviation from periodicity may be caused by the fact that the lattice may be at a temperature higher than absolute zero, in which case the thermal vibrations which are set up introduce a slight aperiodicity, the instantaneous positions of the atoms no longer coinciding exactly with the periodic lattice of their equilibrium positions. This situation may be discussed in terms of a picture in which the electrons are scattered by phonons. Deviations from periodicity may also be caused by the presence of impurity atoms as well as by lattice vacancies, interstitial atoms, dislocations and other structural defects of the crystal lattice. These effects result in the phenomena of lattice scattering, impurity scattering, and defect scattering which were discussed in the preceding chapter. Discussion of the quantitative aspects of these processes will be postponed to a later chapter.

These scattering processes prevent the Zener oscillation behavior from ever occurring in practice.[4] What actually happens in all experimentally realizable situations is that an electron starting from O in Figure 8.6 is accelerated as described by the theory of Sections 8.2–8.4 until it arrives at some point A a small distance along the parabolic lower portion of the curve. Its behavior in this range is that of a free electron of mass m^* as given by (8.4-8). At point A it is scattered by one of the mechanisms described previously, arriving *on the average* back at O, whereupon the cycle is repeated. The behavior is then just that of a free particle subject to the scattering processes discussed above and in the last chapter, leading to Ohm's law, the thermal conductivity, the thermoelectric effect, and all the other results of the free-electron theory of metals, the only difference being that the electron mass is replaced by the effective mass throughout. The distance OA along the ε *versus* k curve in Figure 8.6 is exaggerated for the sake of clarity, and in most cases arising in practice would be very much smaller.

For a band containing a relatively small number of electrons, the current obtained when a small voltage is impressed is most simply given by (8.5-3), summed over all the electrons in the band. If the band is nearly full, however, and there are only a few empty states (which in the equilibrium state will be concentrated near the top of the band[5]) the current equation (8.5-3) is best expressed in another way, by writing (8.5-3) in the form

$$I = -\frac{e}{V}\sum_i v_i = -\frac{e}{V}\left[\sum_j v_j - \sum_k v_k\right] = +\frac{e}{V}\sum_k v_k. \qquad (8.5\text{-}4)$$

Here the sum over i represents the sum over all velocity states occupied by electrons, the sum over j represents the sum over all velocity states in the band and the sum over k represents the sum over all *unoccupied* velocity states. As we have seen before in connection with (8.5-3), the sum over j, taken over all states in the band, must vanish. The remaining sum over the unoccupied states corresponds to a current which could be produced by a corresponding number of *positive* charge carriers! It is possible (and advantageous, as we shall see) to express the current from an almost full band as a

---

[4] A quantitative discussion of the Zener oscillation is given by E. Spenke, *Electronic Semiconductors*, McGraw-Hill Book Co., Inc., New York (1958), pp. 229, 257.

[5] In a nearly empty band, the electrons occupy the lowest available energy states in the absence of thermal excitation. This is also true in a nearly filled band, the result being that the unoccupied states are at the *top* of the band. In an almost completely filled band, then, the empty states "gravitate" upward on a diagram such as that of Figure 8.5, in which the ordinate represents the energy of an electron.

current derived from the motion of a comparatively small number of empty electronic states or *holes*, which behave like positive particles, rather than a very large number of electrons. The velocity associated with a hole is that which an electron would have if it were to occupy the empty energy state, which is ordinarily near the top of the energy band. But since the ε *versus* k relation there is concave downward, $d^2\varepsilon/dk^2$ is negative, giving a negative electron effective mass from (8.4-8). A particle with negative effective mass experiences an acceleration in a direction opposite to that of the applied force. A negative particle with negative effective mass would thus be accelerated in the same direction as the applied field, and would thus exhibit the same dynamical behavior as a *positive particle of positive mass*. We may thus regard the situation in a nearly filled band as one involving a relatively small number of positive particles of positive mass, which we shall refer to as holes, whose velocities and momenta are those corresponding to the unoccupied electronic states in the band. We shall see in a later chapter that in certain materials the physical nature and dynamical behavior of holes is very easily visualized in terms of defects in the electronic valence bonds which connect nearest neighbor atoms and provide the cohesive forces which hold the crystal together. We have not discussed in detail the behavior of electrons and holes under the influence of magnetic force, but it can be shown[6] that they move as would negative or positive particles of effective mass m^* as given by (8.4-8) under the influence of the usual magnetic force $q(\mathbf{v} \times \mathbf{B})/c$.

8·6   THE FREE-ELECTRON APPROXIMATION

In the foregoing sections we have seen how a particular periodic potential model (shown in Figure 8.2) led to the formation of allowed energy bands and an ε *versus* k relation of the form shown in Figure 8.4. The form of this relationship was such that the quantity $\hbar k$ could be advantageously regarded as a "crystal momentum," in terms of which the dynamics of electrons in applied force fields could be described quite simply. The resulting picture of electronic behavior was quite similar to the free-electron picture of Chapter 7, except that an effective mass had to be substituted for the free-electron mass, and that the charge carriers in a nearly filled band had to be viewed as positive holes. We have emphasized that these qualitative conclusions are independent of the precise form of the potential function, as long as it is periodic; the potential function of Figure 8.2 was adopted only for illustrative purposes, since it allowed an exact solution of Schrödinger's equation.

In actual crystals the potential function which is used must be somehow related to the actual potential experienced by an electron due to the ion cores and all the other electrons of the crystal. An exact solution of this problem, even in the one-electron approximation, is impossible; it is therefore customary to approach the problem from the viewpoint of either the *free-electron approximation* or the *tight-binding approximation*, whichever seems most appropriate to the particular situation at hand.

In the free-electron approximation, the total energy of the electron is assumed

---

[6] R. A. Smith, *Wave Mechanics of Crystalline Solids*, John Wiley and Sons, New York (1961), p. 458 ff.

to be always large compared to the periodic potential energy. Under these conditions the allowed bands will be broad and the forbidden energy regions quite narrow. These circumstances are never perfectly realized in any actual crystal, because the potential always goes to $-\infty$ at the ion cores, but for the outermost electrons in many simple metals, including the alkali metals, the requirements are met fairly well over most of the volume of the crystal.

Adopting for simplicity a one-dimensional crystal potential model, and writing the periodic potential $V(x)$ in the form

$$-\frac{2mV(x)}{\hbar^2} = \gamma f(x), \tag{8.6-1}$$

where $f(x)$ has the periodicity of the lattice, the Schrödinger equation (8.3-1) may be written

$$\frac{d^2\psi}{dx^2} + (k_0^2 + \gamma f(x))\psi(x) = 0. \tag{8.6-2}$$

where k_0 is related to the total energy ε by

$$\varepsilon = \frac{\hbar^2 k_0^2}{2m}. \tag{8.6-3}$$

Since $f(x)$ is periodic, it can be expressed as a Fourier series of the form

$$f(x) = \sum_{n=-\infty}^{\infty} c_n e^{-2\pi i n x/a} \tag{8.6-4}$$

with

$$c_n = \frac{1}{a} \int_0^a f(x) e^{2\pi i n x/a} \, dx, \tag{8.6-5}$$

where a is the lattice constant of the crystal. It is obvious from (8.6-1) that the coefficients c_n are related to the Fourier expansion coefficients of the periodic potential itself by

$$V_n = -\frac{\hbar^2 \gamma}{2m} c_n, \tag{8.6-6}$$

where V_n refer to the Fourier expansion coefficients of $V(x)$.

Since the wave functions must have the Bloch form $e^{ikx} u_k(x)$, and since $u_k(x)$ must be a periodic function of period a, expressible as a Fourier series, we may write

$$u_k(x) = \sum_{n=-\infty}^{\infty} b_n e^{-2\pi i n x/a} \tag{8.6-7}$$

and

$$\psi(x) = e^{ikx} u_k(x) = e^{ikx} \sum_n b_n e^{-2\pi i n x/a}. \tag{8.6-8}$$

For perfectly free electrons $\gamma = 0$ and the solutions become free-particle wave functions of the form

$$\psi(x) = b_0 e^{ik_0 x}. \tag{8.6-9}$$

In this limit $u_k(x) \to b_0$ and $k \to k_0$, so that in the general case, we must expect that as γ approaches zero, all the b_n *except* b_0 approach zero. We are thus led to write an approximate expression for the wave function of the form

$$\psi(x) = b_0 e^{ikx} + \gamma \left[e^{ikx} \sum_{n \neq 0} b_n e^{-2\pi inx/a} \right], \tag{8.6-10}$$

which we may expect to be valid for *small* values of γ (that is, for the conditions outlined above under which the free-electron approximation is valid). It is easy to see that this wave function has the Bloch form, with u_k given by

$$u_k(x) = b_0 + \gamma \sum_{n \neq 0} b_n e^{-2\pi inx/a}. \tag{8.6-11}$$

Substituting (8.6-10) into Schrödinger's equation (8.6-2), one may obtain

$$b_0(k_0^2 - k^2)e^{ikx} + \gamma \sum_{n \neq 0} [(k_0^2 - k_n^2)b_n + b_0 c_n]e^{ik_n x}$$

$$+ \gamma^2 \sum_{n \neq 0} \sum_{n' \neq 0} b_{n'} \cdot c_n e^{i\left(k - \frac{2\pi n}{a} - \frac{2\pi n'}{a}\right)x} = 0 \tag{8.6-12}$$

where

$$k_n = k - \frac{2\pi n}{a}. \tag{8.6-13}$$

Since we are interested in the solution in the limit where $\gamma \to 0$, we may as a first approximation neglect the γ^2 term in (8.6-12); then, multiplying the equation through by $e^{-ik_m x}$, integrating over the unit cell from $x = 0$ to $x = a$, we find

$$b_0(k_0^2 - k^2)\int_0^a e^{2\pi imx/a}\, dx + \gamma \sum_{n \neq 0} \left\{ [(k_0^2 - k_n^2)b_n + b_0 c_n] \right.$$

$$\left. \times \int_0^a e^{2\pi i(m-n)x/a}\, dx \right\} = 0. \tag{8.6-14}$$

If $m = 0$, the second integral vanishes for all values of n in the summation, whereby

$$b_0(k_0^2 - k^2)a = 0 \quad \text{or} \quad k = k_0, \tag{8.6-15}$$

while if $m \neq 0$, the first integral vanishes, and the second gives zero except when $n = m$, in which case one obtains

$$\gamma[(k_0^2 - k_m^2)b_m + b_0 c_m]a = 0$$

or

$$b_m = \frac{b_0 c_m}{k_m^2 - k_0^2} = \frac{b_0 c_m}{k_m^2 - k^2}. \tag{8.6-16}$$

To this order of approximation, then, from (8.6-15) and (8.6-3) the relation between ε and k is the same as that for a free particle. The wave function is obtained by substituting the values (8.6-16) for b_n into (8.6-10), whence

$$\psi(x) = b_0 e^{ikx}\left[1 - \gamma \sum_{n \neq 0} \frac{c_n}{k^2 - k_n^2} e^{-2\pi inx/a}\right]. \tag{8.6-17}$$

According to (8.6-15) the first-order correction to the free-particle energy arising from the periodic potential is zero. A second-order energy correction can be obtained, however, by retaining the γ^2 term in (8.6-12). Multiplying (8.6-12) by e^{-ikx} and integrating from $x = 0$ to $x = a$, one may obtain

$$b_0(k_0^2 - k^2) + \gamma \sum_{n \neq 0} [(k_0^2 - k_n^2)b_n + b_0 c_n]\int_0^a e^{-2\pi inx/a}\,dx$$

$$+ \gamma^2 \sum_{n \neq 0}\sum_{n' \neq 0} b_{n'}c_n \int_0^a e^{-2\pi i(n+n')x/a}\,dx = 0. \tag{8.6-18}$$

The first integral above is zero for all allowed values of n, while the second integral is zero unless $n = -n'$. The equation thus reduces to

$$b_0(k_0^2 - k^2)a + \gamma^2 \sum_{n' \neq 0} b_{n'}c_{-n'}a = 0. \tag{8.6-19}$$

If $f(x)$ is given as a Fourier series by (8.6-4), then, substituting $-n$ for n,

$$f(x) = \sum_{n = -\infty}^{\infty} c_{-n}e^{2\pi inx/a}, \tag{8.6-20}$$

while, taking the complex conjugate of both sides of (8.6-4) and noting that since $f(x)$ is a real function, $f^*(x) = f(x)$, one may also write

$$f(x) = \sum_{n = -\infty}^{\infty} c_n^* e^{2\pi inx/a}. \tag{8.6-21}$$

Since the Fourier coefficients associated with the representation of a given function are unique, we must have

$$c_{-n} = c_n^*. \tag{8.6-22}$$

Using the first approximation expression (8.6-16) for the coefficients b_n and the relation (8.6-22), (8.6-19) can be written in the form

$$k_0^2 = k^2 + \gamma^2 \sum_{n \neq 0} \frac{c_n^* c_n}{k^2 - k_n^2} = k^2 + \gamma^2 \sum_{n \neq 0} \frac{c_n^* c_n}{k^2 - \left(k - \dfrac{2\pi n}{a}\right)^2}. \tag{8.6-23}$$

Using (8.6-3) and (8.6-6) to express this result in terms of ε and V_n, it is easy to show that

$$\varepsilon = \frac{\hbar^2 k^2}{2m} + \sum_{n \neq 0} \frac{|V_n|^2}{\left(\frac{\hbar^2 k^2}{2m}\right) - \frac{\hbar^2}{2m}\left(k - \frac{2\pi n}{a}\right)^2} . \tag{8.6-24}$$

These solutions are satisfactory only so long as k^2 does *not* approach one of the values k_n^2. If $k^2 \cong k_n^2$ for some value of n, one of the quantities $k^2 - k_n^2$ in the denominators of the summation terms in (8.6-17) will approach zero, and that term will remain quite large despite the fact that γ is very small. In this case the assumed expression (8.6-10) is not a good approximation for the wave function, and another form must be used. If $k^2 \cong k_n^2$, then

$$k = \pm k_n = \pm k \mp \frac{2\pi n}{a} . \tag{8.6-25}$$

For the upper sign, this equation gives $n = 0$, which is excluded from (8.6-17) in any case and hence is of no particular interest; for the lower sign one obtains

$$k = \frac{n\pi}{a} \tag{8.6-26}$$

which, since n may take on both positive and negative values, refers to all the band edge points. We shall have to use a separate treatment for values of k in the neighborhood of these points.

Suppose now that $k \cong n\pi/a$, whence $k_n = k - (2\pi n/a) \cong -n\pi/a$. Then $k_n^2 \cong k^2$ and b_n will be very large, so that in (8.6-17) the quantity γb_n will no longer be small even though γ is small. In this case we may write the wave function, approximately, as

$$\psi(x) = b_0 e^{ikx} + \gamma b_n e^{ikx} e^{-2\pi i n x/a} = b_0 e^{ikx} + \gamma b_n e^{ik_n x}, \tag{8.6-27}$$

regarding the other terms in the sum of (8.6-17) as negligible compared to the nth. Since $k_n = -k$ at the band edge, this wave function (8.6-27) is a superposition of a wave propagating along the positive x-axis and one propagating in the opposite direction; it has thus the character of a standing wave. This physical situation is in accord with the view that the electron wave function undergoes Bragg reflection at $k = \pm n\pi/a$. If the wave function (8.6-27) is substituted into Schrödinger's equation (8.6-2), using (8.6-4), one obtains

$$b_0(k_0^2 - k^2)e^{ikx} + \gamma b_n(k_0^2 - k_n^2)e^{i\left(k - \frac{2\pi n}{a}\right)x} + \gamma b_0 \sum_{n' \neq 0} c_{n'} e^{i\left(k - \frac{2\pi n'}{a}\right)x}$$

$$+ \gamma^2 b_n \sum_{n' \neq 0} c_{n'} e^{i\left(k - \frac{2\pi n}{a} - \frac{2\pi n'}{a}\right)x} = 0. \tag{8.6-28}$$

Multiplying (8.6-28) through by e^{-ikx} and integrating from $x = 0$ to $x = a$, it is easily seen that the second and third terms contribute nothing and that a contribution from the fourth term is obtained only when $n' = -n$; the final result of this operation is then

$$(k_0^2 - k^2)b_0 + \gamma^2 c_n^* b_n = 0. \tag{8.6-29}$$

Likewise, multiplying (8.6-28) through by $e^{-ik_n x} = e^{-i\left(k - \frac{2\pi n}{a}\right)x}$ and integrating as above it can be shown that

$$c_n b_0 + (k_0^2 - k_n^2)b_n = 0. \tag{8.6-30}$$

The two above equations may be regarded as a set of homogeneous equations determining b_0 and b_n. As such, solutions other than $b_0 = b_n = 0$ exist only if the determinant of the system vanishes, that is, only if

$$\begin{vmatrix} k_0^2 - k^2 & \gamma^2 c_n^* \\ c_n & k_0^2 - k_n^2 \end{vmatrix} = (k_0^2 - k^2)(k_0^2 - k_n^2) - \gamma^2 c_n^* c_n = 0. \tag{8.6-31}$$

This equation is quadratic in k_0^2, and may be solved in the usual way to give

$$k_0^2 = \tfrac{1}{2}[(k^2 + k_n^2) \pm \sqrt{(k^2 - k_n^2)^2 + 4\gamma^2 c_n^* c_n}], \tag{8.6-32}$$

which, using (8.6-3), (8.6-16) and (8.6-13), can be expressed as

$$\varepsilon(k) = \frac{\hbar^2}{4m}\left[k^2 + \left(k - \frac{2\pi n}{a}\right)^2 \pm \sqrt{\left(k^2 - \left(k - \frac{2\pi n}{a}\right)^2\right)^2 + \left(\frac{4m|V_n|}{\hbar^2}\right)^2} \right]. \tag{8.6-33}$$

At the band edge, $k = k_n = n\pi/a$, and at these points (8.6-33) reduces to

$$\varepsilon = \varepsilon_n \pm |V_n| \tag{8.6-34}$$

where

$$\varepsilon_n = \frac{\hbar^2}{2m}\left(\frac{n\pi}{a}\right)^2 \tag{8.6-35}$$

represents the free-particle energy associated with the band edge points.

From the results of Section 8.3, we know that at the band edge points $k = \pm n\pi/a$, internal Bragg reflection takes place, and is accompanied by a discontinuity or "energy gap" in the ε versus k curve.[7] Accordingly, Equation (8.6-34) should be interpreted as implying an energy gap, or forbidden band, of width $2|V_n|$, where V_n is the nth Fourier coefficient in the Fourier series expansion of the periodic lattice potential. For values of k greater in magnitude than $n\pi/a$ we should expect $\varepsilon(k)$ to be greater than the value given by (8.6-34), and to approach the free electron value predicted by (8.6-15) where k^2 differs from k_n^2 by an amount sufficient to render γb_n small enough so that the treatment leading to Equations (8.6-16) and (8.6-24) is legitimate. Likewise, for values of k smaller in magnitude than $n\pi/a$, $\varepsilon(k)$ must be expected to be less than

---

[7] The fact that the electron group velocity is indeed zero at these points, as one would expect for the standing waves which are always associated with Bragg reflection can be proved from (8.6-33); the details of this proof are assigned as an exercise.

the value given by (8.6-34) and again to approach the solutions obtained previously when k^2 differs from k_n^2 by a sufficient amount. These conditions will be satisfied if in (8.6-33) the plus sign is chosen for $k > n\pi/a$ and the minus sign for $k < n\pi/a$. The solution (8.6-15), or the better second approximation (8.6-24), in the regions between the band edges, coupled with the solution (8.6-33) for the regions close to band edges (the signs being chosen as described above) lead to a relation between ε and k such as that illustrated in Figure 8.7.

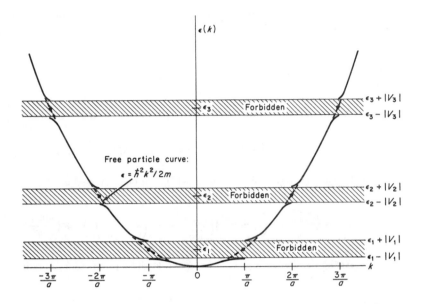

FIGURE 8.7. Schematic representation of the ε *versus* k relation for the free electron approximation as obtained from (8.6-24) and (8.6-33).

It is apparent from this figure that the ε *versus* k curves are approximately parabolic in form near the band edge points. The effective mass can be established from (8.4-8), by direct differentiation of (8.6-33). This calculation is very involved, and it is somewhat easier to proceed by setting

$$k = \frac{n\pi}{a} + k', \tag{8.6-36}$$

where k' is small compared with π/a, in (8.6-33). After a little algebra, one may then write (8.6-36) in the form

$$\varepsilon(k) = \frac{\hbar^2}{2m}\left[\left(\frac{n\pi}{a}\right)^2 + k'^2 + \frac{m\Delta\varepsilon}{\hbar^2}\sqrt{1 + 4k'^2\left(\frac{n\pi}{a}\right)^2\left(\frac{\hbar^2}{m\Delta\varepsilon}\right)^2}\right] \tag{8.6-37}$$

where

$$\Delta\varepsilon = 2|V_n| \tag{8.6-38}$$

represents the width of the forbidden energy region at $k = n\pi/a$. For points near the bottom of the upper band, k' is very small and the radical in (8.6-37) may be expanded by the binomial theorem to give, approximately,

$$\varepsilon(k) = \varepsilon_n + \frac{1}{2}\Delta\varepsilon + \frac{\hbar^2 k'^2}{2m}\left(1 + \frac{4\varepsilon_n}{\Delta\varepsilon}\right), \qquad (8.6\text{-}39)$$

where ε_n is given by (8.6-35). Differentiating twice, and using (8.4-8) it is easily seen that

$$m^* = \frac{\hbar^2}{d^2\varepsilon/dk^2} = \frac{\hbar^2}{d^2\varepsilon/dk'^2} = \frac{m}{1 + \dfrac{4\varepsilon_n}{\Delta\varepsilon}}. \qquad (8.6\text{-}40)$$

It can be shown in a somewhat similar fashion that the effective mass for holes near the top of the lower band is the same as the value for electrons given by (8.6-40).

$8 \cdot 7$ THE TIGHT-BINDING APPROXIMATION

In the free-electron approximation, the potential energy of the electron was assumed to be small compared to its total energy. This led to be an approximate treatment in which allowed and forbidden energy bands were found, the width of the forbidden bands being small compared to that of the allowed bands in the limit in which the initial assumption was well satisfied. The tight binding approximation proceeds from the opposite point of view, namely that the potential energy of the electron accounts for nearly all of the total energy; in this instance the allowed energy bands are narrow in comparison with the forbidden bands. In the free-electron model, the atoms of the crystal are assumed to be so close together that the wave functions for electrons on neighboring atoms overlap to a large extent. There is thus a strong interaction between neighboring atoms, and the allowed energy states of the resulting crystal bear little resemblance to the atomic wave functions of the individual atoms of which the crystal is composed. The tight binding approximation, however, is based on the assumption that the atoms of the crystal are so far apart that the wave functions for electrons associated with neighboring atoms overlap only to a small extent. The interaction between neighboring atoms will in this case be relatively weak, and the wave functions and allowed energy levels of the crystal as a whole will be closely related to the wave functions and energy levels of isolated atoms. The question of which of the two approximations is correct in any given situation, of course, depends upon the particular material at hand. In some substances the free-electron approximation is quite good, while for others the tight binding approximation is more nearly correct; there are also crystals where neither is very good, the situation being intermediate between the two extreme cases.

The crystal wave functions in the tight-binding approximation are based upon the wave functions of isolated atoms. If the potential function associated with an isolated atom is $V_0(\mathbf{r})$, then the solutions of the Schrödinger equation

$$\mathcal{H}_0\psi_0 = -\frac{\hbar^2}{2m}\nabla^2\psi_0 + V_0(\mathbf{r})\psi(\mathbf{r}) = \varepsilon_0\psi_0(\mathbf{r}) \qquad (8.7\text{-}1)$$

represent the electronic wave functions of the atom. Let us assume that the ground state wave function ψ_0 is nondegenerate and corresponds to a ground state energy ε_0. If a large number of atoms of this type are combined into a periodic lattice in such a way that the value of the potential in the regions about each individual atom where the ground-state wave function is large is not much affected by the presence of the neighboring atoms, then the crystal wave function (for these most tightly bound states, at any rate) can be written as a linear superposition of atomic wave functions of the form

$$\psi(\mathbf{r}) = \sum_n a_n\psi_0(\mathbf{r} - \mathbf{r}_n), \qquad (8.7\text{-}2)$$

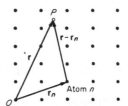

FIGURE 8.8. Vector geometry used in the calculation of Section 8.7.

the vectors r and r_n being related as shown in Figure 8.8. The sum is taken over all atoms of the crystal, which for simplicity we shall assume to be infinite in extent. Since all atoms of the crystal are equivalent, all the coefficients a_n must have the same absolute value; they must therefore be expressible in the form $ae^{i\phi_n}$, where a and ϕ_n are real quantities. For the same reason, the phase difference ϕ between each pair of neighboring atoms along any given crystal axis must be the same (although it is *not* necessarily the same along all three axes). This leads to the choice of the phase factor $\phi_n = \mathbf{k} \cdot \mathbf{r}_n$ with $\mathbf{k}$ a constant vector. Finally, if the functions ψ_0 are normalized such that their integral taken over all space is unity, and if we wish to preserve this normalization, we must take $a = 1$. We may then write (8.7-2) in the form[8]

$$\psi_\mathbf{k}(\mathbf{r}) = \sum_n e^{i\mathbf{k}\cdot\mathbf{r}_n}\psi_0(\mathbf{r} - \mathbf{r}_n). \qquad (8.7\text{-}3)$$

These wave functions satisfy the Schrödinger equation for the entire crystal, with the periodic potential function of Figure 8.9, namely

$$\mathcal{H}\psi_\mathbf{k} = \left(-\frac{\hbar^2}{2m}\nabla^2 + V(\mathbf{r})\right)\psi_\mathbf{k} = \varepsilon\psi_\mathbf{k}. \qquad (8.7\text{-}4)$$

---

[8] Equation (8.7-3) can be shown to have the form of a Bloch function such as (8.2-14). The details of the proof are assigned as an exercise.

The total Hamiltonian $\mathcal{H}$ can be written as the sum of two parts,

$$\mathcal{H} = \mathcal{H}_0 + \mathcal{H}' \tag{8.7-5}$$

with

$$\mathcal{H}_0 = -\frac{\hbar^2}{2m}\nabla^2 + V_0(\mathbf{r} - \mathbf{r}_n) \tag{8.7-6}$$

and

$$\mathcal{H}' = V(\mathbf{r}) - V_0(\mathbf{r} - \mathbf{r}_n). \tag{8.7-7}$$

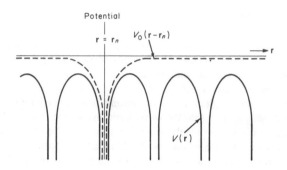

FIGURE 8.9.   Potential functions used in the calculations of Section 8.7.

From (8.7-1) it is clear that $\mathcal{H}_0\psi_0 = \varepsilon_0\psi_0$, whereby, using (8.7-3),

$$\mathcal{H}_0\psi_k = \sum_n e^{i\mathbf{k}\cdot\mathbf{r}_n}\mathcal{H}_0\psi_0(\mathbf{r} - \mathbf{r}_n) = \varepsilon_0 \sum_n e^{i\mathbf{k}\cdot\mathbf{r}_n}\psi_0(\mathbf{r} - \mathbf{r}_n) = \varepsilon_0\psi_k. \tag{8.7-8}$$

Now, according to (4.14-5) the energy ε can be found by evaluating the expectation value of the Hamiltonian operator, whence, using (8.7-5) and (8.7-8),

$$\varepsilon = \frac{\int_v \psi_k^*(\mathcal{H}_0 + \mathcal{H}')\psi_k\,dv}{\int_v \psi_k^*\psi_k\,dv} = \varepsilon_0 + \frac{\int_v \psi_k^* \sum_n e^{i\mathbf{k}\cdot\mathbf{r}_n}[V(\mathbf{r}) - V_0(\mathbf{r} - \mathbf{r}_n)]\psi_0(\mathbf{r} - \mathbf{r}_n)\,dv}{\int_v \psi_k^*\psi_k\,dv}.$$

$$\tag{8.7-9}$$

Provided that the small overlapping of atomic wave functions centered on different lattice sites is neglected, the normalization properties of the functions ψ_0 and the definition (8.7-3) require that $\int_v \psi_k^*\psi_k\,dv$ be equal to N, the number of atoms in the crystal. Utilizing this fact, and substituting the value given by (8.7-3) for ψ_k^* in the numerator of (8.7-9), we obtain

$$\varepsilon = \varepsilon_0 + \frac{1}{N}\sum_n\left\{\sum_m e^{i\mathbf{k}\cdot(\mathbf{r}_n - \mathbf{r}_m)}\int_v \psi_0^*(\mathbf{r} - \mathbf{r}_m)[V(\mathbf{r}) - V_0(\mathbf{r} - \mathbf{r}_n)]\psi_0(\mathbf{r} - \mathbf{r}_n)\,dv\right\}. \tag{8.7-10}$$

It is not difficult to see that due to the periodicity of the crystal, and due to the summation over all values of m, *each term* of the summation over n yields the same value. The value of the sum over n is then just the value of any term of the summation (say, for simplicity, that for which $n = 0$) times the number of terms in the sum, which is N. We may then write (8.7-10) as

$$\varepsilon(\mathbf{k}) = \varepsilon_0 + \sum_m e^{-i\mathbf{k}\cdot\mathbf{r}_m} \int_v \psi_0^*(\mathbf{r} - \mathbf{r}_m)[V(\mathbf{r}) - V_0(\mathbf{r})]\psi_0(\mathbf{r})\, dv, \qquad (8.7\text{-}11)$$

where the sum is taken over all atoms of the crystal. However, since the wave function ψ_0 ordinarily falls off very rapidly with distance, and since the magnitudes of the integrals in (8.7-11) are governed essentially by the amount of overlap between two wave functions centered on atoms separated by distances r_m, the contributions of the terms in the summation decrease very rapidly as r_m increases. It is therefore usually quite a good approximation to consider nearest neighbor terms only. Accordingly we shall neglect all terms beyond nearest neighbor contributions in (8.7-11). We shall also assume that the wave functions ψ_0 are *spherically symmetric*; under these conditions all nearest neighbor contributions will be the same. Our treatment is thus restricted to the case where the ground state electronic configuration in the isolated atom is that of an *s*-state, as it is, for example, for the alkali metals, although the extension to *p*-states is quite straightforward and can be accomplished without much difficulty.[9]

For the case $m = 0$, the integral in (8.7-11) becomes

$$\int_v \psi_0^*(\mathbf{r})[V(\mathbf{r}) - V_0(\mathbf{r})]\psi_0(\mathbf{r})\, dv = -\alpha, \qquad (8.7\text{-}12)$$

while for the nearest neighbor atoms

$$\int_v \psi_0^*(\mathbf{r} - \mathbf{r}_m)[V(\mathbf{r}) - V_0(\mathbf{r})]\psi_0(\mathbf{r})\, dv = -\beta, \qquad (8.7\text{-}13)$$

where $\mathbf{r}_m$ is a vector connecting the atom at the origin with a nearest neighbor atom. Since the actual calculation of the overlap integrals is quite tedious we shall be content with expressing our results in terms of these quantities, which are defined above as $-\alpha$ and $-\beta$. Under these circumstances, equation (8.7-11) may be written

$$\varepsilon(\mathbf{k}) = \varepsilon_0 - \alpha - \beta \sum_m e^{-i\mathbf{k}\cdot\mathbf{r}_m}, \qquad (8.7\text{-}14)$$

where the summation is taken only over nearest neighbor atoms. In the case of a *simple cubic* crystal, the components of the vector $\mathbf{r}_m$ may be expressed as

$$\mathbf{r}_m = (\pm a, 0, 0),\ (0, \pm a, 0),\ (0, 0, \pm a), \qquad (8.7\text{-}15)$$

---

[9] See for instance N. F. Mott and H. Jones, *Theory of Metals and Alloys*, Dover Publications, New York (1958), p. 70.

where a is the lattice constant. Equation (8.7-14) then gives

$$\varepsilon(\mathbf{k}) = \varepsilon_0 - \alpha - 2\beta(\cos k_x a + \cos k_y a + \cos k_z a). \tag{8.7-16}$$

From this equation it is clear that there is a *range* of allowed energy values, corresponding to the various values which k_x, k_y, and k_z may have. The allowed states thus form an *energy band* in much the same way as did those which were discussed in connection with the free electron approximation.

The minimum value of the energy, according to (8.7-16), is at the point $k_x = k_y = k_z = 0$, that is, at the origin in a plot involving orthogonal coordinates (k_x, k_y, k_z) in $\mathbf{k}$-space. The maximum values occur at the corners of a cube in $\mathbf{k}$ space whose coordinates are $\left(\pm \dfrac{\pi}{a}, \pm \dfrac{\pi}{a}, \pm \dfrac{\pi}{a} \right)$, all permutations of $+$ and $-$ signs between the three coordinates being taken to generate the eight corner points. At these sites each cosine term above takes on the value -1. The difference in energy between the maximum energy points and the minimum energy is, from (8.7-16), 12β, and this is the width of the energy band corresponding to the related s-state of the isolated atom. If the nearest neighbor atoms are quite far apart, then the overlap of the wave functions $\psi_0(\mathbf{r})$ and $\psi_0(\mathbf{r} - \mathbf{r}_m)$ as expressed by the product $\psi_0(\mathbf{r})\psi_0(\mathbf{r} - \mathbf{r}_m)$ in the integral (8.7-13) becomes very small, as does the value of β as given by that equation. In the limit where the atomic distances become large (and it is in this limit that the approximations upon which the tight binding calculation are based become really good), the allowed energy bands become narrow, approaching a single discrete energy value, corresponding to the atomic s-state energy level, as the separation becomes infinite. This variation of band width with atomic spacing is shown schematically in Figure 8.10.

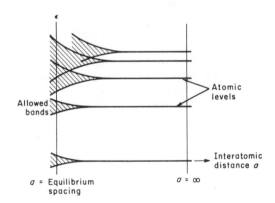

FIGURE 8.10. Schematic representation of the splitting of discrete atomic energy levels into bands as isolated atoms are assembled into a crystal lattice.

For an electron moving in the x-direction, $k_y = k_z = 0$. If the momentum is sufficiently small that $k_x \ll \pi/a$ the electron will be near the bottom of the energy band. Under these conditions, approximating $\cos k_x a$ in (8.7-16) by the first two terms of its series expansion, (8.7-16) be‿omes

$$\varepsilon(k_x) = \varepsilon_0 - \alpha - 2\beta\left(3 - \frac{k_x^2 a^2}{2}\right). \tag{8.7-17}$$

From this it is clear that the ε *versus* k_x relation is parabolic, and the electron therefore behaves essentially as a free electron. The quantity $\hbar k_x$ may, as usual, be interpreted as the x-component of a crystal momentum vector $\mathbf{p} = \hbar k$. The effective mass is given by

$$m^* = \frac{\hbar^2}{d^2\varepsilon/dk_x^2} = \frac{\hbar^2}{2\beta a^2}. \qquad (8.7\text{-}18)$$

From the symmetry of equation (8.7-16) it is evident that the effective mass associated with motion in the y- and z-directions for small momenta will be the *same* as that given by (8.7-18), and the effective mass will therefore be isotropic or independent of direction. This is not always the case, however, especially in noncubic crystals; the effective mass is then often anisotropic and must be regarded as a tensor. This point will be discussed more fully in the next section. From (8.7-18) it would appear that as the lattice distance a increases the effective mass decreases; exactly the *opposite* is true, however, because as a increases the overlap integral β decreases much more rapidly, so that the product βa^2 decreases rapidly with increasing lattice distances. In the extreme tight binding limit, then, according to (8.7-18) an electron becomes very "heavy," which is merely an expression of the fact that it is not easily transferred from one atom to another under these circumstances.

8·8 DYNAMICS OF ELECTRONS IN TWO- AND THREE-DIMENSIONAL LATTICES; CONSTANT ENERGY SURFACES AND BRILLOUIN ZONES

In this section we shall discuss in detail the behavior of electrons in two- and three-dimensional periodic lattices, using wherever possible the example of the tight binding approximation calculation for the simple cubic structure as an illustration of the general principles which are involved. It is easily seen that the dynamical equations of Section 8.4 can be extended in a straightforward manner to a three-dimensional situation, whereupon one obtains

$$\mathbf{v}_g = \frac{1}{\hbar} \nabla_k \varepsilon \qquad (8.8\text{-}1)$$

for the electron velocity, where ∇_k refers to the gradient operator in the orthogonal $\mathbf{k}$-space whose coordinates are (k_x, k_y, k_z). One may then write

$$\frac{d\mathbf{v}_g}{dt} = \frac{1}{\hbar}\frac{d}{dt}\nabla_k \varepsilon. \qquad (8.8\text{-}2)$$

If $\mathbf{A}$ is a vector whose components are functions of $k_x, k_y,$ and k_z, then each component A_α ($\alpha = x, y, z$) must satisfy the relation

$$\frac{dA_\alpha}{dt} = \frac{\partial A_\alpha}{\partial k_x}\frac{dk_x}{dt} + \frac{\partial A_\alpha}{\partial k_y}\frac{dk_y}{dt} + \frac{\partial A_\alpha}{\partial k_z}\frac{dk_z}{dt} = (\nabla_k A_\alpha) \cdot \frac{d\mathbf{k}}{dt}. \qquad (8.8\text{-}3)$$

The vector dA/dt may thus be expressed as the scalar product of the *tensor*[10] $\nabla_k A$ and the vector dk/dt, thus

$$\frac{dA}{dt} = (\nabla_k A) \cdot \frac{dk}{dt},$$ (8.8-4)

each component of the tensor product above leading to an equation of the form of (8.8-3). Applying this formula to (8.8-3), letting $\mathbf{k} = \mathbf{p}/\hbar$, one obtains

$$\frac{dv_g}{dt} = \frac{1}{\hbar^2} \nabla_k (\nabla_k \varepsilon) \cdot \frac{d\mathbf{p}}{dt} = \frac{1}{\hbar^2} \nabla_k (\nabla_k \varepsilon) \cdot \mathbf{F}.$$ (8.8-5)

This equation is essentially Newton's force equation, $dv_g/dt = F/m^*$. It is evident, however, that the reciprocal of the effective mass is a *tensor* of the form

$$\left(\frac{1}{m^*}\right) = \frac{1}{\hbar^2} \nabla_k (\nabla_k \varepsilon).$$ (8.8-6)

The elements of this tensor are easily seen to be

$$\left(\frac{1}{m^*}\right)_{\alpha\beta} = \frac{1}{\hbar^2} \frac{\partial^2 \varepsilon}{\partial k_\alpha \partial k_\beta},$$ (8.8-7)

and from this it is apparent that the reciprocal effective mass tensor is symmetric, that is, that

$$(1/m^*)_{\beta\alpha} = (1/m^*)_{\alpha\beta}.$$ (8.8-8)

The effective mass must in general be regarded as a tensor because the curvature of the $\varepsilon(\mathbf{k})$ relation is not necessarily the same along all possible directions in **k**-space; a different effective mass may thus be required for each possible direction.

For the simple cubic crystal of Section 8.7, for values $ka \ll 1$, equation (8.7-16) may be written in the form

$$\varepsilon(\mathbf{k}) = \varepsilon_0 - \alpha - 6\beta + \beta a^2 (k_x^2 + k_y^2 + k_z^2).$$ (8.8-9)

It is evident from this that all the off-diagonal elements of the effective mass tensor (8.8-7) are zero, and that each of the elements on the diagonal is just equal to $\hbar^2/2\beta a^2$, whereby the tensor reduces to $\hbar^2/2\beta a^2$ times the unit tensor $[\delta_{\alpha\beta}]$. The effective mass is clearly isotropic under these circumstances, and may be represented as the *scalar* quantity $\hbar^2/2\beta^2$ in agreement with the conclusions of the previous section.

Let us now investigate the form of the surfaces of constant energy in **k**-space for the simple cubic structure discussed in the tight-binding approximation in the preceding section. The equation of these surfaces is simply (8.7-16) with $\varepsilon(\mathbf{k})$ being regarded as a

---

[10] A brief outline of the aspects of tensor analysis which are essential for an understanding of the concepts discussed in this book is given in Appendix B.

parametric constant. In connection with (8.7-16), it was mentioned that the *minimum* value of ε was obtained at the point $k_x = k_y = k_z = 0$ where all the cosine terms attain the value unity. For values of ε slightly greater than this minimum, the cosines must be *near* unity, and thus the arguments $k_x a$, $k_y a$, and $k_z a$ must all be very small. In this case (8.7-16) reduces to the form written above as (8.8-9), which can be slightly re-arranged to read

$$k_x^2 + k_y^2 + k_z^2 = \frac{\varepsilon(\mathbf{k})}{\beta a^2} - \frac{\varepsilon_0 - \alpha - 6\beta}{\beta a^2} = \text{const.} \qquad \text{(8.8-10)}$$

This is clearly the equation of a sphere in k-space. The spherical form of the surfaces of constant energy is indicative of the fact that electrons in this region of $\mathbf{k}$-space behave like free electrons; it will be noted from (5.2-18) that for free particles the surface of constant energy in momentum space is spherical. As the value of ε increases, the radius of the spherical constant energy surface increases, until the values of k associated with points on the surface becomes so great that the approximation $ka \ll 1$ is no longer valid. The more general equation (8.7-16) must then be used to plot the constant energy surfaces, and as ε increases the surfaces which are so obtained are illustrated in Figure 8.11 (a–f). With increasing ε the sphere expands, becomes some-what distorted, as at (*b*), and at $\varepsilon = \varepsilon_1 + 4\beta$ (where ε_1 is the value of ε at the minimum point, or $\varepsilon_0 - \alpha - 6\beta$) assumes the quasi-octahedral form shown at (*c*). At this point the energy surface touches the limiting cube of side $2\pi/a$, about which we shall say more later. As ε increases further, the surface assumes a shape like that shown in (*d*), and when $\varepsilon = \varepsilon_1 + 8\beta$ takes on the form shown in (*e*). At this point the surface splits up into eight sheets, each of which approaches a corner of the cube as ε approaches its maximum value for the band, $\varepsilon_1 + 12\beta$. These eight branches are shaped approximately like octants of a sphere in the limit where ε approaches this maximum value. The surface of constant energy always has cubic symmetry in this example. The intersections of the constant energy surface with the $k_x k_y$-plane are shown in Figure 8.12; these curves can also be regarded as the curves of constant energy for a *two-dimensional* crystal with a square lattice whose interatomic spacing is a. It is clear from (8.7-16) that all states within the band must have energy between ε_1 and $\varepsilon_1 + 12\beta$, and that all possible states within this energy range are represented by values of (k_x, k_y, k_z) lying between the limits $\left(\pm \dfrac{\pi}{a}, \pm \dfrac{\pi}{a}, \pm \dfrac{\pi}{a}\right)$.

If there are a given number of electrons belonging to this energy band, and if the temperature is at absolute zero, the electronic energy states will all be occupied up to a given Fermi energy, in accord with the Pauli exclusion principle. The electron distri-bution will therefore lie within a volume in $\mathbf{k}$-space which is enclosed by a surface of constant energy, the value of the energy on the surface being the Fermi energy ε_f. This surface is often referred to as the *Fermi surface*. From the discussion of Section 5.5, we have already seen that the Fermi surface for free electrons is a sphere. For electrons in a periodic crystal potential, the Fermi surface may also be approximately spherical *if the band is nearly empty or nearly full*; this fact is clearly illustrated in Figure 8.11 (a) and (f) for the example which has been discussed in detail. At tempera-tures in excess of absolute zero, some of the electrons will be excited to states of energy higher than ε_f, lying outside the Fermi surface, and there will be some unoccupied states within the Fermi surface. The transition between full and empty states will then no longer be a sharp one. Nevertheless, if the temperature is small compared to the

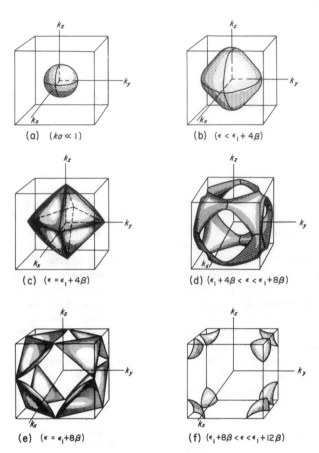

FIGURE 8.11. Successive configurations of the Fermi surface
for a simple cubic crystal with spherically symmetric wave functions in the approximation where only nearest neighbor interactions are considered. As the electron population in the band
increases, thus raising the Fermi level from the bottom of the
band to the top, the Fermi surface goes from the nearly spherical
form shown at (a) through the complex intermediate shapes (b)–
(e) to the cube-minus-sphere configuration in the nearly filled
Brillouin zone shown at (f).

Fermi temperature, the region of transition will be quite thin and the essential validity
of the Fermi surface picture will be unimpaired.

Let us now examine what happens at points on the surface of the limiting cube
enclosing the volume of k-space which the electronic states belonging to this band may
occupy. For simplicity, we shall initially confine our attention to the $k_x k_y$-plane only;
the solutions in this plane are related to a two-dimensional square crystal lattice.
Consider a point on the upper (or lower) edge of the square region shown in Figure
8.12; the k_y coordinate of such a point is π/a, while the k_x coordinate may have any

arbitrary value, say k_{x0}. The electron wavelength λ will then be given by

$$\lambda = \frac{2\pi}{k} = \frac{2\pi}{\sqrt{k_{x0}^2 + \dfrac{\pi^2}{a^2}}} \tag{8.8-11}$$

while the sine of the glancing angle θ related to a set of crystal planes parallel to the x-axis will be

$$\sin\theta = \frac{k_y}{\sqrt{k_x^2 + k_y^2}} = \frac{\pi/a}{\sqrt{k_{x0}^2 + \dfrac{\pi^2}{a^2}}}. \tag{8.8-12}$$

From this it is clear that the Bragg condition (2.1-1) with $n = 1$ is satisfied by the electron wave function at all points on this part of the boundary. In the same way, it is easy to verify that the Bragg condition (related now to a glancing angle referred to an equivalent set of planes parallel to the y-axis) is satisfied at all points on the right-hand (or left-hand) edge of the bounding square. The electron, regarded as a wave, is internally diffracted by the lattice whenever its momentum vector touches this square boundary. Points in **k**-space beyond the limits of this boundary correspond to points *outside* the lowest energy band, belonging to other bands which arise from atomic states of higher energy than the ground state. The limits of the regions of **k**-space occupied by the higher bands are determined by the orientation of boundaries along which higher order Bragg reflections ($n > 1$) occur. There is always an energy gap between the bands, as illustrated in Figure 8.10; for example, whenever one crosses the edge of the square region enclosed by the lines $k_x = \pm\pi/a$, $k_y = \pm\pi/a$, then a discontinuity in energy accompanied by internal Bragg reflection must occur. The interior of the region contains the *totality* of states belonging to the lowest energy band.

It will be noted that the vector **k** has the same dimensions as the *reciprocal* lattice. In Section 2.6 it has already been shown that the Bragg reflection condition can be written as

$$2\mathbf{k} \cdot \mathbf{G} + G^2 = 2(k_x G_x + k_y G_y + k_z G_z) + G_x^2 + G_y^2 + G_z^2 = 0 \tag{8.8-13}$$

where **k** is the propagation vector and where $\mathbf{G}/2\pi$ is a vector from the origin to any point of the *reciprocal* lattice. For a two-dimensional square lattice of points whose lattice spacing is a, the reciprocal lattice is a square lattice of lattice spacing $1/a$. If the origin is taken for simplicity at a lattice point, then the vectors **G** have the form

$$\mathbf{G}_{\mu\nu} = \frac{2\pi}{a}(\mu\mathbf{i}_x + \nu\mathbf{i}_y) \tag{8.8-14}$$

where μ and ν are *integers* and where $\mathbf{i}_x$ and $\mathbf{i}_y$ are unit vectors in the k_x and k_y directions, respectively. Using this form for the vector **G**, (8.8-13) becomes

$$\mu k_x + \nu k_y = -\frac{\pi}{a}(\mu^2 + \nu^2). \tag{8.8-15}$$

This equation represents a family of straight lines in the $k_x k_y$-plane. Their k_x-intercepts and k_y intercepts are

$$k_x \text{ intercepts: } -\frac{\pi}{a}\frac{\mu^2 + v^2}{\mu}$$

$$(8.8\text{-}16)$$

$$k_y \text{ intercepts: } -\frac{\pi}{a}\frac{\mu^2 + v^2}{v}$$

Each of these lines represents a boundary along which a Bragg reflection takes place, and which may form the boundary between states belonging to different energy bands. If we take $\mu = 0$, $v = \pm 1$, and $\mu = \pm 1$, $v = 0$ in (8.8-15) we obtain the four lines $k_x = \pm \pi/a$ and $k_y = \pm \pi/a$ which bound the region shown in Figure 8.12, whose

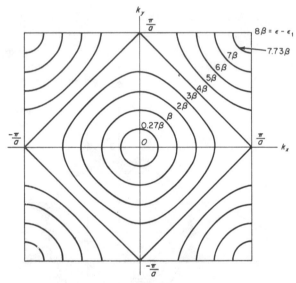

FIGURE 8.12.   Constant energy contours in the Brillouin zone for a two-dimensional square lattice using spherically symmetric wave functions and nearest neighbor interactions only. The same pattern is formed by the traces of the Fermi surfaces of Figure 8.11 for the three-dimensional lattice upon the $k_x k_y$-plane.

interior contains all the states in the lowest energy band of the crystal. This region of k-space is called the *first Brillouin zone* (this terminology has already been introduced in connection with the one-dimensional calculations discussed in Section 3.3). If we now consider $\mu = \pm 1$, $v = \pm 1$, we obtain the four lines $k_x \pm k_y = 2\pi/a$ and $k_x \pm k_y = -2\pi/a$. These four lines are plotted in Figure 8.3(a). On these boundaries the Bragg condition (2.1-1) is satisfied with the order of reflection, n, equal to 1. This can be proved by the same methods which were used previously in connection with the boundaries of the first Brillouin zone. The region enclosed between the first Brillouin zone and these four intersecting lines contains all the states in the second energy band (which would be related to the second lowest atomic energy level if the tight-binding approximation were valid). It is referred to as the second Brillouin zone. It has the same area as the first zone, and indeed the four separate sections of this zone can be translated into the interior of the first zone by moving the sections parallel to the axes by distances which are integral multiples of $2\pi/a$. The four sections then fit together like the pieces of a picture puzzle to form a square region which is an exact replica

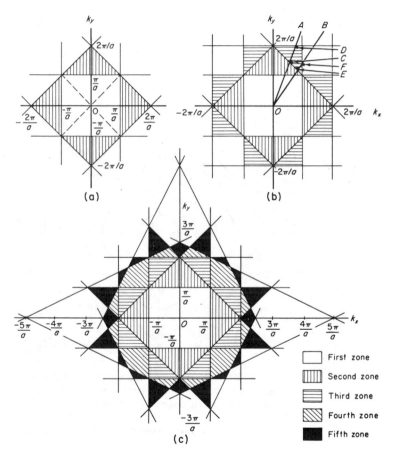

FIGURE 8.13. Construction of Brillouin zones for a two-dimensional square lattice. (a) First and second zones, (b) Third zone construction, (c) First five zones.

of the first zone, as shown by the dotted lines in Figure 8.13(a). The latter representation of the second zone is simply the *reduced zone* representation of Section 8.5. The justification of such a transformation with reference to the Kronig-Penney model has been discussed in that section.

More precisely, if

$$\psi(\mathbf{r}) = e^{i\mathbf{k}\cdot\mathbf{r}}u(\mathbf{r}) = e^{i(k_x x + k_y y + k_z z)}u(\mathbf{r}) \tag{8.8-17}$$

is a wave function of a system, satisfying Schrödinger's equations and all relevant boundary conditions, and if we define

$$\mathbf{k}' = \mathbf{k} - \mathbf{i}_x\left(\frac{2n_x\pi}{a}\right) - \mathbf{i}_y\left(\frac{2n_y\pi}{a}\right) - \mathbf{i}_z\left(\frac{2n_z\pi}{a}\right), \tag{8.8-18}$$

where $\mathbf{i}_x$, $\mathbf{i}_y$, and $\mathbf{i}_z$ are unit vectors along the k_x, k_y, k_z directions, and where n_x, n_y, n_z are integers, then the *same* wave function can be expressed as

$$\psi(\mathbf{r}) = e^{i\mathbf{k}'\cdot\mathbf{r}}[e^{2n_x\pi x/a}e^{2n_y\pi y/a}e^{2n_z\pi z/a}u(\mathbf{r})]$$

$$= e^{i\mathbf{k}'\cdot\mathbf{r}}v(\mathbf{r}), \tag{8.8-19}$$

where $v(\mathbf{r})$ is *again* a periodic function having the periodicity of the lattice. The wave function can thus be written in the Bloch form in terms of either $\mathbf{k}$ or the transformed wave vector $\mathbf{k}'$. It should be noted that (8.8-18) may be written equally well as

$$\mathbf{k}' = \mathbf{k} - \mathbf{G} \tag{8.8-20}$$

where $\mathbf{G}/2\pi$ is a vector connecting two lattice points of the reciprocal lattice.[11] The transformation (8.8-18) or (8.8-20) may always be used to arrive at a reduced zone representation in which the transformed zone occupies the same region of $\mathbf{k}$-space as the first zone. The situation is quite similar to that encountered in Section 3.3 in the case of mechanical vibrations of a linear chain of discrete point masses, and discussed in connection with Figure 3.8.

The higher zones may be constructed by an extension of the procedure used to arrive at the form of the first and second zones. In (8.8-15), if we take $\mu = 0$, $\nu = \pm 2$ or $\mu = \pm 2$, $\nu = 0$, the four lines $k_x = \pm 2\pi/a$, $k_y = \pm 2\pi/a$ are obtained, as shown in Figure 8.13(b). Again, it is clear that Bragg reflection takes place along these lines, and one is initially tempted to assign all the space outside the second zone and within the area enclosed by these four lines to the third Brillouin zone. This is not correct, however, for this area is much larger than that of the first or second zones, and each of the Brillouin zones must encompass the same area in $\mathbf{k}$-space. As a matter of fact there are *additional* Bragg reflection lines which run through this area and form boundaries which divide this space into the third, fourth, fifth, and sixth zones. The criterion for which zone is which is simply this; in travelling along a general radial line which does *not* go through any intersection of Bragg reflection lines, starting from the origin, one must pass through the first, second, third, fourth ⋯ etc., zones successively, and each successive Bragg reflection line which is crossed forms the boundary along the radial path between a zone and the next highest neighboring zone. Thus, progressing from the origin along line OA, one enters the third zone at C and

---

[11] As a matter of fact, expression (8.8-20) is really the fundamental definition of this transformation, while (8.8-18) refers specifically to such a transformation in the particular case of the simple cubic lattice.

leaves it at D, the next Bragg reflection line intersecting OA; beyond D one enters the fourth zone. Traveling along the line OB, however, one enters the third zone at E and *leaves it at* F, which lies on the line $k_x = \pi/a$ which is now the third Bragg reflection line from the origin along path OB. Beyond F along line OB lies a part of the fourth zone. The third zone thus consists of the eight triangular segments shown in Figure 8.13(b). Again, it is clear that these eight segments can be assembled to form a square inside the first zone if they are subjected to appropriate translations according to (8.8-18). The boundaries of the fourth zone are formed, in part, by the lines for which $\mu = \pm 1$, $v = \pm 2$, and $\mu = \pm 2$, $v = \pm 1$ (all possible permutations of plus and minus begins being taken). A family of eight lines (such as $k_x \pm 2k_x = -5\pi/a$) is obtained, leading to boundaries for the fourth and fifth zones as shown in Figure 8.13(c), by the procedures outlined previously. Obviously this method of calculation can be extended *ad infinitum* to find as many zones as desired. In constructing the boundaries of the Brillouin zones in this manner, it is usually advantageous to begin by considering $\mathbf{G}$ vectors whose magnitudes (as obtained from (8.8-14) or its equivalent in another crystal structure) are as small as possible, then proceeding to the next

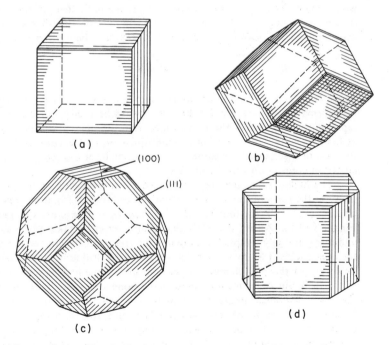

(a)

(100)   (b)

(111)

(c)   (d)

FIGURE 8.14. First Brillouin zone configuration for (a) simple cubic lattice, (b) body-centered cubic lattice, (c) face-centered cubic lattice, and (d) hexagonal lattice.

shortest set of $\mathbf{G}$ vectors, then to the next shortest, and so on. Since the Bragg reflection condition implies the formation of standing waves at normal incidence, the group velocity and thus the gradient of ε normal to the boundary of any Brillouin zone must *vanish*, that is

$$\left(\frac{\partial\omega}{\partial k}\right)_{\perp\,\text{to bdry.}} = \mathbf{n}\cdot\nabla_k\omega = \frac{1}{\hbar}\,\mathbf{n}\cdot\nabla_k\varepsilon = 0 \qquad (8.8\text{-}21)$$

at a zone boundary, if $\mathbf{n}$ is a unit vector normal to the boundary. It is possible to verify this general conclusion from the results of the Kronig-Penney potential model, the free-electron approximation and the tight binding approximation. The details are assigned as exercises.

The extension of these results to three-dimensional lattices is wholly straightforward. In this instance, of course, the Brillouin zones are three-dimensional regions of $\mathbf{k}$-space bounded by planes along which the Bragg condition is satisfied. The equations of these plane boundaries are obtained from equation (8.8-13) just as for the two-dimensional case. Again, each zone is of equal volume and the separate pieces of higher zones can be assembled to form a replica of the first zone by proper application of the transformation given by (8.8-20). It is obvious that the first Brillouin zone for a simple cubic lattice is a cube of side $2\pi/a$. The form of the first Brillouin zone for the face-centered and body-centered cubic lattices and for the hexagonal structure are shown in Figure 8.14. The zone for the b.c.c. lattice is a rhombic dodecahedron, all of whose faces are $\{110\}$ planes in $\mathbf{k}$-space. The zone for the f.c.c. lattice is a tetrakaidecahedron, a polyhedron of 14 faces having the $\{100\}$ and $\{111\}$ orientation. The Fermi surface in all cases must have all the symmetry properties of the related Brillouin zone, which are in turn related to the symmetry properties of the crystal lattice.

8·9   INSULATORS, SEMICONDUCTORS, AND METALS

In Section 8.5, it was demonstrated that no electric current could arise either from empty bands or from completely filled bands. Any electrical conductivity exhibited by a crystal, due to the motion of free electrons, must therefore arise from the motion of electrons in energy bands which are only *partially* filled. This observation forms the basis for the distinction between insulators, metallic conductors and semiconductors.

In an insulator the number of electrons in the crystal is just sufficient to completely fill a number of energy bands. Above these bands in energy is a series of completely empty bands, but between the full and empty bands is a forbidden energy region so wide that it is virtually impossible at physically realizable temperatures to thermally excite a significant number of electrons across this region from the top of the highest filled band to the bottom of the lowest empty band. All the bands are then either full or empty, and no free-electron current can flow. This situation is illustrated in Figure 8.15(a).

If the energy gap $\Delta\varepsilon$ between the full and empty bands in this type of a crystal is quite small, then there will be appreciable statistical probability that electrons will be excited thermally from states near the top of the filled band across the gap to states near the bottom of the empty band. A limited number of free electrons will be available for conduction of electrical currents in the almost empty upper band, and, in addition, the empty electronic states which are left behind near the top of the lower band allow this band to contribute to electric current flow by the mechanism of *hole*

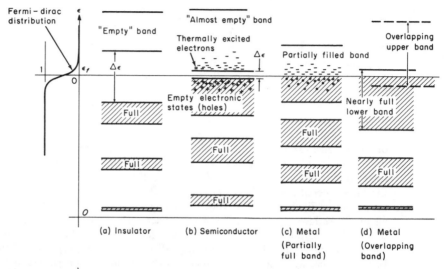

FIGURE 8.15. Energy band diagrams for (a) insulator, (b) semiconductor, (c) metallic conductor (partially filled band) and (d) metallic conductor (overlapping bands).

conduction. A material of this sort is called a *semiconductor*; its electrical conductivity is ordinarily much smaller than that of a metal due to the limited concentration of free electrons and holes, and is in addition strongly temperature dependent, rising rapidly with increasing temperature because the probability of thermal excitation rises with temperature. The electrical conductivity of a semiconductor is also, of course, a function of the energy gap $\Delta\varepsilon$. The distinction between insulators and semiconductors is one of degree only; all semiconductors become ideal insulators as the temperature approaches absolute zero, because then the probability associated with thermal excitation becomes vanishingly small. Likewise, at sufficiently high temperatures (which often cannot be realized experimentally because the crystal melts or vaporizes), all insulators must exhibit semiconductive behavior. The band structure of a semiconductor is shown in Figure 8.15(b).

If the number of electrons in the crystal does not suffice to completely fill the uppermost energy band, but leaves it instead only partially full, then a great number of them can behave as free electrons and can serve as charge or current carriers. This situation is illustrated in Figure 8.15(c). Such a crystal will then exhibit all the characteristic properties of a metallic conductor, for example, high electrical and thermal conductivity and high optical reflectivity.

The determination of the number of electronic states available in a given energy band of a particular substance is not a completely straightforward procedure, and in general involves the use of quantum mechanical calculational techniques which are beyond the scope of this work. If one were to assume that the tight-binding approximation were always valid, then the energy bands would always arise from the energy levels of the isolated atoms, the wave functions for a given band would always consist of a linear combinations of the wave functions of the corresponding electronic energy state of the isolated atoms, and the degeneracy factor associated with a given energy level in the energy band would be the same as that of the corresponding level in the free atom. Accordingly, there would be a $1s$ band corresponding to the $1s$ level, a $2s$ band, a $2p$ band, a $3s$ band, a $3p$ band, a $3d$ band, and so on. The number of levels in each band would be N, according to Equation (8.2-20), and since the degeneracy of each level is the same as that of the corresponding atomic level, the number of electronic quantum states (spin degeneracy included) in each band would be as shown in Table 8.1.

TABLE 8.1.

Band	No. of states
$1s$	$2N$
$2s$	$2N$
$2p$	$6N$
$3s$	$2N$
$3p$	$6N$
$3d$	$10N$

Consider now an alkali metal, such as sodium. Sodium has two $1s$ electrons, two $2s$ electrons, six $2p$ electrons, and a single $3s$ electron. In a crystal of N atoms, there are then $2N$ $1s$ electrons, $2N$ $2s$ electrons, $6N$ $2p$ electrons and N $3s$ electrons. Comparing these figures with those given in Table 8.1, we conclude that the $1s$, $2s$, and $2p$ bands are completely filled, but that the $3s$ band, which contains $2N$ states is only half-full. We should therefore expect sodium—along with all the other alkali metals—to be a metallic conductor, as indeed it is. On the other hand, if we adhere strictly to this approach, we must expect magnesium, whose electronic structure is $1s^2)2s^22p^6)3s^2$, to be an insulator or semiconductor, since now the $3s$ band will be filled. The fact that magnesium (and the other alkaline earth elements) are *metals* is due to the fact that the $3s$ and $3p$ bands *overlap* in energy, and some of the electrons "spill over" from the $3s$ band to the $3p$, leaving the former not quite full, and rendering the latter partially occupied, as shown in Figure 8.15(d). As a matter of fact, the $3s$ and $3p$ wave functions for these bands become strongly intermixed with one another, so that it is not even strictly correct to refer to the bands in that way. The simple tight-binding approach, in effect, is no longer a very good approximation to describe the behavior of s and p electrons in the outer valence shell of most metallic substances; this mixing of s and p bands for the valence electrons is the rule rather than the exception. In silicon, for example, the $3s$ and $3p$ states combine into two nonoverlapping bands each containing $4N$ states, so that the $4N$ valence electrons completely occupy the lower band to form the semiconductor band configuration of Figure 8.15(b).

The overlapping of bands occurs despite the fact that there is an energy gap at all points on the boundary of the Brillouin zone which separates the two bands. Using the square zone of Figure 8.12 as an example, it is quite easy to see how this comes about. The Fermi surface in that figure first touches the zone boundary along the

k_x- and k_y-axes at an energy $\varepsilon = \varepsilon_1 + 4\beta$, but the Brillouin zone and the corresponding energy band are not completely filled until $\varepsilon = \varepsilon_1 + 8\beta$. It may happen that the energy gap between this band and the next, in the direction of the k_x- or k_y-axis may be *less* than 4β, the additional energy required to completely fill the lower band. If this is so, then electrons will spill over into the upper band along the k_x- and k_y-directions *before* the corners of the original zone are completely filled, resulting in the overlapping-band situation referred to previously. This state of affairs is illustrated in Figure 8.16. It

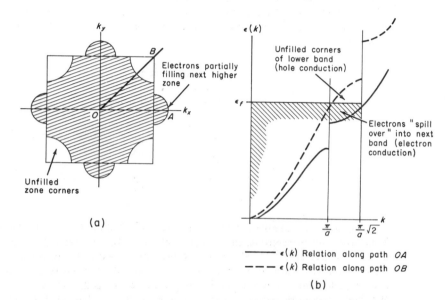

FIGURE 8.16. (a) An instance where the lowest energy states in the higher band (at the centers of the sides of the square zone, on the k_x- and k_y-axes) are lower in energy than the highest energy states in the lower band (at the corners of the zone). Under these circumstances electrons start to occupy the higher band before the lower one is completely filled. (b) Schematic representation of the ε *versus* k relation along the direction OA (solid curves) and OB (dashed curves) illustrating in a slightly different way the situation described in (a).

should be emphasized that when this situation occurs, the simple tight-binding assumption of atomic wave functions associated with a single level is no longer valid, and the energy bands must be recalculated assuming a mixture of wave functions associated with both bands. It will also be noted that in such a situation both electron conduction by the electrons which spill over into the higher band and hole conduction by unfilled states in the upper part of the lower band take place simultaneously. The Fermi surface then has *two* separate branches, one for electrons in the outer zone and one for holes in the original zone. If the hole effective mass is less than the electron effective mass, this situation will lead to a *positive* Hall coefficient.[12] We are thus led to an

---

[12] We shall see how this comes about in detail in a later chapter.

understanding of one of the most puzzling discrepancies in the original free electron theory of metals.

We see that the quantum theory enables us to understand in general terms why the electrical properties of metallic conductors and insulators are so vastly different, even though a rather involved wave-mechanical calculation is often required to establish with accuracy the energy surfaces and wave functions associated with the valence electron bands in any particular substance.

8·10 THE DENSITY OF STATES FUNCTION AND PHASE CHANGES IN BINARY ALLOYS

In Chapter 5, the density of states function for free particles was derived as Equation (5.2-22). It would appear intuitively obvious that the density of states function for electrons in crystals should be the same as that obtained previously, except possibly that the effective mass rather than the inertial mass might be involved. We shall see shortly that this is indeed true, provided that the constant energy surfaces are spherical. If the constant energy surfaces deviate from the spherical form, however, the density of states function becomes more complex.

To begin with, it is important to note that the allowed values for the crystal momentum $\hbar\mathbf{k}$, when periodic boundary conditions are imposed, according to (8.2-20) are the *same* as the allowed values of the intertial momentum of a free particle, as given by (5.2-15). The volume of momentum space per quantum state is thus the same as before, and the task of determining the density of states may be accomplished, as before, by finding the number of quantum states dv between the constant energy surfaces corresponding to energies ε and $\varepsilon + d\varepsilon$, computing the volume of momentum space between the two surfaces, and dividing by $\frac{1}{8}h^3$, the volume per quantum state. The volume element of momentum space can be written

$$dv_p = p^2 \sin\theta \, dp \, d\theta \, d\phi = \frac{h^3}{8\pi^3} k^2 \sin\theta \, dk \, d\theta \, d\phi, \qquad (8.10\text{-}1)$$

whereby the number of states within that volume is just

$$\frac{dv_p}{\frac{1}{8}h^3} = \frac{1}{4\pi^3} k^2 \sin\theta \, dk \, d\theta \, d\phi = \frac{1}{4\pi^3} \frac{dk}{d\varepsilon} \cdot k^2 \sin\theta \, d\varepsilon \, d\theta \, d\phi. \qquad (8.10\text{-}2)$$

The number of states $dv = g(\varepsilon) \, d\varepsilon$ in the energy range $d\varepsilon$ about ε is obtained by integrating over the polar angles θ and ϕ in **k**-space. Thus,

$$g(\varepsilon) = \frac{1}{4\pi^3} \int \int \frac{k^2 \sin\theta \, d\theta \, d\phi}{(d\varepsilon/dk)} = \frac{1}{4\pi^3} \int_S \frac{dS_k}{(d\varepsilon/dk)}, \qquad (8.10\text{-}3)$$

where dS_k represents the surface area element $k^2 \sin\theta \, d\theta \, d\phi$ on the constant energy surface within the Brillouin zone. It is clear that if the function $\varepsilon(\mathbf{k})$ is known, the density of states can in principle be calculated from (8.10-3). Unfortunately, however, the mathematical form of $\varepsilon(\mathbf{k})$ is in most cases of practical interest so complex that the integration can be done only by numerical methods. For the simple cubic lattice in the tight binding approximation, the $\varepsilon(\mathbf{k})$ function for an s-band is given by (8.7-16), and the corresponding density of states function, as given by (8.10-3), is illustrated in Figure 8.17.

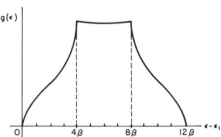

FIGURE 8.17. Density of states curve for the simple cubic lattice in the tight binding approximation, as calculated from (8.10-3) and (8.7-16).

Near the bottom of the energy band for this particular example, the form of $\varepsilon(\mathbf{k})$ is given by (8.8-9); we have then for this region

$$\varepsilon(\mathbf{k}) \cong \varepsilon_1 + \beta a^2 k^2 \qquad \text{and} \qquad \frac{d\varepsilon}{dk} = 2\beta a^2 k. \qquad (8.10\text{-}4)$$

Since ε is independent of θ and ϕ under these circumstances, (8.10-3) can be integrated over these angles to give

$$g(\varepsilon) = \frac{1}{4\pi^3} k^2 \frac{dk}{d\varepsilon} \int_0^{2\pi} \int_0^{\pi} \sin \theta \, d\theta \, d\phi = \frac{k^2}{\pi^2} \frac{dk}{d\varepsilon}, \qquad (8.10\text{-}5)$$

or, using (8.10-4)

$$g(\varepsilon) = \frac{\sqrt{\varepsilon - \varepsilon_1}}{2\pi^2 (\beta a^2)^{3/2}}. \qquad (8.10\text{-}6)$$

Using (8.7-18) to express βa^2 in terms of the effective mass, this can be written as

$$g(\varepsilon) = \frac{8\sqrt{2\pi}}{h^3} m^{*3/2} \sqrt{\varepsilon - \varepsilon_1} \qquad (\varepsilon - \varepsilon_1 \ll 12\beta), \qquad (8.10\text{-}7)$$

which, provided that the origin of energy is taken at the bottom of the band, is the usual parabolic free-electron density of states expression with the inertial mass replaced by the effective mass m^*. In a somewhat similar fashion, it can be shown that for energies near the *top* of the band, where the constant energy surfaces are again spherical in shape, we may write

$$g(\varepsilon) = \frac{8\sqrt{2\pi}}{h^3} m^{*3/2} \sqrt{\varepsilon_1 + 12\beta - \varepsilon} \qquad (\varepsilon_1 + 12\beta - \varepsilon \ll 12\beta). \qquad (8.10\text{-}8)$$

This density of states function vanishes at the top of the band, where $\varepsilon = \varepsilon_1 + 12\beta$, and increases parabolically for decreasing energies. It may thus be regarded as a free-particle density of states function for hole conduction in a nearly filled band. The free-electron density of states functions (8.10-7) and (8.10-8) have been obtained with reference to a specific example, but precisely the same expressions may be shown to result for *any* cubic structure, whether or not the tight-binding approximation is applicable. In metals of the transition group, a rather narrow band arising from atomic $3d$-states, which will accommodate $10N$ electrons, overlaps the $4s$ valence electron band, giving rise to a density of states curve for the valence electrons such as that shown schematically in Figure 8.18. This rather peculiar density of states curve is responsible for many of the characteristic properties of these metals.

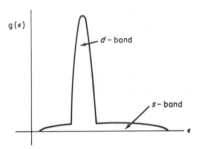

FIGURE 8.18. Overlapping s- and d-bands characteristic of the transition metals.

The form of the density of states curve and the concept of the Fermi surface for valence electrons within the Brillouin zone has been used to explain certain phase changes which are associated with binary alloys of copper, silver and gold with divalent or trivalent metals. Pure copper, silver and gold are all monovalent elements which have the face-centered cubic structure. If one begins with pure silver, for example, and adds increasing amounts of cadmium, the f.c.c. structure, which is called the α-phase, is retained up to a certain concentration of cadmium (called the α-limit), at which point a phase change occurs, and the crystal structure changes from f.c.c. to b.c.c., the latter being referred to as the β-phase. If the concentration of cadmium is further increased, the β-phase b.c.c. structure is stable over a certain range of concentration, after which another transition point (called the β-limit) is reached and the β-phase is transformed into a γ-phase whose structure is a very complex cubic arrangement with some 52 atoms per unit cell. At still higher cadmium concentrations a γ-limit is reached and a hexagonal close-packed η phase is formed. Alloys of the other noble metals and alloys made with other multivalent additives follow the same general pattern of phase changes, although the α-, β-, and γ-limits are different in each individual system.

Hume-Rothery[13] pointed out that the α-, β-, and γ-limits in these alloy systems always occur at certain well-defined valence electron per atom ratios. Thus the β-limit seemed always to be reached at the point where there were just about 3/2 valence electrons for each atom in the crystal, and the γ-limit appeared to be reached at a valence electron per atom ratio of about 21/13. This empirical observation was explained on fundamental grounds by Jones.[14] To begin with, one must note that for

[13] W. Hume-Rothery, *The Metallic State*, Clarendon Press, Oxford (1931), p. 328.
[14] H. Jones, *Proc. Roy. Soc.* **144**, 225 (1934); **147**, 396 (1934). See also N. Mott and H. Jones, *op. cit.*, p. 170.

the pure monovalent f.c.c. metal in the α-phase, the (reduced) Brillouin zone for the valence electrons is as shown in Figure 8.14(c). Since the valence electrons in the noble metals are s-electrons, we may expect that this Brillouin zone will contain $2N$ electronic states. For the pure metal the zone is just half full, because the total number of valence electrons is then N. For simplicity, we shall assume, as Jones did, that the Fermi surface within the zone is spherical at all times.[15] On the basis of this simple assumption, one may calculate the density of states curve for the Brillouin zone quite simply; the result is shown schematically in Figure 8.19. The density of states function

FIGURE 8.19. Picture of the density of states curve for the simple cubic lattice arising from the idea of a spherical Fermi surface expanding within the Brillouin zone, as employed by Jones.

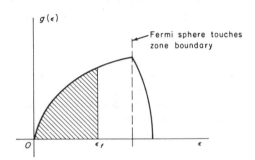

is simply a free-electron parabola, *until the Fermi sphere touches the zone boundary.* From that point on, however, the boundaries of the zone cut off parts of the Fermi sphere, and a given increase in the radius (corresponding to a given increment of energy) encompasses only a relatively small number of states in the corners of the Brillouin zone. The density of states curve thus decreases rapidly, going to zero when the radius of the Fermi sphere equals the radius of the circumscribed sphere. The same general behavior can be noted in the density of states curve of Figure 8.17, which is associated with a simple cubic lattice.

　　When atoms of a multivalent metal are alloyed with the monovalent α-phase, the number of free electrons per atom is increased, the Fermi level is raised, and the radius of the Fermi sphere in the Brillouin zone associated with the valence electrons increases. Beyond the point where the Fermi sphere touches the zone boundary, a large increase in electronic energy takes place with each small increment in free-electron concentration, thus with each small increment in concentration of the multivalent additive. At this point it will result in a *lower* total electronic energy if the crystal undergoes a phase change to a lattice whose Brillouin zone will accommodate a *larger* inscribed sphere, and indeed this is what happens. The f.c.c. α-phase changes to the b.c.c. β-phase, whose Brillouin zone [Figure 8.14(b)], although of the same volume, does in fact have a larger inscribed sphere. Upon further addition of the multivalent alloying element, the radius of the Fermi sphere increases further, finally touching the surface of the β-phase Brillouin zone. Again a phase change takes place, the Brillouin zone of the resulting γ-phase being a complex polyhedron of 36 faces

---

[15] This assumption is much better here than it would have been in the case of the simple cube whose Fermi surfaces are illustrated in Figure 8.11, because of the fact that the bounding Brillouin zone is more nearly spherical in shape. In any case the exact form of the Fermi surface is not particularly important for energies *larger* than that required to make the Fermi surface barely touch the zone boundary.

which is more nearly spherical[16] in form than the rhombic dodecahedron associated with the β-phase and whose inscribed sphere is therefore still larger.

It is possible, on the basis of these arguments to make quantitative estimates of the α-, β-, and γ-limits. In the α phase, the Fermi sphere first reaches the zone boundary along the {111} directions in k-space. The (111) plane forming part of the Brillouin zone polyhedron has [according to (8.8-13)] the equation

$$k_x + k_y + k_z = 3\pi/a. \tag{8.10-9}$$

At the point of contact, in the center of the (111) face, $k_x = k_y = k_z = \pi a$, whereby the radius of the Fermi sphere at contact, k_c, must be given by

$$k_c = \sqrt{k_x^2 + k_y^2 + k_z^2} = \pi\sqrt{3}/a. \tag{8.10-10}$$

The volume of the Fermi sphere, under these conditions, is $4\pi k_c^3/3$, or $4\sqrt{3}\pi(\pi/a)^3$, while it is easily shown that the total volume of the zone is just half the volume of a cube of edge $4\pi/a$, or $32(\pi/a)^3$. The ratio of the volume of the inscribed sphere to that of the zone is then the ratio of these two numbers, or $\pi\sqrt{3}/8$. The average number of valence electrons per atom is just the number per atom when the zone is filled (i.e., 2) times the fraction of the number of states within the zone which are occupied. The number of valence electrons per atom for the α-limit is thus $\pi\sqrt{3}/4$ or 1.362. The values for the β- and γ-limits can be worked out using the same general procedures. The values obtained, compared with Hume-Rothery's empirically derived values are shown in Table 8.2.

TABLE 8.2.

Electron-atom ratio

Phase limit	Hume-Rothery	Jones
α	—	1.362
β	1.5	1.480
γ	1.615	1.538

Naturally, since the Fermi surfaces cannot be perfectly spherical, the Jones estimates must be only approximate, but they are found nevertheless to agree quite well with experimentally determined phase limits. The phase diagrams of more complex alloy systems can also often be understood in terms of electronic energy surfaces and the density of states function within the Brillouin zone, but the details of calculation and the resulting answers are inevitably more involved.

---

[16] See F. Seitz, *Modern Theory of Solids*, McGraw-Hill Book Company, Inc., New York (1940), Figure 13, p. 433.

EXERCISES

1. Show that if in the potential model of Figure 8.2, used in connection with the Kronig-Penney calculation, V_0 is allowed to become infinitely large and b is allowed to approach zero, such that the product $P = -\beta^2 ab/2$ remains fixed, then equation (8.3-12) becomes

$$P \frac{\sin \alpha a}{\alpha a} + \cos \alpha a = \cos ka.$$

If a plot of the function on the left-hand side is made, it is seen to behave in a manner similar to (8.3-15) and (8.3-16) as shown in Figure 8.3. The quantity P may be regarded as the "scattering power" of a single potential spike.

2. Show that in the case discussed in Exercise 1 above, the quantity $d\varepsilon/dk$ vanishes at the band edges ($k = n\pi/a$).

3. Show that for the one-dimensional free-electron approximation discussed in Section 8.6, $d\varepsilon/dk$ vanishes at the band edges.

4. Show that for the tight binding approximation in the simple cubic lattice, discussed in Section 8.7, $\mathbf{n} \cdot \nabla_k \varepsilon$ vanishes at all points on the cubical boundaries of the first Brillouin zone.

5. Show explicitly that the polyhedra shown in Figure 8.14(b) and (c) are the correct forms for the first Brillouin zone of the b.c.c. and f.c.c. lattices, respectively.

6. Find the $\varepsilon(\mathbf{k})$ relation for an s-band (spherically symmetric wave functions) in the tight binding approximation for a body-centered cubic crystal. Consider nearest neighbor overlap integrals only. Plot the forms of the constant energy surfaces for several energies within the zone. Show that these surfaces are spherical, as for free electrons, for energies near the bottom of the band. Show that $\mathbf{n} \cdot \nabla_k \varepsilon$ vanishes on the zone boundaries.

7. Obtain all the results required in Exercise 6 above for an s-band in a face-centered cubic crystal in the tight binding approximation, considering overlap of nearest neighbor wave functions only.

8. Make a diagram [similar to Figure 8.13(c)] in the $\mathbf{k}$-plane, of the first five Brillouin zones of a two-dimensional lattice defined by the basis vectors $\mathbf{a} = a\mathbf{i}_x$, $\mathbf{b} = 3a\mathbf{i}_y$. *Note:* The formulas (2.5-5) cannot be used to define a two-dimensional reciprocal lattice, because the cross product cannot be defined in a two-dimensional space. Proceed from the fundamental definitions (2.5-1) and (2.5-2).

9. Show explicitly from the results of Section 8.7 that the density of states near the top of the energy band for the simple cubic crystal which was discussed in detail in that section is as given by Equation (8.10-8).

10. Show that the limiting valence electron per atom ratio for the β-phase (b.c.c.) of as binary alloy of the type discussed in Section 8.10 is 1.489. You may make the assumption of the Jones calculation.

11. Show that the wave function (8.7-3) which was used in the tight binding approximation calculation can be expressed as a Bloch function of the form (8.2-14).

GENERAL REFERENCES

W. Hume-Rothery, *The Metallic State*, Clarendon Press, Oxford (1931).

H. Jones, *The Theory of Brillouin Zones and Electronic States in Crystals*, North Holland Publishing Co., Amsterdam (1960).

N. F. Mott and H. Jones, *Theory of Metals and Alloys*, Dover Publications, Inc., New York (1958).

S. Raimes, *The Wave Mechanics of Electrons in Metals*, Interscience Publishers, New York (1961).

F. Seitz, *Modern Theory of Solids*, McGraw-Hill Book Co., Inc., New York (1940).

R. A. Smith, *Wave Mechanics of Crystalline Solids*, John Wiley & Sons, Inc., New York (1961).

E. Spenke, *Electronic Semiconductors*, McGraw-Hill Book Co., Inc., New York (1958).

A. H. Wilson, *Theory of Metals*, 2nd Edition, Cambridge University Press, New York (1953).

CHAPTER 9

UNIFORM ELECTRONIC SEMICONDUCTORS IN EQUILIBRIUM

9·1 SEMICONDUCTORS

In the preceding chapters our main concern has been to understand the electrical and thermal properties of metallic conductors and of insulators, and to understand why some substances are extremely good electrical conductors and others are extremely poor ones. These objectives have now been accomplished in a general way, and from this point on we shall turn our attention toward a rather detailed examination of the properties of semiconductors. There are three reasons why this is a useful thing to do. First, it is usually possible to use Maxwell-Boltzmann statistics in discussing the statistical mechanics of charge carriers in semiconductors. This means that it is possible to obtain exact analytical solutions for many problems in semiconductors which can be solved only by approximate or numerical methods for metals, where Fermi-Dirac statistics must be used. An understanding of such effects in semiconductors may furnish valuable physical insight into their counterparts in metallic substances. In short, it is possible to understand electronic processes more easily and more deeply in semiconductors than in any other class of crystals. Second, the technology of crystal growth has made it possible to produce crystals of certain semiconducting substances (notably germanium and silicon) of fantastic purity and crystalline perfection, far in excess of what can be obtained at present for metals and insulators. The existence of these nearly perfect crystals makes it possible for the experimentalist to observe electronic transport properties and thermoelectric and galvanomagnetic effects with ease and precision, secure in the knowledge that what he is trying to observe will not be obscured by effects caused by impurities or structural imperfections in the crystal lattice. We may thus easily observe and interpret phenomena in semiconductors which would be difficult or impossible either to measure accurately or to explain quantitatively in metallic substances. Finally, semiconductors have acquired over the last two decades an enormous technological importance. They have come to be used in making all sorts of electronic devices, including rectifiers, transistors, photocells, voltage regulators, parametric amplifiers, and switching devices. An appreciation of the fundamental electronic transport properties in these substances is of direct application in understanding and analyzing the operation of all the device structures of contemporary technical importance.

A semiconductor is a crystalline substance having an energy band structure in which a band of electronic states, completely filled at zero temperature, is separated from one which is completely empty at absolute zero by a narrow region of forbidden energies. This band structure is shown schematically in Figure 9.1(a). At absolute zero the semiconductor is a perfect insulator, since there are no partially filled bands. At higher temperatures, however, a few electrons from the valence band may acquire enough random thermal energy to be excited across the forbidden band to become conduction electrons in the heretofore empty *conduction* band. The empty states left behind in the lower or *valence* band also contribute to the conductivity, behaving like

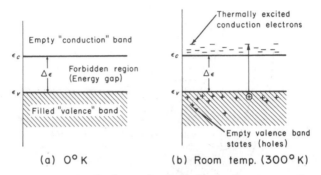

(a) 0° K   (b) Room temp. (300° K)

FIGURE 9.1. Conduction and valence bands of a pure semiconductor (a) at absolute zero, (b) at room temperature, showing thermally excited electrons and holes.

positively charged holes. It is clear that the number of conduction electrons and the number of holes must rise with increasing temperature, and hence the electrical conductivity likewise increases with temperature.

The physical mechanism of electron and hole conduction in *covalent* semiconductors such as carbon (diamond), germanium, and silicon, which form crystals having the diamond structure of Figure 1.9 can be understood from Figure 9.2. This figure shows the calculated energy band structure of diamond, plotted against the interatomic spacing, as in Figure 8.10. The corresponding diagrams for silicon and germanium are quite similar. When N isolated atoms are assembled into a crystal, the $2s$ and $2p$ atomic levels first broaden into energy bands; as the interatomic spacing decreases these bands broaden further, and eventually overlap. As the interatomic distance becomes still smaller, the continuum of what were originally $2s$ and $2p$ states splits once more into two bands, each of which, however, now contain precisely $4N$ states. At the equilibrium interatomic distance a, these bands are separated by an "energy gap" or forbidden region of width $\Delta\varepsilon$. Since the electronic structure of carbon is $1s^2)2s^22p^2$, there are $4N$ valence electrons available, which exactly suffice to fill the lower of the two bands, and form the valence band of the crystal. The upper band becomes the conduction band. The forbidden energy region under normal conditions has a width of about 7 eV in diamond, 1.2 eV in silicon and 0.7 eV in germanium. Due to thermal expansion of the lattice, this "energy gap" has a weak dependence on temperature; it is clear from Figure 9.2 that the energy gap will decrease as the crystal expands. It is also evident from this diagram that $\Delta\varepsilon$ will be a function of pressure, becoming larger as the interatomic spacing is reduced by the application of pressure by hydrostatic or other means. Beyond the point A in Figure 9.2, where the $2p$ and $2s$ bands overlap, the s and p character of the electronic states is lost; in the valence band of the semiconductor, therefore, the electron wave functions are a mixture of s and p atomic wave functions.

The electrons in the valence band are the electrons which form the tetrahedrally disposed electron-pair covalent bonds between the atoms in Figure 1.9. The thermal excitation of an electron from the valence band to the conduction band corresponds physically to the removal of an electron from a covalent pair bond by thermal agitation

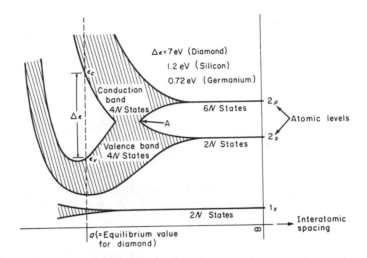

FIGURE 9.2. The bands arising from the $1s$, $2s$, and $2p$ atomic states of carbon (diamond) as a function of interatomic distance. [After W. Shockley, *Electrons and Holes in Semiconductors*, D. van Nostrand Co., Inc., New York (1950).]

of the lattice. The electron then becomes a free electron, outside the covalent lattice bond scheme, and is available to function as a charge carrier to conduct current through the crystal. This process is depicted schematically in Figure 9.3(a), though it must be remembered that the bond scheme shown there is a two-dimensional rendering of what in actuality is a three-dimensional network of tetrahedral bonds. The excitation of an electron leaves behind a localized defect in the covalent bond structure of the crystal, which can be identified as a valence band state which could be occupied by an electron but is actually empty. This defect is simply the "hole" of Section 8.5. Both the free electron and the hole are migratory; the free electron may wander about within the crystal in a random fashion, impelled by thermal energy it may acquire from the lattice. The hole, likewise, may migrate by virtue of the fact that an electron in a covalent bond pair adjacent to the hole may very easily move into the hole, completing the bond pair at the former hole site, but transferring the location of the hole to the site whence the electron came. This hole migration may also be caused by thermal agitation of the lattice. Free electrons and holes will also move in response to an electric field, and may then give rise to a macroscopic current flow through the crystal. This situation is illustrated in Figure 9.3(b). Here *all* the electrons in the crystal, in both valence band and conduction band, are subjected to a force $-e\mathbf{E}$, acting to the right in the diagram. The free electrons move to the right, causing a conventional current flow to the left in view of their negative charge. In addition to this, an electron in a covalent bond pair adjacent to a hole may move to the right into the empty electron site associated therewith, the hole then moving to the left, to the site whence this electron came; this process may now be repeated, the net result being a transfer of an electron to the right, accompanied by the motion of the hole to the

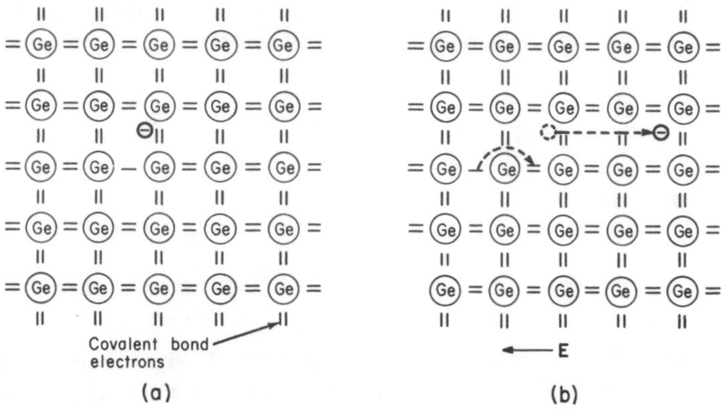

FIGURE 9.3. (a) A free electron and a hole arising from thermal ionization of an electron originally in a covalent bond. (b) Motion of free electron (to the right) and hole (to the left) when an electric field is applied as shown.

left, hence in the direction in which a *positively* charged particle would move under the influence of the applied field. The net electron current to the right again causes a conventional current flow to the left, which might as well be represented as a current of *positive holes* in that direction. Current flow may thus arise both from the movement of free electrons and from the migration of valence electrons into and out of empty states in the valence band, which we prefer to think of as the migration of positive holes.

9·2 INTRINSIC SEMICONDUCTORS AND IMPURITY SEMICONDUCTORS

A semiconductor in which holes and electrons are created solely by thermal excitation across the energy gap is called an *intrinsic* semiconductor. Holes and electrons which are created in this manner are often referred to as *intrinsic charge carriers* and the conductivity arising from such carriers is sometimes termed *intrinsic conductivity*. In an intrinsic semiconductor the concentrations of electrons and holes must always be *the same*, since the thermal excitation of an electron inevitably creates one and only one hole.

The population of holes and electrons in an intrinsic semiconductor is described statistically in terms of the Fermi-Dirac distribution function and the density of states functions for the valence and conduction bands. Since the bottom of the conduction band and the top of the valence band exhibit an essentially parabolic dependence of ε on k, the behavior of electrons and holes in these regions is essentially that of free particles, with appropriate effective mass factors. Electrons and holes are only very rarely excited into regions of the conduction and valence bands where their properties may depart from free-particle behavior at physically realizable temperatures, and so the effects of such excitations may for practical purposes be neglected. The density of states functions to be used, then, are essentially those for free particles, the density of states in the conduction band being given by

$$g_c(\varepsilon)\, d\varepsilon = \frac{8\sqrt{2\pi}}{h^3}\, m_n^{*3/2}\sqrt{\varepsilon - \varepsilon_c}\, d\varepsilon \qquad (\varepsilon > \varepsilon_c) \qquad (9.2\text{-}1)$$

and the density of states in the valence band by

$$g_v(\varepsilon)\, d\varepsilon = \frac{8\sqrt{2\pi}}{h^3}\, m_p^{*3/2}\sqrt{\varepsilon_v - \varepsilon}\, d\varepsilon, \qquad (\varepsilon < \varepsilon_v) \qquad (9.2\text{-}2)$$

where m_n^* is the effective mass for electrons in the conduction band, and m_p^* the effective mass for holes in the valence band.[1] The density of states in the forbidden region $\varepsilon_v < \varepsilon < \varepsilon_c$ is, of course, zero. A plot of the density of states curve for an intrinsic semiconductor is shown in Figure 9.4. If m_p^* and m_n^* are precisely equal, the

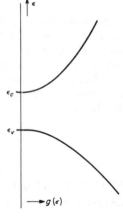

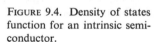

FIGURE 9.4. Density of states function for an intrinsic semi-conductor.

Fermi energy must lie exactly in the center of the forbidden region. This is true because otherwise the population of electrons in the conduction band and holes in the valence band, obtained by integrating the product of the density of states function and the probability factor $f_0(\varepsilon)$ for electrons in the conduction band or $1 - f_0(\varepsilon)$ for holes in the valence band would not be equal. The situation is represented diagrammatically in Figure 9.5(a). If m_p^* and m_n^* are not equal (and this is usually the case), the

---

[1] The subscripts n and p refer to *negative* and *positive* charge carriers. This convention will be used consistently.

Fermi energy must be adjusted upwards or downwards a bit, away from the precise center of the energy gap, to equalize the population integrals, and hence must lie *near* but not *at* the center of the forbidden region. This state of affairs is shown in Figure 9.5(b). We shall investigate the statistics of holes and electrons in some detail in a later section.

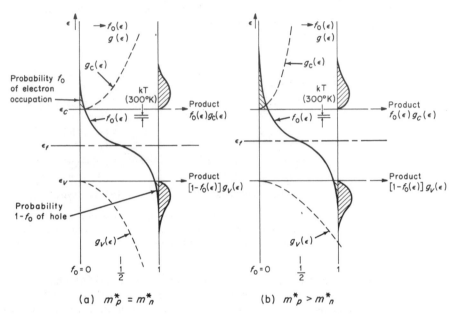

FIGURE 9.5.   Distribution function, Fermi level, density of states function and electron and hole populations for an intrinsic semiconductor (a) where $m_p^* = m_n^*$ and (b) where $m_p^* > m_n^*$. The "spread" of the Fermi distribution is exaggerated here for the sake of illustrative clarity; in reality at 300°K the Fermi distribution would be much more like the "step" function which is found for $T = 0$.

It is quite easy to introduce very small amounts of substances such as arsenic, antimony or other elements belonging to group V of the periodic table into otherwise pure crystals of silicon or germanium as substitutional impurities, that is, as impurity atoms which occupy lattice sites which would normally be occupied by atoms of the covalent semiconductor. The group V atoms have five valence electrons apiece. Four of these are used in forming covalent electron pair bonds with neighboring semi-conductor atoms; the fifth is bound to the impurity atom only by electrostatic forces which are quite weak, and it can therefore be easily ionized by thermal agitation of the lattice at ordinary temperatures to provide an *extra* conduction electron. The impurity atom which is left behind then becomes a positive ion, which, however, is *immobile* in view of the fact that it is strongly bound to four neighboring atoms by the usual covalent bonds. The situation is illustrated in Figure 9.6(a). In crystals containing this type of impurity there are more electrons than holes (although *some* holes are still present because hole-electron pairs are still created thermally on occasion). Such crystals are termed *n*-type semiconductors, the terminology arising from the fact that

most of the charge carriers are *negative* electrons. The component of electrical conductivity arising from the impurity atoms is called *impurity conductivity*, and a substance most of whose charge carriers originate from impurity atoms is called an
impurity semiconductor. The substitutional group V atoms are often called *donor*
atoms, because they each donate an additional free electron to the crystal. It takes so
little energy (as we shall soon see) to ionize a typical group V donor impurity atom in
Si and Ge that essentially *all* group V impurities in these materials are ionized at
temperatures above about 20°K.

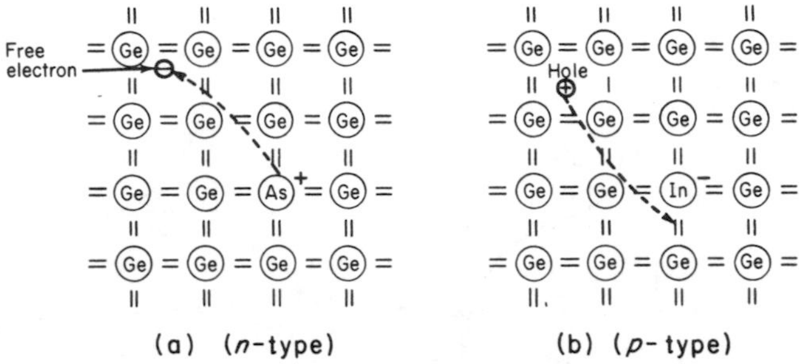

(a) (*n*-type)   (b) (*p*-type)

FIGURE 9.6.   (a) Free electron arising from ionization of a substitutional arsenic impurity atom; (b) free hole arising from ionization
of a substitutional indium impurity atom.

If, instead of group V atoms, group III impurity atoms (Al, Ga, In, and so forth)
are introduced substitutionally into the lattice, a rather different effect is observed.
These atoms have only three valence electrons, which are used in forming covalent
pair bonds to three neighboring atoms, but the fourth bond of necessity lacks an
electron. There is, in effect, an extra *hole* built into the covalent bond structure at the
impurity atom. This hole can easily migrate away from the impurity site, because an
extra electron from a neighboring covalent bond can migrate to the impurity site and
fill in the fourth electron pair bond there (which, of course, puts a *negative* charge on
the impurity atom); the hole is then associated with a neighboring atom, indistinguishable from a thermally created hole. This state of affairs is illustrated in Figure 9.6(b).
The energy required for the migration of the hole away from the impurity site is of the
order of the energy required to remove the extra electron from a donor atom. Therefore, except at very low temperatures, all the holes will be migratory and all the group
III impurity atoms will have the character of immobile negative ions. In crystals
containing predominantly this type of impurity there are more holes than electrons,
although there will always be some electrons arising from thermal excitation. Crystals
of this sort are called *p*-type semiconductors, because the charge carriers are for the
most part *positive*. The substitutional group III atoms are usually termed *acceptor*
atoms, because they readily accept an electron from the covalent bond structure, free-
ing a mobile hole. When impurities of either type are present in a semiconductor
crystal, the conductivity is invariably *greater* than that associated with a pure or
intrinsic semiconductor at the same temperature, due to the extra charge carriers
contributed by the impurity atoms, and, in general, the greater the impurity concentration, the greater the conductivity.

The statistical picture of *n*-type and *p*-type semiconductors is characterized by the presence of the Fermi level above (for *n*-type) or below (for *p*-type) the position associated with the pure or intrinsic crystal. In an *n*-type crystal, for example, there cannot be more electrons than holes *unless* the Fermi level is adjusted *upward* from the

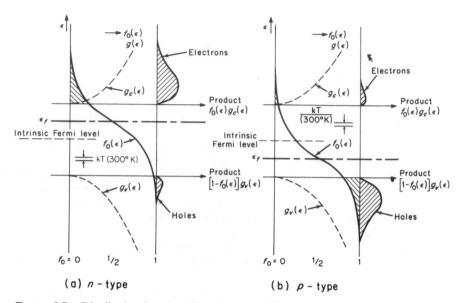

(a) *n* - type   (b) *p* - type

FIGURE 9.7.   Distribution function, Fermi level, and electron and hole populations for (a) an *n*-type impurity semiconductor and (b) a *p*-type impurity semiconductor. Again, as in Figure 9.5, the "spread" of the Fermi distribution is exaggerated for illustrative purposes.

intrinsic position, and vice-versa for *p*-type, as shown in Figure 9.7. As the temperature and impurity concentration are varied, the position of the Fermi level changes in a rather complex way, about which we shall have more to say later.

9.3   STATISTICS OF HOLES AND ELECTRONS—THE CASE OF THE INTRINSIC SEMICONDUCTOR

Under conditions of thermal equilibrium, the number of electrons dn_0 per unit volume having energy in a range $d\varepsilon$ about ε in the conduction band of *any* semiconductor, intrinsic or impurity, is, according to the results of Chapter 5

$$dn_0 = f_0(\varepsilon)g_c(\varepsilon)\,d\varepsilon = \frac{8\sqrt{2}\pi}{h^3}\,m_n^{*3/2}\,\frac{\sqrt{\varepsilon - \varepsilon_c}\,d\varepsilon}{1 + e^{\frac{\varepsilon - \varepsilon_f}{kT}}} \qquad (9.3\text{-}1)$$

where $f_0(\varepsilon)$ represents the equilibrium Fermi distribution function (5.5-20) and $g_c(\varepsilon)$ is the density of states factor (9.2-1). Now $\Delta\varepsilon$, the width of the forbidden energy region is typically of the order of 1 eV, and kT at room temperature (300°K) is only about 1/40 eV. If the Fermi energy is *within the forbidden energy region, several kT units away from the edge of the conduction band* (that is, if $\varepsilon_c - \varepsilon_f \gg kT$), then for all energies belonging to the conduction band, the exponential factor in the denominator of the Fermi distribution function is much greater than unity. We may then represent the Fermi distribution function in the conduction band as

$$f_0(\varepsilon) \cong e^{-(\varepsilon - \varepsilon_f)/kT} \qquad (9.3\text{-}2)$$

which is a distribution function of essentially Maxwellian form. The situation here is substantially that which was discussed in Chapter 5 in connection with Equation (5.5-27). Since we have already shown in Figure 9.5 that the Fermi energy for an intrinsic semiconductor must lie well within the forbidden region of energy, and since $kT \ll \Delta\varepsilon$ at all accessible temperatures as pointed out above, the condition $\varepsilon_c - \varepsilon_f \gg kT$ will always be satisfied in practice, and we shall always be able to use (9.3-2) for intrinsic semiconductors rather than the more complicated expression (5.5-20). As a matter of fact, in most cases involving impurity semiconductors we shall find that this condition is also satisfied, allowing us to use (9.3-2) also for impurity semiconductors. It is clear, for example, that this simplification would be valid for the impurity semiconductors of Figure 9.7. In both Figures 9.5 and 9.7 it is easily seen that the exponential "tail" of the Fermi distribution function is the only part of that function which overlaps the conduction and valence bands. The part of the Fermi function in the neighborhood of the Fermi energy, which is invariably the part which causes mathematical difficulties, coincides in all these cases with a region where the density of available electronic states is *zero*. We shall refer to the simplifying approximation (9.3-2) as the *Boltzmann approximation*.

Using this approximation in (9.3-1) and integrating to get the total number of electrons per unit volume in the conduction band, we find

$$n_0 = \int_{\varepsilon_c}^{\infty} dn_0 = \frac{8\sqrt{2\pi}}{h^3} m_n^{*3/2} e^{\varepsilon_f/kT} \int_{\varepsilon_c}^{\infty} \sqrt{\varepsilon - \varepsilon_c}\, e^{-\varepsilon/kT}\, d\varepsilon. \qquad (9.3\text{-}3)$$

Actually, the limits of integration should be from ε_c to the top of the conduction band, wherever that may be; however, since $f_0(\varepsilon)$ falls off very rapidly with increasing ε (at least for reasonable values of temperature) most of the electrons are concentrated in states near the bottom of the conduction band, and so long as the top of the band is many kT units higher in energy than ε_c, it makes little difference whether we integrate to the top of the band or to infinity. Making the substitution

$$x_c = \frac{\varepsilon - \varepsilon_c}{kT} \qquad (9.3\text{-}4)$$

the integral (9.3-3) can be expressed in the form

$$n_0 = \frac{8\sqrt{2\pi}}{h^3} (m_n^* kT)^{3/2} e^{\frac{\varepsilon_f - \varepsilon_c}{kT}} \int_0^{\infty} x_c^{1/2} e^{-x_c}\, dx_c. \qquad (9.3\text{-}5)$$

According to (5.4-4), the value of the integral above is $\Gamma(3/2) = \sqrt{\pi}/2$. Substituting this value for the integral, the final result may be written

$$n_0 = U_c e^{-\frac{\varepsilon_c - \varepsilon_f}{kT}} \tag{9.3-6}$$

where

$$U_c = 2(2\pi m_n^* kT/h^2)^{3/2}. \tag{9.3-7}$$

The numerical value of the quantity U_c can be written (using cgs units) as

$$U_c = (4.83 \times 10^{15})(m_n^*/m_0)^{3/2} T^{3/2} cm^{-3}, \tag{9.3-8}$$

where m_0 is the inertial mass of the electron. For $m_n^* = m_0$ and $T = 300°K$, this gives $U_c = 2.51 \times 10^{19}$ cm^{-3}.

The number of holes p_0 per unit volume in the valence band of a semiconductor can be found in a similar manner. In this case we may write as dp_0, the number of holes in the valence band in a range of energy $d\varepsilon$ about ε, the expression

$$dp_0 = f_{p0}(\varepsilon)g_v(\varepsilon)\, d\varepsilon = [1 - f_0(\varepsilon)]g_v(\varepsilon)\, d\varepsilon. \tag{9.3-9}$$

The quantity $f_{p0}(\varepsilon)$ is the probability that a hole will be associated with a quantum state of energy ε, in other words, that such a state will be unoccupied. This probability is just one minus the probability of occupation $f_0(\varepsilon)$. Since $f_0(\varepsilon)$ is given by (5.5-20) it is easily seen that

$$f_{p0}(\varepsilon) = 1 - f_0(\varepsilon) = 1 - \frac{1}{1 + e^{(\varepsilon - \varepsilon_f)/kT}} = \frac{1}{1 + e^{(\varepsilon_f - \varepsilon)/kT}}. \tag{9.3-10}$$

Again, if the Fermi energy is several kT units above the edge of the valence band, that is, if $\varepsilon_f - \varepsilon_v \gg kT$, the exponential factor in the denominator of the expression (9.3-10) will be much larger than unity for all values of ε in the valence band, whereby one may set

$$f_{p0}(\varepsilon) \cong e^{-(\varepsilon_f - \varepsilon)/kT}. \tag{9.3-11}$$

This is the Boltzmann approximation for holes in the valence band. Using (9.3-11) and the density of states function (9.2-2) in (9.3-9) and integrating over the valence band we obtain

$$p_0 = \frac{8\sqrt{2}\pi}{h^3} m_p^{*3/2} e^{-\varepsilon_f/kT} \int_{-\infty}^{\varepsilon_v} \sqrt{\varepsilon_v - \varepsilon}\; e^{\varepsilon/kT}\, d\varepsilon. \tag{9.3-12}$$

Again, the lower limit of the integral has been taken as $-\infty$ rather than the bottom of the valence band; this simplification can be justified by essentially the same argument which was made in connection with (9.3-3) for the conduction band. Making the substitution

$$x_v = \frac{\varepsilon_v - \varepsilon}{kT} \tag{9.3-13}$$

the integral can be expressed in the form

$$p_0 = \frac{8\sqrt{2\pi}}{h^3}(m_p^* kT)^{3/2} e^{-(\varepsilon_f - \varepsilon_v)kT} \int_0^\infty x_v^{1/2} e^{-x_v}\, dx_v, \qquad (9.3\text{-}14)$$

which can be evaluated as a Γ-function as above to give

$$p_0 = U_v e^{-(\varepsilon_f - \varepsilon_v)/kT} \qquad (9.3\text{-}15)$$

with

$$U_v = 2(2\pi m_p^* kT/h^2)^{3/2} \qquad (9.3\text{-}16)$$

The numerical value of U_v may be obtained from (9.3-8), substituting m_p^* in the formula for m_n^*.

The product $n_0 p_0$ can easily be shown to be a function *only* of the energy gap $\Delta\varepsilon$, the effective masses, and the temperature, *independent* of the Fermi level or impurity content. Thus, multiplying the expressions (9.3-6) and (9.3-15) for n_0 and p_0, one may obtain

$$n_0 p_0 = U_c U_v e^{-(\varepsilon_c - \varepsilon_v)/kT} = U_c U_v e^{-\Delta\varepsilon/kT}. \qquad (9.3\text{-}17)$$

Using (9.3-7) and (9.3-16) this can be written in the form

$$n_0 p_0 = 4\left(\frac{2\pi (m_p^* m_n^*)^{1/2} kT}{h^2}\right)^3 e^{-\Delta\varepsilon/kT}. \qquad (9.3\text{-}18)$$

For a given semiconducting substance, the effective masses and the energy gap $\Delta\varepsilon$ are fixed; hence the product $n_0 p_0$ in a given material must be a function only of temperature. This is essentially a mass-action law governing the relative concentrations of holes and electrons in a given material. If the semiconductor is in the pure or intrinsic state, then the concentrations of holes and electrons must be equal, as we have already seen, because the only holes and electrons which can be present are those which are generated in pairs by direct thermal excitation of valence band electrons. In such a material, then

$$p_0 = n_0 = n_i(T) \qquad (9.3\text{-}19)$$

where $n_i(T)$ stands simply for the number of holes or electrons per unit volume in an *intrinsic* sample of the semiconductor in question at temperature T. In a sample of the same substance which is *not* intrinsic but rather an *impurity* semiconductor, the number of holes and electrons is no longer equal; nevertheless, according to (9.3-17) *the product $n_0 p_0$ must be the same for this material as for the intrinsic substance.* We may then write, from (9.3-19)

$$n_0 p_0 = n_i^2(T) \qquad (9.3\text{-}20)$$

with

$$n_i(T) = 2\left(\frac{2\pi (m_p^* m_n^*)^{1/2} kT}{h^2}\right)^{3/2} e^{-\Delta\varepsilon/2kT}, \qquad (9.3\text{-}21)$$

as given by (9.3-18). According to these results, the concentration of carriers in an intrinsic semiconductor is a strong function of temperature, increasing rapidly as the temperature rises, and is likewise strongly dependent upon the energy gap $\Delta\varepsilon$, decreasing rapidly as $\Delta\varepsilon$ increases.

The mass action relation (9.3-17) or (9.3-20) is strongly reminiscent of the laws governing the relative concentrations of ions in substances which are weakly ionized, for example the H^+ and OH^- ions in aqueous systems. As a matter of fact, semiconductors can be understood as an example of the classical Arrhenius theory of weakly ionized electrolytes in terms of a system in which a covalent bond dissociates into an electron and a hole with dissociation energy $\Delta\varepsilon$.

In a purely intrinsic sample, where the hole and electron concentrations are equal, we may equate expressions (9.3-6) and (9.3-15), take the logarithm of both sides of the resulting equation, and solve for the Fermi energy, obtaining

$$\varepsilon_f = \tfrac{1}{2}(\varepsilon_v + \varepsilon_c) + kT \ln\sqrt{U_v/U_c} \qquad \text{(intrinsic).} \qquad (9.3\text{-}22)$$

Since U_v and U_c are given by (9.3-16) and (9.3-7), we may express this result in the form

$$\varepsilon_{fi} = \tfrac{1}{2}(\varepsilon_v + \varepsilon_c) + kT \ln(m_p^*/m_n^*)^{3/4} \qquad \text{(intrinsic),} \qquad (9.3\text{-}23)$$

using the symbol ε_{fi} to denote the Fermi energy of an intrinsic semiconductor. It is clear from this that in an intrinsic semiconductor the Fermi energy is displaced from the center of the energy gap by only a relatively small amount of energy as given by the second term on the right-hand side of (9.3-23). If $m_p^* = m_n^*$, the Fermi level is exactly at the midpoint of the forbidden region, as indicated previously.

9·4 IONIZATION ENERGY OF IMPURITY CENTERS

A donor center in a semiconductor consists of a fixed ion of charge $+e$ to which an electron is loosely bound. If the binding energy is small enough, the "orbit" of the electron will be quite large—so large, indeed, compared to the interatomic spacing that many atoms will be encompassed by the path of the electron about the donor ion.[2] Under these circumstances it will be approximately correct to regard the electron as being immersed in a uniform polarizable medium whose dielectric constant κ is the macroscopic dielectric constant of the semiconductor crystal, since the electrostatic force between electron and donor ion will then be modified, on the average, by the polarization of many intervening atoms. The picture is similar to that of a hydrogen atom in a uniform continuous medium of dielectric constant κ, if one can envision such a situation. This picture will be valid as long as κ is of sufficient magnitude so that the Bohr orbits are large compared with the interatomic spacing.

We may now work out the size of the orbits and associated energy levels according

---

[2] Although descriptive of the situation, this is not very good wave mechanical language; it would be more accurate to say that the wave function of the electron extends over many interatomic distances, so that many atoms are included within the region where the probability density is large.

to the Bohr theory discussed in Section 4.5. It would really be more accurate to use the wave mechanics of the hydrogen atom for these calculations, but since we know that the Bohr theory gives the correct energy levels, and since the Bohr radii and the wave mechanical average electron-nuclear distances are almost the same, we are confident that the simpler Bohr theory will yield the right answers for this very similar situation. The quantization condition for the orbital angular momentum holds in both cases, so that Equation (4.5-2) is still correct provided the electron mass is replaced by m_n^*. On the other hand, due to the polarization of the crystal, the electrostatic force between the donor ion and the bound electron is reduced by a factor κ, becoming $e^2/\kappa r_n^2$, so that Equation (4.5-3) should now be written

$$e^2/\kappa r_n^2 = m_n^* v_n^2/r_n = m_n^* r_n \omega_n^2. \tag{9.4-1}$$

Equations (9.4-1) and (4.5-2) may now be solved as simultaneous equations for r_n and ω_n giving [in analogy to (4.5-4) and (4.5-5)],

$$r_n = \frac{n^2 \hbar^2 \kappa}{m_n^* e^2} \tag{9.4-2}$$

and

$$\omega_n = \frac{m_n^* e^4}{n^3 \hbar^3 \kappa^2} \tag{9.4-3}$$

The kinetic and potential energies may now be evaluated as in (4.5-6) and (4.5-7), recalling, however, that the potential energy is now given by $-e^2/\kappa r_n$, whereby the total energy is

$$\varepsilon_n = \varepsilon_k + \varepsilon_p = -\frac{m_n^* e^4}{2n^2 \kappa^2 \hbar^2} . \tag{9.4-4}$$

The Bohr theory for the hydrogen atom is expressed by the above equations with $m_n^* = m_0$ and $\kappa = 1$. It is evident from (9.4-2) that the Bohr orbits for the donor center are larger than the hydrogen orbits by a factor κ and also by a factor m_0/m_n^*. For n-type germanium, $\kappa = 16$ and $m_n^* \cong m_0/4$, so that the radius of the first Bohr orbit is 64 times the radius of the first Bohr orbit of hydrogen. The value for hydrogen is 0.528 Å, so that the value for the donor atom in germanium is about 34 Å. This is indeed considerably greater than the interatomic distance (which is about 2.44 Å for germanium), hence this picture of the donor center is in this instance to a large extent physically justified.

The ionization energy of the donor center is, from (9.4-4), smaller than that of the hydrogen atom by a factor $1/\kappa^2$ and also by a factor m_n^*/m_0. For germanium this amounts to $(1/256) \cdot (1/4) = 1/1024$. The energy of ionization for the ground state of a donor atom would then be expected to be about 1000 times less than that of the ground state of the hydrogen atom (13.6 eV). We arrive thus at the figure 0.013 eV for the energy of ionization of a donor atom in germanium. This value is in good agreement with the experimentally determined values obtained from measurements of carrier concentration as a function of temperature and from infrared optical absorption as given in Table 9.1. The higher values of ionization energy observed for silicon are accounted for partially by the smaller dielectric constant ($\kappa = 12$) for that material, and partly by the higher effective masses for electrons in silicon. The characteristics of acceptor centers may be understood along the same lines of reasoning, using a picture involving a negatively charged acceptor ion and a positively charged hole in a uniform dielectric medium.

TABLE 9.1.

Ionization Energies for Donor and Acceptor Centers in Ge and Si

Impurity	Ge	Si
P	0.012 eV	0.045 eV
As	0.0127	0.05
Sb	0.010	0.039
B	0.0104	0.045
Al	0.0102	0.06
Ga	0.0108	0.07
In	0.0112	0.16

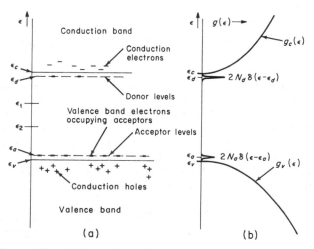

FIGURE 9.8. (a) Energy band diagram of an impurity semiconductor, showing donor and acceptor levels, (b) corresponding density of states curve.

The electrons contributed by donor atoms can thus be construed to originate from localized *donor states* which lie within the forbidden energy gap, a few hundredths of an electron volt *below* the conduction band. Likewise the holes contributed by acceptors can be visualized as being created when electrons which would normally occupy states near the top of the valence band are promoted into initially empty *acceptor levels* which lie in energy a few hundredths of an electron volt above the valence band. This scheme of *donor and acceptor levels* is shown in Figure 9.8(a), and the related density of states picture is shown in Figure 9.8(b).

9·5   STATISTICS OF IMPURITY SEMICONDUCTORS

Before discussing the actual statistical mechanics of impurity semiconductors it is necessary to examine a subtle feature of the occupation statistics of donor and acceptor levels. Confining our discussion for the moment to donor levels, one might be tempted to conclude that the occupation probability associated with such levels would be given simply by the equilibrium Fermi function (5.5-20) with $\varepsilon = \varepsilon_d$. This is not quite true, however, because of the spin degeneracy of the donor levels. There are really *two* quantum states associated with each donor impurity level, corresponding to the two allowed spin orientations of the electron on the donor atom. But as soon as one of these states is occupied occupancy of the other one is precluded, since the valency requirements of the donor ion are satisfied by one and only one electron. This situation alters the statistical problem which leads to the distribution function.

Consider a system having energy levels of this type, as shown in Figure 9.9. We

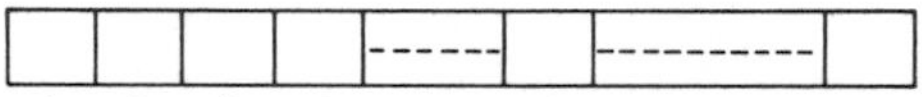

Energy level no.	1	2	3	4	---	i	---	n
Energy	ϵ_1	ϵ_2	ϵ_3	ϵ_4	---	ϵ_i	---	ϵ_n
Degeneracy	g_1	g_2	g_3	g_4	---	g_i	---	g_n
No. of electrons	N_1	N_2	N_3	N_4	---	N_i	---	N_n

FIGURE 9.9.   Energy levels and notation used in the calculations of Section 9.5.

shall suppose that the electrons occupying the levels are indistinguishable particles, and that there are N_i electrons in the ith level, whose degeneracy is g_i. The number of ways of inserting the first electron into the quantum states belonging to the ith level is g_i; the number of ways of inserting the second is $g_i - 2$, since the occupancy of the first state by the first electron precludes the occupancy of the state of opposite spin; the number of ways of inserting the third is $g_i - 4$, and so on. For the ith the number will be $g_i - 2N_i + 2$. The total number of ways of arranging N_i indistinguishable electrons in g_i states under these circumstances is then

$$\frac{g_i(g_i - 2)(g_i - 4) \cdots (g_i - 2N_i + 2)}{N_i!} = \frac{2^{g_i/2}(g_i/2)!}{2^{\frac{g_i}{2} - N_i} N_i! \left(\frac{g_i}{2} - N_i\right)!}. \qquad (9.5\text{-}1)$$

The factor $N_i!$ in the denominator is included because those distributions which are identical except for the permutation of the electrons among themselves must not be counted as distinct distributions, since the electrons are indistinguishable one from another. The total number of ways of realizing a distribution wherein there are $N_1, N_2 \cdots N_n$ electrons in levels $1, 2, \cdots n$ is just the product of factors of the form (9.5-1) over the levels of the system. This is just the quantity $Q(N_1 N_2 \cdots N_n)$ of Chapter 5 [see, for example, (5.5-3)]. For this system, then, we have

$$Q_d(N_1, N_2 \cdots N_n) = \prod_{i=1}^{n} \frac{2^{N_i}(g_i/2)!}{N_i!\left(\frac{g_i}{2} - N_i\right)!}. \qquad (9.5\text{-}2)$$

The resulting distribution function is found by maximizing $\ln Q_d$ as before, using the method of Lagrangean multipliers developed in Chapter 5, the result being

$$N_j = \frac{\frac{1}{2}g_j}{1 + \frac{1}{2}e^{(\varepsilon_j - \varepsilon_f)/kT}}, \tag{9.5-3}$$

where ε_f is the Fermi energy, defined in the usual way. For donor levels, the number of donor atoms is just half the number of spin states, or $\frac{1}{2}g_j$, so that for this system, we may write

$$n_d = \frac{N_d}{1 + \frac{1}{2}e^{(\varepsilon_d - \varepsilon_f)/kT}} \tag{9.5-4}$$

where N_d is the concentration of donor impurity atoms, ε_d is the energy of the donor levels, and n_d is the number of electrons per unit volume occupying the donor levels (thus the concentration of *unionized* donors). The Fermi function is clearly modified by the presence of the factor 1/2 before the exponential term in the denominator.

In a generally similar way one may show that if N_a is the concentration of acceptor impurity atoms and ε_a the energy of the acceptor levels, then the concentration of holes associated with acceptor atoms (thus the density of unionized acceptor sites) p_a will be

$$p_a = \frac{N_a}{1 + \frac{1}{2}e^{(\varepsilon_f - \varepsilon_a)/kT}}. \tag{9.5-5}$$

The density of states associated with the donor and acceptor levels may be represented in terms of Dirac δ-functions by

$$g_d(\varepsilon) = 2N_d\delta(\varepsilon - \varepsilon_d) \tag{9.5-6}$$

and

$$g_a(\varepsilon) = 2N_a\delta(\varepsilon - \varepsilon_a). \tag{9.5-7}$$

These equations express the fact that there are no donor or acceptor levels at energies other than ε_d and ε_a, and that the total number of donor and acceptor quantum states per unit volume is $2N_d$ and $2N_a$, respectively.

If the Boltzmann approximation is valid for *both* the conduction band and the donor levels (that is, if the Fermi level is several kT units below ε_d), the exponential factor in the denominator of the expression in (9.5-4) is much greater than unity, and hence the latter may be neglected; Equation (9.5-4) may then be written as

$$n_d = 2N_d e^{-(\varepsilon_d - \varepsilon_f)/kT}. \tag{9.5-8}$$

Since the concentration of electrons in the conduction band is given by (9.3-6) we may write the ratio of n_d, the number of electrons associated with unionized donors, to $n_0 + n_d$, the total number of free and loosely bound electrons as

$$\frac{n_d}{n_0 + n_d} = \frac{1}{1 + \dfrac{U_c}{2N_d}e^{-(\varepsilon_c - \varepsilon_d)/kT}}. \tag{9.5-9}$$

Now $\varepsilon_c - \varepsilon_d$ is the donor ionization energy, which is ordinarily of the order of or smaller than kT; the exponential factor above is therefore of the order of unity. If $N_D \ll \frac{1}{2} U_c$ the ratio of the number of electrons on unionized donors to the total number will be very small. Since at $300°K$ U_c is of the order of 10^{19} cm^{-3}, this condition is satisfied at that temperature for all donor impurity concentrations which are small compared to this figure. In such instances, it is clear that the donors are *almost completely ionized*, and it is usually convenient and accurate to proceed on the assumption that their ionization is complete, setting n_d equal to zero. At very low temperatures the exponential factor in (9.5-9) may be appreciably smaller than unity, and the criterion for complete donor ionization must be written in the more general form

$$N_d \ll \tfrac{1}{2} U_c e^{-(\varepsilon_c - \varepsilon_d)/kT}. \tag{9.5-10}$$

In a similar fashion, provided the Boltzmann approximation for holes is valid both in the valence band and at the acceptor levels, one may show that the acceptor levels are practically completely ionized whenever

$$N_a \ll \tfrac{1}{2} U_v e^{-(\varepsilon_a - \varepsilon_v)/kT}. \tag{9.5-11}$$

Unless the temperature is very low, this amounts to the requirement $N_a \ll \frac{1}{2} U_v$. In most circumstances, the Fermi level will be within the forbidden gap, many kT units away from both acceptor and donor levels, and both acceptor and donor impurities which may be present will be almost completely ionized.

In a semiconductor at equilibrium there must be either a thermal hole or a positively charged donor ion for every free electron, and a thermal electron or a negatively charged acceptor ion for every free hole. The entire crystal must thus be electrically neutral. This electrical neutrality condition can be expressed by equating the algebraic sum of all negative and positive charges to zero, whereby (recalling that the concentration of *ionized* donors is $N_d - n_d$ and of ionized acceptors $N_a - p_a$),

$$p_0 - n_0 + N_d - N_a + p_a - n_d = 0. \tag{9.5-12}$$

For all but the lowest temperatures and the highest values of impurity concentration, the Boltzmann approximation will be valid for both conduction and valence bands and for both donor and acceptor levels, whereby the concentrations of unionized donors and acceptors, p_a and n_d, may be neglected in (9.5-12), giving

$$p_0 - n_0 + N_d - N_a = 0 \tag{9.5-13}$$

or, using (9.3-6) and (9.3-15),

$$U_v e^{-(\varepsilon_f - \varepsilon_v)/kT} - U_c e^{-(\varepsilon_c - \varepsilon_f)/kT} + (N_d - N_a) = 0. \tag{9.5-14}$$

If we let

$$\alpha = e^{\varepsilon_f/kT}, \qquad \beta_c = e^{-\varepsilon_c/kT} \quad \text{and} \quad \beta_v = e^{\varepsilon_v/kT}, \tag{9.5-15}$$

then (9.5-14) can be written in the form

$$\alpha^2 - \frac{N_d - N_a}{U_c \beta_c} \alpha - \frac{U_v \beta_v}{U_c \beta_c} = 0. \tag{9.5-16}$$

Solving this quadratic equation for α, and taking the logarithm of the result, one may obtain

$$\ln \alpha = \frac{\varepsilon_f}{kT} = \ln\left[\frac{N_d - N_a}{2U_c \beta_c} + \sqrt{\left(\frac{N_d - N_a}{2U_c \beta_c}\right)^2 + \frac{U_v \beta_v}{U_c \beta_c}}\right]. \tag{9.5-17}$$

We must choose the positive sign of the radical in (9.5-17) since when $N_d = N_a = 0$, $e^{\varepsilon_f/kT}$ must be a *positive* quantity. This can be put in a somewhat more satisfying form by noting that

$$\ln(a + \sqrt{a^2 + x^2}) = \ln x + \sinh^{-1} \frac{a}{x}. \tag{9.5-18}$$

Using this relation, and substituting the values given by (9.5-15) for β_v and β_c, (9.5-17) can be expressed as

$$\varepsilon_f = \tfrac{1}{2}(\varepsilon_v + \varepsilon_c) + kT \ln\left(\frac{m_p^*}{m_n^*}\right)^{\tfrac{3}{4}} + kT \sinh^{-1}\left(\frac{N_d - N_a}{2\sqrt{U_c U_v}e^{-\Delta\varepsilon/2kT}}\right). \tag{9.5-19}$$

The first two terms on the right-hand side of this equation are seen from (9.3-23) to represent the Fermi level ε_{fi} for an intrinsic semiconductor, and since from (9.3-17) and (9.3-20) $n_i = \sqrt{U_c U_v}\, e^{-\Delta\varepsilon/2kT}$, we may write (9.5-19) finally as

$$\varepsilon_f = \varepsilon_{fi} + kT \sinh^{-1}\left(\frac{N_d - N_a}{2n_i}\right). \tag{9.5-20}$$

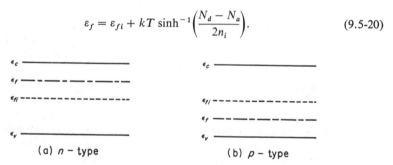

(a) n – type   (b) p – type

FIGURE 9.10. Relative positions of band edges, intrinsic Fermi level, and actual Fermi level in (a) n-type and (b) p-type semiconductors.

This expression gives the Fermi level for an impurity semiconductor in the range over which the Boltzmann approximation is satisfied and the donors and acceptors can be regarded as completely ionized. The quantities n_0, p_0, n_d and p_a can easily be obtained from (9.3-6), (9.3-15), (9.5-4), and (9.5-5) once the Fermi level has been found.

Since $\sinh^{-1} x$ is positive for $x > 0$ and negative for $x < 0$, it is easily seen that $\varepsilon_f > \varepsilon_{fi}$ for n-type semiconductors ($N_d - N_a > 0$), and $\varepsilon_f < \varepsilon_{fi}$ for p-type semiconductors ($N_d - N_a < 0$) as illustrated in Figure 9.10. If the number of donors and acceptors are equal, the inverse hyperbolic sine function vanishes and the material behaves exactly like an intrinsic semiconductor so far as electron and hole populations are concerned. Under these conditions the n- and p-type impurities are said to be fully *compensated*. If the net impurity density $|N_d - N_a|$ is much larger than n_i, the number of thermally excited carriers will be small compared with the total number; in this case the argument of the inverse hyperbolic sine function in (9.5-20) is very large. Since $\sinh^{-1} x \cong \pm \ln |2x|$ for large values of x, it is clear that under these circumstances (9.5-20) becomes

$$\varepsilon_f = \varepsilon_{fi} \pm kT \ln \frac{|N_d - N_a|}{n_i}, \qquad (9.5\text{-}21)$$

the plus sign being used for n-type material ($N_d > N_a$), the minus sign for p-type ($N_a > N_d$). A semiconducting material of this type is said to be a strongly *extrinsic* semiconductor, and the range of applicability of (9.5-21) is often referred to as the *extrinsic* range.

The density of carriers in the conduction and valence bands can, of course, be found by substituting the expression (9.5-17) for the Fermi level into (9.3-6) and (9.3-15). These quantities may also be determined from the electrical neutrality condition (9.5-13), with the help of (9.3-20). For example, substituting $p_0 = n_i^2/n_0$ into (9.5-13), we obtain

$$n_0^2 - (N_d - N_a)n_0 - n_i^2 = 0 \qquad (9.5\text{-}22)$$

which may be solved for n_0, giving

$$n_0 = \tfrac{1}{2}(N_d - N_a) + \sqrt{\tfrac{1}{4}(N_d - N_a)^2 + n_i^2}. \qquad (9.5\text{-}23)$$

In solving the quadratic equation, it is necessary to choose the positive sign for the radical, since the result must reduce to $+n_i$ for $N_d = N_a = 0$. Likewise, substituting $n_0 = n_i^2/p_0$ into the charge neutrality equation, we can show that

$$p_0 = -\tfrac{1}{2}(N_d - N_a) + \sqrt{\tfrac{1}{4}(N_d - N_a)^2 + n_i^2}. \qquad (9.5\text{-}24)$$

From these formulas, it is clear that when $N_d - N_a = 0$, $n_0 = p_0 = n_i$. This is the case of complete compensation (or, if $N_d = N_a = 0$, of an intrinsic semiconductor). If $N_d - N_a > 0$, $n_0 > p_0$ as expected for an n-type impurity semiconductor; if $N_d - N_a < 0$, $p_0 > n_0$ as required for a p-type semiconductor. For strongly *extrinsic* materials these formulas become much simpler. For example, if $N_d - N_a \gg n_i$, as it is for a strongly n-type sample, then

$$n_0 \cong N_d - N_a \qquad \text{and} \qquad p_0 \cong \frac{n_i^2}{N_d - N_a} \qquad (9.5\text{-}25)$$

while for a strongly p-type material where $N_d - N_a$ is large and negative,

$$p_0 \cong N_a - N_d \qquad \text{and} \qquad n_0 \cong \frac{n_i^2}{N_a - N_d}. \qquad (9.5\text{-}26)$$

For the sake of complete generality we have assumed from the beginning that *both* donor and acceptor impurities are present. In many cases, however, especially those involving crystals into which impurities of one type have been intentionally incorporated, the concentration of the *other* impurity species may be completely neglected and indeed set equal to zero in all the formulas of this section.

9·6   CASE OF INCOMPLETE IONIZATION OF IMPURITY LEVELS (VERY LOW TEMPERATURE)

At very low temperatures there may not be sufficient thermal energy available to maintain complete ionization of donor and acceptor impurity levels, as determined by the criteria (9.5-10) and (9.5-11). At zero absolute temperature, for example, in an *n*-type semiconductor, the donor levels must be completely occupied and the conduction band completely devoid of electrons. Both these conditions cannot be realized unless the Fermi level lies *between* the donor levels and the conduction band, as depicted in Figure 9.11. In this case the Boltzmann approximation can no longer be used to

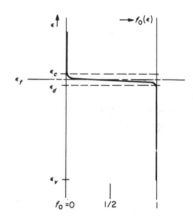

FIGURE 9.11.   Distribution function in an *n*-type semiconductor at a very low temperature.

describe the occupation probability associated with the donor levels, since the Fermi level is now above rather than below the donor energy levels. The Boltzmann approximation can, however, still be used for the electrons in the conduction band, because the Fermi level will remain many kT units below the bottom of the conduction band; it should be recalled in this connection that kT is only about 0.001 eV at 10°K.

We shall treat in detail the case of a purely *n*-type semiconductor where $N_a = 0$, and we shall assume that T is so low that the concentration of thermally created holes in the valence band is negligible. Since $N_a = 0$, of course, p_a must also vanish. Setting $N_a = p_a = p_0 = 0$ in (9.5-12), one may write

$$n_0 - N_d + n_d = 0. \tag{9.6-1}$$

Using (9.3-6) and (9.5-4) to express n_0 and n_d in this equation, we obtain

$$U_c e^{-(\varepsilon_c - \varepsilon_f)/kT} - N_d\left(1 - \frac{1}{1 + \frac{1}{2}e^{(\varepsilon_d - \varepsilon_f)/kT}}\right) = 0. \qquad (9.6\text{-}2)$$

However, since

$$1 - \frac{1}{1 + \frac{1}{2}e^{(\varepsilon_d - \varepsilon_f)/kT}} = \frac{1}{1 + 2e^{-(\varepsilon_d - \varepsilon_f)/kT}}, \qquad (9.6\text{-}3)$$

(9.6-2) can be written as

$$U_c e^{-(\varepsilon_c - \varepsilon_f)/kT} + 2U_c e^{-(\varepsilon_c + \varepsilon_d - 2\varepsilon_f)/kT} - N_d = 0. \qquad (9.6\text{-}4)$$

Letting

$$\beta_c = e^{-\varepsilon_c/kT}, \qquad \beta_d = e^{-\varepsilon_d/kT} \qquad \text{and} \qquad \alpha = e^{\varepsilon_f/kT}, \qquad (9.6\text{-}5)$$

(9.6-4) becomes

$$\alpha^2 + \frac{\alpha}{2\beta_d} - \frac{N_d}{2U_c\beta_c\beta_d} = 0. \qquad (9.6\text{-}6)$$

This equation can be solved for α to yield

$$\alpha = -\frac{1}{4\beta_d} + \sqrt{\frac{1}{16\beta_d^2} + \frac{N_d}{2U_c\beta_c\beta_d}}. \qquad (9.6\text{-}7)$$

The plus sign must be chosen in order that $\alpha = e^{\varepsilon_f/kT}$ be positive and thus ε_f be real. Now, by definition

$$\ln(x + \sqrt{x^2 + a^2}) = \ln a + \sinh^{-1}\frac{x}{a}. \qquad (9.6\text{-}8)$$

Taking the logarithm of both sides of (9.6-7), using (9.6-8) and reverting to original notation with (9.6-5), one may finally obtain

$$\varepsilon_f = \frac{1}{2}(\varepsilon_c + \varepsilon_d) + \frac{kT}{2}\ln\frac{N_d}{2U_c} - kT\sinh^{-1}\left(\sqrt{\frac{U_c}{8N_d}}\,e^{-\Delta\varepsilon_i/2kT}\right), \qquad (9.6\text{-}9)$$

where $\Delta\varepsilon_i = \varepsilon_c - \varepsilon_d$ is the donor ionization energy. The density of free electrons in the conduction band may now be calculated from (9.3-6).

At absolute zero, (9.6-9) gives $\varepsilon_f = \frac{1}{2}(\varepsilon_c + \varepsilon_d)$, which means that the Fermi level is midway between the donor levels and the conduction band. As the temperature increases, the Fermi level first increases slightly (remaining nevertheless below ε_c) and then decreases, moving down through the donor levels toward the center of the gap. For temperatures such that the Fermi level is several kT units below the donor energy ε_d at which point the donors are almost completely ionized, Equation (9.6-9) above reduces, approximately, to (9.5-21). The proof of this statement is assigned as an exercise. It can be shown in a similar way that for a p-type semiconductor under these same circumstances, the Fermi level at zero temperature lies midway between the acceptor levels and the valence band, and with increasing temperature moves first downward a bit, then upward, approaching the value given by (9.5-21) as the acceptor levels become fully ionized. In both n-type and p-type semiconductors the Fermi level approaches the *intrinsic* value ε_{fi} as given by (9.3-23) for sufficiently high temperatures, since at some high temperature the number n_i of thermally excited carriers will be far in excess of the number $|N_d - N_a|$ contributed by donor and acceptor impurities. Beyond this point the behavior of the material will in every way approximate that of an intrinsic semiconductor. In some instances, however, this transition temperature may exceed the melting point of the semiconductor. The Fermi level, as determined by (9.6-9) in the low temperature range and by (9.5-20) at higher temperatures is shown as a function of temperature for both n-type and p-type materials in Figure 9.12.

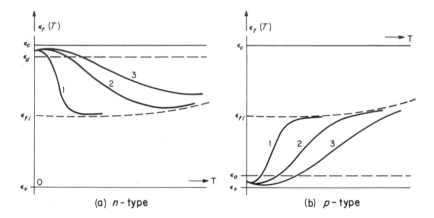

FIGURE 9.12.   Variation of Fermi energy with temperature for (a) n-type and (b) p-type semiconductors, of various impurity densities. The curves marked "1" correspond to relatively low impurity densities, those marked "2" to intermediate densities, and those marked "3" to relatively high impurity densities.

9.7   CONDUCTIVITY

The current density carried by holes and by electrons is equal to the product of their respective charge densities and velocities. Thus, the current density, or the charge density per unit time transported by holes and electrons is

$$\mathbf{I}_n = \rho_n \mathbf{v}_n = -en\mathbf{v}_n \qquad (9.7\text{-}1)$$

and

$$\mathbf{I}_p = \rho_p \mathbf{v}_p = ep\mathbf{v}_p \qquad (9.7\text{-}2)$$

where ρ_n and ρ_p refer to the charge densities associated with the electron and hole densities n and p, and $\mathbf{v}_n$ and $\mathbf{v}_p$ are the average vector velocities of electrons and holes, respectively. Since the average electron velocity is $-\mu_n\mathbf{E}$ and the average hole velocity is $\mu_p\mathbf{E}$, where μ_n and μ_p are the electron and hole mobilities and $\mathbf{E}$ is the field, we may write (9.7-1) and (9.7-2) as

$$\mathbf{I}_n = ne\mu_n\mathbf{E} \tag{9.7-3}$$

and

$$\mathbf{I}_p = pe\mu_p\mathbf{E}, \tag{9.7-4}$$

whereby the total electrical current density $\mathbf{I}$ may be expressed as

$$\mathbf{I} = e(n\mu_n + p\mu_p)\mathbf{E} = \sigma\mathbf{E}. \tag{9.7-5}$$

The electrical conductivity σ is then given by

$$\sigma = e(n\mu_n + p\mu_p). \tag{9.7-6}$$

The mobilities μ_n and μ_p in the above formulas are defined as

$$\mu_n = \frac{e\bar{\tau}_n}{m_n^*} \quad \text{and} \quad \mu_p = \frac{e\bar{\tau}_p}{m_p^*} \tag{9.7-7}$$

where $\bar{\tau}_n$ and $\bar{\tau}_p$ are weighted averages of the relaxation times τ_n for electrons and τ_p for holes over the Maxwell-Boltzmann distribution, calculated as prescribed by Equation (7.3-12).

It should be noted also that n and p are the actual instantaneous values of hole and electron concentration, which are not necessarily identical with the *equilibrium* hole and electron densities n_0 and p_0. It is, however, so far assumed that no *gradients* of carrier density (which would give rise to *diffusion currents*) are present. If the carrier densities are those associated with the equilibrium state, of course, we shall find that

$$\sigma_0 = e(n_0\mu_n + p_0\mu_p) \tag{9.7-8}$$

is the corresponding value of conductivity. It is necessary to make this distinction because it is quite possible to create carrier densities in excess of the equilibrium values in semiconductors, as we shall see later.

For an intrinsic semiconductor with carrier densities equal to the equilibrium values, $n_0 = p_0 = n_i$ and (9.7-8) becomes

$$\sigma_0 = en_i(\mu_n + \mu_p) = en_i\mu_p(b + 1) \tag{9.7-9}$$

where b is defined as the ratio of electron to hole mobility, that is,

$$b = \mu_n/\mu_p. \tag{9.7-10}$$

Using (9.3-21), (9.7-9) can be written in the form

$$\sigma_0 = 2e\mu_p(b + 1)\left(\frac{2\pi(m_p^* m_n^*)^{\frac{1}{2}}kT}{h^2}\right)^{\frac{3}{2}} e^{-\Delta\varepsilon/2kT}. \tag{9.7-11}$$

Since the mobility μ_p usually has a temperature dependence which largely cancels the $T^{\frac{3}{2}}$ temperature variation of the term preceding the exponential factor, and since b is not strongly temperature dependent, the above variation of σ_0 as a function of $1/T$ is essentially exponential. A semilogarithmic plot of σ_0 versus $1/T$ thus yields a straight line of slope $\Delta\varepsilon/2k$, and the value of the energy gap $\Delta\varepsilon$ may be rather accurately determined by determining the slope of an experimentally measured plot of $\ln \sigma_0$ versus $1/T$ in the intrinsic range.

For a semiconductor which is not necessarily intrinsic, but whose hole and electron densities correspond to the equilibrium values n_0, p_0, the conductivity is given by (9.7-8). Using (9.7-10) to eliminate μ_n, and then substituting the expressions (9.5-23) and (9.5-24) for n_0 and p_0, we may write the conductivity as

$$\sigma_0 = e\mu_p \left(bn_0 + p_0\right) = e\mu_p \left[\frac{1}{2}(b-1)(N_d - N_a) + (b+1)\sqrt{\frac{1}{4}(N_d - N_a)^2 + n_i^2}\right]. \tag{9.7-12}$$

Consider for the moment an n-type semiconductor, for which $N_d - N_a$ is positive. At low temperatures, $n_i \ll N_d - N_a$, whereupon, according to (9.7-12), $\sigma_0 \cong (N_d - N_a)be\mu_p$ $= (N_d - N_a)e\mu_n$, which is independent of temperature except insofar as μ_n may depend on T. The semiconductor is then a strongly extrinsic material, most of the carriers being contributed by impurity atoms. As the temperature is increased, n_i will increase rapidly, according to (9.3-21), and at length n_i will equal, then exceed, $N_d - N_a$. For temperatures such that $n_i \gg N_d - N_a$, (9.7-12) gives $\sigma_0 \cong e\mu_p n_i(b + 1)$, the same value given by (9.7-9) for an intrinsic semiconductor, and the conductivity then rises rapidly with increasing temperature due to the rapid increase of n_i. The sample is now an essentially intrinsic semiconductor. The transition between these two extremes of behavior occurs at the temperature where $n_i = N_d - N_a$. As the impurity density $N_d - N_a$ increases this transition temperature increases, and for different semiconducting substances of equal impurity density the transition temperature increases as the band gap $\Delta\varepsilon$ increases, because the temperature required to generate a given density of intrinsic carriers increases with $\Delta\varepsilon$. The same qualitative behavior may be shown to follow from (9.7-12) for p-type materials, wherein $N_d - N_a$ is negative. A semilogarithmic plot of conductivity versus $1/T$ as predicted by (9.7-12) for several samples of varying impurity content is shown in Figure 9.13.

At extremely low temperatures (less than about 30°K) there is not sufficient thermal energy available to maintain the ionization of donor and acceptor levels. In this temperature range the conductivity again falls rapidly with decreasing temperature as free electrons and holes are "frozen out" of the conduction and valence bands into donor and acceptor levels. In crystals of quite high impurity density the wave functions of electrons on neighboring donor atoms or of holes on neighboring acceptor atoms may overlap appreciably. The result of this is to broaden the donor or acceptor level into a narrow energy band, just as atomic energy levels broaden into a band, in the tight binding approximation, as the electronic wave functions of neighboring atoms

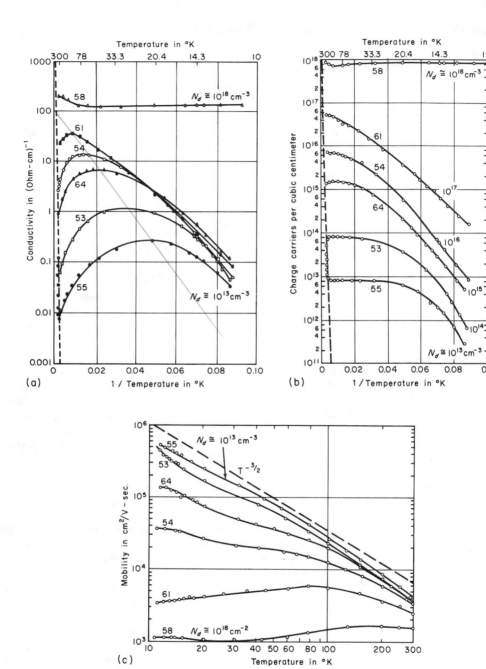

FIGURE 9.13.   (a) Electrical conductivity, (b) carrier concentration and (c) electron mobility for a series of n-type germanium samples covering a wide range of donor impurity content. [After E. M. Conwell, *Proc. I.R.E.*, **40**: 1327 (1952).]

begin to overlap. Since there are twice as many states in this band as there are electrons (or holes), a small residual *impurity band conductivity* may be observed in the partially filled impurity band in the extreme low temperature region where electrons and holes are "frozen out" of the conduction and valence bands into the donor and acceptor levels.

9·8 THE HALL EFFECT AND MAGNETORESISTANCE

In Section 7.6 the Hall effect in free-electron gases was discussed for the case of electrons or holes all having the same x-component of velocity. This analysis leads to the expression

$$E_y = RI_0B_0 \qquad\qquad (9.8\text{-}1)$$

for the transverse Hall field, where I_0 is the current density, B_0 is the magnetic induction and R is a constant of proportionality called the Hall coefficient. The orientation of the sample and of the current and fields is as shown in Figure 7.3. In the simple case discussed in Section 7.6, where the charge carriers are assumed to be electrons, the expression

$$R = -\frac{1}{n_0 ec} \qquad \text{(electrons)} \qquad\qquad (9.8\text{-}2)$$

was obtained for the Hall coefficient. It is easy to see from the analysis of Section 7.6 that if the charge carriers are positive holes whose concentration is p_0, the expression for the Hall coefficient would be

$$R = \frac{1}{p_0 ec} \qquad \text{(holes)}. \qquad\qquad (9.8\text{-}3)$$

We have already pointed out that the effect of the *distribution* of velocities has been neglected in obtaining these results. We must now attempt to include the effect of the velocity distribution in redetermining the Hall coefficient for a semiconductor. In addition, we must account for the fact that in a semiconductor positive and negative charge carriers may be present *simultaneously*.

Beyond these effects which are of immediate interest, there are two other complications with which we shall eventually be concerned. Up to this point, in all the transport and carrier population calculations we have made, we have always gone under the assumption that the surfaces of constant energy in momentum space or k-space were spherical. This assumption arises naturally when the minimum value of energy in the reduced Brillouin zone occurs at the central point $k = 0$. But there are substances, including germanium and silicon (conduction bands only), in which the energy extremum occurs not at the origin, but at a set of crystallographically equivalent points elsewhere in the Brillouin zone. In such instances, the constant energy surfaces in the neighborhood of the energy minima may be ellipsoidal, rather than spherical. The effect of this is to introduce an anisotropy into the effective mass associated with each ellipsoid; the carriers, although behaving as free particles in the sense that the variation of ε with $\mathbf{k}$ is parabolic for any direction of motion, have a different effective mass along each such direction. Although the anisotropy in the transport properties vanishes when one sums over the energy ellipsoids associated with all the energy minimum points, the procedure for computing density of states factors and averaging transport properties over velocities must be modified to account for this rather different configuration of constant energy surfaces.

In addition to this, the *valence* bands of germanium and silicon really consist of two distinct overlapping bands with each of which is associated a different effective mass. Although the constant energy surfaces for both of these bands are roughly spherical near the maximum point at $k = 0$, there are actually two species of holes ("light" holes and "heavy" holes corresponding to the two effective masses) present simultaneously.

Any theory of transport processes which seriously attempts to explain all the very abundant experimental data in this field relating to germanium and silicon must of necessity take into account both these added complications. But to begin with, we shall have enough to do in extending the picture of the Hall effect and related phenomena to cover electrons and holes simultaneously and to include the effect of the distribution of velocities. We shall therefore continue to assume that we are dealing with a simple semiconductor with spherical energy surfaces in both conduction and valence band, each having only one variety of electrons or holes. We shall later indicate how our results may be extended to allow for the two additional difficulties discussed above.

We shall begin by considering only a single type of carrier (we shall choose holes to work with) and incorporating the effect of the velocity distribution. The most obvious way of proceeding is from the Boltzmann equation with the Lorentz force term (7.6-2). It is simpler, however, and quite instructive to use an alternative method[3] which begins with the equation of motion for a hole in the presence of both a steady electric field $\mathbf{E}$ and a steady magnetic induction $\mathbf{B}_0$. This equation can be written

$$\mathbf{F} = m_p^* \frac{d\mathbf{v}}{dt} = e\mathbf{E} + \frac{e}{c}\mathbf{v} \times \mathbf{B}_0. \qquad (9.8\text{-}4)$$

If we assume that the sample, current, and magnetic field are oriented as shown in Figure 7.3, then, clearly $\mathbf{B}_0$ has only a z-component, and (9.8-4) reduces to two equations for the x- and y-velocity components, which have the form

$$dv_x/dt = \frac{eE_x}{m_p^*} + \omega_0 v_y \qquad (9.8\text{-}5)$$

$$dv_y/dt = \frac{eE_y}{m_p^*} - \omega_0 v_x, \qquad (9.8\text{-}6)$$

where

$$\omega_0 = eB_0/m_p^* c. \qquad (9.8\text{-}7)$$

---

[3] See, for example, H. Brooks, *Advances in Electronics and Electron Physics*. New York: Academic Press, Inc. (1955), Vol. VII, p. 127 ff.

This set of equations can be most easily solved by multiplying (9.8-6) by the imaginary number i and adding the result to (9.8-5). One obtains thereby the single equation

$$\frac{dV}{dt} + i\omega_0 V = \frac{e\mathscr{E}}{m_p^*} \tag{9.8-8}$$

where the complex quantities V and $\mathscr{E}$ are defined by

$$V = v_x + iv_y \quad \text{and} \quad \mathscr{E} = E_x + iE_y. \tag{9.8-9}$$

Equation (9.8-8) is easily solved by multiplying both sides by $e^{i\omega_0 t}$. The left-hand side then can be written as the time derivative of $Ve^{i\omega_0 t}$, and the resulting equation integrated with respect to time to give

$$Ve^{i\omega_0 t} = \frac{e\mathscr{E}}{i\omega_0 m_p^*} e^{i\omega_0 t} + C \tag{9.8-10}$$

where C is a constant of integration. If we let $V_0 = v_{x0} + iv_{y0}$ be the value of V at $t = 0$, C can be evaluated in terms of V_0 from this equation, allowing us to write (9.8-10) as

$$V = V_0 e^{-i\omega_0 t} + \frac{e\mathscr{E}}{i\omega_0 m_p^*}(1 - e^{-i\omega_0 t}). \tag{9.8-11}$$

This expression must first be averaged over the exponential distribution of relaxation times, whereby, in analogy with (7.2-13)

$$\langle V \rangle_{\text{Av over } t} = \frac{\int_0^\infty V(t)e^{-t/\tau_p}\, dt}{\int_0^\infty e^{-t/\tau_p}\, dt} = \frac{1}{1 + i\omega_0 \tau_p}\left[V_0 + \frac{e\mathscr{E}\tau_p}{m_p^*}\right], \tag{9.8-12}$$

where τ_p represents the relaxation time for holes. We must now average this expression over a Maxwell-Boltzmann distribution of velocities. If we assume, as usual, that τ_p is a function of the magnitude v, but that it is independent of *direction*, i.e., isotropic, then the averages associated with the first term in (9.8-12) will always amount to an average of a function only of the magnitude of v times a component v_{x0} ($= v_0 \sin\theta \cos\phi$) or v_{y0} ($= v_0 \sin\theta \sin\phi$). Such an average is always zero, because the integral of the resulting expression over the azimuthal coordinate ϕ always vanishes. We may thus drop this term from our consideration altogether. If we separate the second term above into real and imaginary parts, we may identify the real part with v_x and the imaginary part with v_y, according to (9.8-9). We obtain thus

$$\langle V \rangle_{\text{Av over } t} = \frac{e}{m_p^*}\left[\left(\tau_p - \frac{\omega_0^2 \tau_p^3}{1 + \omega_0^2 \tau_p^2}\right)E_x + \frac{\omega_0 \tau_p^2}{1 + \omega_0^2 \tau_p^2}E_y\right]$$

$$+ i\frac{e}{m_p^*}\left[\frac{\tau_p}{1 + \omega_0^2 \tau_p^2}E_y - \frac{\omega_0 \tau_p^2}{1 + \omega_0^2 \tau_p^2}E_x\right], \tag{9.8-13}$$

and it is this expression which must be averaged over the Boltzmann distribution to obtain average values for v_x and v_y. We saw in Section 7.3, using the Boltzmann equation, that if $\alpha(v)$ is the quantity to be averaged, the proper average to take is

$$\bar{\alpha} = \frac{\langle v^2 \alpha(v) \rangle}{\langle v^2 \rangle}. \tag{9.8-14}$$

In the expression (9.8-13), if there is no magnetic field then $\omega_0 = 0$ and the equation gives simply the velocities v_x and v_y in terms of the electric fields E_x and E_y. In this limit the equation should agree with (7.3-11), and indeed it does, *provided* that τ_p is averaged according to the prescription (9.8-14). As a matter of fact, in the general case *all* the averages must be computed in this way, leading to

$$\overline{V} = \bar{v}_x + i\bar{v}_y \tag{9.8-15}$$

where

$$\bar{v}_x = \mathrm{Re}(\overline{V}) = \frac{e}{m_p^*} \left[\bar{\tau}_p E_x + \omega_0 \overline{\left(\frac{\tau_p^2}{1 + \omega_0^2 \tau_p^2} \right)} E_y - \omega_0^2 \overline{\left(\frac{\tau_p^3}{1 + \omega_0^2 \tau_p^2} \right)} E_x \right] \tag{9.8-16}$$

and

$$\bar{v}_y = \mathrm{Im}(\overline{V}) = \frac{e}{m_p^*} \left[\overline{\left(\frac{\tau_p}{1 + \omega_0^2 \tau_p^2} \right)} E_y - \omega_0 \overline{\left(\frac{\tau_p^2}{1 + \omega_0^2 \tau_p^2} \right)} E_x \right]. \tag{9.8-17}$$

The x- and y-components of the current density may, in the customary way, be expressed as

$$I_x = p_0 e \bar{v}_x \quad \text{and} \quad I_y = p_0 e \bar{v}_y. \tag{9.8-18}$$

In the usual experimental arrangement, as shown in Figure 7.3, the y-component of current is zero, whereby, from (9.8-17)

$$E_y = \omega_0 \frac{\overline{\tau_p^2/(1 + \omega_0^2 \tau_p^2)}}{\overline{\tau_p/(1 + \omega_0^2 \tau_p^2)}} E_x. \tag{9.8-19}$$

Substituting this in (9.8-16) one may obtain finally

$$I_x = \frac{p_0 e^2}{m_p^*} \left[\bar{\tau}_p + \omega_0^2 \left(\frac{\left[\overline{\tau_p^2/(1 + \omega_0^2 \tau_p^2)} \right]^2}{\overline{\tau_p/(1 + \omega_0^2 \tau_p^2)}} - \left[\overline{\tau_p^3/(1 + \omega_0^2 \tau_p^2)} \right] \right) \right] E_x. \tag{9.8-20}$$

If τ_p is independent of v, then $\overline{f(\tau_p)} = f(\tau_p)$ and the above equations reduce to the very simple and intuitive results

$$I_x = \sigma_0 E_x = \frac{p_0 e^2 \tau_p}{m_p^*} E_x \tag{9.8-21}$$

$$E_y = \omega_0 \tau_p E_x = \frac{e B_0 \tau_p}{m_p^* c} \frac{I_x}{\sigma_0} = \frac{1}{p_0 e c} B_0 I_x \tag{9.8-22}$$

which are in agreement with (7.3-14) and with (9.8-1) and (9.8-3). It is clear that the simple expression for the Hall coefficient obtained previously is correct even when there is a Boltzmann distribution of velocities, provided that τ is independent of velocity. It is also evident from (9.8-21) that in this case the electrical conductivity is independent of the magnetic field; there is no *magnetoresistance*.

Usually, however, τ_p does depend on velocity, and then these results are no longer correct. Under such circumstances it is most convenient to make the simplifying assumption that $\omega_0\tau_p \ll 1$. This condition will be satisfied best for small values of the magnetic induction and small values of τ_p which are generally associated with higher temperatures. It is quite generally satisfied for most substances under commonly encountered experimental conditions down to the liquid nitrogen temperature range, although it is also usually quite possible to violate it, if that is what is desired. For example, if $B_0 = 10{,}000$ oersteds, $\tau_p = 10^{-12}$ sec (corresponding to a mobility of the order of 2000) and $m_p^* = m_0$, then $\omega_0\tau_p \cong 0.2$. The requirement would then be marginally satisfied, but higher fields or larger relaxation times would result in its violation. If the condition is assumed to hold, the quantity $\omega_0^2\tau_p^2$ in (9.8-20) can be neglected throughout in comparison with unity. This equation may then be written [expressing ω_0 by (9.8-7)] as

$$I_x = \sigma_0 E_x\left[1 - \frac{e^2 B_0^2}{m_p^{*2}c^2}\frac{\overline{\tau_p^3}\,\overline{\tau_p} - (\overline{\tau_p^2})^2}{(\overline{\tau_p})^2}\right] \tag{9.8-23}$$

where $\sigma_0 = p_0 e^2 \overline{\tau}_p/m_p^*$. It is apparent now that the electrical conductivity is dependent on the magnetic field; this effect is called *magnetoresistance*. It can be shown that the second term in brackets above is always positive, so that the conductivity is invariably *reduced* by the magnetic field. Equation (9.8-23) is more commonly written in a slightly different way; solving for E_x in terms of I_x and noting that $(1 - \omega_0^2\tau_p^2)^{-1} \cong 1 + \omega_0^2\tau_p^2$ for $\omega_0\tau_p \ll 1$, we may obtain

$$E_x = \rho_0 I_x\left[1 + \frac{e^2 B_0^2}{m_p^{*2}c^2}\frac{\overline{\tau_p^3}\,\overline{\tau_p} - (\overline{\tau_p^2})^2}{(\overline{\tau_p})^2}\right] = (\rho_0 + \Delta\rho)I_x, \tag{9.8-24}$$

where $\rho_0 = 1/\sigma_0$ is the zero-field resistivity. The *magnetoresistance coefficient* is then defined as

$$\frac{\Delta\rho}{\rho_0 B_0^2} = \frac{e^2}{m_p^{*2}c^2}\frac{\overline{\tau_p^3}\,\overline{\tau_p} - (\overline{\tau_p^2})^2}{(\overline{\tau_p})^2}. \tag{9.8-25}$$

It is clear from (9.8-24) that the small-field longitudinal magnetoresistance predicted by these equations is proportional to the *square* of the magnetic induction B_0. Physically, the magnetoresistance effect is due to the fact that the magnetic field deflects the particles upward or downward along the y-axis in a direction normal to the direction of the current vector. The resulting trajectories are curved rather than straight, and the average drift distance along the current flow direction between successive collisions is thereby reduced.

The Hall field E_y, using the same approximation in (9.8-19), becomes

$$E_y = \omega_0 \frac{\overline{\tau_p^2}}{\overline{\tau_p}} E_x. \tag{9.8-26}$$

The field E_x in this equation may be expressed with good accuracy as I_x/σ_0, the magnetoresistive term in (9.8-24) being small compared to the zero-field term for $\omega_0\tau_p \ll 1$. Substituting this value, along with the expressions for σ_0 and ω_0 into (9.8-26), we find that

$$E_y = RB_0I_x \qquad (9.8\text{-}27)$$

where [recalling (9.8-14)] the Hall coefficient R is given by

$$R = \frac{1}{p_0ec}\frac{\overline{\tau_p^2}}{(\overline{\tau}_p)^2} = \frac{1}{p_0ec}\frac{\langle v^2\rangle\langle v^2\tau_p^2\rangle}{\langle v^2\tau_p\rangle^2}. \qquad (9.8\text{-}28)$$

The Hall coefficient is modified by the factor $(\overline{\tau_p^2})/(\overline{\tau}_p)^2$ from the value given by the simple theory of Chapter 7. It is evident also that the value given by (9.8-27) is a small-field value, good only so long as $\omega_0\tau_p \ll 1$; should $\omega_0\tau_p$ equal or exceed unity, the magnetoresistance term in (9.8-24) would become comparable in magnitude to the zero-field term and we could no longer replace E_x by I_x/σ_0 in (9.8-26). We should have to express E_x as I_x/σ, where σ has a strong magnetic field dependence. The Hall coefficient would then exhibit a significant dependence upon the magnetic field. The extension of these analyses to the case where $\omega_0\tau$ is no longer small compared with unity involves extensive numerical calculations and will not be attempted here. In the above analysis, we have assumed that the charge carriers are holes. The result for an n-type specimen, wherein the charge carriers are electrons, can be obtained by replacing e with $-e$ throughout. It should be noted that when this is done the characteristic frequency (9.8-7) becomes $-eB_0/m_n^*c$, and therefore one must *also* replace ω_0 with $-\omega_0$ throughout. In this manner, one may readily ascertain that when the carriers are electrons, the sign of the Hall coefficient is changed, but the form of the conductivity and magnetoresistance coefficients are unaltered.

If the dominant scattering process is acoustical mode lattice scattering, then, according to the results of Section 7.5, the mean free path will be essentially independent of velocity, whereby $\tau_p(v) = \lambda_p/v$. It is then clear that $\langle v^2\tau_p^2\rangle = \langle\lambda_p^2\rangle = \lambda_p^2$; also, from (7.3-12) and (7.3-18) we see that $\langle v^2\rangle = 3kT/m$ and $\langle v^2\tau_p\rangle = \lambda_p\langle v\rangle = \lambda_p\overline{c}$, where $\overline{c}$ is given by (5.4-20). Substituting these values into (9.8-28), we obtain $\overline{\tau_p^2}/(\overline{\tau}_p)^2 = 3\pi/8$, and the Hall coefficient then becomes

$$R = \frac{3\pi}{8}\frac{1}{p_0ec} \qquad (p\text{-type})$$

$$= -\frac{3\pi}{8}\frac{1}{n_0ec} \qquad (n\text{-type}). \qquad (9.8\text{-}29)$$

The magnetoresistance coefficient (9.8-25) may be evaluated in this case using the same general approach. One may in fact show that when acoustical mode phonon scattering is the dominant mechanism, (9.8-23) becomes

$$I_x = \sigma_0E_x\left[1 - \frac{e^2B_0^2}{m_p^{*2}c^2}\frac{\lambda^2}{kT}\left(\frac{4\text{-}\pi}{8}\right)\right].$$

The details of deriving this result are assigned as an exercise for the student.

It would seem, from this expression, that the magnetoresistance might become very large at low temperatures, and this is indeed possible. Under these circumstances, however, the contribution of small amounts of impurity scattering, which causes a magnetoresistance effect proportional to T^3 becomes important and in fact represents the limiting factor as the absolute temperature approaches zero.

Consider now a semiconductor crystal which is not strongly extrinsic, but which contains both free electrons and holes in significant concentrations. In this case, the holes and electrons can be considered separately by exactly the same methods used before in deriving equations (9.8-16) and (9.8-17). The hole and electron velocity components are then expressed as equations having the form of (9.8-16) and (9.8-17), the velocity components $\bar{v}_{px}$ and $\bar{v}_{py}$ for the holes being identical with those equations, while the electron velocity components $\bar{v}_{nx}$ and $\bar{v}_{ny}$ are given by equations wherein e is replaced by $-e$, m_p^* by m_n^*, τ_p by τ_n (the latter symbol referring to the relaxation time for electrons in the conduction band), and eB_0/m_p^*c by $-eB_0/m_n^*c$. If one makes the further assumption that $\omega_0\tau \ll 1$ for *both* holes and electrons so that the term $\omega_0^2\tau^2$ may be neglected in the denominators of the averages, then one may write for the current components the expressions

$$I_x = -n_0 e\bar{v}_{nx} + p_0 e\bar{v}_{px} = \frac{n_0 e^2}{m_n^*}\left[E_x\bar{\tau}_n - E_y\omega_{on}\overline{\tau_n^2} - E_x\omega_{on}^2\overline{\tau_n^3}\right]$$

$$+ \frac{p_0 e^2}{m_p^*}\left[E_x\bar{\tau}_p + E_y\omega_{op}\overline{\tau_p^2} - E_x\omega_{op}^2\overline{\tau_p^3}\right], \qquad (9.8\text{-}30)$$

and $I_y = -n_0 e\bar{v}_{ny} + p_0 e\bar{v}_{py} = \dfrac{n_0 e^2}{m_n^*}(E_y\bar{\tau}_n + E_x\omega_{on}\overline{\tau_n^2}) + \dfrac{p_0 e^2}{m_p^*}(E_y\bar{\tau}_p - E_x\omega_{op}\overline{\tau_p^2})$, (9.8-31)

where $\qquad\qquad\qquad \omega_{on} = \dfrac{eB_0}{m_n^*c} \qquad$ and $\qquad \omega_{op} = \dfrac{eB_0}{m_p^*c}.$ $\qquad\qquad$ (9.8-32)

Using (9.7-7) to eliminate the effective masses and rearranging the terms, the above equations may be put in the form

$$I_x = n_0 e\mu_n\left[E_x - E_y\omega_{on}(\overline{\tau_n^2}/\bar{\tau}_n) - E_x\omega_{on}^2(\overline{\tau_n^3}/\bar{\tau}_n)\right]$$

$$+ p_0 e\mu_p\left[E_x + E_y\omega_{op}(\overline{\tau_p^2}/\bar{\tau}_p) - E_x\omega_{op}^2(\overline{\tau_p^3}/\bar{\tau}_p)\right], \qquad (9.8\text{-}33)$$

and $\qquad I_y = E_x\left((n_0 e\mu_n\omega_{on}(\overline{\tau_n^2}/\bar{\tau}_n) - p_0 e\mu_p\omega_{op}(\overline{\tau_p^2}/\bar{\tau}_p)\right) + E_y(n_0 e\mu_n + p_0 e\mu_p).$ (9.8-34)

Since the Hall voltage is measured under open-circuit conditions, $I_y = 0$, as before, whereby (9.8-34) reduces to

$$E_y = \frac{p_0\mu_p\omega_{op}\dfrac{\overline{\tau_p^2}}{\bar{\tau}_p} - n_0\mu_n\omega_{on}\dfrac{\overline{\tau_n^2}}{\bar{\tau}_n}}{n_0\mu_n + p_0\mu_p} \cdot E_x \qquad (9.8\text{-}35)$$

By substituting this expression back into (9.8-33), we may after some algebra obtain

$$
I_x = \sigma_0 E_x \left[1 - \frac{n_0\mu_n\omega_{on}^2 \dfrac{\overline{\tau_n^3}}{\overline{\tau}_n} + p_0\mu_p\omega_{op}^2 \dfrac{\overline{\tau_p^3}}{\overline{\tau}_p}}{n_0\mu_n + p_0\mu_p} - \left(\frac{n_0\mu_n\omega_{on}\dfrac{\overline{\tau_n^2}}{\overline{\tau}_n} - p_0\mu_p\omega_{op}\dfrac{\overline{\tau_p^2}}{\overline{\tau}_p}}{n_0\mu_n + p_0\mu_p} \right)^2 \right]. \quad (9.8\text{-}36)
$$

where σ_0, as usual, is given by (9.7-8). The magnetoresistance term is now much more complex, but is nevertheless small compared to the zero-field value provided that $\omega_{on}\tau_n \ll 1$ and $\omega_{op}\tau_p \ll 1$. The Hall field is given by (9.8-35), and the small-field Hall coefficient can be obtained by replacing E_x by I_x/σ_0, as before, the difference between σ and σ_0 as represented by the magnetoresistance term being negligible under these conditions. Expressing ω_{on} and ω_{op} by (9.8-32) and eliminating the effective masses in favor of the mobilities and relaxation times by (9.7-7), we obtain

$$
E_y = RB_0I_x = \frac{1}{ec} \frac{p_0 \dfrac{\overline{\tau_p^2}}{(\overline{\tau}_p)^2} - b^2 n_0 \dfrac{\overline{\tau_n^2}}{(\overline{\tau}_n)^2}}{(bn_0 + p_0)^2} B_0 I_x. \quad (9.8\text{-}37)
$$

where, as always, $b = \mu_n/\mu_p$.

This expression gives the Hall coefficient in a nonextrinsic semiconductor. It is seen to reduce to (9.8-29) when $p_0 \gg n_0$ or $n_0 \gg p_0$. If the dominant scattering process for both electrons and holes is acoustical mode phonon scattering[4] then $\overline{\tau_n^2}/(\overline{\tau}_n)^2 = \overline{\tau_p^2}/(\overline{\tau}_p)^2 = 3\pi/8$ as shown previously and (9.8-37) then yields

$$
R = \frac{3\pi}{8ec} \frac{p_0 - b^2 n_0}{(bn_0 + p_0)^2}. \quad (9.8\text{-}38)
$$

For an intrinsic semiconductor $p_0 = n_0 = n_i$, and (9.8-38) reduces to

$$
R = \frac{3\pi}{8} \frac{1}{n_i ec} \frac{1 - b}{1 + b} \qquad \text{(Intrinsic)}. \quad (9.8\text{-}39)
$$

For $p_0 \gg n_0$, the Hall coefficient, as given by (9.8-37), is positive; for $n_0 \gg p_0$ it is negative. A reversal of the sign of the Hall coefficient takes place for values of p_0 and n_0 such that the numerator of (9.8-37) vanishes. If the conditions under which (9.8-38) is valid are satisfied, this reversal of the sign of the Hall field occurs when $p_0 = b^2 n_0$. Since $n_0 p_0 = n_i^2$, this amounts simply to $p_0 = bn_i$ or $n_0 = n_i/b$. If $b \neq 1$, the reversal point will not coincide exactly with the intrinsic point. For example, for $b > 1$, there will be a range of carrier concentrations wherein the sign of the Hall coefficient is negative despite the fact that the sample is p-type!

The Hall effect, aside from any fundamental interest it may excite, is important because it affords a practical way of telling whether a sample is n-type or p-type and determining the concentration of carriers. When combined with measurements of conductivity, the Hall effect can also be used to determine the mobility of the charge

---

[4] Despite the fact that formula (9.8-38) is very frequently used to analyze Hall measurements in nonextrinsic materials, these conditions are often not strictly fulfilled—usually because of the presence of optical mode phonon scattering.

carriers. In samples which are highly *extrinsic* and in which the predominant scattering mechanism is acoustical mode phonon scattering, it is clear from (9.8-29) that a measurement of R can be used directly to obtain n_0 or p_0, and the conductivity type follows from the sign of R. If other scattering mechanisms are important, then, of course, in order to obtain an absolute measurement of carrier concentration, the quantity $\overline{\tau^2}/(\bar{\tau})^2$ associated with the dominant scattering mechanism must be known, and in order to evaluate this one must know how τ varies as a function of velocity. However, the factor $\overline{\tau^2}/(\bar{\tau})^2$ is not much greater nor less than unity for any known scattering process which occurs in actual practice, so that a measurement of carrier concentration which is accurate within about a factor of two can always be obtained whatever the scattering mechanisms are, and this is often all that is required.

The mobility of the charge carriers in extrinsic samples can be obtained by eliminating the carrier concentration between the expression (9.8-28) and the equation $\sigma_0 = p_0 e \mu_p$ which results from (9.7-8) when the concentration of the minority carriers (electrons in this example) is neglected. The result, for a p-type semiconductor, is

$$\mu_p = cR\sigma_0 \frac{(\bar{\tau}_p)^2}{\overline{\tau_p^2}}. \tag{9.8-40}$$

For an n-type extrinsic semiconductor, one may likewise show that

$$\mu_n = -cR\sigma_0 \frac{(\bar{\tau}_n)^2}{\overline{\tau_n^2}}. \tag{9.8-41}$$

The product $\pm cR\sigma_0$ is often defined as the *Hall Mobility*, thus

$$\mu_{Hn} = -cR\sigma_0 \qquad (n\text{-type})$$

$$\mu_{Hp} = cR\sigma_0 \qquad (p\text{-type}). \tag{9.8-42}$$

Equations (9.8-40) and (9.8-41) can then be written

$$\mu_p = \mu_{Hp}((\bar{\tau}_p)^2/\overline{\tau_p^2}) \tag{9.8-43}$$

and

$$\mu_n = \mu_{Hn}((\bar{\tau}_n)^2/\overline{\tau_n^2}). \tag{9.8-44}$$

The quantities μ_p and μ_n are the true *drift mobilities* or *conductivity mobilities*; the Hall mobility on the other hand is simply a quantity with the dimensions of mobility which is easily obtained from quantities which can be quite simply measured experimentally. The factor $(\bar{\tau})^2/\overline{\tau^2}$ must be known to convert Hall mobility into the true drift mobility. Since this factor is usually not far from unity, however, the Hall mobility gives a rough indication of the true mobility in most cases which arise in connection with experiment. Experimental measurements of the carrier concentration and electron Hall mobility for n-type germanium samples as a function of temperature are illustrated in Figure 9.13 for a wide range of donor impurity concentrations.

For nonextrinsic samples, one proceeds in much the same way. From (9.8-37) or (9.8-38) one may eliminate either n_0 or p_0 by means of the mass action relation $n_0 p_0 = n_i^2$. The resulting equation, expressing n_0 or p_0 as a function of R is of the fourth degree and cannot be treated analytically in a simple way, but may be solved in each instance by approximate or numerical methods. In addition, there is an ambiguity in that there are two real roots rather than one. This can be seen from Figure 9.14, where the Hall coefficient is plotted as a function of p_0; for a given value of R there are always *two* possible values of p_0. In practice, however, it is usually fairly easy to choose which root is the right one by making a series of Hall effect

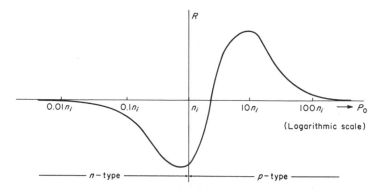

FIGURE 9.14.   Hall coefficient of a semiconductor crystal as a function of hole concentration.

measurements over a wide range of temperatures. In addition, in order to obtain the carrier concentrations by the above procedure, it is necessary to know the mobility ratio b. This may be ascertained by making separate measurements of Hall mobility in highly extrinsic n- and p-type samples, or from the results of excess carrier drift measurements, which will be discussed later. If n_0 and p_0 are ascertained from the Hall coefficient in this way, and if the ratio $\mu_n/\mu_p = b$ is known, the mobilities μ_n and μ_p may be determined from conductivity data, by means of (9.7-8).

9·9   CYCLOTRON RESONANCE AND ELLIPSOIDAL ENERGY SURFACES

The electrons and holes in semiconductors may be made to exhibit a resonance in the presence of a large constant magnetic field (which we shall assume to lie along the z-axis) and a small transverse oscillating radiofrequency or microwave electromagnetic field whose electric vector lies in the xy-plane. The resulting hole or electron orbits are circular at resonance, and energy is absorbed from the rf-field every half cycle, as in a cyclotron. The situation is, in fact, precisely analogous to what happens to nuclear charged particles in a cyclotron, and for this reason the phenomenon is termed *cyclotron resonance*.

    Let us initially investigate the case of spherical energy surfaces and a single scalar effective mass. The Lorentz force on a hole (and we shall consider holes in a p-type sample in this specific calculation) is given by (9.8-4). If the effective mass m_p^* is independent of direction, and if the electric vector of the oscillating field, whose frequency is ω, is assumed to vibrate along the x-direction, then

$$E_x = E_0 e^{i\omega t}; \qquad E_y = E_z = 0 \tag{9.9-1}$$

while
$$B_z = B_0; \qquad B_x = B_y = 0, \tag{9.9-2}$$

and the components of the Lorentz force give for the equations of motion

$$F_x = m_p^* \frac{d^2 x}{dt^2} = \frac{e}{c} v_y B_0 + e E_0 e^{i\omega t} \tag{9.9-3}$$

$$F_y = m_p^* \frac{d^2 y}{dt^2} = -\frac{e}{c} v_x B_0 \tag{9.9-4}$$

$$F_z = m_p^* \frac{d^2 z}{dt^2} = 0. \tag{9.9-5}$$

In writing these equations it has been assumed that the magnetic vector of the rf-field is negligible compared with the constant field $\mathbf{B}_0$. The z-component equation (9.9-5) tells us merely that the particle moves with constant velocity along the z-direction; this is of no particular interest and we need trouble ourselves no further about it. The x- and y-component equations may be written

$$\frac{d^2 x}{dt^2} = \omega_0 \frac{dy}{dt} + \frac{e E_0}{m_p^*} e^{i\omega t} \tag{9.9-6}$$

$$\frac{d^2 y}{dt^2} = -\omega_0 \frac{dx}{dt} \tag{9.9-7}$$

where ω_0 is the very same frequency defined by (9.8-7) in connection with the Hall effect.

If oscillatory solutions of the form $x = x_0 e^{i\omega t}$ and $y = y_0 e^{i\omega t}$ are assumed these equations reduce to

$$i\omega\omega_0 y_0 + \omega^2 x_0 = -e E_0 / m_p^* \tag{9.9-8}$$

and
$$-\omega^2 y_0 + i\omega\omega_0 x_0 = 0 \tag{9.9-9}$$

which can be solved for the amplitudes x_0 and y_0 to give

$$\dot{x}_0 = \frac{e E_0 / m_p^*}{\omega_0^2 - \omega^2} \tag{9.9-10}$$

$$y_0 = \frac{i\omega_0}{\omega} x_0 = \frac{i\omega_0 e E_0 / m_p^*}{\omega(\omega_0^2 - \omega^2)}. \tag{9.9-11}$$

When $\omega = \omega_0 = e B_0 / m_p^* c$ resonance occurs and the amplitudes become very large.

The resonance frequency ω_0 is often referred to as the *cyclotron frequency*. Clearly, if ω_0 is measured experimentally and if the magnetic induction B_0 is accurately known, the effective mass m_p^* can be determined. In practice, the cyclotron resonance effect affords one of the best methods of measuring effective masses. From (9.9-10) and (9.9-11), it can be seen that as $\omega \to \omega_0$

$$y_0 \cong ix_0 = x_0 e^{i\pi/2} \tag{9.9-12}$$

whereby $$x(t) = x_0 e^{i\omega_0 t} \tag{9.9-13}$$

$$y(t) = y_0 e^{i\omega_0 t} = x_0 e^{i(\omega_0 t + \pi/2)}. \tag{9.9-14}$$

At resonance, then, $x(t)$ and $y(t)$ are orthogonal harmonic vibrations of equal amplitude which differ in phase by $90°$. It is easily seen that the resultant particle trajectory, representing the hole orbit, is circular.

In practice, the resonance is usually observed by measuring the Q-factor of a microwave resonant cavity as the magnetic field B_0 (and hence the frequency ω_0) is varied. When ω_0 equals the microwave excitation frequency, a sharp decrease in the Q of the cavity is obtained, because at resonance a great deal of electromagnetic field energy is absorbed when carriers are excited to large resonance amplitudes and then are scattered by impurities or lattice vibrations. It is worth noting that in this experiment one wishes to make the quantity $\omega_0\tau$ much *larger* than unity, so that the carriers can be excited through several complete orbits before being scattered. Under these conditions they are capable of absorbing a great deal of microwave energy and transforming it, when scattering takes place, into vibrations of the crystal lattice. The experiment is usually performed at low temperature, so as to obtain a large value of τ, and with as large a static field B_0 as is possible consistent with the requirement that ω_0 be an experimentally accessible microwave frequency. In the case of Hall effect measurements, on the other hand, the objective is often to make $\omega_0\tau$ much *less* than unity so as to simplify the interpretation of the data.

When the cyclotron resonance effect in silicon and germanium was studied experimentally, it was found for both substances that there were *more* than two resonances (one for electrons, one for holes) as predicted by the simple theory discussed above.[5,6] Moreover, it was observed that some of the cyclotron resonance frequencies underwent significant changes as the orientation of the crystal with respect to the static magnetic field was altered, while others were practically independent of sample orientation. These effects were finally explained using a model in which the constant energy surfaces for electrons in **k**-space are ellipsoidal rather than spherical, and in which the presence of three separate valence bands, two of which are degenerate at $k = 0$, are considered. The constant energy surfaces for holes in these valence bands are roughly spherical, although since $\partial^2 \varepsilon / \partial k^2$ differs for the different valence bands there is more than one possible effective mass for holes. The various features of this model have been verified not only by the agreement with experimental data which has been obtained, but also by quantum-mechanical calculations of the energy band structure which begin, more or less, from first principles.[7,8,9]

[5] G. Dresselhaus, A. F. Kip and C. Kittel, *Phys. Rev.*, **92**, 827 (1953).
[6] R. N. Dexter, H. J. Zeiger and B. Lax, *Phys. Rev.*, **104**, 637 (1956).
[7] F. Herman, *Phys. Rev.*, **93**, 1214 (1954); *Proc. Inst. Radio Engrs.*, **43**, 1703 (1955).
[8] D. P. Jenkins, *Physica*, **20**, 967 (1954).
[9] E. M. Conwell, *Proc. Inst. Radio Engrs.*, **46**, 1281 (1958).

We shall first consider the matter of ellipsoidal energy surfaces, and investigate how the results of the cyclotron resonance experiment are modified when the energy surfaces have this form. Up to this point we have always assumed that the minimum energy point in the reduced Brillouin zone occurs at $k = 0$, in the center of the zone, as illustrated in Figure 8.6. In this case an electron near the energy minimum exhibits the dynamical behavior of a free electron, leading to constant energy surfaces which, within this region of k-space, are approximately spherical, as shown in Figure 8.11(a).

It has been found, however, that the minimum value of energy within the reduced zone need not necessarily be at $k = 0$, but may lie elsewhere within the zone. This is indeed the observed state of affairs for electrons in the conduction band of germanium and silicon. Since the diamond lattice is really two interpenetrating face-centered cubic lattices separated along the cube diagonal by a distance $a\sqrt{3}/4$ (where a is the cube edge), the primitive unit cell for this lattice is the same as that shown in Figure 1.5, except that there is a second atom within the cell. The reciprocal lattice is thus the same as that for the f.c.c. structure, and hence the Brillouin zone has the same shape as the f.c.c. zone shown in Figure 8.14(c). A plot of ε versus k along the (100) direction in this Brillouin zone for electrons in the conduction band of silicon is shown in Figure 9.15(a). The smallest value of ε is reached at a value of k which is about

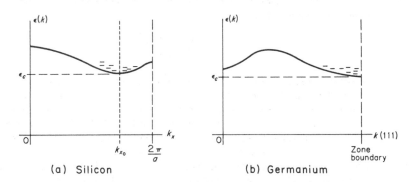

(a) Silicon   (b) Germanium

FIGURE 9.15.   The ε versus k relation for (a) the conduction band of silicon, plotted along the k_x-direction, (b) the conduction band of germanium, plotted along a (111) direction in k-space.

$0.8(2\pi/a)$. This is the absolute minimum value of ε within the Brillouin zone; curves of ε versus k along other paths may exhibit minimum values of ε, but none so low as the value ε_c in the figure. For energies greater than ε_c, one may construct constant energy surfaces about the minimum energy point. It is clear that the energy versus crystal momentum curve shown in Figure 9.15(a) is roughly parabolic about the minimum point, corresponding to free electron behavior with an appropriate effective mass which we shall call $m_\parallel^*$. If one were to plot the ε versus k curve along some path passing through the minimum point, but *perpendicular* to the k_x axis (for example a line parallel to the k_y axis) one would also obtain a curve which would be parabolic about the minimum point, but there is now no aspect of crystal symmetry which would require that the *curvature* in this direction be the same as it is along the k_x direction. (This situation is illustrated in Figure 9.16.) As a result the effective mass related to changes in the momentum component along this direction has a value which is different from the value discussed previously, and which we shall denote as $m_\perp^*$. It turns out, as we shall soon see, that requirements of crystal symmetry demand that the effective mass have the same value $m_\perp^*$ for *any* direction in **k**-space normal to the k_x-direction for this example.

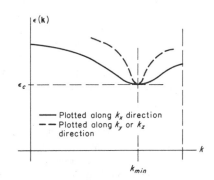

FIGURE 9.16.   The ε *versus* k relation for the conduction band of silicon plotted along the k_x-direction (solid curve) and along a line normal to the k_x-axis which passes through the point $k_{\min}$ (dashed curve). The different curvatures at the minimum point reflect the anisotropy of the effective mass.

The energy in excess of ε_c represents the kinetic energy of a free electron in the conduction band, whereby

$$\varepsilon - \varepsilon_c = \frac{(p_x - p_{x0})^2}{2m_\parallel^*} + \frac{(p_y^2 + p_z^2)}{2m_\perp^*} \qquad (9.9\text{-}15)$$

where $p_{x0} = \hbar k_{x0}$ represents the value of p_x at the bottom of the band, as shown in Figure 9.15(a). According to this the surface of constant energy for an electron of energy ε slightly greater than ε_c is seen to be an *ellipsoid of revolution* whose center is at $k_x = k_{x0}$, $k_y = k_z = 0$. Since, however, the six $\langle 100 \rangle$ directions of a cubic lattice are crystallographically equivalent, the ε *versus* k curve along any of these six directions must be the same. There must then be *six* equivalent energy minima along the six $\langle 100 \rangle$ directions, and associated with each of these minima there must be a family of ellipsoidal constant energy surfaces similar to those described by (9.9-15). This situation is illustrated in Figure 9.17(a). It is now clear why the ellipsoids have to be ellipsoids of revolution; as we saw in the previous chapter, the constant energy surfaces must have all the symmetry properties of the Brillouin zone, which in turn has all the symmetry properties of a cubic crystal. If the ellipsoids were not ellipsoids of revolution, the family of constant energy surfaces would no longer be invariant under all the operations for which a cubic crystal remains invariant.

The situation for electrons in the conduction band of germanium is somewhat similar, except that there are a set of eight equivalent minimum energy points which lie along the $\langle 111 \rangle$ directions in **k**-space at the intersection of those directions with the surface of the Brillouin zone. These minima thus lie at the centers of the hexagonal faces of the Brillouin zone. The constant energy surfaces are again ellipsoidal in form, but since the energy minima are at the zone boundary, and since there is a rather large forbidden energy region which must be surmounted before electrons can be excited outside the Brillouin zone in these directions, there are eight half-ellipsoids which extend into the interior of the Brillouin zone along the $\langle 111 \rangle$ directions, as shown in Figure 9.17(b). For many purposes these may be considered to be equivalent to four full ellipsoids. In both cases there is an effective mass *tensor* associated with *each*

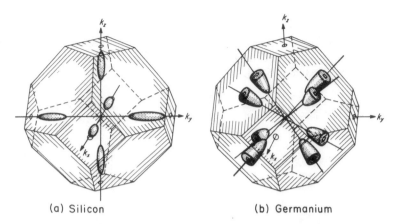

(a) Silicon   (b) Germanium

FIGURE 9.17. Ellipsoidal constant energy surfaces in k-space for (a) silicon and
(b) germanium. In silicon the major axes of the ellipsoids are along {100} directions,
while in germanium, since the energy minimum is on the zone boundary, the constant
energy surfaces form eight half-ellipsoids whose major axes lie along {111} directions.

ellipsoid whose elements are described by (8.8-7). In calculating the transport properties and other physical characteristics of crystals in which the energy surfaces are
ellipsoidal, one usually calculates the effect due to one ellipsoid (which is generally
anisotropic) and then sums over all ellipsoids. In cubic crystals, the transport properties are found to be isotropic after the summation is carried out, despite the fact that
anisotropies are associated with individual ellipsoids. In both germanium and silicon
the longitudinal effect mass $m_\parallel^*$ is much larger than the transverse mass $m_\perp^*$, the ellipsoids thus being quite long and thin. The actual values, as determined by cyclotron
resonance experiments,[5] are given in Table 9.2.

TABLE 9.2.

Effective Masses of Electrons in Germanium and Silicon

	$m_\parallel^*$	$m_\perp^*$	"mass ratio" $m_\parallel^*/m_\perp^*$
Germanium	1.64 m_0	0.0819 m_0	20.0
Silicon	0.98 m_0	0.19 m_0	5.2

Let us now investigate the cyclotron resonance effect in materials where the energy
surfaces are ellipsoidal. To begin with we shall calculate the resonance frequency
associated with a *single* ellipsoid, which we shall assume to be symmetric about the
k_z-axis and to be centered on the point k_{z0}, as shown in Figure 9.18. The magnetic
induction $\mathbf{B}_0$ makes an angle θ with the major axis of the ellipsoid, and for convenience
we shall assume that the oscillating electric field vector has only a y-component.
The choice of this coordinate system is made only for the purposes of the present

calculation, and we shall see when we are finished that the only coordinate of physical importance is the angle θ between the field and the longitudinal axis of the ellipsoid.

Under these circumstances, the equation of motion (for a particle of positive charge e) may be written in terms of the tensor (8.8-6) which represents the reciprocal of the effective mass as

$$\frac{d\mathbf{v}}{dt} = \left(\frac{1}{\underline{\mathbf{m}}^*}\right) \cdot \mathbf{F} = \left(\frac{1}{\underline{\mathbf{m}}^*}\right) \cdot \left(e\mathbf{E} + \frac{e}{c}\,\mathbf{v} \times \mathbf{B}\right). \tag{9.9-16}$$

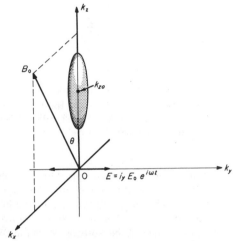

FIGURE 9.18. Vector geometry used for the calculation of Section 9.9.

The relation between ε and $\mathbf{k}$ for the case of the ellipsoidal energy surface shown in Figure 9.18 must, according to (9.9-15) be

$$\varepsilon - \varepsilon_c = \hbar^2 \left[\frac{k_x^2 + k_y^2}{2m_\perp^*} + \frac{(k_z - k_{z0})^2}{2m_\parallel^*}\right]. \tag{9.9-17}$$

The tensor components $(1/m^*)_{\alpha\beta}$ may now be found according to (8.8-7). In this case, $\partial^2 \varepsilon / \partial k_\alpha \partial k_\beta = 0$ for $\alpha \neq \beta$, so that all the off-diagonal elements of the tensor are zero. The diagonal elements are easily evaluated, the result being

$$(1/m^*)_{xx} = (1/m^*)_{yy} = 1/m_\perp^* \quad \text{and} \quad (1/m^*)_{zz} = 1/m_\parallel^*. \tag{9.9-18}$$

The effective mass tensor then has the form

$$\left(\frac{1}{\underline{\mathbf{m}}^*}\right) = \begin{bmatrix} \dfrac{1}{m_\perp^*} & 0 & 0 \\[2ex] 0 & \dfrac{1}{m_\perp^*} & 0 \\[2ex] 0 & 0 & \dfrac{1}{m_\parallel^*} \end{bmatrix} \tag{9.9-19}$$

and
$$\left(\frac{1}{\underline{\underline{m}}^*}\right) \cdot F = \begin{bmatrix} \dfrac{1}{m_\perp^*} & 0 & 0 \\[2ex] 0 & \dfrac{1}{m_\perp^*} & 0 \\[2ex] 0' & 0 & \dfrac{1}{m_\parallel^*} \end{bmatrix} \cdot \begin{bmatrix} F_x \\[1ex] F_y \\[1ex] F_z \end{bmatrix} = \begin{bmatrix} F_x/m_\perp^* \\[1ex] F_y/m_\perp^* \\[1ex] F_z/m_\parallel^* \end{bmatrix}. \qquad (9.9\text{-}20)$$

Inserting the Lorentz force components for F_x, F_y, and F_z and equating to the components of dv/dt as directed by (9.9-16), the equation of motion can be written

$$m_\perp^* \frac{dv_x}{dt} = \frac{e}{c}(v_y B_z - v_z B_y) + eE_x$$

$$m_\perp^* \frac{dv_y}{dt} = \frac{e}{c}(v_z B_x - v_x B_z) + eE_y \qquad (9.9\text{-}21)$$

$$m_\parallel^* \frac{dv_z}{dt} = \frac{e}{c}(v_x B_y - v_y B_x) + eE_z.$$

Noting that $B_x = B_0 \sin\theta$, $B_z = B_0 \cos\theta$, $B_y = 0$, $E_y = E_0 e^{i\omega t}$, $E_x = E_z = 0$, these equations may be stated as

$$dv_x/dt = \omega_\perp v_y \cos\theta$$

$$dv_y/dt = \omega_\perp v_z \sin\theta - \omega_\perp v_x \cos\theta + \frac{eE_0}{m_\perp^*} e^{i\omega t} \qquad (9.9\text{-}22)$$

$$dv_z/dt = -\omega_\parallel v_y \sin\theta$$

where
$$\omega_\perp = \frac{eB_0}{m_\perp^* c} \qquad \text{and} \qquad \omega_\parallel = \frac{eB_0}{m_\parallel^* c}. \qquad (9.9\text{-}23)$$

As before, one may assume oscillatory solutions of the form $x(t) = x_0 e^{i\omega t}$, $y(t) = y_0 e^{i\omega t}$, and $z(t) = z_0 e^{i\omega t}$, substitute these into the equations of motion (9.9-22) and solve for the amplitudes to obtain

$$x_0 = \frac{eE_0}{m_\perp^*} \frac{i\omega_\perp \cos\theta}{\omega(\omega^2 - \omega_\perp^2 \cos^2\theta - \omega_\perp \omega_\parallel \sin^2\theta)}$$

$$y_0 = -\frac{eE_0}{m_\perp^*} \frac{1}{\omega^2 - \omega_\perp^2 \cos^2\theta - \omega_\perp \omega_\parallel \sin^2\theta} \qquad (9.9\text{-}24)$$

$$z_0 = -\frac{eE_0}{m_\perp^*} \frac{i\omega_\parallel \sin\theta}{\omega(\omega^2 - \omega_\perp^2 \cos^2\theta - \omega_\perp \omega_\parallel \sin^2\theta)}.$$

These amplitudes become very large when

$$\omega = \sqrt{\omega_\perp^2 \cos^2 \theta + \omega_\perp \omega_\parallel \sin^2 \theta}, \qquad (9.9\text{-}25)$$

and this is accordingly the cyclotron resonance frequency for the particular ellipsoid which we are considering. The resonance frequency depends only upon the magnetic field, the effective masses and the angle θ between the field $\mathbf{B}_0$ and the major axis of the ellipsoid. For ellipsoidal energy surfaces, it is clear that the resonance frequency is a *strong function of the orientation of the sample relative to the magnetic field*. At the resonant frequency the particle orbits can be shown from (9.9-24) to be elliptical in form and to lie in a plane perpendicular to the field $\mathbf{B}_0$. The above results have been derived for positive particles of charge e, but for electrons we need only replace e by $-e$ (in which case we must also replace $\omega_\parallel$ by $-\omega_\parallel$ and $\omega_\perp$ by $-\omega_\perp$). The resonant frequency is obviously unaltered by these substitutions; the only difference is that electrons traverse their orbits in the opposite sense from that found for positive particles.

In the case of silicon, if the field $\mathbf{B}_0$ is oriented along the z-direction, then $\theta = 0$ for the two ellipsoids whose major axes are along the k_z direction and $\theta = 90°$ for the other four ellipsoids. There will then, according to (9.9-25), be two superimposed resonance peaks at $\omega = \omega_\perp$ from the first two ellipsoids and four superimposed resonances at $\omega = \sqrt{\omega_\perp \omega_\parallel}$ from the other four. Assuming that the number of electrons belonging to each ellipsoid is equal, which should be true in equilibrium, the second resonance, involving four ellipsoids should have twice the absorptive strength of the first, which involves only two. If the field $\mathbf{B}_0$ is otherwise oriented, there may be as many as three resonance peaks, corresponding to the three possible angles between $\mathbf{B}_0$ and the ellipsoid axes. This theory of cyclotron resonance with ellipsoidal energy surfaces serves to account for all the experimentally observed resonances which are strongly affected by sample orientation, and allows a determination of both $m_\perp^*$ and $m_\parallel^*$ from the experimentally determined cyclotron resonance frequencies, once a correct model for the orientation of the ellipsoids is proposed.

The cyclotron resonance data for holes in the valence bands of silicon and germanium do not exhibit the features characteristic of the effect with ellipsoidal energy surfaces. Although there are multiple resonance peaks for holes in both materials, the characteristic changes of resonance frequency with sample orientation which are observed for ellipsoidal energy surfaces are either absent or much less pronounced. The picture of the valence band which seems to fit these (and other) data best is that of two separate bands which are degenerate at the energy maximum at $k = 0$, but which have different effective masses. There is also a third "split-off" band with a maximum energy (at $k = 0$) somewhat below the top of the valence band. This state of affairs is shown in Figure 9.19. There are simultaneously present light and heavy holes (belonging to the l and h bands in the figure), and a different cyclotron resonance effect for each type can be observed. Since the energy surfaces associated with these bands are approximately spherical, the orientation dependence of the cyclotron resonance effect is much less pronounced than that which is associated with the ellipsoidal surfaces of the conduction band. Nevertheless, the energy surfaces are only roughly spherical; they are somewhat distorted from a truly spherical form, even for small values of k. This "warping" must be taken into account to explain all the details of the cyclotron resonance data, especially in the case of the heavy hole band, where the departure from sphericity is most pronounced.

The energy difference between the maximum point of the lower-lying "split-off" band and the maximum of the valence band ($\varepsilon_v - \varepsilon_s$ in Figure 9.19) amounts to about 0.28 eV in germanium and 0.035 eV in silicon. This energy difference, in the case of germanium, is sufficiently great that under ordinary conditions the population of holes in the lower band will be negligible. In silicon, however, except at low temperatures, a significant number of holes will be found in this band. The splitting of the bands is caused by the interaction of the electron spins of the valence electrons with

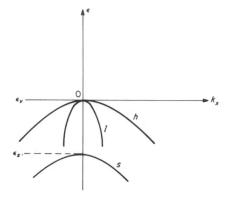

FIGURE 9.19. The three distinct branches of the valence bands of silicon and germanium.

the magnetic moment arising from their orbital motion. If it were not for this *spin-orbit coupling* the energy difference $\varepsilon_v - \varepsilon_s$ would be zero and all three bands would coalesce at the point $k = 0$. The effective masses associated with the three varieties of holes are given in Table 9.3.

TABLE 9.3.

Effective Masses of Holes in Germanium and Silicon

	m_l^*	m_h^*	m_s^*
Germanium	0.044 m_0	0.28 m_0	0.077 m_0
Silicon	0.16 m_0	0.49 m_0	0.245 m_0

The existence of a hole as one particular species is in general limited to a single mean free time. For example, a hole may easily be scattered from the l band to the h-band or vice versa by any of the scattering processes in which it may normally be involved. For this reason it is not possible to detect the existence of the separate varieties of holes in any measurement (such as conductivity or carrier drift) which involves only the average behavior of the carrier over relatively long distances or times, within which many collisions may occur. The results of such a measurement will necessarily only depend upon the properties of the carrier *averaged* over a long time interval during which the identities of holes in the three bands may be frequently interchanged. In the cyclotron resonance experiment, where one measures a resonance which is excited in a time *short* compared to the mean free time ($\omega_0\tau \gg 1$), the presence of the several varieties of holes is easily exhibited. The same remark can be made with reference to the electrons in the various equivalent energy minima of the conduction band; the normal scattering processes may readily transfer electrons from one ellipsoid to another (*intervalley* scattering), so that in a time long compared to the relaxation time, the identity of an electron as belonging to one particular ellipsoid is irretrievably lost.

9·10 DENSITY OF STATES, CONDUCTIVITY AND HALL EFFECT WITH COMPLEX ENERGY SURFACES

In a situation where there are two parabolic bands, degenerate at $k = 0$, with spherical energy surfaces, giving rise to two different species of carriers, the relative populations of the two species and the density of states factors can be easily calculated. We shall discuss the case of a valence band of this sort, in which there are present simultaneously light holes (effective mass m_l^*) and heavy holes (effective mass m_h^*). In this instance we may write the total hole concentration p_0 as the sum of the light hole concentration p_l and the heavy hole concentration p_h, whereby, from (9.3-15) and (9.3-16),

$$p_0 = p_l + p_h = U_{vl}e^{-(\varepsilon_f - \varepsilon_v)/kT} + U_{vh}e^{-(\varepsilon_f - \varepsilon_v)/kT}$$

$$= 2\left(\frac{2\pi kT}{h^2}\right)^{3/2}\left(m_l^{*3/2} + m_h^{*3/2}\right)e^{-(\varepsilon_f - \varepsilon_v)/kT} \tag{9.10-1}$$

where U_{vl} and U_{vh} are factors of the form of (9.3-16) for light and heavy holes. This can be written in the form

$$p_0 = 2\left(\frac{2\pi m_{ds}^* kT}{h^2}\right)^{3/2} e^{-(\varepsilon_f - \varepsilon_v)/kT}, \tag{9.10-2}$$

as if there were a single species of carrier, if we define the *density of states equivalent effective mass* m_{ds}^* as

$$m_{ds}^{*3/2} = m_l^{*3/2} + m_h^{*3/2}. \tag{9.10-3}$$

This is an approximation to the actual state of affairs in the valence band of silicon and germanium, although it is a rather rough one, since the effect of the "split-off" band is entirely neglected and since the actual energy surfaces are by no means perfectly spherical. The conductivity follows from (9.10-2) directly, by the use of (9.7-8).

For a set of ellipsoidal energy surfaces such as those discussed in connection with the conduction band of germanium or silicon, one may calculate the density of states factor $g(\varepsilon)$ by an extension of the procedure developed in Section 5.2. The equation (9.9-17) for one of the ellipsoids may be reduced by the transformation

$$p'_x = p_x\sqrt{m_0/m^*_\perp} = \hbar k_x\sqrt{m_0/m^*_\perp}$$

$$p'_y = p_y\sqrt{m_0/m^*_\perp} = \hbar k_y\sqrt{m_0/m^*_\perp}$$

$$p'_z = (p_z - p_{z0})\sqrt{m_0/m^*_\parallel} = \hbar(k_z - k_{z0})\sqrt{m_0/m^*_\parallel} \tag{9.10-4}$$

to that of a sphere in $\mathbf{p}'$-space.

$$\varepsilon - \varepsilon_c = \frac{p'^2_x + p'^2_y + p'^2_z}{2m_0} = \frac{p'^2}{2m_0}. \tag{9.10-5}$$

The volume of $\mathbf{p}'$-space in the spherical shell bounded by the radii p' and $p' + dp'$ for a single ellipsoid may now be calculated exactly as in Section 5.2 for spherical energy surfaces, but since there is more than one ellipsoidal constant energy surface, the total volume of $\mathbf{p}'$-space corresponding to energies in the range ε to $\varepsilon + d\varepsilon$ will be obtained by multiplying the volume found for a single surface by the number of equivalent ellipsoids. If this number is ν, then the corresponding volume $dV_{p'}$ of $\mathbf{p}'$-space is

$$dV_{p'} = 4\sqrt{2\pi}\nu m_0^{3/2}\sqrt{\varepsilon - \varepsilon_c}\, d\varepsilon. \tag{9.10-6}$$

But, from (9.10-4) the volume elements $dV_p = dp_x dp_y dp_z$ and $dV_{p'} = dp_{x'} dp_{y'} dp_{z'}$ are related by

$$dV_{p'} = \frac{m_0^{3/2}}{(m^{*2}_\perp m^*_\parallel)^{1/2}}\, dV_p, \tag{9.10-7}$$

whereby the volume dV_p of $\mathbf{p}$-space within the energy range ε to $\varepsilon + d\varepsilon$ becomes

$$dV_p = 4\sqrt{2\pi}\nu(m^{*2}_\perp m^*_\parallel)^{1/2}\sqrt{\varepsilon - \varepsilon_c}\, d\varepsilon. \tag{9.10-8}$$

Dividing this by the volume $h^3/2$ assigned to a single quantum state gives the number of quantum states in this energy range which, by definition, is $g(\varepsilon)\, d\varepsilon$, whereby

$$g(\varepsilon)\, d\varepsilon = \frac{8\sqrt{2\pi}\nu(m^{*2}_\perp m^*_\parallel)^{1/2}}{h^3}\sqrt{\varepsilon - \varepsilon_c}\, d\varepsilon. \tag{9.10-9}$$

If, once more, we define a *density of states equivalent effective mass* m^*_{ds} as

$$m^{*3/2}_{ds} = (m^{*2}_\perp m^*_\parallel)^{1/2}, \tag{9.10-10}$$

Equation (9.10-9) can be written (except for the factor ν) in the same form (5.2-22) as the usual density of states expression for the case where the energy surfaces are spherical. By a straightforward application of the procedures by which (9.3-6) and (9.3-7) were obtained, it is easily seen that we must obtain

$$n_0 = 2\nu\left(\frac{2\pi m^*_{ds}kT}{h^2}\right)^{3/2} e^{-(\varepsilon_c - \varepsilon_f)/kT} \tag{9.10-11}$$

with m^*_{ds} as given by (9.10-10).

In calculating the conductivity of a semiconductor in which the constant energy surfaces are ellipsoidal, we must first calculate the current or conductivity contribution from a single ellipsoid and then sum over all the ellipsoids to obtain the total conductivity. Since the effective mass associated with each individual ellipsoid is a tensor quantity, so also is the conductivity. In general, the equations relating the current density and electric field may be written as tensor relations of the form

$$\mathbf{I} = \underline{\sigma} \cdot \mathbf{E} \tag{9.10-12}$$

and
$$\mathbf{E} = \underline{\rho} \cdot \mathbf{I} \tag{9.10-13}$$

where $\underline{\sigma}$ is the *conductivity tensor* and $\underline{\rho}$ is a resistivity tensor. Since, clearly, from (9.10-12) and (9.10-13),

$$\underline{\rho} \cdot \mathbf{I} = \underline{\rho} \cdot \underline{\sigma} \cdot \mathbf{E} = \mathbf{E}, \tag{9.10-14}$$

we must have

$$\underline{\rho} \cdot \underline{\sigma} = \underline{1}, \tag{9.10-15}$$

where $\underline{1}$ is the unit tensor, whose elements are $\delta_{\alpha\beta}$. The resistivity tensor is thus the inverse of the conductivity tensor, that is,

$$\underline{\rho} = \underline{\sigma}^{-1}. \tag{9.10-16}$$

The elements of the conductivity tensor $\sigma^{(i)}$ related to the ith energy ellipsoid may in analogy with (7.3-15) be written

$$\sigma_{\alpha\beta}^{(i)} = n^{(i)} e^2 \bar{\tau} \left(\frac{1}{m^*} \right)_{\alpha\beta} \tag{9.10-17}$$

where $n^{(i)}$ is the density of charge carriers associated with that ellipsoid. The total conductivity may then be expressed as

$$\underline{\sigma} = \sum_i \underline{\sigma}^{(i)}, \tag{9.10-18}$$

the sum being taken over all ellipsoids. Using (8.8-7) to express the tensor components, (9.10-17) can be written as

$$\underline{\sigma}^{(i)} = \frac{n^{(i)} e^2 \bar{\tau}}{\hbar^2} \begin{bmatrix} \dfrac{\partial^2 \varepsilon^{(i)}}{\partial k_x^2} & \dfrac{\partial^2 \varepsilon^{(i)}}{\partial k_x \, \partial k_y} & \dfrac{\partial^2 \varepsilon^{(i)}}{\partial k_x \, \partial k_z} \\[2ex] \dfrac{\partial^2 \varepsilon^{(i)}}{\partial k_y \, \partial k_x} & \dfrac{\partial^2 \varepsilon^{(i)}}{\partial k_y^2} & \dfrac{\partial^2 \varepsilon^{(i)}}{\partial k_y \, \partial k_z} \\[2ex] \dfrac{\partial^2 \varepsilon^{(i)}}{\partial k_z \, \partial k_x} & \dfrac{\partial^2 \varepsilon^{(i)}}{\partial k_z \, \partial k_y} & \dfrac{\partial^2 \varepsilon^{(i)}}{\partial k_z^2} \end{bmatrix} \tag{9.10-19}$$

where $\varepsilon^{(i)}(\mathbf{k})$ is the energy expressed as a function of $\mathbf{k}$ for the ith ellipsoid. This treatment assumes, of course, that the relaxation time τ is the same for all possible directions in $\mathbf{k}$-space, which may not be true in the most general case.

In the case of silicon, the major axes of the ellipsoids lie along the coordinate axes of k-space; for the two ellipsoids whose major axes lie along the k_z-axis, in the [001] and [00$\bar{1}$] directions the function $\varepsilon^{(001)}(\mathbf{k})$ is given by (9.9-17), leading to tensor components (9.9-18) for the effective mass tensor. The conductivity tensor for these ellipsoids then becomes

$$\underline{\sigma}^{(001)} = \underline{\sigma}^{(00\bar{1})} = n^{(001)}e^2\bar{\tau}\begin{bmatrix} \dfrac{1}{m_\perp^*} & 0 & 0 \\ 0 & \dfrac{1}{m_\perp^*} & 0 \\ 0 & 0 & \dfrac{1}{m_\parallel^*} \end{bmatrix} \tag{9.10-20}$$

The equations for the other ellipsoids may be written and the tensor components obtained in the same way. For the two ellipsoids whose major axes lie along the k_x-axis, then

$$\underline{\sigma}^{(100)} = \underline{\sigma}^{(\bar{1}00)} = n^{(100)}e^2\bar{\tau}\begin{bmatrix} \dfrac{1}{m_\parallel^*} & 0 & 0 \\ 0 & \dfrac{1}{m_\perp^*} & 0 \\ 0 & 0 & \dfrac{1}{m_\perp^*} \end{bmatrix}, \tag{9.10-21}$$

while for the two ellipsoids whose major axes lie along the k_y-axis,

$$\underline{\sigma}^{(010)} = \underline{\sigma}^{(0\bar{1}0)} = n^{(010)}e^2\bar{\tau}\begin{bmatrix} \dfrac{1}{m_\perp^*} & 0 & 0 \\ 0 & \dfrac{1}{m_\parallel^*} & 0 \\ 0 & 0 & \dfrac{1}{m_\perp^*} \end{bmatrix} \tag{9.10-22}$$

Since all six ellipsoids are equivalent energy minima, $n^{(100)} = n^{(\bar{1}00)} = n^{(010)} = n^{(0\bar{1}0)} = n^{(001)} = n^{(00\bar{1})} = n_0/6$. Performing the summation indicated in (9.10-18) and utilizing this fact, one may obtain

$$\underline{\sigma}_0 = n_0 e^2\bar{\tau}\begin{bmatrix} \dfrac{1}{3}\left(\dfrac{2}{m_\perp^*} + \dfrac{1}{m_\parallel^*}\right) & 0 & 0 \\ 0 & \dfrac{1}{3}\left(\dfrac{2}{m_\perp^*} + \dfrac{1}{m_\parallel^*}\right) & 0 \\ 0 & 0 & \dfrac{1}{3}\left(\dfrac{2}{m_\perp^*} + \dfrac{1}{m_\parallel^*}\right) \end{bmatrix}$$

$$= n_0 e^2\bar{\tau}\frac{1}{3}\left(\frac{2}{m_\perp^*} + \frac{1}{m_\parallel^*}\right)\underline{1} \tag{9.10-23}$$

for the total conductivity. Since $\underline{\sigma}$ is a scalar multiple of the unit tensor $\underline{1}$, the *total* conductivity is *isotropic* (albeit the conductivity associated with a single ellipsoid is not). The scalar magnitude of the conductivity can be represented as

$$\sigma_0 = \frac{n_0 e^2 \bar{\tau}}{m_c^*} \tag{9.10-24}$$

where m_c^* is a *conductivity effective mass* defined by

$$\frac{1}{m_c^*} = \frac{1}{3}\left(\frac{2}{m_\perp^*} + \frac{1}{m_\parallel^*}\right). \tag{9.10-25}$$

It can be shown that exactly the same result will be obtained for *any* set of ellipsoidal energy surfaces whatever, so long as the configuration is invariant under all of the symmetry operations under which a cubic crystal is invariant. In fact, it can be proved that the conductivity of all cubic crystals must be isotropic, whatever specific form the surfaces of constant energy take. In particular, Equations (9.10-23) through (9.10-25) are valid *also* for the conduction electrons in germanium, where the constant energy ellipsoids lie along the $\langle 111\rangle$ directions. It is instructive to work out the conductivity tensor for germanium; this problem is assigned as an exercise for the reader.

The resistivity tensor $\underline{\rho}$ is a matrix of components inverse to the conductivity matrix. The elements of the matrix which is the inverse of a matrix $\underline{A}$ with elements $a_{\alpha\beta}$ are given by the formula

$$a_{\alpha\beta}^{-1} = \frac{\text{cof}\,(a_{\beta\alpha})}{\Delta(a)}, \tag{9.10-26}$$

where $\text{cof}\,(a_{\beta\alpha})$ is the cofactor[10] of the element $a_{\beta\alpha}$ and $\Delta(a)$ is the determinant of the coefficients of $\underline{A}$. The elements of ρ are thus readily seen to be

$$\underline{\rho} = \underline{\sigma}^{-1} = \frac{1}{\Delta(\sigma)}\begin{bmatrix} \sigma_{yy}\sigma_{zz} - \sigma_{yz}^2 & \sigma_{xz}\sigma_{yz} - \sigma_{xy}\sigma_{zz} & \sigma_{xy}\sigma_{yz} - \sigma_{xz}\sigma_{yy} \\ \sigma_{yz}\sigma_{xz} - \sigma_{xy}\sigma_{zz} & \sigma_{xx}\sigma_{zz} - \sigma_{xz}^2 & \sigma_{xz}\sigma_{xy} - \sigma_{xx}\sigma_{yz} \\ \sigma_{xy}\sigma_{yz} - \sigma_{xz}\sigma_{yy} & \sigma_{xz}\sigma_{xy} - \sigma_{xx}\sigma_{yz} & \sigma_{xx}\sigma_{yy} - \sigma_{xy}^2 \end{bmatrix} \tag{9.10-27}$$

where

$$\Delta(\sigma) = \sigma_{xx}(\sigma_{yy}\sigma_{zz} - \sigma_{yz}^2) - \sigma_{xy}(\sigma_{xy}\sigma_{zz} - \sigma_{yz}\sigma_{xz})$$

$$+ \sigma_{xz}(\sigma_{xy}\sigma_{yz} - \sigma_{yy}\sigma_{xz}). \tag{9.10-28}$$

It is readily seen that if $\underline{\sigma}$ has the form (9.10-23), the resistivity tensor is isotropic and can be represented by a scalar quantity whose magnitude is simply the reciprocal of the conductivity (9.10-24).

We shall illustrate the computation of the Hall effect in a material with ellipsoidal energy surfaces by considering in detail the Hall effect in a material where the major axes of the ellipsoids extend along the k_x-, k_y- and k_z-axes (as, for example, in silicon),

---

[10] See, for example, L. A. Pipes, *Applied Mathematics for Engineers and Physicists*. New York: McGraw-Hill (1946), p. 71.

in the special case where the magnetic field is in the z-direction and the current is flowing in the x-direction. We shall also assume that only one species of carrier is present. Under these conditions the inverse effective mass tensors for the various ellipsoids have the form shown in Equations (9.10-20, 21, and 22), and in particular, the tensor equation of motion for carriers of charge e in the (100) and ($\bar{1}$00) ellipsoids can be written, in analogy with (9.8-4), as

$$
\left(\frac{1}{\underline{\mathbf{m}}^*}\right) \cdot \underline{\mathbf{F}} =
\begin{bmatrix}
\dfrac{1}{m_\parallel^*} & 0 & 0 \\
0 & \dfrac{1}{m_\perp^*} & 0 \\
0 & 0 & \dfrac{1}{m_\perp^*}
\end{bmatrix}
\begin{bmatrix}
eE_x + \dfrac{e}{c}(v_yB_z - v_zB_y) \\
eE_y + \dfrac{e}{c}(v_zB_x - v_xB_z) \\
eE_z + \dfrac{e}{c}(v_xB_y - v_yB_x)
\end{bmatrix}
=
\begin{bmatrix}
dv_x/dt \\
dv_y/dt \\
dv_z/dt
\end{bmatrix}.
\qquad (9.10\text{-}29)
$$

Performing the indicated matrix multiplication, writing out the component equations of motion, and noting that $B_x = B_y = 0$, while $B_z = B_0$, we may obtain

$$
\frac{dv_x}{dt} = \frac{eE_x}{m_\parallel^*} + \omega_\parallel v_y
$$

$$
\frac{dv_y}{dt} = \frac{eE_y}{m_\perp^*} - \omega_\perp v_x
$$

$$
\frac{dv_z}{dt} = \frac{eE_z}{m_\perp^*},
\qquad (9.10\text{-}30)
$$

where $\omega_\parallel$ and $\omega_\perp$ are as given by (9.9-23).

If the first of these equations is multiplied by $\sqrt{\omega_\perp}$ and added to $i\sqrt{\omega_\parallel}$ times the second, the resulting equation can be expressed in the form

$$
\frac{dZ}{dt} + i\sqrt{\omega_\parallel \omega_\perp} Z = e\left(\frac{\sqrt{\omega_\perp}}{m_\parallel^*} E_x + \frac{i\sqrt{\omega_\parallel}}{m_\perp^*} E_y\right)
\qquad (9.10\text{-}31)
$$

where Z is the complex quantity

$$
Z = v_x\sqrt{\omega_\perp} + iv_y\sqrt{\omega_\parallel}.
\qquad (9.10\text{-}32)
$$

Equation (9.10-31) is clearly seen to have exactly the same form as (9.8-8). All the steps involving the solution of this equation and the averaging over path lengths and velocity distribution, leading to Equations (9.8-16) and (9.8-17) may now be repeated, starting with (9.10-31). The physical justification of each step is the same as before. In this manner we may show that

$$
\bar{v}_x^{(100)} = \bar{v}_x^{(\bar{1}00)} = \frac{1}{\sqrt{\omega_\perp}}\,\mathrm{Re}(\bar{Z})
$$

$$
= e\left[\frac{\bar{\tau}E_x}{m_\parallel^*} + \frac{\omega_\parallel}{m_\perp^*}\overline{\left(\frac{\tau^2}{1 + \omega_\parallel\omega_\perp\tau^2}\right)}E_y - \frac{\omega_\parallel\omega_\perp}{m_\parallel^*}E_x\overline{\left(\frac{\tau^3}{1 + \omega_\parallel\omega_\perp\tau^2}\right)}\right]
\qquad (9.10\text{-}33)
$$

while $\quad \bar{v}_y^{(100)} = \bar{v}_y^{(\bar{1}00)} = \dfrac{1}{\sqrt{\omega_{\parallel}}} \operatorname{Im}(\bar{Z})$

$$= e\left[\frac{1}{m_{\perp}^*}\left(\overline{\frac{\tau}{1+\omega_{\parallel}\omega_{\perp}\tau^2}}\right)E_y - \frac{\omega_{\perp}}{m_{\parallel}^*}\left(\overline{\frac{\tau^2}{1+\omega_{\parallel}\omega_{\perp}\tau^2}}\right)E_x\right]. \quad (9.10\text{-}34)$$

If we confine our results to the small field Hall coefficient, we may assume $\omega_{\parallel}\omega_{\perp}\tau^2 \ll 1$ and neglect the $\omega_{\parallel}\omega_{\perp}\tau^2$ terms in the denominators in Equations (9.10-33) and (9.10-34). Now the current densities may be obtained from the velocities as before, whence, for the ith ellipsoid,

$$I_x^{(i)} = p_0^{(i)}e\bar{v}_x^{(i)} \quad \text{and} \quad I_y^{(i)} = p_0^{(i)}e\bar{v}_y^{(i)}, \quad (9.10\text{-}35)$$

while, summing over all ellipsoids, the total current density components are

$$I_x = \sum_i p_0^{(i)}e\bar{v}_x^{(i)} \quad \text{and} \quad I_y = \sum_i p_0^{(i)}e\bar{v}_y^{(i)}. \quad (9.10\text{-}36)$$

In the steady state, there can be no current in the y-direction, so that from (9.10-36) we must have

$$\sum_i p_0^{(i)}e\bar{v}_y^{(i)} = 0. \quad (9.10\text{-}37)$$

As mentioned previously the concentrations $p_0^{(i)}$ for all six ellipsoids are the same and equal to $p_0/6$. The expression for $\bar{v}_y^{(100)}$ and $\bar{v}_y^{(\bar{1}00)}$ is given above as (9.10-34). For the (010) and (0$\bar{1}$0) ellipsoids, it is easily established by using the appropriate mass tensor in (9.10-29) that the equations of motion for the x- and y-components of velocity are the same as those given in (9.10-30) except that $m_{\parallel}^*$ and $m_{\perp}^*$, and $\omega_{\parallel}$ and $\omega_{\perp}$ are interchanged. The expression for $\bar{v}_y^{(010)}$ and $\bar{v}_y^{(0\bar{1}0)}$ is then just (9.10-34) with these quantities interchanged. In the case of the (001) and (00$\bar{1}$) ellipsoids, it is clear, following the same procedure, that the equations of motion for the x- and y-components of velocity are the same as those in (9.10-30) except that $m_{\parallel}^*$ and $\omega_{\parallel}$ do not appear at all; they are replaced by $m_{\perp}^*$ and $\omega_{\perp}$ in all instances. The expression for $\bar{v}_y^{(001)}$ and $\bar{v}_y^{(00\bar{1})}$ is then obtained from (9.10-34) by replacing $m_{\parallel}^*$ and $\omega_{\parallel}$ by $m_{\perp}^*$ and $\omega_{\perp}$. Making all these substitutions, performing the summation indicated in (9.10-37) and solving for E_y in terms of E_x, we may obtain

$$E_y = \frac{eB_0}{c}\left(\frac{\overline{\tau^2}}{\bar{\tau}}\right)\frac{2m_{\perp}^* + m_{\parallel}^*}{m_{\perp}^{*2}m_{\parallel}^*\left(\dfrac{2}{m_{\perp}^*}+\dfrac{1}{m_{\parallel}^*}\right)}E_x. \quad (9.10\text{-}38)$$

Neglecting magnetoresistance effects, which are not important for low magnetic field-strengths, we may express E_x in terms of I_x, according to (9.10-24), as

$$E_x = I_x/\sigma_0 = \frac{3I_x}{p_0e^2\bar{\tau}\left(\dfrac{2}{m_{\perp}^*}+\dfrac{1}{m_{\parallel}^*}\right)}. \quad (9.10\text{-}39)$$

Substituting this into (9.10-38) we may write, finally

$$E_y = RB_0 I_x \tag{9.10-40}$$

with

$$R = \frac{1}{p_0 ec} \frac{\overline{\tau^2}}{(\overline{\tau})^2} \frac{3(2m_\perp^* + m_\parallel^*)}{m_\perp^{*2} m_\parallel^* \left(\dfrac{2}{m_\perp^*} + \dfrac{1}{m_\parallel^*}\right)^2}. \tag{9.10-41}$$

This is often written in a slightly different form; if one defines K as the ratio of the longitudinal to transverse mass, whereby

$$K = m_\parallel^*/m_\perp^*, \tag{9.10-42}$$

then (9.10-41) can be expressed as

$$R = \frac{1}{p_0 ec} \frac{\overline{\tau^2}}{(\overline{\tau})^2} \frac{3K(K + 2)}{(2K + 1)^2}. \tag{9.10-43}$$

Although we assumed particularly simple directions for the sample current and for the magnetic field, had we taken the trouble to work out the result for arbitrary direction of current flow or field orientation, we should have found the small-field Hall coefficient (9.10-41) or (9.10-43) to be independent of the orientation of current and field direction with respect to the energy ellipsoids. The small field Hall coefficient may thus be shown to be *isotropic*, although this is not true if the magnetic field is so large that $\omega_\parallel \omega_\perp \tau^2$ is no longer small compared to unity. The above results were obtained also for a particularly simple set of energy ellipsoids, but they can be shown to hold for *any* set of ellipsoidal energy surfaces so long as the set has cubic symmetry. These results were, of course, worked out for charge carriers having a charge $+e$. For n-type semiconductors (and n-type Si and Ge are the most common examples of substances with ellipsoidal energy surfaces) one must replace e by $-e$ and p_0 by n_0 in (9.10-41) and (9.10-43). If Equation (9.10-43) is multiplied by σ_0, and if the quantity $cR\sigma_0$ is, as before, defined as the "Hall mobility" μ_H, it is clear that the ratio of the Hall mobility to the true drift mobility μ is given by

$$\frac{\mu_H}{\mu} = \frac{\overline{\tau^2}}{(\overline{\tau})^2} \frac{3K(K + 2)}{(2K + 1)^2} \tag{9.10-44}$$

for a semiconductor with ellipsoidal constant energy surfaces.

A complete discussion of other galvanomagnetic effects and of magnetoresistance for semiconductors having ellipsoidal energy surfaces is somewhat beyond the scope of this work and will be omitted. It should be noted particularly, however, that the magnetoresistance is not isotropic under these conditions, and that measurements of magnetoresistance as a function of sample orientation and magnetic field direction can provide a rich source of experimental data against which any proposed model of the energy surfaces may be checked in detail.

9·11 SCATTERING MECHANISMS AND MOBILITY OF CHARGE CARRIERS

In semiconductors, as in metals, the most important scattering processes involve interactions of electrons (or holes) with lattice vibrations and with impurity atoms. In relatively pure crystals, or at relatively high temperatures, the former interaction is ordinarily predominant, while at low temperatures or in impure specimens, the latter may be more important. There is a broad range of intermediate conditions where both processes are significant.

The effect of any particular scattering interaction i may be evaluated by calculating the mean free time τ_i associated with this process. The overall free time τ may then be expressed by

$$\frac{1}{\tau} = \sum_i \frac{1}{\tau_i} \tag{9.11-1}$$

as shown in Section 7.5. It is clear from t...is that the scattering process which leads to the shortest free time τ_i is the dominant one. Once the mean free time τ is known, the mobility may easily be expressed as $e\bar{\tau}/m^*$.

Unfortunately, the calculation of the free time associated with any of the scattering processes which are of importance in semiconductors is a lengthy and complex problem, the mathematical details of which are in some respects beyond the level which has been set for the present treatment. We must, therefore, limit ourselves to a rather crude and qualitative treatment of lattice scattering, although we shall attempt to give a fairly complete treatment of impurity scattering along the lines which were first taken by Conwell and Weisskopf.[11]

At low temperatures the thermal energy available to excite optical-mode lattice vibrations is quite limited, and the most important lattice scattering process is the scattering of charge carriers by acoustical mode lattice vibrations. We shall consider only the scattering of carriers by longitudinal mode vibrations; the reason why the effect of transverse modes is negligible will become clear in the course of the discussion.

The passage of a longitudinal vibration through a crystal gives rise to alternate regions of compression and extension of the crystal lattice. When the crystal is compressed, the positions of the energy bands are altered in such a manner that the width of the forbidden gap is increased (see Figure 9.2); in the condition of extension the forbidden gap width decreases. This effect causes a local variation in the energy associated with the conduction band edge, as shown in Figure 9.20(a), which, to simplify the calculation, may be replaced by the stepwise variation shown in Figure 9.20(b) without introducing much error. There is a similar variation in the energy of the valence band edge. One may then calculate the reflection probability for an electron incident upon a single "step" of Figure 9.20(b).

To do this, an electron incident upon such a step of height $\delta\varepsilon_c$ is represented by a free particle wave function of positive momentum of the form

$$\psi_1(x) = Ae^{ik_0x}. \tag{9.11-2}$$

---

[11] E. Conwell and V. F. Weisskopf, *Phys. Rev.*, **77**, 388 (1950).

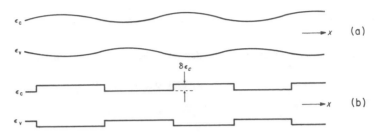

FIGURE 9.20. The sinusoidal variation of conduction and valence band energies in a semiconductor crystal caused by the compressional and extensional forces associated with a longitudinal thermal vibration. (b) A "square wave" variation which approximates the one shown in (a) and which simplifies the calculation of relaxation time.

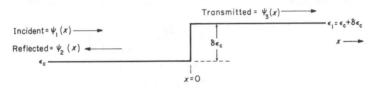

FIGURE 9.21. The "reflection" of an electron from a potential barrier of height $\delta\varepsilon_c$.

The electron may be "reflected" at the step, or it may be transmitted, as illustrated in Figure 9.21. Accordingly, there may be a component of the wave function associated with a negative momentum in the region $(x < 0)$, representing the reflected electron, which must have the form

$$\psi_2(x) = Be^{-ik_0x}. \tag{9.11-3}$$

In the region beyond the barrier $(x > 0)$ the wave function representing an electron transmitted over the barrier is

$$\psi_3(x) = Ce^{ik_1x}. \tag{9.11-4}$$

These are all free particle wave functions satisfying Schrödinger's equation for constant potential. Since the momentum $\hbar k_0$ for the reflected component is assumed to be the same as the incident particle momentum, the collision interaction is assumed to be perfectly elastic. Actually, however, since the lattice wave as represented by the energy barrier is moving with the velocity of sound, there is a Doppler effect which renders the reflected momentum different from the incident momentum. Since the velocity of sound is small compared to the average thermal velocity of the particle the momentum shift is only a small fraction of the incident particle momentum and may safely be neglected in this calculation. From (4.9-8) it is clear that if the energy of the incident electron is ε_0 and that of the "transmitted" electron is ε_1, then

$$\varepsilon_0 = \hbar^2 k_0^2/2m_n^* \quad \text{and} \quad \varepsilon_1 = \varepsilon_0 - \delta\varepsilon_c = \hbar^2 k_1^2/2m_n^* \tag{9.11-5}$$

whereby
$$-\delta\varepsilon_c = \varepsilon_1 - \varepsilon_0 = \frac{\hbar^2}{2m_n^*}(k_1^2 - k_0^2). \qquad (9.11\text{-}6)$$

Since the wave function and its derivative must be continuous at $x = 0$, we may write

$$\psi_1(0) + \psi_2(0) = \psi_3(0) \qquad (9.11\text{-}7)$$

and
$$\psi_1'(0) + \psi_2'(0) = \psi_3'(0).$$

Substituting the values given by (9.11-2, 3 and 4) into these equations, it is easy to show that

$$\frac{B}{A} = \frac{k_0 - k_1}{k_0 + k_1} \quad \text{and} \quad \frac{C}{A} = \frac{2k_0}{k_0 + k_1}. \qquad (9.11\text{-}8)$$

The transmission and reflection probabilities R and T are given by

$$R = \frac{\psi_2^*\psi_2}{\psi_1^*\psi_1} = \left(\frac{k_0 - k_1}{k_0 + k_1}\right)^2 \qquad (9.11\text{-}9)$$

$$T = 1 - R = \frac{4k_0 k_1}{(k_0 + k_1)^2} \qquad (9.11\text{-}10)$$

If the step height $\delta\varepsilon_c$ is small, which we shall assume is the case, then, according to (9.11-6), $k_1 \cong k_0$, and (9.11-9) can, with the help of this approximation and (9.11-6), be written as

$$R \cong \left(\frac{m_n^*\delta\varepsilon_c}{2\hbar^2 k_0^2}\right)^2. \qquad (9.11\text{-}11)$$

The quantity $\delta\varepsilon_c$ can be related to the compressional or extensional strain $\delta V/V_0$ to a first approximation by a linear relation of the form

$$-\delta\varepsilon_c = \Xi_c \cdot \frac{\delta V}{V_0}, \qquad (9.11\text{-}12)$$

where the *deformation potential constant* Ξ_c represents the shift of the conduction band edge per unit dilational strain.

Since the shear distortions associated with transverse vibrational modes produce no first-order change in volume, no change in ε_c results from these vibrations, and their effect in scattering charge carriers is negligible. The passage of a longitudinal elastic wave through the crystal, however, is accompanied by coherent regions of compression or extension which are *of the order of l/2 in linear extent*, where l is the wavelength of the disturbance. If a volume element V_0 of this linear extent is subjected to a dilatational stress producing a maximum pressure δp and a volume change δV, it is easily seen that the stored strain energy is $-\frac{1}{2}\delta p\,\delta V$, and if the source of the strain energy is thermal, its magnitude must be proportional to kT; in other words,

$$\delta\varepsilon = -\tfrac{1}{2}\delta p\,\delta V = ckT \qquad (9.11\text{-}13)$$

where c is a constant.[12] But since the compressibility β is defined as

$$\beta = \frac{1}{V_0}\frac{\delta V}{\delta p}, \tag{9.11-14}$$

we may express δp in terms of β by this equation and substitute the result into (9.11-13) to obtain

$$\delta\varepsilon = \frac{1}{2}\frac{(\delta V)^2}{\beta V_0} = ckT \tag{9.11-15}$$

whereby

$$\frac{(\delta V)^2}{V_0^2} = \frac{2c\beta kT}{V_0}. \tag{9.11-16}$$

Substituting this result into (9.11-12), and using the resulting expression for $\delta\varepsilon_c^2$ in (9.11-11), R may be written as

$$R = \frac{m_n^{*2}}{2\hbar^4 k_0^4}\frac{c\beta kT}{V_0}\Xi_c^2. \tag{9.11-17}$$

Now, in a distance δx, the probability of scattering (reflection) is $\delta x/\lambda_n$ where λ_n is the mean free path. In going a distance $l/2$ (the linear dimension of the volume V_0) the probability of reflection is approximately the value given as R above. Then $l/2\lambda_n$ may be equated to R, and one may solve for λ_n to obtain

$$\lambda_n = \frac{\hbar^4 k_0^4 l V_0}{m_n^{*2}c\beta kT\Xi_c^2} = \frac{\hbar^4}{8m_n^* c\beta kT\Xi_c^2}. \tag{9.11-18}$$

The second expression follows from the first by noting that $V_0 = l^3/8$ and $k_0 = 2\pi/l$. Making the assumption that λ_n is independent of velocity, the approximate validity of which has been discussed in Section 7.5, λ_n and the mean free time are related by (7.3-18). Using this equation, and Equation (5.4-20), $\bar{\tau}_n$ can be expressed in the form

$$\bar{\tau}_n = \frac{\hbar^4}{3\sqrt{8\pi}m_n^{*3/2}c\beta(kT)^{3/2}\Xi_c^2}. \tag{9.11-19}$$

This calculation is clearly quite crude; in particular, no attempt has been made to evaluate c. It clearly illustrates, nevertheless, the principles involved and also exhibits the dependence of $\bar{\tau}_n$ and thus the mobility on temperature, effective mass and deformation potential constant. A full quantum mechanical treatment[13] yields the result

$$\bar{\tau}_n = \frac{\sqrt{8\pi}}{3}\frac{\hbar^4 c_{ll}}{m_n^{*3/2}(kT)^{3/2}\Xi_c^2}, \tag{9.11-20}$$

---

[12] This constant has to do primarily with the fraction of all the possible normal modes of vibration which contribute to a dilatational strain of the type discussed here. It is clearly independent of temperature, effective mass and deformation potential constant.

[13] See, for example, W. Shockley, *Electrons and Holes in Semiconductors*. New York: D. van Nostrand, Inc. (1950), p. 539.

where c_{ll} is the elastic constant for a longitudinal extension in the [110] direction. The corresponding mobility is

$$\mu_n = \frac{e\bar{\tau}}{m_n^*} = \frac{\sqrt{8\pi}}{3} \frac{e\hbar^4 c_{ll}}{m_n^{*5/2}(kT)^{3/2}\Xi_c^2}. \tag{9.11-21}$$

If the constant energy surfaces are ellipsoidal rather than spherical, it is clear [from (9.11-5)] that the effective mass m_n^* in (9.11-17, 19 and 20) should be replaced by the density of states effective mass (9.10-10). The additional factor of m_n^* which appears in (9.11-21) must naturally in this case be the conductivity effective mass (9.10-25), so that the factor $m_n^{*5/2}$ in the denominator of (9.11-21) will then become $m_{ds}^{*3/2}m_c^*$. The above calculations refer specifically to electrons in the conduction band, but the procedure for holes in the valence band is essentially the same, and the results are the same as those given above, with the electron masses and conduction band deformation potential constant replaced by their respective valence band analogues.

According to (9.11-20) and (9.11-21), the mobility should vary as $T^{-3/2}$ and as $m_n^{*-5/2}$ when acoustical mode lattice scattering is the dominant scattering interaction. If the ratio of the electron to hole mobility as expressed by (9.11-21) and its analogue for holes is calculated, one may easily show that

$$\frac{\mu_n}{\mu_p} = (m_n^*/m_p^*)^{-5/2}(\Xi_c/\Xi_v)^{-2}. \tag{9.11-22}$$

Since Ξ_c and Ξ_v are usually about the same, it is often quoted as a rough rule of thumb that the ratio of mobilities is the inverse ratio of the effective masses to the 5/2 power.

Unfortunately, the experimental results are only in rough agreement with the predictions of (9.11-21). The temperature variation of the mobility in the range of temperatures where lattice scattering predominates, in particular, is usually rather stronger than the $T^{-3/2}$ variation found by this analysis. Average values of the measured mobility at 300°K (where lattice scattering is dominant in all but very impure samples) and the measured temperature variation of mobility for holes and electrons in germanium and silicon are shown in Table 9.4. In materials with energy surfaces which are essentially spherical, such as p-type Si and Ge, the effect of scattering by *optical* mode lattice vibrations is important, especially at higher temperatures.[14]

TABLE 9.4.

Measured Mobility Data for Silicon and Germanium

	μ_n	Temp. var. of μ_n	μ_p	Temp. var. of μ_p
Si	1600	$T^{-2.5}$	500	$T^{-2.3}$
Ge	3800	$T^{-1.66}$	1900	$T^{-2.3}$

In cases where the energy surfaces are ellipsoidal, for example, n-type Ge and Si, effects arising from the scattering of electrons from one ellipsoid to another (inter-valley scattering), which have not been considered in the above calculation are also of

---

[14] H. Ehrenreich and A. W. Overhauser, *Phys. Rev.*, **104**, 649 (1956).

importance. In this process when an electron is transferred from one ellipsoid to another, conservation of momentum requires the generation or absorption of a phonon whose energy is of the same order of that of the electron itself, and hence these scattering events cannot be regarded as even approximately elastic. A detailed consideration of this type of scattering by Herring[15] has shown that the observed temperature dependence of mobility in both n-type Ge and n-type Si can be explained on the basis of intervalley scattering, although, of course, the optical mode contribution is no doubt significant also. In addition, the effect of transverse vibrational modes is much greater when the band structure is of the "many valley" form,[16] and hence cannot be entirely neglected for n-type Ge and Si. For a more detailed discussion of these matters, the excellent articles by Herring[15] and Blatt[17] are recommended.

The calculation of the relaxation time for impurity scattering is based upon the theory of scattering of charged particles by the Coulomb potential of nuclei which was originally developed by Rutherford to explain the scattering of α-particles.[18,19] In a semiconductor crystal which contains substitutional impurities, the Coulomb potential due to the charged donor and acceptor ions serve to deflect the paths of electrons and holes much as the potential of a heavy nucleus will deflect an α-particle in the Rutherford scattering experiment.

If the differential cross section $\sigma(\theta)$ is defined such that $\sigma(\theta)\, d\Omega$ is the differential area of the incident particle beam which is scattered through an angle θ into an element of solid angle $d\Omega$, then it may be shown by either classical or quantum mechanics[19] that

$$\sigma(\theta)\, d\Omega = \left(\frac{Ze^2}{2\kappa m^* v_0^2}\right)^2 \frac{1}{\sin^4 \dfrac{\theta}{2}}\, d\Omega \tag{9.11-23}$$

and that

$$\tan\frac{\theta}{2} = \frac{Ze^2}{\kappa a m^* v_0^2}. \tag{9.11-24}$$

The scattering geometry is shown in Figure 9.22 for a repulsive Coulomb interaction (hole and donor ion) and in Figure 9.23 for an attractive Coulomb interaction (electron and donor ion); the above equations are applicable to both cases. The quantity κ is the dielectric constant of the substance, v_0 is the initial (and final) velocity and a is the "impact parameter," or the perpendicular distance between the scattering center and the projection of the initial line of approach of the particle.

For a single incident particle, the number of collisions which it may make per unit time into a solid angle $d\Omega$ is readily seen to be $Nv_0\sigma(\theta)\, d\Omega$, where N is the number of scattering impurities per unit volume. Since energy is conserved in the scattering interaction, the magnitudes of the initial and final momenta of the scattered particle are the same; since the direction is changed by the scattering angle θ, however,

[15] C. Herring, *Bell Syst. Tech. J.*, **34**, 237 (1955).

[16] C. S. Smith, *Phys. Rev.*, **94**, 42 (1954).

[17] F. J. Blatt, *Theory of Mobility of Electrons in Solids* in *Solid State Physics, Advances in Research and Applications*. New York: Academic Press (1957), Vol. 4, p. 199. See particularly p. 332 ff.

[18] E. Rutherford, *Phil. Mag.*, **21**, 669 (1911).

[19] R. B. Leighton, *Principles of Modern Physics*. New York: McGraw-Hill (1959), p. 485 ff.

there is an accompanying change in the forward x-component of momentum of

$$\Delta p_x = p_0(1 - \cos \theta). \tag{9.11-25}$$

Now, from (9.11-23), it is clear that most of the collisions produce only a *small* angle of deflection. The relaxation time, however, is the time required, on the average, for the forward velocity of the particle to be reduced to *zero*, and this clearly is much

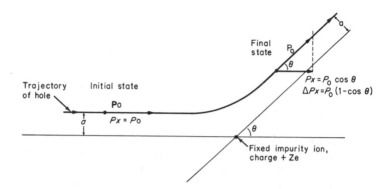

FIGURE 9.22.   The trajectory of a hole which is scattered by Coulomb interaction with a fixed scattering center of charge $+ Ze$.

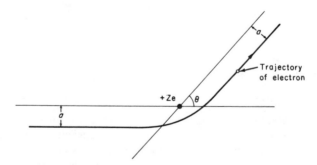

FIGURE 9.23.   The trajectory of an electron scattered by a fixed scattering center of charge $+ Ze$.

longer than the average time between such scattering events. As a matter of fact, from (9.11-25), it can be seen that the *effective* number of velocity destroying collisions is *less* than the actual total number of scatterings by a factor $(1 - \cos \theta)$. The equivalent number of velocity destroying collisions per unit time involving an angle of deflection θ into a solid angle $d\Omega$ is thus not $Nv_0\sigma(\theta)\, d\Omega$ but rather

$$dn_{\text{eff}} = Nv_0\sigma(\theta)(1 - \cos \theta)\, d\Omega = d(1/\tau). \tag{9.11-26}$$

This quantity is just the reciprocal of the relaxation time for particles of this specific initial velocity undergoing collisions involving this particular scattering angle. The total relaxation time may be calculated by integrating over the solid angle and averaging over velocities.

Since the scattering is independent of the azimuthal angle ϕ about the polar axis, $d\Omega = 2\pi \sin\theta\, d\theta$, and integrating (9.11-26) over the solid angle, one may obtain

$$\frac{1}{\tau(v_0)} = 2\pi N v_0 \int_{\theta_0}^{\pi} \sigma(\theta)(1 - \cos\theta) \sin\theta\, d\theta = 2\pi N v_0 \left(\frac{Ze^2}{2\kappa m^* v_0^2}\right)^2 \int_{\theta_0}^{\pi} \frac{(1 - \cos\theta) \sin\theta}{\sin^4 \dfrac{\theta}{2}}\, d\theta$$

$$= -16\pi N v_0 \left(\frac{Ze^2}{2\kappa m^* v_0^2}\right)^2 \ln\sin\frac{\theta_0}{2} = 8\pi N v_0 \left(\frac{Ze^2}{2\kappa m^* v_0^2}\right)^2 \ln\left[1 + \left(\frac{\kappa m^* v_0^2}{2Ze^2 N^{1/3}}\right)^2\right].$$

$$(9.11\text{-}27)$$

In performing the integration it is best to express all trigonometric functions in terms of the argument $\theta/2$. The upper limit of the integral must be $\theta = \pi$, which corresponds to a "direct hit" which reflects the incident particle back upon its original path. For such a collision, the impact parameter a is zero. The lower limit would be $\theta = 0$ for an isolated scattering center (which would give rise to an infinite integrated cross-section for scattering); but in the crystal the influence of an individual scattering center extends outwards only to a distance of the order of the mean spacing between the scattering ions. It is therefore reasonable to "cut off" the scattering from a particular ion when the impact parameter a equals $d/2$, where $d = N^{-1/3}$ is the mean distance between impurity atoms. The lower limit of the integral is then at

$$a = \frac{1}{2N^{1/3}} \qquad (9.11\text{-}28)$$

which, according to (9.11-24), corresponds to

$$\theta = \theta_0 = 2 \tan^{-1}\left(\frac{2Ze^2 N^{1/3}}{\kappa m^* v_0^2}\right). \qquad (9.11\text{-}29)$$

The above calculations permit one to express $\tau(v_0)$ in the form

$$\tau(v_0) = \frac{\kappa^2 m^{*2} v_0^3}{2\pi Z^2 e^4 N} \frac{1}{\ln\left[1 + \left(\dfrac{\kappa m^* v_0^2}{2Ze^2 N^{1/3}}\right)^2\right]}. \qquad (9.11\text{-}30)$$

It is now necessary to calculate $\bar{\tau}$ by the usual averaging process, whereby

$$\bar{\tau} = \frac{\langle v_0^2 \tau(v_0)\rangle}{\langle v_0^2\rangle} = \frac{\kappa^2 m^{*2}}{2\pi Z^2 e^4 N} \frac{\displaystyle\int_0^\infty \frac{v_0^7 e^{-m^* v_0^2/2kT}\, dv_0}{\ln[1 + (\kappa m^{*2} v_0^2/2Ze^2 N^{1/3})^2]}}{\displaystyle\int_0^\infty v_0^4 e^{-m^* v_0^2/2kT}\, dv_0}. \qquad (9.11\text{-}31)$$

This integral cannot be exactly evaluated by analytical means, but if one observes that the logarithmic term in the denominator varies only very slowly with respect to v_0 as compared with the multiplying factor $v_0^7 e^{-m^* v_0{}^2/2kT}$, an approximate value can be obtained by assuming that the logarithmic term has a constant value which is equal to the value it attains when the multiplying factor is a maximum. It is easily ascertained that this maximum value is reached for $v_0 = \sqrt{7kT/m^*}$. Setting v_0 equal to this quantity in the logarithmic term, removing it from the integral, and evaluating the remaining integrals by referring to Table 5.1, we may obtain

$$\bar{\tau} = \frac{8\kappa^2 (kT)^{\frac{3}{2}}(2m^*)^{\frac{1}{2}}}{\pi^{\frac{3}{2}} Z^2 e^4 N \ln\left[1 + (7\kappa kT/2Ze^2 N^{\frac{1}{3}})^2\right]} \tag{9.11-32}$$

whereby the mobility will be given by

$$\mu = \frac{e\bar{\tau}}{m^*} = \frac{8\sqrt{2}\kappa^2 (kT)^{\frac{3}{2}}}{\pi^{\frac{3}{2}} Z^2 e^3 m^{*\frac{1}{2}} N \ln\left[1 + (7\kappa kT/2Ze^2 N^{\frac{1}{3}})^2\right]}. \tag{9.11-33}$$

This method of calculation was developed originally by Conwell and Weisskopf[11,20] and (9.11-33) is often called the Conwell-Weisskopf formula. The same calculation was done in a somewhat more refined way by Brooks and Herring;[21] their result is similar, except that the slowly varying logarithmic term is somewhat different. The temperature and concentration dependences are the same in both cases. In the above calculation, the averaging was done assuming spherical energy surfaces. The computation for ellipsoidal surfaces is much more complex and will not be considered here; the magnitude of the relaxation time will be affected but not the temperature and concentration dependences. We have assumed throughout that the impurity atoms were completely ionized; at very low temperatures this is no longer true. Under these circumstances the quantity N in the above equations must be taken to represent the concentration of *ionized* impurity centers only, as given by (9.5-4) or (9.5-8). This introduces an additional temperature dependence, but it is significant only at very low temperatures.

According to Equation (9.11-33), when ionized impurity scattering is the dominant scattering process, the mobility should be found to be proportional to $T^{3/2}$ and to exhibit an inverse dependence on the concentration of impurities. Although it is rather difficult to obtain specimens in which impurity scattering alone determines the mobility, these predictions have been verified at least approximately by experiment. A plot of mobility *versus* impurity concentration for germanium is shown in Figure 9.24. For low impurity concentrations, the dominant scattering interaction is lattice scattering, which is independent of impurity content, and the mobility [as given, for example, by (9.11-21)] is likewise independent of impurity density. At high concentrations of impurity atoms, the ionized impurity scattering becomes the most important

---

[20] There appears to be an error in the original Conwell-Weisskopf article which has persisted in most of the discussions of this subject by other authors. Conwell and Weisskopf state in their original paper [their Equation (18)] that the maximum of the function multiplying the logarithmic term in the integrand in the numerator of (9.11-31) occurs at $v^2 = 6kT/m$, while a simple calculation suffices to show that it really occurs at $7\ kT/m$. This accounts for the factor 7/2 in the argument of the logarithmic term of (9.11-33) in place of the factor 3 which is in the original Conwell-Weisskopf formula. The correction to the original result is clearly quite small, and does not affect its general validity.

[21] H. Brooks, *Phys. Rev.*, **83**, 879 (1951).

scattering process and the mobility decreases with increasing impurity density as predicted by the Conwell-Weisskopf formula (9.11-33). In the intermediate range, both scattering interactions are important and the overall mobility μ must be expressed as $\mu^{-1} = \mu_I^{-1} + \mu_l^{-1}$ where μ_I is the mobility attributable to ionized impurity scattering alone and μ_l that which would be found were lattice scattering the only important process.

In addition to lattice scattering and impurity scattering, certain other scattering interactions are of occasional importance. Neutral impurity centers as well as ionized impurity centers may give rise to appreciable scattering if present in sufficient numbers.

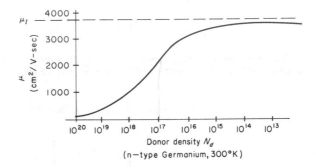

FIGURE 9.24. The effect of impurity concentration upon the room-temperature mobility of electrons in n-type germanium, according to the Conwell-Weisskopf theory of impurity scattering. [After E. M. Conwell, *Proc. I.R.E.*, **40**: 1327 (1952).]

A treatment of this subject has been given by Erginsoy.[22] Neutral impurity scattering is found to be independent of temperature, and hence may be of importance at very low temperatures, when lattice scattering is negligible and most of the impurities are unionized. Furthermore, vacancies, interstitial atoms, dislocations, grain boundaries, and sample surfaces can scatter holes and electrons, although in many cases the scattering attributable to these agencies is negligible in comparison with lattice or impurity scattering.

The account of the behavior of bulk semiconductors given in this chapter is by no means complete. Indeed, we have not really touched upon thermal or optical properties at all, and for lack of space we have had to omit an account of the very interesting effects which may be observed in semiconductors under hydrostatic or uniaxial stresses.[23] It is hoped, however, that the present treatment has illuminated some of the very basic and important areas and has developed methods of general usefulness in investigating some of the subjects which have not been discussed in detail.

————————
[22] C. Erginsoy, *Phys. Rev.*, **79**, 1013 (1950).
[23] R. W. Keyes, "The Effects of Elastic Deformation on the Electrical Conductivity of Semiconductors," in *Solid State Physics, Advances in Research and Applications*. New York: Academic Press (1960), Vol. 11, p. 149.

EXERCISES

1. Make plots, on semilogarithmic graph paper, of the intrinsic carrier concentration n_i as a function of $1/T$ for (a) Germanium (b) Silicon (c) InSb ($\Delta\varepsilon = 0.22$ eV, $m_n^* = 0.013\ m_0$, $m_p^* = 0.18\ m_0$) (d) GaAs ($\Delta\varepsilon = 1.5$ eV, $m_n^* = 0.1\ m_0$, $m_p^* = 0.4\ m_0$). Be sure your plots show clearly the variation of n_i with T over the range from 10–500°K.

2. Consider a two-dimensional intrinsic semiconductor for which, in analogy with two-dimensional metallic systems, the density of states functions are

$$g_n(\varepsilon)\,d\varepsilon = (4\pi m_n^*/h^2)\,d\varepsilon \qquad \varepsilon > \varepsilon_c$$

$$g(\varepsilon) = 0 \qquad \varepsilon_v < \varepsilon < \varepsilon_c$$

$$g_p(\varepsilon)\,d\varepsilon = (4\pi m_p^*/h^2)\,d\varepsilon \qquad \varepsilon < \varepsilon_v$$

Show, using Fermi statistics only, *without* the use of the Boltzmann approximation that for the case where $m_p^*/m_n^* = 2$ the position of Fermi level may be expressed as

$$\varepsilon_c - \varepsilon_f = \frac{1}{2}\Delta\varepsilon - \frac{kT}{2}\ln\frac{8}{3} - kT\ln\cos\frac{\Phi}{3},$$

where
$$\Phi = \tan^{-1}\left[\frac{32}{27}\left(e^{\,\Delta\varepsilon/kT}\right) - 1\right]^{1/2}$$

3. Make plots of the conductivity of both p-type and n-type germanium as a function of net impurity concentration $|N_d - N_a|$ for values of $|N_d - N_a|$ ranging from zero to 10^{17} cm^{-3}, at temperatures of 80°K, 300°K and 500°K. Assume that μ_n varies as $T^{-1.65}$ and μ_p as $T^{-2.3}$. Use logarithmic graph paper.

4. For what value of carrier concentrations n_0, p_0 is the conductivity of a semiconductor a *minimum*? What value of net impurity content $|N_d - N_a|$ does this imply?

5. For a semiconductor with spherical energy surfaces in both conduction and valence bands, but with *two* species of holes, show that the small field Hall coefficient can be written as

$$R = \frac{p_{01}\mu_{p_1}^2\,\dfrac{\overline{\tau_{p_1}^2}}{(\overline{\tau_{p_1}})^2} + p_{02}\mu_{p_2}^2\,\dfrac{\overline{\tau_{p_2}^2}}{(\overline{\tau_{p_2}})^2} - n_0\mu_n^2\,\dfrac{\overline{\tau_n^2}}{(\overline{\tau_n})^2}}{ec(n_0\mu_n + p_{01}\mu_{p_1} + p_{02}\mu_{p_2})^2}$$

where p_{01}, p_{02} are the concentrations, μ_{p_1}, μ_{p_2} the mobilities, and τ_{p_1}, τ_{p_2} the relaxation times for the two species of holes.

6. Find, for each of the constant energy surfaces of germanium, the individual conductivity tensors $\sigma_{\alpha\beta}^{(i)}$ referred to the cube axes as coordinate axes. In germanium the constant-energy surfaces are four ellipsoids whose major axes lies along the four $\langle 111 \rangle$ directions. Show that the result (9.10-23) is obtained by summing over all ellipsoids.

7. Show that if the static field is in the [100] direction, only a single cyclotron resonance peak at the frequency $\omega = (eB/c)[(m_\parallel^* + 2m_\perp^*)/3m_\parallel^* m_\perp^{*2}]^{\frac{1}{2}}$ will be observed for germanium. What resonances would be observed with the static field in the [111] direction?

8. Using the laws of classical mechanics, derive the Rutherford scattering formula (9.11-23) and the associated relation (9.11-24).

9. A semiconductor specimen is determined to have a resistivity of 12.5 ohm-cm. The specimen is 1 cm long by 5 mm wide by 1 mm thick. A current of 1 ma. is made to flow

along the longest dimension of the sample and a Hall voltage of 5 mV is measured across the 5 mm width of the sample, when a magnetic field of 2000 gauss is used. What is the carrier concentration and the Hall mobility of the carriers in the sample? You may assume that the sample is an extrinsic one, that the energy surfaces are spherical, and that acoustical mode lattice scattering is the dominant scattering process.

10. Fill in the details involving the maximization of expression (9.5-2) by the method of Lagrangean multipliers, thus completing the derivation of (9.5-3).

11. Show that (9.3-6) and (9.3-15) can be transformed so as to read

$$n_0 = n_i e^{(\epsilon_f - \epsilon_{fi})/kT}$$

$$p_0 = n_i e^{(\epsilon_{fi} - \epsilon_f)/kT}.$$

12. Explain *physically* the principal features of the experimental curves shown in Figure 9.13, and discuss the relationships between the three separate plots.

13. Show that in the case where acoustical mode lattice scattering is the dominant scattering mechanism, Equation (9.8-23) may be expressed in the form shown at the bottom of page 286.

14. In the case where ionized impurity scattering is the dominant mechanism, it is evident from Equation (9.11-30) that the dependence of the relaxation time on velocity can be expressed as $\tau(v) = \alpha v^3$ where α is constant, provided that the weak velocity dependence of the logarithmic term is neglected. Show, in this instance, that the longitudinal magnetoresistance Equation (9.8-23) may be written in the form

$$I_x = \sigma_0 E_x \left[1 - \frac{e^2 B_0^2}{m_p^{*2} c^2} \left(\frac{2kT}{m_p^*} \right) \left(120 - \frac{\pi (315)^2}{(64)^2} \right) \right].$$

GENERAL REFERENCES

A. C. Beer, "Galvanomagnetic Effects in Semiconductors"; Supplement 4 in *Solid State Physics, Advances in Research and Applications*, Academic Press, Inc., New York (1963).

F. J. Blatt, "Theory of Mobility of Electrons in Solids," in *Solid State Physics, Advances in Research and Applications*, Vol. 4, Academic Press, Inc., New York (1957).

H. Brooks, "Theory of the Electrical Properties of Germanium and Silicon," in *Advances in Electronics and Electron Physics*, Vol. 7, Academic Press, Inc., New York (1955).

E. Burstein and P. H. Egli, "The Physics of Semiconductor Materials," in *Advances in Electronics and Electron Physics*, Vol. 7, Academic Press, Inc., New York (1955).

H. Y. Fan, "Valence Semiconductors, Germanium and Silicon," in *Solid State Physics, Advances in Research and Applications*, Vol. 1, Academic Press, Inc., New York (1955).

R. W. Keyes, "The Effect of Elastic Deformation on the Electrical Conductivity of Semiconductors," in *Solid State Physics, Advances in Research and Applications*, Vol. 11, Academic Press, Inc., New York (1960).

W. Shockley, *Electrons and Holes in Semiconductors*, D. van Nostrand Co., Inc., New York (1950).

R. A. Smith, *Semiconductors*, Cambridge University Press, New York (1961).

CHAPTER 10

EXCESS CARRIERS IN SEMICONDUCTORS

10·1 INTRODUCTION

In a metal it is practically impossible to alter the bulk concentration of free charge carriers in any way whatever. Excess carriers can be introduced, but only by inducing an electrical charge on the specimen. Under these conditions, the excess charge density resides essentially at the surface of the metal, the bulk concentration of charge carriers being completely unaffected. No change in the bulk transport properties can be observed in this situation.

In a semiconductor, however, it is possible to profoundly alter the charge carrier concentration in the bulk material *without* the introduction of any significant electric charge density. This can be done because it is possible for two types of charge carriers, holes and electrons, to be present simultaneously; if excess holes and electrons are introduced in *pairs*, large deviations from the thermal equilibrium values of bulk carrier concentration can be created without the formation of any net charge density. The formation of these carrier densities in excess of equilibrium is accompanied by a significant modulation of the bulk conductivity of the material, as determined by (9.7-6). It is this possibility which permits the use of semiconductor elements as electronic devices such as rectifiers, transistors and switching units. By *extracting* electrons and holes in pairs one may obtain bulk carrier concentrations *less* than the equilibrium values quite as easily as excess concentrations are created by the introduction or injection of charge carrier pairs.

Excess carriers can be created in semiconductors by illuminating the material with light of frequency such that the photon energy $\hbar\omega_0$ equals or exceeds the forbidden gap energy $\Delta\varepsilon$. Under these conditions the incident photons have enough energy to break the covalent electron pair lattice bonds, liberating free electrons and leaving holes behind at the excitation sites. The excess electrons and holes which are thus created in pairs contribute to the conductivity of the crystal, which is then increased under illumination, the increase being proportional to the light intensity. This phenomenon is known as photoconductivity and is quite characteristic of all semiconductors. Since a photon is absorbed in the creation of each electron-hole pair, light of wavelength sufficiently short that the photon energy exceeds $\Delta\varepsilon$ will be quite strongly absorbed within the crystal; on the other hand, light of longer wavelength, whose photon energy is less than $\Delta\varepsilon$ cannot create electron-hole pairs and is absorbed hardly at all. The absorption spectrum of a semiconductor is therefore always

320

characterized by a very rapid change of absorption coefficient at a wavelength corresponding to photon energy $\Delta\varepsilon$, the crystal being strongly absorptive for shorter wavelengths and nearly transparent for longer wavelengths, as shown in Figure 10.1(a). The region of sharp transition from opaqueness to transparency is called the *absorption edge*. In germanium the absorption edge occurs in the near infrared at about $\lambda = 1.75\ \mu$, corresponding to $\Delta\varepsilon = 0.7$ eV, in silicon at about 1.13 μ, corresponding to

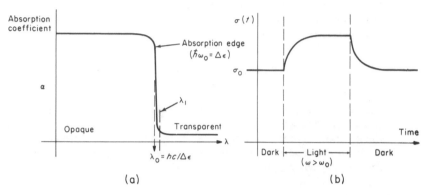

(a)   (b)

FIGURE 10.1.   (a) Absorption spectrum of a typical semiconductor in the neighborhood of the fundamental absorption edge, (b) Photoconductive response of a semiconductor crystal to light of wavelength sufficiently short to excite excess electron-hole pairs.

$\Delta\varepsilon = 1.12$ eV.   The forbidden energy gap $\Delta\varepsilon$ can be accurately determined from the absorption spectrum by measuring the wavelength at which the absorption edge occurs. The photoconductive response of a semiconductor to light of wavelength short enough to create electron-hole pairs is shown in Figure 10.1(b).

Excess carriers can also be created in semiconductor crystals by high energy particle beams (electrons, protons, α-particles), X-rays, γ-radiation, and by suitably biased metal-semiconductor contacts or p-n junctions, about which we shall have more to say later.

10·2 TRANSPORT BEHAVIOR OF EXCESS CARRIERS; THE CONTINUITY EQUATIONS

In any semiconductor crystal electron-hole pairs are continually being generated by thermal or other means and are continually recombining. In thermal equilibrium the only generation process is thermal generation, and the rate at which pairs are generated and the rate at which they recombine must be equal; otherwise, the concentration of electrons and holes would build up or decline as a function of time. The equilibrium thermal generation rate g_0 (cm^{-3} sec^{-1}) is the number of electron-hole pairs generated per unit volume per unit time from thermal breakage of covalent bonds. This quantity is a function of temperature and certain crystal parameters, but is independent of electron and hole concentration. The recombination rate is related to the mean time which elapses between the generation of an electron or hole and its subsequent re-combination. This quantity is called the mean carrier *lifetime*. If there were only one carrier of a given species per unit volume of the crystal at any given time, then the number of recombination events per unit volume per second involving that type of carrier would be $1/\tau$, where τ is the lifetime of that carrier species.[1] When there are n carriers per unit volume, then the recombination rate (number of recombination events per unit volume per unit time) is n times as great or n/τ. In thermal equilibrium, the concentration of electrons is n_0 and the concentration of holes is p_0; since the generation and recombination rates of holes and of electrons must be equal under these circumstances, then we must have

$$g_{0n} = \frac{n_0}{\tau_{n_0}} \quad \text{and} \quad g_{0p} = \frac{p_0}{\tau_{p_0}} \tag{10.2-1}$$

where g_{0n} and g_{0p} are the equilibrium thermal generation rates for electrons and holes, respectively, and where τ_{n_0} and τ_{p_0} are the average lifetimes for the respective carrier species under equilibrium conditions. In addition, since thermal generation always results in the production of electron-hole *pairs*, one hole being generated for each electron, g_{0n} and g_{0p} must be the same, whereby

$$g_{0n} = \frac{n_0}{\tau_{n_0}} = \frac{p_0}{\tau_{p_0}} = g_{0p}. \tag{10.2-2}$$

The equality of electron and hole generation rates as well as the equality of electron and hole recombination rates must be preserved even in systems which are not in a steady-state condition, because an electron generation (or recombi-nation) is inevitably accompanied by a hole generation (or recombination). In a non-steady-state system, however, the generation and recombination rates for elec-trons (or holes) may be different. In any event, we may *always* write

$$g_n = g_p \quad \text{and} \quad \frac{n}{\tau_n} = \frac{p}{\tau_p}, \tag{10.2-3}$$

where g_n and g_p are the actual generation rates, n and p the local concentrations and τ_n and τ_p the lifetime of electrons and holes, respectively. In these equations the generation rates g_n and g_p as well as the concentrations n and p and the lifetimes τ_n and τ_p must in general be regarded as *functions of space coordinates and time* within the crystal.

Consider now a region of the crystal of dimensions dx, dy, dz, as shown in Figure 10.2, and suppose that particle flux densities $\mathbf{J}_p$ and $\mathbf{J}_n$ of holes and electrons, respec-tively, are flowing into and out of this region. The x-component of $\mathbf{J}_p$ at $x + dx$ can

---

[1] We have heretofore used the symbols τ_n, τ_p and τ to refer to relaxation times connected with scattering processes. Unfortunately, the *same* symbols are *universally* used in the literature to repre-sent recombination lifetimes as well. Since this practice has been sanctified by many years of usage, and since we do not wish to adopt a notation which is very complex, we shall also use the same symbols for both sets of quantities. From this point forward, however, we shall understand that τ, τ_n and τ_p always refer to *recombinative lifetimes* unless otherwise specified.

be expressed in terms of $\mathbf{J}_p$ at x by making a Taylor expansion of $J_{px}(x + dx)$; in the limit where dx becomes small only the first two terms will be significant, whereby

$$J_{px}(x + dx) = J_{px}(x) + \frac{\partial J_{px}}{\partial x}\, dx. \qquad (10.2\text{-}4)$$

The net increase in the number of holes within the region per unit time arising from a difference in the x-components of $\mathbf{J}_p$ at the two faces $ABCD$ and $EFGH$ is then

$$[J_{px}(x) - J_{px}(x + dx)]\, dy\, dz = -\frac{\partial J_p}{\partial x}\, dx\, dy\, dz. \qquad (10.2\text{-}5)$$

There will be similar terms arising from the difference of the y-components of $\mathbf{J}_p$ at

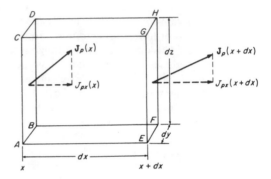

FIGURE 10.2. Carrier flow into and out of a small volume element of the crystal.

the faces $ACGE$ and $BDHF$, and from the difference of the z-components of $\mathbf{J}_p$ at $ABFE$ and $CDHG$. The total net increase per unit time in the number of holes within the region arising from these terms is $-\nabla \cdot \mathbf{J}_p\, dx\, dy\, dz$. If generation and recombination processes are active within the volume, there will be $g_p\, dx\, dy\, dz$ holes generated and $(p/\tau_p)\, dx\, dy\, dz$ holes lost by recombination per unit time within the region of interest. The algebraic sum of all these quantities represents the total net increase of the number of holes inside the volume $dx\, dy\, dz$ per unit time, which is clearly $(\partial p/\partial t)\, dx\, dy\, dz$. Taking the algebraic sum of the various contributions, equating it to $(\partial p/\partial t)\, dx\, dy\, dz$ and canceling the volume element on both sides of the equation, one may easily show that

$$-\nabla \cdot \mathbf{J}_p + g_p - \frac{p}{\tau_p} = \frac{\partial p}{\partial t}, \qquad (10.2\text{-}6)$$

while a similar calculation for electrons within the same volume element will obviously yield

$$-\nabla \cdot \mathbf{J}_n + g_n - \frac{n}{\tau_n} = \frac{\partial n}{\partial t}. \qquad (10.2\text{-}7)$$

Equations (10.2-6) and (10.2-7) are referred to as the equations of continuity for electrons and holes. The solutions of these equations, under appropriate boundary conditions, describe the distribution of electron and hole concentration as a function of space coordinates and time, and furnish a complete description of the transport behavior of electrons and holes in the semiconductor under nonequilibrium conditions. In order to arrive at explicit solutions of the continuity equations, it is necessary to express the current in terms of the concentration. This is easily done by writing the particle flux density as the sum of a diffusion flux density and a drift current density arising from any electric field which might be present. We may in this way write

$$\mathbf{J}_p = -D_p \nabla p + p\mu_p \mathbf{E} \qquad (10.2\text{-}8)$$

and
$$\mathbf{J}_n = -D_n \nabla n - n\mu_n \mathbf{E}. \qquad (10.2\text{-}9)$$

The origin of the current component due to the field in the above equations is quite clear;[2] the diffusion term states, in effect, that whenever there is a *gradient* of concentration ∇p, a net current of particles flows from regions of high concentration to regions of lower concentration, proportional to the gradient of concentration at any point. This diffusive flow is analogous to the flow of heat in the presence of a temperature gradient. The constant of proportionality is called the *diffusion coefficient* or *diffusivity*. That the diffusive component of current should have this form can be proved rigorously by the methods developed in Chapter 7 starting with the Boltzmann equation, assuming the relaxation time approximation (7.2-4). We shall not reproduce the details of this proof here, but it is assigned as an exercise for the reader; it affords good practice in utilizing the transport theory discussed previously. In the case where the mean free path is velocity-independent, which is quite a good approximation in semiconductors where lattice scattering is dominant, it turns out that the diffusion coefficients for electrons and holes can be expressed as

$$D_n = \lambda_n \bar{c}_n / 3 \qquad \text{and} \qquad D_p = \lambda_p \bar{c}_p / 3 \qquad (10.2\text{-}10)$$

where λ_n and λ_p are electron and hole mean free paths and $\bar{c}_n$ and $\bar{c}_p$ are the mean thermal velocities for electrons and holes, respectively. In the circumstances under which (10.2-10) is valid, the mean free paths and relaxation times for electrons and holes are related by (7.3-18). If the equations (10.2-10) are expressed in terms of relaxation times, and if the thermal velocities are represented by (5.4-20), using the appropriate effective masses, it can readily be seen with the aid of (7.3-13) that the diffusion coefficients can be represented in terms of the mobilities as

$$D_n = \frac{\mu_n kT}{e} \qquad \text{and} \qquad D_p = \frac{\mu_p kT}{e}. \qquad (10.2\text{-}11)$$

These relations between diffusion coefficients and mobilities are known as the *Einstein relations*. Although derived above under fairly restrictive assumptions, they can be shown to be true in all systems which obey Boltzmann statistics. As a matter of fact,

---

[2] It is assumed in (10.2-8) and (10.2-9) that no *magnetic* fields are present, or at least that their effect on the carriers is small compared to that of the electric field. The symbols $\mathbf{J}_n$ and $\mathbf{J}_p$ will be used to represent particle flux density, while $\mathbf{I}_n (= -e\mathbf{J}_n)$ and $\mathbf{I}_p (= e\mathbf{J}_p)$ will be used for electrical current density.

if an appropriate numerical constant is included, they may be shown to apply to Fermi systems as well.

If the current equations (10.2-8) and (10.2-9) are substituted into the continuity equations (10.2-6) and (10.2-7), the latter can be written as

$$D_p \nabla^2 p - \mu_p \nabla \cdot (p\mathbf{E}) + g_p - \frac{p}{\tau_p} = \frac{\partial p}{\partial t} \qquad (10.2\text{-}12)$$

and

$$D_n \nabla^2 n + \mu_n \nabla \cdot (n\mathbf{E}) + g_n - \frac{n}{\tau_n} = \frac{\partial n}{\partial t}. \qquad (10.2\text{-}13)$$

The divergence terms in the above equations can be transformed to a slightly more tractable form by the use of the vector identity

$$\nabla \cdot (\phi \mathbf{A}) = \mathbf{A} \cdot \nabla \phi + \phi \nabla \cdot \mathbf{A} \qquad (10.2\text{-}14)$$

which is true for any vector $\mathbf{A}$ and any scalar function ϕ. Also, it is advantageous to express the generation rates as the sum of the thermal generation rate plus the rate at which carriers are generated in *excess* of the thermal generation rate, hence

$$g_n = g_{0n} + g_n' \qquad \text{and} \qquad g_p = g_{0p} + g_p', \qquad (10.2\text{-}15)$$

where g_n' and g_p' are the excess generation rates. The thermal generation rates may now be expressed in terms of equilibrium concentrations and lifetimes by (10.2-1). After making these transformations, the continuity equations (10.2-12) and (10.2-13) take the form

$$D_p \nabla^2 p - \mu_p (\mathbf{E} \cdot \nabla p + p \nabla \cdot \mathbf{E}) + g_p' - \left(\frac{p}{\tau_p} - \frac{p_0}{\tau_{p0}} \right) = \frac{\partial p}{\partial t} \qquad (10.2\text{-}16)$$

$$D_n \nabla^2 n + \mu_n (\mathbf{E} \cdot \nabla n + n \nabla \cdot \mathbf{E}) + g_n' - \left(\frac{n}{\tau_n} - \frac{n_0}{\tau_{n0}} \right) = \frac{\partial n}{\partial t}. \qquad (10.2\text{-}17)$$

The field $\mathbf{E}$ can be expressed as the sum of the applied field and the *internal* field which arises because of the fact that the diffusing particles are charged, whereby

$$\mathbf{E} = \mathbf{E}_{\text{int}} + \mathbf{E}_{\text{app}}. \qquad (10.2\text{-}18)$$

The internal field has its origin in the fact that the mean free paths (and thus the diffusion coefficients) of electrons and holes may be different. If the electrons and holes could move entirely independently of each other, the faster diffusing species would tend to go ahead of the slower diffusing species, leaving the latter behind altogether. When this actually begins to happen, however, since the diffusing particles are charged, a separation of positive and negative charges begins to occur, and an internal electric field which tends to retard the faster diffusing particles and to drag the slower ones on more rapidly is set up. This is the field $\mathbf{E}_{\text{int}}$ of Equation (10.2-18). Eventually this field becomes strong enough to counteract entirely any tendency for the positive and negative charge distributions to separate further; the positive and

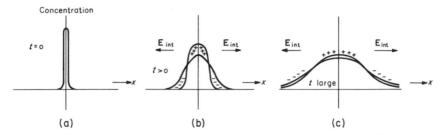

FIGURE 10.3. Successive stages in the diffusive spreading of a distribution of excess electron-hole pairs, which initially has the form of a δ-function, in an intrinsic semiconductor.

negative particle distributions then diffuse *together*, the effective diffusion coefficient being larger than that of the more slowly diffusing species, but smaller than that of the more rapidly diffusing type. This phenomenon, which is known as *ambipolar* diffusion, is illustrated in Figure 10.3. As we shall soon see, a similar effect alters the character of the drift motion which takes place when an applied electric field is present.

In Equations (10.2-16) and (10.2-17), it is clear from the preceding discussion that there are really *three* unknown quantities, n, p, and $\mathbf{E}$. To arrive at a complete solution for these quantities another equation is needed. The set of equations may be completed by writing Poisson's equation relating the electric field and the net charge density $e(p - n + N_d - N_a + p_a - n_d)$. According to this,

$$\nabla \cdot \mathbf{E} = \frac{4\pi\rho}{\kappa} = \frac{4\pi e(p - n + N_d - N_a + p_a - n_d)}{\kappa} \qquad (10.2\text{-}19)$$

where κ is the dielectric constant. We should note that since the applied field $\mathbf{E}_{app}$ has no internal sources or sinks, its divergence is zero, whence from (10.2-18)

$$\nabla \cdot \mathbf{E} = \nabla \cdot \mathbf{E}_{int}. \qquad (10.2\text{-}20)$$

Equations (10.2-16), (10.2-17), and (10.2-19) now form a set of three equations which can in principle be solved for the three unknowns n, p, and $\mathbf{E}_{int}$.

Unfortunately, there seems to be no way of arriving at the solution of these equations in a straightforward analytical manner. We must, therefore, make some reasonable physical approximation which will allow us to arrive at a solution which, although not exact, will be adequate for most cases of practical importance. The approximation which we shall use for this purpose is the *electrical neutrality* condition, or *charge balance* assumption. We shall assume in (10.2-16) and (10.2-17) that the excess electron density $\delta n = n - n_0$ is just balanced by an excess hole density $\delta p = p - p_0$. It is clear that this cannot be precisely correct, for then no internal field would ever be set up, and it is the internal field arising from the basic disparity in electron and hole mean free paths which tends to keep the diffusing electron and hole density distributions together to begin with. However, such a small difference between electron and hole density (compared with the excess carrier densities themselves) is ordinarily required to create this internal field, that the charge balance approximation is usually a very good one insofar as the carrier densities of Equations (10.2-16) and (10.2-17) are concerned. We shall *not*, however, expect to use the charge neutrality assumption in (10.2-19). On the contrary, we may *calculate* the field $\mathbf{E}_{int}$ from

(10.2-16) and (10.2-17) using the charge neutrality assumption, since there will then be only the two unknowns, δp and $\mathbf{E}_{int}$, in these two equations. This calculated value of $\mathbf{E}_{int}$ may then be used in (10.2-19) to determine the source density (thus the disparity in electron and hole densities) required to maintain the field. If the calculated source density is only a small fraction of the excess electron and hole densities calculated from (10.2-16) and (10.2-17) we may be confident of the consistency of our calculation and the validity of the assumptions upon which it is based; otherwise, we must conclude that our results are inconsistent with our original hypotheses and our calculations are in error. We shall see later that the former state of affairs is usually the one which is encountered in actual practice.[3]

Following this line of reasoning, then, we shall assume initially that

$$\delta n = n - n_0 = p - p_0 = \delta p. \qquad (10.2\text{-}21)$$

We shall limit ourselves to a discussion of homogeneous samples, where the impurity density is uniform throughout. For such samples n_0 and p_0 are constants and the gradients and time derivatives of n and p are simply equal to the gradients and time derivatives of δn and δp, respectively. Also, from (10.2-3) it is readily seen that the generation and recombination terms in (10.2-12) are equal to those in (10.2-13). The generation and recombination terms in (10.2-16), which arise directly from the corresponding terms in (10.2-12) are therefore equal to the generation and recombination terms in (10.2-17) which arise in the same way from those terms in (10.2-13). Employing these results in (10.2-16) and (10.2-17) it may easily be shown that

$$D_p\nabla^2(\delta p) - \mu_p(\mathbf{E}\cdot\nabla(\delta p) + p\nabla\cdot\mathbf{E}) + g' - \left(\frac{p_0 + \delta p}{\tau_p} - \frac{p_0}{\tau_{p0}}\right) = \frac{\partial(\delta p)}{\partial t} \qquad (10.2\text{-}22)$$

$$D_n\nabla^2(\delta p) + \mu_n(\mathbf{E}\cdot\nabla(\delta p) + n\nabla\cdot\mathbf{E}) + g' - \left(\frac{p_0 + \delta p}{\tau_p} - \frac{p_0}{\tau_{p0}}\right) = \frac{\partial(\delta p)}{\partial t} \qquad (10.2\text{-}23)$$

where

$$g_n' = g_p' = g'. \qquad (10.2\text{-}24)$$

The term involving $\nabla\cdot\mathbf{E}$ may be eliminated by multiplying (10.2-22) by $n\mu_n$, (10.2-23) by $p\mu_p$ and adding the two equations (noting from (10.2-21) that $n - p = n_0 - p_0$) to obtain

$$\frac{n\mu_n D_p + p\mu_p D_n}{n\mu_n + p\mu_p}\nabla^2(\delta p) - \frac{\mu_n\mu_p(n_0 - p_0)}{n\mu_n + p\mu_p}\mathbf{E}\cdot\nabla(\delta p) + g' - \left(\frac{p_0 + \delta p}{\tau_p} - \frac{p_0}{\tau_{p0}}\right) = \frac{\partial(\delta p)}{\partial t}.$$

$$(10.2\text{-}25)$$

Now in the coefficient of the first term above the mobilities can be expressed in terms of the diffusion coefficients by the Einstein relations (10.2-11). Equation (10.2-25) may thus be written as

$$D^*\nabla^2(\delta p) - \mu^*\mathbf{E}\cdot\nabla(\delta p) + g' - \frac{\delta p}{\tau} = \frac{\partial(\delta p)}{\partial t} \qquad (10.2\text{-}26)$$

---

[3]This line of approach for the semiconductor case was first taken by W. van Roosbroeck [*Phys. Rev.*, **91**, 282 (1953)], but similar methods were used much earlier to treat problems dealing with the transport of electrons and ions in gas discharges, which are quite similar.

where
$$D^* = \frac{(n + p)D_n D_p}{n D_n + p D_p} = \frac{(n_0 + p_0 + 2\delta p)D_n D_p}{(n_0 + \delta p)D_n + (p_0 + \delta p)D_p} \qquad (10.2\text{-}27)$$

$$\mu^* = \frac{(n_0 - p_0)\mu_n \mu_p}{n \mu_n + p \mu_p} = \frac{(n_0 - p_0)\mu_n \mu_p}{(n_0 + \delta p)\mu_n + (p_0 + \delta p)\mu_p} \qquad (10.2\text{-}28)$$

and where τ is an "excess carrier lifetime" defined by

$$\frac{\delta p}{\tau} = \frac{p_0 + \delta p}{\tau_p} - \frac{p_0}{\tau_{p_0}} = \frac{n_0 + \delta p}{\tau_n} - \frac{n_0}{\tau_{n_0}}. \qquad (10.2\text{-}29)$$

The excess concentration of the other carrier species, δn, is, of course, equal to δp in accord with the charge balance assumption (10.2-21).

Equation (10.2-26) is the *ambipolar* transport equation which is obeyed by the excess electron and hole density.[4] The ambipolar diffusion coefficient D^* and the ambipolar mobility μ^* are seen from (10.2-27) and (10.2-28) to be, in general, dependent upon the concentration δp of excess carriers. It is impossible to solve (10.2-26) under these circumstances, and one must in general proceed from this point using approximate or numerical methods. However, if the excess carrier density δp is much less than the larger of the two quantities (n_0, p_0), D^* and μ^* are substantially constant and analytical solutions of (10.2-26) are quite easily obtained. We shall restrict ourselves to the discussion of situations where this condition is fulfilled; we shall refer to this state of affairs as the *low level* case.

Let us now turn our attention to the case of strongly extrinsic n-type material, where $n_0 \gg p_0$ or δp. Suppose, for example, we have in equilibrium 10^{15} electrons and 10^{11} holes per cm^3. There will be a certain equilibrium electron lifetime τ_{n_0} and an equilibrium hole lifetime τ_{p_0}, which will be related by (10.2-2) or (10.2-3). It is clear that τ_{p_0} is much smaller (by a factor of 10^4 in this case) than τ_{n_0}. Suppose now that a uniform excess density of electron-hole pairs is created such that $\delta p = \delta n = 10^{11}$ cm^{-3}. The concentration of electrons is now $n_0 + \delta n = 1.0001 \times 10^{15}$ cm^{-3}, an increase of 0.01 percent over the equilibrium density, while the concentration of holes is 2×10^{11}, double the equilibrium density. Since the density of electrons *relative to the equilibrium density* is hardly changed, the probability per unit time that any given hole will encounter an electron is practically unchanged. Therefore, the *hole lifetime* will be essentially unaffected, and will be practically independent of δp for such small values of δp. On the other hand, since the density of holes has doubled, the probability that any given electron will encounter a hole is essentially doubled, and the electron lifetime will be reduced to about half what it was in the equilibrium state. The electron lifetime must thus vary rapidly with δp even though that quantity may be quite small. Clearly the situation will be just the opposite in strongly extrinsic p-type material. In general, we observe that the lifetime of the *minority* carrier is essentially independent of δp for small values of δp while the majority carrier lifetime is not. For a strongly n-type semiconductor, then, $\tau_p = \tau_{p_0}$ for $\delta p \ll n_0$, and the excess carrier lifetime τ, as

---

[4] This equation *resembles* the continuity equation for independently diffusing particles, which would have the form (10.2-13), but it is not exactly the same, the term $\nabla \cdot (p\mathbf{E})$ being replaced by $\mathbf{E} \cdot \nabla p$. If the field is constant the terms are the same, but otherwise the equations and their solutions will be different.

defined by (10.2-29) reduces to

$$\tau = \tau_{p_0} \qquad \text{(strongly } n\text{-type material)}. \qquad (10.2\text{-}30)$$

Likewise, for a strongly extrinsic p-type crystal, it is readily seen that (10.2-29) reduces to

$$\tau = \tau_{n_0} \qquad \text{(strongly } p\text{-type material)}. \qquad (10.2\text{-}31)$$

For the low-level case, then, the excess carrier lifetime τ simply reduces to the lifetime of the *minority* carrier. The terms *excess carrier lifetime* and *minority carrier lifetime* are often used interchangeably, despite the fact that the two are not always the same. In particular, when the conditions for the low level case are not fulfilled, or when the crystal is nearly intrinsic, the excess carrier lifetime τ differs from the minority carrier lifetime. It is the excess carrier lifetime which is ordinarily measured in experiments designed to detect excess carrier photoconductivity.

Under low-level conditions, it is clear from (10.2-27) and (10.2-28) that in the case of a strongly extrinsic n-type semiconductor D^* and μ^* reduce to D_p and μ_p, and for strongly extrinsic p-type material, D^* and μ^* reduce to D_n and $-\mu_n$. Combining these results with those expressed in (10.2-30) and (10.2-31), we may write for the extrinsic n-type samples

$$D_p \nabla^2(\delta p) - \mu_p \mathbf{E} \cdot \nabla(\delta p) + g' - \frac{\delta p}{\tau_{p_0}} = \frac{\partial(\delta p)}{\partial t} \qquad (n_0 \gg p_0, \delta p) \quad (10.2\text{-}32)$$

and for extrinsic p-type samples

$$D_n \nabla^2(\delta n) + \mu_n \mathbf{E} \cdot \nabla(\delta n) + g' - \frac{\delta n}{\tau_{n_0}} = \frac{\partial(\delta n)}{\partial t} \qquad (p_0 \gg n_0, \delta n). \quad (10.2\text{-}33)$$

In either case the concentration of the other species is obtained from the relation $\delta p = \delta n$. These equations look very much like continuity equations for *minority* carrier flow. The transport and recombination parameters are in all cases those of the minority carrier. It is often stated that in the above situation one writes down and solves the equation for diffusive flow of *minority* carriers, although this is not precisely true: Equations (10.2-32) and (10.2-33) contain within them in addition an implicit assumption about the behavior of the *majority* carriers.

In a sample for which n_0 and p_0 are not vastly different, and where δp is much less than the *smaller* of (n_0, p_0), the ambipolar transport coefficients, (10.2-27) and (10.2-28), are constant and can be written as

$$D^* = \frac{(n_0 + p_0)D_n D_p}{n_0 D_n + p_0 D_p} = \frac{(p_0^2 + n_i^2)D_n D_p}{n_i^2 D_n + p_0^2 D_p} \qquad (10.2\text{-}34)$$

and

$$\mu^* = \frac{(n_0 - p_0)\mu_n \mu_p}{n_0 \mu_n + p_0 \mu_p} = \frac{(n_i^2 - p_0^2)\mu_n \mu_p}{n_i^2 \mu_n + p_0^2 \mu_p}. \qquad (10.2\text{-}35)$$

The ambipolar diffusion coefficient approaches D_p for $p_0 \ll n_i$ and D_n for $p_0 \gg n_i$.

For intermediate values of p_0, D^* lies somewhere between these two extreme values. Likewise for $p_0 \ll n_i$, μ^* approaches μ_p, and for $p_0 \gg n_i$, μ^* approaches $-\mu_n$, lying between these limits for intermediate values of p_0. For material which is precisely intrinsic, $p_0 = n_i$ and the above equations reduce to

$$D^* = D_i^* = \frac{2D_n D_p}{D_n + D_p} \quad \text{and} \quad \mu_i^* = 0, \qquad (10.2\text{-}36)$$

the continuity Equation (10.2-26) then becoming

$$D_i^* \nabla^2(\delta p) + g' - \frac{\delta p}{\tau} = \frac{\partial(\delta p)}{\partial t}. \qquad (10.2\text{-}37)$$

Actually, for precisely intrinsic material, it is easily seen from (10.2-27) and (10.2-28) that this is the correct continuity equation for δp regardless of how large δp may be compared with n_0 or p_0. In the same way it may be seen that this continuity equation is *approximately* correct for near-intrinsic material whenever the density of excess carriers δp is greatly in *excess* of the equilibrium majority carrier density since then μ^* becomes very small. As we shall see in a later chapter, this fact greatly simplifies the analysis of semiconductor device structures under high-current conditions. The fact that μ^* approaches zero in intrinsic samples does not mean that electrons and holes do not acquire a drift velocity when an electric field is applied, but only that an excess carrier concentration distribution does not drift under these circumstances. Electrons and holes drift into and out of such a distribution, although the distribution itself may remain stationary, or nearly so.

We must be careful to note that the electric field $\mathbf{E}$ in Equations (10.2-26), (10.2-32), and (10.2-33) refers to the total electric field as given by (10.2-18), which *includes* the internal field set up by the ambipolar diffusion and drift of the charged particles of the system. We may derive an expression for this field by noting that the total electrical current density $\mathbf{I}$ may be written, with the aid of (10.2-8), (10.2-9), and (10.2-21), as

$$\mathbf{I} = e(\mathbf{J}_p - \mathbf{J}_n) = \sigma\mathbf{E} + e(D_n - D_p)\nabla(\delta p), \qquad (10.2\text{-}38)$$

where σ is the conductivity as given by (9.7-5). Solving for the field, using the Einstein relations to express the diffusion constants in terms of mobilities, and introducing the mobility ratio b, we may readily obtain

$$\mathbf{E} = \frac{\mathbf{I}}{\sigma} - \frac{kT}{e}\frac{b-1}{nb+p}\nabla(\delta p) = \mathbf{E}_{app} + \mathbf{E}_{int}. \qquad (10.2\text{-}39)$$

The first term above is clearly the field arising from the application of an external potential source; the second represents the internal field. It is obvious that the internal field, as expressed in the second term, vanishes if there is no gradient of carrier concentration, and also if the hole and electron mobilities are identical ($b = 1$). In the latter instance, (10.2-27) and (10.2-28) give $D^* = D$ and $\mu^* = [(n_0 - p_0)/(n + p)]\mu$, where D and μ are the common hole and electron diffusion coefficient and mobility. Using these values, one may proceed to solve (10.2-26) using for $\mathbf{E}$ the value of the applied field, wasting no further concern about the internal field.

For materials in which the electron and hole mobilities are different, it is clear that the internal field will contribute to the total field $\mathbf{E}$ of (10.2-26). We shall see, however, that in the great majority of cases, this contribution can be neglected. The reason for this is that, aside from the effects which have been embodied in the modified transport coefficients D^* and μ^*, the internal field ordinarily exerts only a small *direct* influence upon the spatial distribution of excess carriers, compared to that of diffusion and applied fields. This can be demonstrated by a comparison of the solutions of (10.2-26) where there is diffusion but no field, and when there is a field, but essentially no diffusion. In the former case (assuming a one-dimensional geometry in which δp varies only along the x-direction, that the excess bulk generation rate g' is zero, and that steady-state conditions prevail, so that $\partial(\delta p)/\partial t = 0$) Equation (10.2-26) reduces to

$$\frac{d^2(\delta p)}{dx^2} - \frac{\delta p}{D^*\tau} = 0. \tag{10.2-40}$$

If we seek only solutions which approach zero as x becomes large, we must write

$$\delta p = Ae^{-x/(D^*\tau)^{1/2}} \tag{10.2-41}$$

as the solution to (10.2-40). On the other hand, if there is a constant field E, and if diffusive transport may be neglected (10.2-26) may be written, approximately, as

$$\frac{d(\delta p)}{dx} = -\frac{\delta p}{\mu^* E\tau}, \tag{10.2-42}$$

whose solution is

$$\delta p = A'e^{-x/\mu^* E\tau}. \tag{10.2-43}$$

From these equations it is evident that there is a characteristic length $(D^*\tau)^{1/2}$ associated with purely diffusive transport in the steady state, and another characteristic length $\mu^* E\tau$ in this case where the effect of a field is predominant. If the field is zero, the latter length is likewise zero, and it is quite evident from the preceding results that the distribution of excess carriers resulting from purely diffusive transport will not be much affected by the presence of an electric field unless the field is sufficiently large that $\mu^* E\tau$ is comparable in magnitude with the characteristic diffusion length $(D^*\tau)^{1/2}$. In other words, if

$$\mu^* E\tau \ll \sqrt{D^*\tau}, \quad \text{or} \quad E \ll \sqrt{D^*/\mu^{*2}\tau} \tag{10.2-44}$$

then the effect of the electric field on the distribution of excess carrier concentration will be negligible. Inserting the values of D^* and μ^* from (10.2-27) and (10.2-28) and the value of the internal field as given by (10.2-39) into this expression, we may obtain as a condition on the concentration gradient that

$$|\nabla(\delta p)| \ll \frac{1}{\sqrt{D^*\tau}} \frac{(n + p)(nb + p)}{(b - 1)(n_0 - p_0)} \tag{10.2-45}$$

in order that the effect of the internal field be neglected in (10.2-26). In practice one might proceed initially by neglecting the internal field and calculating δp from (10.2-26) substituting for E the applied field. The gradient of δp could then be calculated from the resulting solution and checked against (10.2-45) to insure that that condition is satisfied by the solution everywhere within the region of interest. If (10.2-45) were violated, then the solution would have to be rejected, and a new calculation made in which the internal field is incorporated from the outset. In purely diffusive systems, we shall see that the solution often has the form $\delta p = \text{(const)}\, e^{-x/L}$ where L is approximately (if not precisely) equal to $(D^*\tau)^{1/2}$. In such instances the gradient of δp is given by $-\delta p/L$, and it is readily seen from the form of (10.2-45) that the condition expressed by this equation is never seriously violated regardless of what values n_0, p_0, δp and b may have. Larger values of $\nabla(\delta p)$ may be obtained if an applied field is present, but in such instances, although condition (10.2-45) may no longer be satisfied, it will generally be found that the internal field which is generated is now small compared to the *applied* field which is necessary to maintain this state of affairs. One may thus arrive at the conclusion that it is usually a very good approximation to regard the field E in Equation (10.2-26) as the *applied* field only, neglecting the explicit effect of the internal field. It should be noted, however, that the effect of the internal field is *implicitly* embodied in the modified transport coefficients D^* and μ^*.

The continuity Equations (10.2-16) and (10.2-17) lead directly to the single differential Equation (10.2-26) for δp when the electrical neutrality assumption is made. When an expression for δp has been obtained as a solution of (10.2-26), the validity of the electrical neutrality assumption may easily be checked by computing the *departure* from neutrality required to produce the internal field (10.2-39) and comparing the disparity in charge densities thus obtained with the total excess carrier density obtained from the solution of (10.2-26). Using (9.5-12), we may write Poisson's Equation (10.2-19) in terms of excess densities δn and δp; using (10.2-20) and (10.2-39) to express the electric field, we obtain

$$\nabla \cdot \mathbf{E}_{\text{int}} = \frac{4\pi e(\delta p - \delta n)}{\kappa} = \frac{kT}{e}\frac{b-1}{nb+p}\nabla^2(\delta p). \tag{10.2-46}$$

The ratio of the disparity in densities $\delta p - \delta n$ required to set up the internal field to the actual density δp may then be written as

$$\frac{\delta p - \delta n}{\delta p} = -\frac{\kappa kT}{4\pi e^2}\frac{b-1}{nb+p}\frac{\nabla^2(\delta p)}{\delta p}. \tag{10.2-47}$$

The expression on the right-hand side of this equation must be much less than unity for the charge balance assumption (10.2-21) to be valid. This condition may be expressed in a slightly different way by multiplying numerator and denominator above by n_i. The condition for validity of the charge balance assumption then becomes

$$\left|\frac{\delta p - \delta n}{\delta p}\right| = L_{D_i}^2\frac{(b-1)n_i}{nb+p}\frac{\nabla^2(\delta p)}{\delta p} \ll 1 \tag{10.2-48}$$

where

$$L_{D_i} = \sqrt{\frac{\kappa kT}{4\pi e^2 n_i}} \tag{10.2-49}$$

is a parameter with the dimensions of length which is called the "intrinsic Debye length" associated with the material. At 300°K, the intrinsic Debye length is about 34 μ for silicon and 0.96 μ for germanium. If, as is very frequently the case, the excess concentration density has the form $\delta p = $ (const) $e^{\pm x/L}$ where L is a characteristic decay distance, which in purely diffusive systems has the value $(D^*\tau)^{1/2}$, then $\nabla^2(\delta p)/\delta p = 1/L^2$ and (10.2-48) may be expressed as

$$\left| \frac{\delta p - \delta n}{\delta p} \right| = \frac{L_{D_i}^2}{L^2} \frac{(b-1)n_i}{nb + p} \ll 1. \tag{10.2-50}$$

It is clear from this that the assumption of electrical neutrality should be a good one when the characteristic distance L is much larger than the intrinsic Debye length, and where $nb + p$ is much larger than $(b - 1)n_i$. These conditions are almost always met in intrinsic or extrinsic samples of silicon and germanium unless the excess carrier lifetime τ is extremely short. In the compound semiconductors of the III-V type, such as InSb and GaAs, carrier lifetimes are usually quite short; for these materials the condition (10.2-50) is often, though not always satisfied. In semiinsulating substances where $nb + p$ as well as L may be very small, condition (10.2-50) is violated more often than not, and other methods of analyzing added carrier problems must be used. The situation then is usually quite complex, although a very general theory for the analysis of such problems has been worked out by van Roosbroeck.[5] Since the charge balance assumption is valid for the great majority of cases involving the more familiar semi-conducting materials, we shall not consider explicitly situations where it is violated.

10·3 SOME USEFUL PARTICULAR SOLUTIONS OF THE CONTINUITY EQUATION

We shall now find it advantageous to examine the form which the solution of the continuity equation takes in certain particularly simple and important cases. To begin, let us examine what happens in a very large uniform crystal, whose boundaries may for practical purposes be assumed to be at infinity, in which there is no applied field nor bulk excess electron-hole pair generation. At time $t = 0$ we shall assume that a *uniform* spatial distribution of excess carrier density has somehow been created. Under these circumstances $\nabla(\delta p) = \nabla^2(\delta p) = 0$, and since the effect of diffusion is always to reduce rather than create concentration gradients, these quantities will remain zero at all later times. Equation (10.2-26) then takes the form

$$\frac{d(\delta p)}{dt} = -\frac{\delta p}{\tau}. \tag{10.3-1}$$

If τ is independent of δp this may easily be solved to give

$$\delta p = p - p_0 = Ae^{-t/\tau} \tag{10.3-2}$$

---

[5] W. van Roosbroeck, *Phys. Rev.*, **123**, 474 (1961).

where A is an arbitrary constant, equal here to the value of δp at $t = 0$. The excess carrier density in this case dies away everywhere in an exponential fashion with a time constant τ equal to the excess carrier lifetime.

Another case of great importance is the steady-state solution for an infinite sample wherein excess carriers are created by a uniform plane generation source, which, for convenience, we shall assume coincides with the yz-plane. There is no applied field nor any excess carrier generation in the bulk. Under these conditions, the excess carrier density will vary only along the x-direction. For simplicity, we shall discuss the case of a strongly extrinsic n-type sample in which the excess carrier density is everywhere much less than the majority carrier concentration; under these conditions the transport coefficients D^*, μ^* and τ reduce to the constant coefficients D_p, μ_p and τ_p related to the minority carrier, and (10.2-26) becomes

$$\frac{d^2(\delta p)}{dx^2} - \frac{\delta p}{L_p^2} = 0 \tag{10.3-3}$$

where

$$L_p = \sqrt{D_p \tau_p}. \tag{10.3-4}$$

The solutions to (10.3-3) must everywhere have the general form

$$\delta p = p(x) - p_0 = A e^{x/L_p} + B e^{-x/L_p} \tag{10.3-5}$$

where A and B are arbitrary constants. Since there is a plane source of carriers located at $x = 0$ in the yz-plane, there will be some sort of singularity in the solution there, and it is best to consider the solution for δp in the region $x < 0$ (which we shall call $\delta p_-(x)$), and the solution in the region $x > 0$ (which we shall call $\delta p_+(x)$) separately. Accordingly, then, from (10.3-5), these must have the form

$$\delta p_+(x) = p_+(x) - p_0 = A_+ e^{x/L_p} + B_+ e^{-x/L_p} \quad (x > 0) \tag{10.3-6}$$

and

$$\delta p_-(x) = p_-(x) - p_0 = A_- e^{x/L_p} + B_- e^{-x/L_p} \quad (x < 0). \tag{10.3-7}$$

Since there is a nonzero bulk recombination rate, the excess carrier densities must approach zero far from the source. These boundary conditions require that $A_+ = B_- = 0$. Furthermore, from the geometrical symmetry of the situation it is clear that $\delta p(x)$ must be an even function of x, whereby $A_- = B_+ = A_0$ and

$$\delta p_+(x) = p_+(x) - p_0 = A_0 e^{-x/L_p} \quad (x > 0) \tag{10.3-8}$$

and

$$\delta p_-(x) = p_-(x) - p_0 = A_0 e^{x/L_p} \quad (x < 0). \tag{10.3-9}$$

The excess *electron* concentration, by (10.2-21) is simply equal to the excess hole concentration. The resulting concentration profiles are shown in Figure 10.4. The excess carrier concentration dies off exponentially on either side of the source plane, with a characteristic decay distance of L_p.

The minority carrier diffusion current associated with the excess carrier distribution can be calculated from (10.2-8) to be

$$J_{p+}(x) = -D_p \frac{d(\delta p_+)}{dx} = \frac{A_0 D_p}{L_p} e^{-x/L_p} \quad (x > 0) \tag{10.3-10}$$

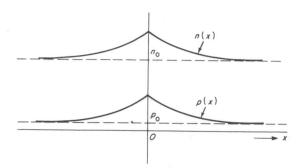

FIGURE 10.4. Steady-state electron and hole concentrations within an infinite crystal in which a plane source of excess carrier generation exists at $x = 0$.

$$J_{p-}(x) = -D_p \frac{d(\delta p_-)}{dx} = -\frac{A_0 D_p}{L_p} e^{x/L_p} \qquad (x < 0). \tag{10.3-11}$$

As x approaches zero through positive values, J_{p+} approaches $A_0 D_p/L_p$, while as x approaches zero through negative values, J_{p-} approaches $-A_0 D_p/L_p$. The difference between these quantities is the total flux strength of the source, or the number of minority carriers per cm^2 per second emitted by the source. In other words

$$F_{0p} = J_{p+}(0) - J_{p-}(0) = \frac{2A_0 D_p}{L_p} \tag{10.3-12}$$

where F_0 is the flux source strength. This equation enables one to express the coefficient A_0 of (10.3-8) and (10.3-9) in terms of the flux strength F_0 when the source is described in terms of the initial emitted flux. For a sample which is not strongly extrinsic but in which the excess carrier density is everywhere sufficiently small, the solutions (10.3-8) and (10.3-9) are, of course, still good, provided that L_p is replaced by $L^* = (D^* \tau)^{1/2}$.

It is of some value in understanding the physical nature of ambipolar charge transport to examine in detail the electron and hole fluxes which are present in this particular example. For this purpose we shall consider a sample in which the equilibrium carrier density is arbitrary rather than an extrinsic n-type material. In general, there will be both hole and electron fluxes arising from diffusion and also fluxes of both carrier species which are due to the presence of the internal field as given by (10.2-39). Noting that in the region $(x > 0)$ $\delta p_+ = A_0 e^{-x/L^*}$, whereby $d(\delta p_+)/dx = -\delta p_+/L^*$, and using the Einstein relations (10.2-11), the hole and electron fluxes due to diffusion may be written as

$$J_{p+}^d = -D_p \frac{dp_+}{dx} = -D_p \frac{d(\delta p_+)}{dx} = \frac{\mu_p kT}{e} \frac{\delta p_+}{L^*} \tag{10.3-13}$$

and

$$J_{n+}^d = -D_n \frac{dn_+}{dx} = -D_n \frac{d(\delta p_+)}{dx} = \frac{\mu_n kT}{e} \frac{\delta p_+}{L^*}. \tag{10.3-14}$$

The fluxes of holes and electrons which are created by the internal field (10.2-39) are

$$J^f_{p+} = p_+ \mu_p E_{int} = \frac{kT}{e} \frac{p_+ \mu_p (b-1)}{n_+ b + p_+} \frac{\delta p_+}{L^*} \tag{10.3-15}$$

and
$$J^f_{n+} = -n_+ \mu_n E_{int} = -\frac{kT}{e} \frac{n_+ \mu_n (b-1)}{n_+ b + p_+} \frac{\delta p_+}{L^*}. \tag{10.3-16}$$

The total hole and electron fluxes may now be found simply by adding the diffusion and internal field components together. The result is

$$J_{p+} = J_{n+} = \frac{\mu_p kT}{e} \frac{b(n_+ + p_+)}{n_+ b + p_+} \frac{\delta p_+}{L^*}. \tag{10.3-17}$$

From these equations, the fraction of the total flux of either species of carrier which is due to diffusion and to the presence of the internal field may be calculated very easily. It is found that

$$\frac{J^d_{p+}}{J_{p+}} = \frac{n_+ b + p_+}{b(n_+ + p_+)} \qquad \frac{J^d_{n+}}{J_{n+}} = \frac{n_+ b + p_+}{n_+ + p_+}$$

$$\frac{J^f_{p+}}{J_{p+}} = \frac{p_+ (b-1)}{b(n_+ + p_+)} \qquad \frac{J^f_{n+}}{J_{n+}} = -\frac{n_+ (b-1)}{n_+ + p_+} \tag{10.3-18}$$

It is most interesting to examine these expressions in three specific cases; extrinsic n-type material, where $n_+ \gg p_+$, extrinsic p-type material, where $p_+ \gg n_+$ and intrinsic, where $n_+ = p_+$. In these three instances the Equations (10.3-18) become

	n-type $(n_+ \gg p_+)$	p-type $(p_+ \gg n_+)$	intrinsic $(n_+ = p_+)$
$\dfrac{J^d_{p+}}{J_{p+}} =$	$1\,;$	$\dfrac{1}{b}\,;$	$\dfrac{b+1}{2b}$
$\dfrac{J^f_{p+}}{J_{p+}} =$	$0\,;$	$1 - \dfrac{1}{b}\,;$	$\dfrac{b-1}{2b}$
$\dfrac{J^d_{n+}}{J_{n+}} =$	$b\,;$	$1\,;$	$\dfrac{b+1}{2}$
$\dfrac{J^f_{n+}}{J_{n+}} =$	$1-b\,;$	$0\,;$	$\dfrac{1-b}{2}.$

$$(10.3\text{-}19)$$

It is clear from these results that in extrinsic samples the flux of *minority* carriers due to the internal field is *negligible in comparison with diffusion flux*. The *majority* carrier flux due to the field, however, is significant. This flux must, in fact, be large enough to assure quasineutrality at all points, and therefore must compensate for the imbalance between minority and majority carrier diffusion fluxes arising from the inequality of electron and hole diffusion coefficients. The situation may be understood by noting from (10.2-39) that in extrinsic samples the magnitude of the external field becomes very small because of the factor $nb + p$ in the denominator, which becomes quite large in extrinsic samples. The majority carrier flux arising from the internal field, however, is still appreciable, since it is proportional to the *product* of the very small internal field and the very large majority carrier density. The minority carrier flux due to the internal field, of course, is extremely small indeed, since both internal field and minority carrier density are small. The diffusion fluxes of majority and minority carriers, on the other hand, are dependent only upon the concentration gradients and the inherent electron and hole diffusion coefficients, and since the concentration gradients of electrons and holes are equal, must be comparable in magnitude, differing only by an amount proportional to the difference between the inherent diffusivities of the two carriers.

An understanding of this situation leads to the explanation of why, in extrinsic samples, the ambipolar diffusion coefficient is simply the inherent diffusion coefficient of the minority carrier. Under such conditions the internal field is small; it may create relatively large majority carrier fluxes, but its influence upon minority carrier flux is negligible. The minority carrier flux is therefore exclusively due to the *diffusion* of the minority carriers and is essentially unaffected by the internal field. The *observed* minority carrier flux is then simply that which would be produced by *independently diffusing minority carriers*. In order that the quasineutrality condition be satisfied, the majority carrier flux must be *the same*. Some of this flux is supplied by the diffusion of the majority carriers, but since the inherent diffusivities of the two carrier species are unequal, there will be an inevitable deficit (or excess). This deficit or excess may be accounted for by *majority* carrier flux whose source is the internal field. The net effect is that electron and hole fluxes are equal to each other and to the flux which would be produced by a distribution of independently diffusing minority carriers; the diffusion coefficient of the excess carrier distribution is thus that associated with the minority carrier. In extrinsic systems, it is therefore usually necessary to consider explicitly only the behavior of the minority carriers. In samples which are intrinsic or nearly intrinsic, of course, the effect of the internal field upon both carrier types is important. This fact is illustrated by (10.3-19) and by the dependence of the ambipolar transport coefficients upon both electron and hole diffusivity or mobility. Although these results have been discussed in connection with a rather restricted specific example, it is clear that the physical principles involved are the same in all systems and that the conclusions which are drawn are of general applicability.

The minority carrier diffusion fluxes (10.3-10) and (10.3-11) are now seen to be the *total* minority carrier fluxes in the case of an *extrinsic* n-type sample. In a crystal which is not extrinsic, the total electron or hole fluxes can be obtained from (10.3-17), by writing $b = \mu_n/\mu_p$ and using the Einstein relations, in the form

$$J_{p+} = J_{n+} = \frac{(n_+ + p_+)D_n D_p}{n_+ D_n + p_+ D_p} \cdot \frac{\delta p_+}{L^*} = -D^* \frac{d(\delta p_+)}{dx}. \tag{10.3-20}$$

In the example at hand, of course, since there is no applied field, the total electrical current $I = e(J_{p+} - J_{n+})$ is zero.

Next let us consider the extension of the preceding solution to the case where there is a *constant* applied field E_0 in the positive x-direction. The differential equation

(10.2-26) then may be written for extrinsic n-type material as

$$\frac{d^2(\delta p)}{dx^2} - \frac{\mu_p E_0}{D_p}\frac{d(\delta p)}{dx} - \frac{\delta p}{L_p^2} = 0. \tag{10.3-21}$$

This equation may be solved using the standard techniques applicable to linear differential equations with constant coefficients to give

$$\delta p(x) = p(x) - p_0 = Ae^{\gamma_p + x/L_p} + Be^{\gamma_p - x/L_p} \tag{10.3-22}$$

where
$$\gamma_{p\pm} = \gamma_p \pm \sqrt{1 + \gamma_p^2} \tag{10.3-23}$$

and
$$\gamma_p = \mu_p E_0 L_p/2D_p = eE_0 L_p/2kT. \tag{10.3-24}$$

This solution is the same as (10.3-5), except that the arguments of the exponential factors are multiplied by the factors γ_{p+} and $-\gamma_{p-}$, which depend upon the electric field. From the definition given above, it is readily seen that whatever value E_0 may have, γ_{p+} must always be positive while γ_{p-} must always be negative; when γ_{p+} and γ_{p-} are plotted as a function of field, the two quantities are represented by the two branches of a hyperbola, as shown in Figure 10.5. For $\gamma_p = 0$, corresponding to zero

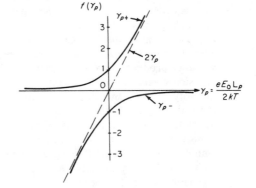

FIGURE 10.5. A plot of the functions γ_{p+} and γ_{p-}.

field, (10.3-22) reduces to Equation (10.3-5) of the preceding section. For large positive values of γ_p (hence for fields such that $E_0 \gg 2kT/eL_p$), it may be shown with the aid of the binomial expansion and (10.3-24) that

$$\gamma_{p-} \cong -\frac{1}{2\gamma_p} = -\frac{L_p}{\mu_p E_0 \tau_p}, \tag{10.3-25}$$

while it is obvious that under these circumstances

$$\gamma_{p+} \cong 2\gamma_p = \mu_p E_0 \tau_p/L_p. \tag{10.3-26}$$

It is easily seen from (10.3-23) that γ_{p+} and γ_{p-} are always related by

$$\gamma_{p+}\gamma_{p-} = -1. \tag{10.3-27}$$

We may now repeat the development of the zero-field case as given by Equations (10.3-6)–(10.3-12), assuming as before two separate expressions for $\delta p_+(x)$ in the region $x > 0$ and $\delta p_-(x)$ in the region $x < 0$, with coefficients A_+, B_+, A_- and B_-, and showing in the same way that A_+ and B_- must be taken as zero in order to insure proper behavior at $\pm \infty$. The resulting solution may be written

$$\delta p_+(x) = p_+(x) - p_0 = A_0 e^{\gamma_p - x/L_p} \qquad (x > 0) \qquad (10.3\text{-}28)$$

$$\delta p_-(x) = p_-(x) - p_0 = A_0 e^{\gamma_p + x/L_p} \qquad (x < 0) \qquad (10.3\text{-}29)$$

where A_0 is an arbitrary constant which expresses the strength of the generation source. The excess electron concentration, as usual, will simply be equal to the excess hole concentration. The concentration profiles are shown in Figure 10.6 for a positive

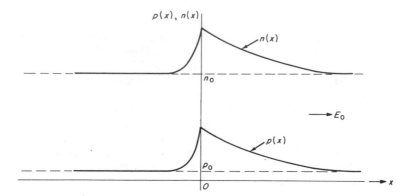

FIGURE 10.6. Steady-state electron and hole concentrations within a crystal in which a plane source of excess carriers exists at $x = 0$, and within which a constant electric field in the x-direction is present.

value of E_0. Since the transport characteristics of the excess carrier distribution are essentially those of the minority carriers, we find that the transport of the excess carriers to the right (in the direction of motion of holes under the applied field) is aided by the action of the field, while the transport of carriers in the opposite direction is retarded. From (10.3-25) one may verify that (10.3-28) approaches the previously obtained result (10.2-43) for large applied fields, as might be expected. The minority carrier particle fluxes $J_{p+}(x)$ and $J_{p-}(x)$ may, as before, be calculated from (10.2-8); contributions are now obtained from both diffusion and electric field terms. Part of the resulting particle flux will arise from the drift of the holes which are present in thermal equilibrium in the applied field, and part from the diffusion and drift of the excess carrier distribution. The flux arising from the latter effect, evaluated at the source plane, is readily shown to be

$$F_{0p} = J_{p+}(0) - J_{p-}(0) = A_0 \left[2\mu_p E_0 + \frac{D_p}{L_p}(\gamma_{p+} - \gamma_{p-}) \right]. \qquad (10.3\text{-}30)$$

This equation is the analog of (10.3-12), which relates the amplitude coefficient A_0

to the total number of carriers per unit area per unit time, F_{0p}, generated at the source plane. When the sample is not strongly extrinsic, and when the added carrier density is sufficiently small, the above solutions will still be correct, provided that D_p, μ_p, L_p and τ_p are replaced by D^*, μ^*, L^* and τ, respectively.

The same remarks regarding the relation of fluxes due to diffusion and to the *internal* fields in connection with the field-free case apply also in the present situation; for example, in an extrinsic sample, the flux of minority carriers due to the internal field is negligible in comparison with the corresponding diffusion flux, while the majority carrier fluxes attributable to either source are of comparable magnitude. There are now, in addition, however, large fluxes of both carrier types, comprising both excess and equilibrium carriers, due to the *applied* field. These fluxes are superposed upon the fluxes arising from diffusion and internal fields; as a result the total electric current density no longer vanishes but is instead represented by σE_{appl}. It must be noted in this connection that the present example deals only with a situation wherein δp is everywhere much less than the equilibrium majority carrier density. Since the total electric current density must be everywhere the same, a situation wherein δp is large enough to appreciably modulate the conductivity would imply a field which could not be constant, but which would instead be large where δp is small and *vice versa*. For small values of δp the same difficulty is still present, but the corresponding variations in the field are so small that the constant-field treatment is a good approximation. There are also, of course, numerous other difficulties which arise when δp becomes comparable to the equilibrium carrier density.

As a final example, let us consider a uniform one-dimensional system in which N hole-electron pairs are generated instantaneously at a point $x = x'$ at time $t = t'$. The sample will be assumed to extend to infinity in both directions along the x-axis, and a constant applied electric field E_0 is assumed to be present. For an extrinsic n-type sample, Equation (10.2-26) then becomes

$$D_p \frac{\partial^2(\delta p)}{\partial x^2} - \mu_p E_0 \frac{\partial(\delta p)}{\partial x} - \frac{\delta p}{\tau_p} = \frac{\partial(\delta p)}{\partial t}. \qquad (10.3\text{-}31)$$

A solution to this equation valid for all $t > t'$ which meets all the requirements of the problem is

$$\delta p(x,t) = \frac{N}{\sqrt{4\pi D_p(t - t')}} \exp\left[-\frac{\{(x - x') - \mu_p E_0(t - t')\}^2}{4D_p(t - t')} - \frac{t - t'}{\tau_p} \right]. \qquad (10.3\text{-}32)$$

The fact that (10.3-32) satisfies the differential Equation (10.3-31) can be established by direct substitution.[6] It is readily established that the initial value $\delta p(x,0)$ is zero everywhere except at $x = x'$, at which point it is infinite. The initial concentration distribution thus corresponds to a Dirac δ-function. For later times the distribution has a Gaussian shape, whose half-width increases with time and whose maximum amplitude decreases with time. The maximum concentration point also moves along the field direction with velocity $\mu_p E_0$. The expression (10.3-32) can be integrated over x between the limits $-\infty$ and ∞ (by letting $u = (x - x') - \mu_p E_0(t - t')$, whence $du = dx$, t being regarded as constant) to obtain the total number of excess holes δP

---

[6] The solution (10.3-32) can be obtained by Fourier integral or Laplace transform techniques.

in the distribution as a function of time. The result is

$$\delta P(t) = \int_{-\infty}^{\infty} \delta p(x,t)\, dt = N e^{-(t-t')/\tau_p}. \tag{10.3-33}$$

The total number of carriers, initially N, thus falls off exponentially with time constant τ_p, as might be expected. The behavior of the excess carrier concentration as a function of x and t is shown in Figure 10.7.

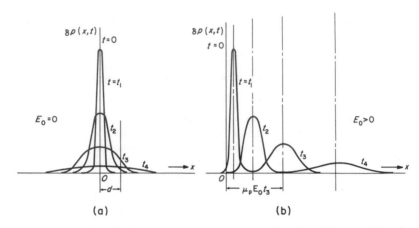

FIGURE 10.7. (a) Diffusive spread of excess carriers injected initially as a δ-function distribution at $x = 0$, with no applied electric field. (b) Diffusive spread and drift of the distribution in (a) with a constant electric field in the x-direction. In both cases $0 < t_1 < t_2 < t_3 < t_4$.

If, now, we revise our definition of N, the initial source strength, by regarding $N(x',t')\, dx'\, dt'$ as the number of excess holes or electrons generated in a time interval dt' about $t = t'$ within a space element dx' about $x = x'$, then the subsequent concentration due to this *element* of generation as a function of x and t will be obtained by simply replacing N in (10.3-32) with $N(x',t')\, dx'\, dt'$; the resulting concentration distribution should then be labeled as a function of x' and t' as well as x and t, that is, it should be written as $\delta p(x,t;x',t')$. The concentration $\delta p(x,t)$ arising from a *known arbitrary distribution of generation* whose strength may vary with space and time coordinates x' and t' may then be calculated by *superposing* solutions of this form, each of which is characterized by the *local* generation strength $N(x',t')\, dx'\, dt'$. This superposition of infinitesimal elements, each with its own local source strength leads to an integral over dx' and dt' of the form

$$\delta p(x,t) = \int_{-\infty}^{t} \int_{-\infty}^{\infty} \frac{N(x',t')}{\sqrt{4\pi D_p(t-t')}} \exp\left[-\frac{\{(x-x') - \mu_p E_0(t-t')\}^2}{4D_p(t-t')} - \frac{t-t'}{\tau_p} \right] dx'\, dt'.$$
$$\tag{10.3-34}$$

The time integral is carried only up to $t' = t$ because, of course, *future* generation does not contribute to the concentration observed at time t. The above solution, naturally, applies only to an infinite sample, in which the concentration distribution satisfies the boundary condition that the concentration must approach zero far from the source. For finite samples, one would first have to find a solution for a δ-function initial source distribution, such as (10.3-32), but satisfying in addition appropriate boundary conditions at the surfaces of the sample.[7] The same superposition procedure could then be used to obtain the solution for an arbitrary distribution of generation, with respect to the space and time coordinates, for samples of this sort.

This procedure can be fairly easily generalized to a three-dimensional system. The solution for an instantaneous generation of $N(\mathbf{r}',t')\,dv'\,dt'$ in the volume element dv' about the point $\mathbf{r} = \mathbf{r}'$ at the time $t = t'$, describing the subsequent concentration distribution arising from this initial generation element in an infinite sample, is

$$\delta p(\mathbf{r},t\,;\mathbf{r}',t') = \frac{N(\mathbf{r}',t')\,dv'dt'}{[4\pi D_p(t-t')]^{\frac{3}{2}}} \exp\left[-\frac{\{(\mathbf{r}-\mathbf{r}') - \mu_p\mathbf{E}_0(t-t')\}^2}{4D_p(t-t')} - \frac{t-t'}{\tau_p} \right]. \quad (10.3\text{-}35)$$

It is possible in this case also to verify that (10.3-35) is a solution to the three-dimensional diffusion equation corresponding to (10.3-31) by direct substitution. The superposition of solutions of the form (10.3-35) may now be made in the same way as before, resulting in an integral over the generation source distribution of the form

$$\delta p(\mathbf{r},t) = \int_{-\infty}^{t} \int_V \frac{N(\mathbf{r}',t')}{[4\pi D_p(t-t')]^{\frac{3}{2}}} \exp\left[-\frac{\{(\mathbf{r}-\mathbf{r}') - \mu_p\mathbf{E}_0(t-t')\}^2}{4D_p(t-t')} - \frac{t-t'}{\tau_p} \right] dv'dt'.$$

We shall find most of the particular solutions of the continuity equation which have been described in this section useful in discussing experimental work involving excess carrier distributions or semiconductor device structures of technological importance.

10·4 DRIFT MOBILITY AND THE HAYNES-SHOCKLEY EXPERIMENT

The Haynes-Shockley experiment was first performed in order to measure accurately and directly the drift mobility of holes and electrons in semiconductor crystals.[8] Although it is of great interest because it was the first really successful *direct* measurement[9] of charge carrier drift velocity, it is also very instructive in illustrating the transport behavior of excess electrons and holes in semiconductors, and also in demonstrating the basic features of transistor action, since the Haynes-Shockley circuit is in effect a type of transistor.

In the Haynes-Shockley experiment, as illustrated in Figure 10.8, excess minority carriers are injected into and collected from the crystal by metallic point probe

---

[7] The solution for a δ-function initial source distribution is ordinarily referred to as the *Green's function* appropriate to a given system and a given set of boundary conditions.

[8] J. R. Haynes and W. Shockley, *Phys. Rev.*, **81**, 835 (1951).

[9] As contrasted with the Hall effect, which allows only an indirect determination of mobility.

contacts. Although we must postpone a detailed account of what happens at a rectifying metal point contact until later, a general idea of their behavior may be obtained
by referring to Figure 10.9. In this illustration a semiconductor sample is depicted,
to which a point probe contact and a large area contact have been made. The large
area contact, which may be a suitably fabricated soldered or alloyed region or simply
a pressure contact acts simply as an *ohmic* contact to the sample, and does not affect

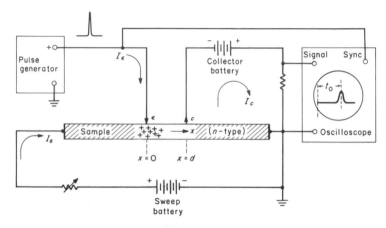

FIGURE 10.8.   A schematic diagram of the Haynes-Shockley experiment.

the distribution of carriers in the sample in any appreciable way. The point probe,
however, acts as a rectifying contact; when biased in such a way as to attract minority
carriers to it, it will act as a collector of minority carriers, depleting the region of the
crystal adjacent to it of these carriers until a steady state is reached wherein minority
carriers diffuse from the interior of the crystal just rapidly enough to supply the deficit
caused by their disappearance at the probe point. Under these conditions (called
reverse bias) this minority carrier current is the only current which can flow through
the probe, and since the supply of minority carriers in the crystal is quite small, the
current flow is severely limited, and is relatively independent of the bias voltage because the supply of minority carriers is unaffected by changing the reverse bias voltage.
This situation is illustrated in Figure 10.9(a) and (c) for n- and p-type semiconducting
materials. When the bias voltage is reversed, so as to attract majority carriers to the
probe, it is found that in addition to the flow of majority carriers into the probe,
excess *minority* carriers are *injected* into the sample at the point contact. In this mode
of operation current flows quite easily, and relatively large currents can be made to
flow with small bias voltages. The current is found to increase approximately exponentially as the bias voltage is increased, and it is not difficult to produce a sufficient
number of injected carriers to modulate the conductivity of the semiconductor quite
strongly.[10] This forward bias condition is illustrated in Figure 10.9(b) and (d). The

---

[10] It should be remembered that an excess concentration of majority carriers must always
accompany the excess minority carrier distribution. The majority carriers necessary to maintain
electrical neutrality are in plentiful supply in the bulk of the crystal, whence they are drawn as the
excess minority carrier distribution is formed. Any resulting deficit in the majority carrier concentration in the bulk is immediately made up by majority carriers drawn from the metallic ohmic contact.

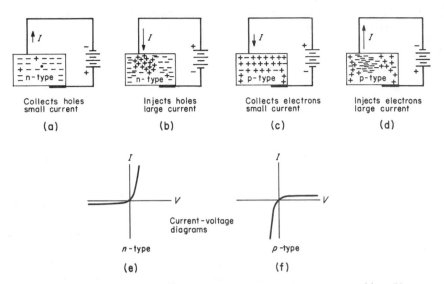

FIGURE 10.9. A point contact semiconductor rectifier (a) *n*-type, reverse bias, (b) *n*-type, forward bias, (c) *p*-type, reverse bias, (d) *p*-type, forward bias. (e) Current-voltage characteristic of *n*-type semi-conductor point contact rectifier, (f) Current-voltage characteristic of *p*-type semiconductor point contact rectifier.

current-voltage relationships for *n*-type and *p*-type crystals are shown in the same set of diagrams at (e) and (f), respectively. The point contact rectification effect, as shown at (e) and (f), in fact, provides quite a simple test for distinguishing between *n*-type and *p*-type semiconductor crystals.

In the Haynes-Shockley experiment a pulse of minority carriers is injected at the emitter probe ε of Figure 10.8. The sample is shaped in the form of a long thin bar, and an electric field is set up by an external battery, which, of course, causes a sweep current I_s to flow in the sample. This electric field sweeps the injected minority carriers down the sample past a second electrode, positioned a known distance d away from the first, which is reverse-biased so that it acts as a collector of minority carriers. When there are no excess carriers present at the collector, the only current flowing in that electrode is a small saturation current which is due to the collection of equilibrium minority carriers by the probe point. As the distribution of injected minority carriers flows past the collector, the concentration of minority carriers in its neighborhood increases, and the number of minority carriers collected there per unit time goes up proportionally. If the oscilloscope trace is initiated at the time when the pulse of carriers is originally injected, nothing will be observed on the screen until the excess carrier distribution has arrived at the collector, at which time the collector current will increase and a signal pulse will be observed on the oscilloscope.

Since the excess carrier concentration as a function of the distance x along the sample is essentially that given by (10.3-32), and shown graphically in Figure 10.7(b) it is clear that the maximum of the concentration distribution moves along with velocity equal to $\mu^* E_0$, where μ^* is the ambipolar mobility and E_0 the applied field. (In (10.3-32) conditions are assumed to be such that D^* and μ^* equal D_p and μ_p, but this may not be generally true.) It is apparent that the time t_0 required for the maximum of the concentration pulse to traverse the known distance d between emitter and collector is just

$$t_0 = \frac{d}{v} = \frac{d}{\mu^* E_0} \tag{10.4-1}$$

giving
$$\mu^* = \frac{d}{t_0 E_0}. \tag{10.4-2}$$

If conditions are such that μ^* equals $-\mu_n$ or μ_p (i.e., if the sample is strongly extrinsic) this experiment allows one to measure accurately the minority carrier drift mobility.

Unfortunately, the pulse that arrives at the collector is not perfectly sharp, but is rather spread out due to diffusion, as illustrated in Figure 10.7(b). This causes no difficulty in the measurements, but leads to certain errors in the interpretation of the data unless great care is taken. The reason for this is that what is actually displayed on the screen of the oscilloscope is *not* a plot of the concentration *versus* distance at some fixed time, but rather a plot of the concentration at some fixed point as a function of time, $\delta p(d,t)$. A measurement to the maximum of this curve does not accurately represent the drift time unless the conditions of the experiment are arranged so that a minimum of diffusive spread takes place as the distribution drifts between emitter and collector. For example, suppose that the applied field were zero. Then the distribution of excess carriers would diffuse as shown in Figure 10.7(a). Even though there is no applied field, however, it is readily seen from the figure that a concentration maximum would be observed as a function of time at a point $x = d$; in the example shown there the maximum would occur at about $t = t_3$. This maximum is observed simply because diffusive transport alone is sufficient to move the excess carriers belonging to the distribution from emitter to collector, and then beyond. This effect leads to the observation of a finite drift time in the absence of a field, which, according to (10.4-2) would correspond to infinite mobility. Obviously, then, if what is *meant* by the drift time t_0 is the time elapsed between injection and the observation of the maximum value attained by $\delta p(d,t)$ as given by (10.3-32) (and which is what is in fact measured from the oscilloscope trace) then Equation (10.4-2) cannot be correct as it stands.

Actually, Equation (10.4-2) is approximately correct so long as the diffusive spread of the carrier distribution is reasonably small during the transit from emitter to collector. These conditions can be realized experimentally by the use of very large applied fields, which, however, lead to serious heating effects unless rather cumbersome pulse techniques are resorted to. Alternatively, one may derive a correct expression for the transit time from (10.4-2) which takes diffusion fully into account, and thus arrive at a modified expression[11] of the form

$$\mu^* = \frac{d}{t_0 E_0} (\sqrt{1 + x^2} - x) \tag{10.4-3}$$

where
$$x = \frac{2kT}{eE_0 d} \left(\frac{t_0}{\tau} + \frac{1}{2}\right). \tag{10.4-4}$$

[11] J. P. McKelvey, *J. Appl. Phys.*, **27**, 341 (1956).

It is necessary, of course, to know the excess carrier lifetime τ to use this formula, but, as we shall see, that quantity is easily determined by a simple independent measurement. The details involved in the calculation of (10.4-3) are assigned as an exercise for the reader.

It will be shown in Chapter 13 how transistor action can be explained and understood on the basis of the Haynes-Shockley experiment. By measuring both the diffusive spread and the transit time, it is possible to determine the diffusion coefficient D^* as well as the drift mobility by the Haynes-Shockley technique. This possibility has been exploited[12] to verify the Einstein relation experimentally in strongly extrinsic samples of germanium.

10·5 SURFACE RECOMBINATION AND THE SURFACE BOUNDARY CONDITION

If the effect of the surfaces of a finite crystal were solely to confine the charge carriers to the interior of the sample, then the boundary condition which the excess charge carrier distribution would have to satisfy at the sample surfaces would simply be that the electron and hole currents must vanish at the surfaces. Unfortunately, the situation is not quite as simple as this, because charge carriers may *recombine* at the surface, by mechanisms which are quite independent of those which regulate charge carrier recombination rates in the interior of the sample. Under these circumstances, the surface acts as a partial absorber for electrons and holes, and there may be net current flow toward the surfaces of the sample.

At first thought one might be tempted to conclude that since electrons and holes may recombine at the surface of a crystal, there should be a deficiency of charge carrier concentration near the surface which would result in a diffusive current of carriers toward the surface *even in the thermal equilibrium state*. This, however, is not true, because in addition to recombination, *thermal generation* of electron-hole pairs takes place at the surface, and in thermal equilibrium the generation rate precisely equals the rate at which carrier pairs recombine at the surface. There is, therefore, no net flux toward the surface, nor any change in concentrations in the surface region under thermal equilibrium conditions.[13] This situation is a specific example of a very general principle of statistical mechanics, called the principle of *detailed balancing* or the principle of *microscopic reversibility*. This principle states that *in the thermal equilibrium condition* any given microscopic process and the reverse process must proceed at the same rate. We have already encountered an example of the validity of this principle when we ascertained that the equilibrium bulk recombination rate of electron-hole pairs and the bulk thermal generation rate were equal. The present

---

[12] Transistor Teachers Summer School (Bell Telephone Laboratories), *Phys. Rev.*, **88**, 1368 (1952).
    [13] This statement is true for the simple model of the surface which we are now discussing. There are changes in the equilibrium carrier concentration near the surface, associated with a surface space charge layer which arises because of the presence of *surface states*, at the actual physical surface of the sample, as we shall see later. The simple viewpoint adopted here is generally adequate for a phenomenological discussion of surface recombination.

situation, in which the surface recombination rate and the surface thermal generation rate are equal at equilibrium, is simply illustrative of the same general law. It should be noted, of course, that the principle of detailed balance applies *only* to the thermal equilibrium state. In the present example, the surface thermal generation rate is a function only of the temperature, and is quite independent of local charge carrier concentrations, while the recombination rate clearly depends directly upon the local carrier concentrations. If the local charge carrier concentration exceeds the thermal equilibrium value, then the surface recombination rate will exceed the surface generation rate, and a net absorption of carriers by the surface will result, which in turn will set up a diffusive flow of carriers toward the surface. A depletion of excess carrier concentration in the neighborhood of the surface will likewise result in a diffusive flow of excess thermally generated carriers away from the surface. Similar remarks can, of course, be made regarding the flow of carriers in the bulk arising from local excesses or deficiencies in bulk carrier concentrations. It is important to realize that these general conclusions are valid independent of the precise mechanisms involved in the generation or recombination processes.

The effect of surface recombination upon the charge carrier distribution within the sample may now be investigated by considering the flux interchange between the surface and interior region of the sample, as illustrated by Figure 10.10. We shall

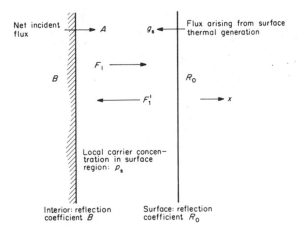

FIGURE 10.10. Flux diagram illustrating flux interchange between bulk and surface.

begin by defining the *surface reflection* coefficient R_0 as the probability that a single carrier in a single collision with the surface will be sent back to the bulk rather than absorbed by recombination. The corresponding recombination probability is thus $1 - R_0$. Likewise, the *bulk reflection coefficient* B will be defined as the probability that a carrier, upon entering the interior of the crystal from the surface, will reappear at the surface in the course of its random wandering through the material before recombining in the interior. These coefficients are assumed to be independent of carrier fluxes or concentrations. We may then consider the system of fluxes set up between surface and interior shown in Figure 10.10. A net flux of particles of magnitude A is assumed to originate from the interior, and the surface generation flux g_s is directed from the surface toward the interior of the crystal. The surface of the crystal in Figure 10.10 is shown separated from the interior to clearly illustrate the flux interchange between the two, but, of course, actually the two regions are physically contiguous. The *total* flux F_1 flowing from interior to surface, and the total reverse flux F_1' can be obtained by observing that F_1 is made up of the incident flux A plus that part of F_1' which is reflected by the bulk, whereby

$$F_1 = A + BF_1'. \tag{10.5-1}$$

Similarly,

$$F_1' = g_s + R_0 F_1. \tag{10.5-2}$$

These two equations can be solved for F_1 and F_1', giving

$$F_1 = \frac{A + Bg_s}{1 - R_0 B} \tag{10.5-3}$$

and

$$F_1' = \frac{g_s + AR_0}{1 - R_0 B}. \tag{10.5-4}$$

Now it is easily proved that for a distribution of free particles which obey Boltzmann statistics, the number of particles per unit time crossing a plane surface of unit area in either direction is just $p\bar{c}/4$ where p is the local particle concentration.[14] In the thermal equilibrium state, the particle concentration is p_0 everywhere and

$$F_1 = F_1' = p_0\bar{c}/4. \tag{10.5-5}$$

This value of F_1 and F_1' may be substituted into (10.5-3) and (10.5-4) and the resulting pair of equations solved for g_s and A_0, which is the value of the incident flux A under conditions of thermal equilibrium, to yield

$$g_s = \frac{p_0\bar{c}}{4}(1 - R_0) \tag{10.5-6}$$

and

$$A_0 = \frac{p_0\bar{c}}{4}(1 - B). \tag{10.5-7}$$

We have seen in the preceding argument that in the thermal equilibrium state F_1 and F_1' have the common value $p\bar{c}/4$. If there is a departure from thermal equilibrium resulting from a diffusive flow set up by a concentration gradient, and if this departure from the equilibrium state is small enough that the Boltzmann distribution is still approximately correct at every point in the system, then F_1 and F_1' will no longer be equal. Under these circumstances, we must regard the positively directed flux arriving at a point x_0 as having originated, on the average, a distance of the order of the mean free path "upstream" and the negatively directed flux arriving there as having originated on the average a distance of the order of the mean free path "downstream."

---

[14] See, for example, Exercise 7 at the end of Chapter 5.

Then the positively directed flux F_1' may be written

$$F_1' = \left[p(x_0) - \alpha\lambda \left(\frac{\partial p}{\partial x}\right)_{x_0} \right] \frac{\bar{c}}{4}. \qquad (10.5\text{-}8)$$

where α is a numerical constant of the order of unity, because the quantity in brackets represents, approximately, the local concentration at the place where the flux arriving at x_0 *originated*. Likewise, the negatively directed flux F_1' may be written as

$$F_1' = \left[p(x_0) + \alpha\lambda \left(\frac{\partial p}{\partial x}\right)_{x_0} \right] \frac{\bar{c}}{4}. \qquad (10.5\text{-}9)$$

It is clear that the *sum* of the fluxes then must be

$$F_1 + F_1' = \frac{p\bar{c}}{2}, \qquad (10.5\text{-}10)$$

where p is the local concentration, independent of the concentration gradient, so long as the distribution function is not seriously disturbed from its equilibrium form.

Using (10.5-10) to express the sum of the fluxes F_1 and F_1' in the surface region, and using the expressions (10.5-3), (10.5-4) and (10.5-6) to represent F_1, F_1' and g_s, we may obtain

$$\frac{p_s\bar{c}}{2} = F_1 + F_1' = \frac{A(1 + R_0)}{1 - R_0 B} + \frac{p_0\bar{c}}{4} \frac{(1 - R_0)(1 + B)}{1 - R_0 B}, \qquad (10.5\text{-}11)$$

where p_s represents the concentration in the neighborhood of the surface. We must, of course, remember that while g_s has the same value as in the thermal equilibrium state, this is *not* true of A, whose magnitude depends upon the nature of the excess carrier distribution which may be present in the bulk. Equation (10.5-11) may, in fact, be solved for A to give

$$A = \frac{p_s\bar{c}}{2} \frac{1 - R_0 B}{1 + R_0} - \frac{p_0\bar{c}}{4} \frac{(1 - R_0)(1 + B)}{1 + R_0}. \qquad (10.5\text{-}12)$$

The *difference* between F_1 and F_1' is the *net* flux of carriers, which, if there is no electric field present, as we shall assume, is simply equal to the net diffusion current $-D_p(\partial(\delta p)/\partial x)$ evaluated at the surfaces. Writing the expression for $F_1 - F_1'$, using (10.5-3) and (10.5-4) to represent the fluxes, and using (10.5-6) and (10.5-12) to represent the values of g_s and A, we may obtain after some tedious but straightforward algebra,

$$F_1 - F_1' = \frac{A(1 - R_0)}{1 - R_0 B} - \frac{g_s(1 - B)}{1 - R_0 B} = \frac{(p_s - p_0)\bar{c}}{2} \frac{1 - R_0}{1 + R_0} = -D_p\left(\frac{\partial(\delta p)}{\partial x}\right)_s. \qquad (10.5\text{-}13)$$

This equation is, in fact, a statement of the surface boundary condition which must be applied to the continuity equation. It can conveniently be written in the form

$$-D_p\left(\frac{\partial(\delta p)}{\partial x}\right)_s = s \cdot (\delta p)_s \qquad (10.5\text{-}14)$$

where

$$s = \frac{\bar{c}}{2}\frac{1 - R_0}{1 + R_0}. \qquad (10.5\text{-}15)$$

The s subscripts indicate that the quantities concerned are evaluated at the surface. The constant s is usually referred to as the *surface recombination velocity*; its relation to the more fundamental reflection coefficient is given by (10.5-15). The surface boundary condition (10.5-14) can be stated in a more general vector form as

$$-D_p[\mathbf{n} \cdot \nabla(\delta p)]_s = s \cdot (\delta p)_s, \qquad (10.5\text{-}16)$$

where $\mathbf{n}$ is a unit outward vector normal to the surface.

If there is no surface recombination, then $R_0 = 1$ and $s = 0$, from (10.5-5). In this case, (10.5-14) gives $\partial(\delta p)/\partial x = 0$, corresponding to the condition of no net diffusive flow to the surface. If carriers inevitably recombine upon striking the surface, then $R_0 = 0$ and $s = \bar{c}/2$; this value is clearly an upper limit for the surface recombination velocity. Nevertheless, it is quite common in this limiting case to set $s = \infty$ in (10.5-14), which is tantamount to taking the excess carrier concentration at the surface $(\delta p)_s$ to be zero. This approximation, which may greatly simplify calculation, can be shown to be a good one provided $D_p/L_p \ll \bar{c}/2$. For germanium at 300°K, $D_p \cong 50$ cm^2/sec, $\bar{c} \cong 10^7$ cm/sec, and L_p is usually greater than 10^{-3} cm; the condition is therefore very well satisfied in this quite typical example. As a matter of fact, circumstances under which the requirement is violated arise only very infrequently, and the adoption of the boundary condition $s = \infty$ or $(\delta p)_s = 0$ in the high surface recombination limit is thus nearly always justified. An example, which shows clearly the origin of the condition discussed above, is assigned as an exercise. In all intermediate cases, where the recombination probability is neither zero nor unity, the boundary condition in the form (10.5-14) must be used. All the above calculations have been carried out with regard to excess holes in n-type material, but, of course, exactly similar considerations apply to excess electrons in a p-type semiconductor.

In germanium, quite high reflection coefficients (~ 0.99999) corresponding to surface recombination velocities of the order of 100 cm/sec may be obtained in samples whose surfaces have been prepared by careful chemical etching. For samples whose surfaces have been heavily damaged by lapping or other abrasive action, there are many dislocations and other lattice imperfections at the surfaces which may act as recombination centers, and the surface recombination velocity may be in excess of 10^5 cm/sec, corresponding to a value of less than 0.99 for the reflection coefficient.[15] In silicon, the surface recombination velocity is usually much higher on etched surfaces, a typical value being perhaps 2000 cm/sec.

---

[15] The reflection coefficient R_0 is usually quite close to unity for germanium and silicon, even in samples whose surfaces are quite heavily abraded. Nevertheless, in many cases, the effect of a recombination velocity of 10^5 cm/sec, corresponding to $R_0 = 0.99$, may be hardly distinguishable from the condition where $s = c/2$ and $R_0 = 0$.

10·6 STEADY-STATE PHOTOCONDUCTIVITY

In order to illustrate the application of some of the principles which were discussed in the preceding sections, we shall now proceed to calculate the steady-state photoconductive response of a uniform semiconductor sample which is illuminated with radiation of wavelength long enough so as to be only slightly absorbed in passing through the crystal, but at the same time sufficiently short to create a measurable concentration of electron-hole pairs. A suitable wavelength would thus lie just on the long wavelength side of the absorption edge, at the position λ_1 in Figure 10.1(a). We shall assume that our sample is in the form of a rectangular solid, of dimensions x_0, y_0, z_0, that the thickness x_0 is much less than the other dimensions y_0 and z_0, and that the illumination is incident along the x-direction, as shown in Figure 10.11. Both

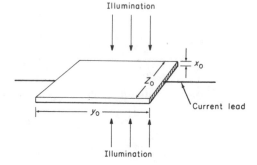

FIG. 10.11. Geometry of dc photoconductivity experiment discussed in Section 10.6.

of the large surfaces of the sample will have been prepared in the same way, so as to produce the same surface recombination velocity s on both sides. If the absorption coefficient is relatively small, the light intensity may be considered to be approximately uniform throughout the crystal, and this will lead to a generation rate of excess carriers g' which is constant and proportional to light intensity. In a large thin sample of the type shown, the excess carrier concentration varies essentially only along the x-direction, allowing one to use the one-dimensional form of the continuity equation. Since there are no fields in the x-direction, and since in the steady-state condition $\partial(\delta p)/\partial t = 0$, Equation (10.2-26) takes the form

$$\frac{d^2(\delta p)}{dx^2} - \frac{\delta p}{L_p^2} = -\frac{g'}{D_p}$$ (10.6-1)

where g' is constant. In writing the equation this way, we are assuming that we are dealing with extrinsic n-type material and that δp is everywhere small compared to the majority carrier density.

The solution to this equation may be expressed as the general solution to the homogeneous Equation (10.3-3) plus a particular solution to (10.6-1). It is clear that $\delta p = g' L_p^2 / D_p = g' \tau_p$ is just such a particular solution to (10.6-1), and that the general solution to (10.6-1) can be written

$$\delta p(x) = A \cosh \frac{x}{L_p} + B \sinh \frac{x}{L_p} + g' \tau_p.$$ (10.6-2)

By the symmetry of the sample geometry shown in Figure 10.11, $\delta p(x)$ must be an even function of x, whereby $B = 0$ and (10.6-2) becomes

$$\delta p(x) = A \cosh \frac{x}{L_p} + g'\tau_p. \tag{10.6-3}$$

Applying the surface boundary condition (10.5-16) at either surface $\left(x = \frac{\pm x_0}{2} \right)$ to this equation, we may evaluate the constant A to obtain

$$A = \frac{-sg'\tau_p}{s \cosh \dfrac{x_0}{2L_p} + \dfrac{D_p}{L_p} \sinh \dfrac{x_0}{2L_p}} \tag{10.6-4}$$

and

$$\delta p(x) = g'\tau_p \left[1 - \frac{s \cosh \dfrac{x}{L_p}}{s \cosh \dfrac{x_0}{2L_p} + \dfrac{D_p}{L_p} \sinh \dfrac{x_0}{2L_p}} \right]. \tag{10.6-5}$$

The concentration profile $\delta p(x)$ is shown in Figure 10.12(a) for several values of s.

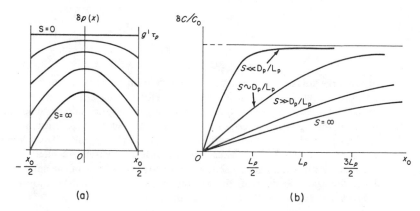

FIGURE 10.12.   (a) Concentration profiles within the sample illustrated in Figure 10.11 for $s = 0$, $s = \infty$ and several intermediate values of s. (b) Photoconductive response $\delta C/C_0$ as a function of thickness x_0 plotted for several values of s.

The change in conductance due to the presence of excess carriers will be proportional to the total number of excess carriers present in the sample. If we denote by δP the total number of excess holes and by δN the total number of excess electrons, then

$$\delta P = \delta N = y_0 z_0 \int_{-x_0/2}^{x_0/2} \delta p(x)\, dx, \tag{10.6-6}$$

where $y_0 z_0$ is the surface area of the illuminated face of the sample. If (10.6-5) is used to represent $\delta p(x)$ in (10.6-6), the integral can easily be evaluated to give

$$\delta P = \delta N = 2y_0 z_0 g' \tau_p \left[\frac{1}{2} x_0 - \frac{sL_p \sinh \dfrac{x_0}{2L_p}}{s \cosh \dfrac{x_0}{2L_p} + \dfrac{D_p}{L_p} \sinh \dfrac{x_0}{2L_p}} \right]. \qquad (10.6\text{-}7)$$

The differential element of conductance between the end electrodes associated with a thin sheet of material of thickness dx is given by

$$dC = \sigma(x) \frac{da}{y_0} = \frac{z_0 \sigma(x)\, dx}{y_0}. \qquad (10.6\text{-}8)$$

But

$$\sigma(x) = \sigma_0 + \delta\sigma(x) = \sigma_0 + e\mu_p \delta p(x)(b + 1), \qquad (10.6\text{-}9)$$

whereby
$$dC = \frac{z_0}{y_0} [\sigma_0 + e\mu_p(b + 1)\delta p(x)]\, dx. \qquad (10.6\text{-}10)$$

Integrating this equation between the limits $x = \pm x_0/2$, one may easily obtain

$$C = \frac{z_0 x_0}{y_0} \sigma_0 + \frac{z_0}{y_0} e\mu_p(b + 1) \int_{-x_0/2}^{x_0/2} \delta p(x)\, dx. \qquad (10.6\text{-}11)$$

But the first term above is the equilibrium conductance ($= \sigma_0 \cdot$ area/length), while in the second, the integral can be expressed in terms of δP by (10.6-6). Making these substitutions (10.6-11) can be written as

$$C = C_0 + \frac{e\mu_p(b + 1)\delta P}{y_0^2} = C_0 + \delta C. \qquad (10.6\text{-}12)$$

The relative conductance change may then be expressed as $\delta C/C_0$, where $C_0 = \sigma_0 z_0 x_0/y_0$, whence

$$\frac{\delta C}{C_0} = \frac{e\mu_p(b + 1)\delta P}{\sigma_0 V}. \qquad (10.6\text{-}13)$$

where δP is given by (10.6-7) and $V = x_0 y_0 z_0$ is the volume of the sample. The photoconductance $\delta C/C_0$ is shown plotted as a function of sample thickness for several values of s in Figure 10.12(b). For $s = 0$, of course, the quantity $\delta C/C_0$ is independent of the sample thickness x_0. For $s > 0$, the photoconductance is very small for thin samples, since then carriers may diffuse very rapidly to the surface and recombine there. For thicker samples, carriers in the interior reach the surface only after much diffusion, and surface recombination has much less effect on the overall conductance. In the limit $x_0 \to \infty$ the effect of the surface becomes negligible. It is clear that if the sample dimensions, D_p, the generation rate g', and *one* of the two quantities (s,L_p) are known, then either s or L_p, whichever is unknown, may be measured by this experiment. If two different samples, identical but for thickness, are used, then both s and L_p can be determined. The chief difficulty with the use of the steady-state photoconductivity effect as a technique for the determination of s or L_p is that the absolute generation rate g' must be known, and this is quite difficult to determine for any given experiment. This difficulty is eliminated if the *transient* photoconductivity is measured instead. The analysis of this effect is given in the next section.

10·7 TRANSIENT PHOTOCONDUCTIVITY; EXCESS CARRIER LIFETIME

In this section we shall consider the transient decay of photoconductivity produced by penetrating radiation in a sample exactly like that which was discussed in the previous section. It will be assumed that the light source has generated a uniform excess carrier density p_1 everywhere within the sample at time $t = 0$, at which instant the exciting radiation is abruptly turned off. The subsequent decay of the excess carrier distribution to equilibrium is then described by (10.2-26), which for the case at hand takes the form

$$D_p \frac{\partial^2(\delta p)}{\partial x^2} - \frac{\delta p(x,t)}{\tau_p} = \frac{\partial(\delta p)}{\partial t}.$$ (10.7-1)

The surface boundary conditions[16] are

$$-D_p\left(\frac{\partial(\delta p)}{\partial x}\right)_{x_0/2} = s\delta p(x_0/2,t) \quad \text{and} \quad D_p\left(\frac{\partial(\delta p)}{\partial x}\right)_{-x_0/2} = s\delta p(-x_0/2,t),$$ (10.7-2)

while in addition, we must require that

$$\delta p(x,0) = p_1 = \text{const}$$ (10.7-3)

and

$$\lim_{t \to \infty} \delta p(x,t) = 0.$$ (10.7-4)

By making the substitution

$$\delta p(x,t) = e^{-t/\tau_p}u(x,t)$$ (10.7-5)

[16] Although the steady state condition was assumed in arriving at the form (10.5-16) for the boundary condition, it is clear that this boundary condition can be used for time-dependent situations as well, provided that the carrier concentration in the surface region does not change much over the time needed for the flux equilibrium between surface and volume to establish itself. We shall always assume that this condition is realized.

Equation (10.7-1) can be transformed into a differential equation for $u(x,t)$ of the form

$$D_p \frac{\partial^2 u}{\partial x^2} = \frac{\partial u}{\partial t}, \tag{10.7-6}$$

while the boundary conditions transform to

$$-D_p \left(\frac{\partial u}{\partial x} \right)_{x_0/2} = su(x_0/2,t) ; \qquad D_p \left(\frac{\partial u}{\partial x} \right)_{-x_0/2} = su(-x_0/2,t) \tag{10.7-7}$$

and $$u(x,0) = p_1 = \text{const} \qquad (-x_0/2 < x < x_0/2). \tag{10.7-8}$$

Now let us seek product solutions of the form

$$u(x,t) = X(x)T(t). \tag{10.7-9}$$

Substituting this form for the solution back into (10.7-6) it is readily seen that

$$\frac{1}{X(x)} \frac{d^2 X}{dx^2} = \frac{1}{D_p T(t)} \frac{dT}{dt} = -\alpha^2 = \text{const.} \tag{10.7-10}$$

Both sides of the equation above must be separately equal to a constant, since only in this way can a function of x alone and a function of t alone be equal for all possible values of x and t. The constant is written as $-\alpha^2$ so that for any real α it will be negative; this is necessary (as will soon become evident) to insure that (10.7-4) be satisfied. For the time dependence of the above equation, it is obvious that

$$\frac{dT}{dt} = -\alpha^2 D_p T(t), \tag{10.7-11}$$

whence $$T(t) = e^{-\alpha^2 D_p t}. \tag{10.7-12}$$

For the spatial dependence, we have

$$\frac{d^2 X}{dx^2} = -\alpha^2 X(x), \tag{10.7-13}$$

whereby $$X(x) = A \cos \alpha x + B \sin \alpha x. \tag{10.7-14}$$

By the symmetry of the problem, it is clear that the spatial dependence of the excess carrier concentration *must always be an even function of x*. A sine function, or any superposition of sine functions, however, is invariably odd, while a cosine or any superposition of cosines is even. It is apparent that this physical requirement can be fulfilled only by a cosine function or by a superposition of cosine functions, and that we must therefore set B equal to zero in (10.7-14). From (10.7-9), (10.7-12) and

(10.7-14), it is evident that a suitable solution may be written as

$$u(x,t) = X(x)T(t) = Ae^{-\alpha^2 D_p t} \cos \alpha x. \tag{10.7-15}$$

This solution, by itself, satisfies the differential equation (10.7-6), but does *not* satisfy the boundary conditions (10.7-7) and (10.7-8). However, we may form a linear superposition of solutions of this form such as

$$u(x,t) = \sum_n u_n(x,t) = \sum_n A_n e^{-\alpha_n D_p t} \cos \alpha_n x, \tag{10.7-16}$$

which still satisfies the differential equation, and we are at liberty to choose the values A_n and α_n so that the boundary conditions are satisfied by the superposition. If *each term* of the summation in (10.7-16) is required to satisfy the surface boundary condition

$$-D_p(\partial u_n/\partial x)_{c_0/2} = su_n(x_0/2,t), \tag{10.7-17}$$

then, of course, the surface boundary conditions (10.7-7) will automatically be satisfied by the superposition.[17] Substituting a solution of the form (10.7-15) for u_n into (10.7-17), we find that in order to satisfy (10.7-17) we must have

$$\alpha_n A_n D_p e^{-\alpha_n^2 D_p t} \sin \frac{\alpha_n x_0}{2} = sA_n e^{-\alpha_n^2 D_p t} \cos \frac{\alpha_n x_0}{2}, \tag{10.7-18}$$

whereby α_n must be chosen in such a way that the equation

$$\text{ctn} \frac{\alpha_n x_0}{2} = \frac{\alpha_n D_p}{s} = \frac{\alpha_n x_0}{2}(2D_p/sx_0) \tag{10.7-19}$$

is satisfied. The roots of this equation may be obtained numerically or graphically as the intersection of the curves $f(\alpha x_0) = \text{ctn}(\frac{1}{2}\alpha x_0)$ and $f(\alpha x_0) = (\frac{1}{2}\alpha x_0)(2D_p/sx_0)$, as shown in Figure 10.13.

Now we must try to satisfy the boundary condition (10.7-8). We shall try to do this by choosing the values A_n so that at $t = 0$ the sum of all the $u_n(x,0)$ add to form a Fourier-like representation of the required initial concentration profile. To accomplish this, suppose that $u(x,0)$ is an *arbitrary* even function[18] $f(x)$, so that

$$u(x,0) = \sum_{n=0}^{\infty} A_n \cos \alpha_n x = f(x). \tag{10.7-20}$$

If both sides of this equation are multiplied by $\cos \alpha_m x$ and integrated over the interval $(-x_0/2 < x < x_0/2)$, the result is

$$\sum_n \int_{-x_0/2}^{x_0/2} A_n \cos \alpha_n x \cos \alpha_m x \, dx = \int_{-x_0/2}^{x_0/2} f(x) \cos \alpha_m x \, dx. \tag{10.7-21}$$

---

[17] Note that if $u(x,t)$ is an even function of x, the satisfaction of one of the boundary conditions (10.7-7) *implies* the satisfaction of the other.

[18] If $f(x)$ were not even, we would have to admit solutions of the form $B_n \exp(-\alpha_n^2 D_p t) \sin \alpha_n x$ into our superposition as well as the cosine function we are using.

If the set of functions $\{\cos \alpha_n x\}$ is orthogonal on the interval $(-x_0/2 < x < x_0/2)$, then all the integrals on the left-hand side of this equation will vanish except that for which $n = m$, which is easily evaluated. Under these circumstances, solving for A_m, one may easily show that

$$A_m = \frac{2\alpha_m}{\alpha_m x_0 + \sin \alpha_m x_0} \int_{-x_0/2}^{x_0/2} f(x) \cos \alpha_m x \, dx. \qquad (10.7\text{-}22)$$

It is clear that since this is a special case of the Sturm-Liouville problem, the functions $\{\cos \alpha_n a\}$ must form an orthogonal set.[19] In any case, it is not difficult to show directly

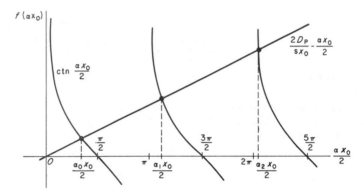

FIGURE 10.13. Diagram illustrating the determination of the roots of (10.7-19) by the graphical solution of the transcendental equation.

that provided the α_n are chosen so as to satisfy (10.7-19) the functions $\{\cos \alpha_n a\}$ are orthogonal. The details of this proof are left as an exercise for the reader.

For the case at hand, we must choose $f(x) = p_1 = \text{const}$, and with this choice, the integral in (10.7-22) can easily be evaluated giving

$$A_n = \frac{4p_1 \sin \dfrac{\alpha_n x_0}{2}}{\alpha_n x_0 + \sin \alpha_n x_0} \qquad (10.7\text{-}23)$$

whereby
$$f(x) = p_1 = 4p_1 \sum_{n=0}^{\infty} \frac{\sin \dfrac{\alpha_n x_0}{2} \cos \alpha_n x}{\alpha_n x_0 + \sin \alpha_n x_0}. \qquad (10.7\text{-}24)$$

Now $u(x,t)$ can be found by substituting the above value for A_n in (10.7-16), and $\delta p(x,t)$

---

[19] R. V. Churchill, *Fourier Series and Boundary Value Problems*. New York: McGraw-Hill (1941), pp. 46–52.

can then be obtained from (10.7-5). The final result is

$$\delta p(x,t) = 4p_1 e^{-t/\tau_p} \sum_{n=0}^{\infty} \frac{\sin \frac{\alpha_n x_0}{2} \cos \alpha_n x}{\alpha_n x_0 + \sin \alpha_n x_0} e^{-\alpha_n^2 D_p t}. \tag{10.7-25}$$

When $s \to 0$, the slope of the straight line shown in Figure 10.13 becomes extremely large, and the eigenvalues $\{\alpha_n a\}$ approach $(0, 2\pi, 4\pi, 6\pi \cdots)$. For this set of values, $\sin(\alpha_n a/2) = 0$ for all α_n, and all the terms in the sum in (10.7-25) vanish, with the exception of the first. In the case of the first term, both numerator and denominator vanish, giving an indeterminate form. Using L'Hôpital's rule, it is readily shown that

$$\lim_{\alpha_0 x_0 \to 0} \frac{\sin \frac{\alpha_0 x_0}{2}}{\alpha_0 x_0 + \sin \alpha_0 x_0} = \lim_{\alpha_0 x_0 \to 0} \frac{\frac{1}{2} \cos \frac{\alpha_0 x_0}{2}}{1 + \cos \alpha_0 x_0} = \frac{1}{4}, \tag{10.7-26}$$

whence $$\delta p(x,t) = p_1 e^{-t/\tau_p} \qquad (s = 0). \tag{10.7-27}$$

The concentration profile for this case is plotted for several values of t in Figure 10.14(a). It is clear that the concentration remains the same at any given time at all points in the sample.

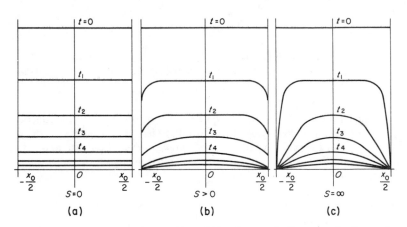

FIGURE 10.14. Concentration profiles within the sample for several values of t, (a) for $s = 0$, (b) for $s > 0$, (c) for $s = \infty$. In all cases $0 < t_1 < t_2 < t_3 < t_4$.

If $s \to \infty$, the slope of the straight line in Figure 10.13 approaches zero, the eigenvalues $\{\alpha_n a\}$ approaching $(\pi, 3\pi, 5\pi, 7\pi, \ldots)$ in the limit. In this case (10.7-25) becomes

$$\delta p(x,t) = 4p_1 e^{-t/\tau_p} \sum_{n=0}^{\infty} \frac{(-1)^n \cos \frac{(2n+1)\pi x}{x_0}}{(2n+1)\pi} \exp\left[-\frac{(2n+1)^2 \pi^2 D_p t}{x_0^2}\right], \tag{10.7-28}$$

which is an ordinary Fourier series. In this instance, when $x = \pm x_0/2$, δp is zero, since $\cos(n + \frac{1}{2})\pi = 0$. The concentration profile for this case is plotted in Figure 10.14(c). For intermediate values of s, the values $\{\alpha_n\}$ must be determined numerically or graphically, and the results, when plotted, yield concentration profiles such as those shown in Figure 10.14(b).

The actual photoconductance can be obtained from $\delta p(x,t)$ by means of (10.6-12) and (10.6-6), whereby

$$\frac{\delta C(t)}{C_0} = \frac{e\mu_p(b+1)}{\sigma_0 V} \cdot y_0 z_0 \int_{-x_0/2}^{x_0/2} \delta p(x,t)\, dx, \qquad (10.7\text{-}29)$$

or, using (10.7-25) and evaluating the integral,

$$\frac{\delta C(t)}{C_0} = \frac{8p_1 e\mu_p(b+1)}{\sigma_0} e^{-t/\tau_p} \sum_{n=0}^{\infty} \frac{\sin^2 \dfrac{\alpha_n x_0}{2}}{\alpha_n x_0(\alpha_n x_0 + \sin \alpha_n x_0)} e^{-\alpha_n^2 D_p t}. \qquad (10.7\text{-}30)$$

This expression is a summation of terms each of which has its own characteristic amplitude and each of which decays exponentially with time with a different time constant. It may be written in the form

$$\frac{\delta C(t)}{C_0} = \sum_m C_m e^{-t/\tau_m} \qquad (10.7\text{-}31)$$

where C_m is the amplitude of the mth decay mode[20] defined by (10.7-30), and where each exponential time constant may be expressed as

$$\frac{1}{\tau_m} = \frac{1}{\tau_p} + \alpha_n^2 D_p. \qquad (10.7\text{-}32)$$

Since α_0 is the smallest member of the set $\{\alpha_n\}$, the time constant τ_0 must be longer than any of the others; furthermore, the form of the amplitude factor C_m multiplying each term is such that the amplitudes of the higher terms are all smaller than that of the first. The result is that the higher order modes die out more rapidly than the zeroth, and after a sufficiently long time the decay can be represented as a simple exponential with the single *principal mode* decay constant τ_0, as shown in Figure 10.15. It is clear that if the logarithm of the photoconductive response is plotted as a function of time, a straight line is obtained after the effects of the higher mode terms in (10.7-30) have died out. The slope of this line is $1/\tau_0$, where

$$\frac{1}{\tau_0} = \frac{1}{\tau_p} + \alpha_0^2 D_p. \qquad (10.7\text{-}33)$$

There are two terms on the right-hand side of this equation; the first represents the effect of bulk recombination upon the observed principal mode time constant, the

---

[20] The index m is adopted here in order to avoid possible confusion with the electron lifetime τ_n.

second the effect of surface recombination. The *observed lifetime* τ_0 is shorter than the true bulk lifetime τ_p unless the surface recombination velocity is zero, in which case, as we have already noted, $\alpha_0 = 0$. Although it is ordinarily impossible to reduce s to zero, experimentally, the same effect can be produced, as is evident from Figure 10.13, by arranging the conditions of the experiment so that $2D_p/sx_0$ is much less than unity. This may be accomplished by using a very thick sample. Under these circumstances carriers generated in the interior must diffuse a long way to the surface, whose role is then much less important than for a thin sample, and the observed lifetime τ_0 approximates the true bulk lifetime τ_p.

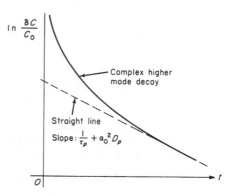

FIGURE 10.15. Logarithm of the transient photoconductive response *versus* time.

If the true bulk lifetime τ_p and the observed lifetime τ_0 are both known, then the surface recombination velocity can be evaluated by solving (10.7-33) for α_0, inserting the value so obtained into (10.7-19) and solving for s. The result is

$$s = \sqrt{D_p\left(\frac{1}{\tau_0} - \frac{1}{\tau_p}\right)} \tan \frac{x_0}{2} \sqrt{\frac{1}{D_p}\left(\frac{1}{\tau_0} - \frac{1}{\tau_p}\right)}. \qquad (10.7\text{-}34)$$

This formula may also be used to obtain the true bulk lifetime if the observed lifetime and surface recombination velocity are known.

If both τ_p and s are unknown to begin with, they may be determined by working first with a thick sample whose surface is treated in such a way as to minimize surface recombination; the transient photoconductive decay associated with this type of sample will approximate τ_p. The thick sample may then be cut into thin ones, the surfaces of these being treated so as to obtain the surface conditions under which the surface recombination velocity is to be measured. The decay constant τ_0 is then measured and s follows directly from (10.7-34), using the value for τ_p obtained previously. Alternatively, two samples of different thickness can be cut from material of *uniform* lifetime. The surfaces of both samples are treated so as to produce the conditions under which a measurement of s is desired, and the observed lifetime associated with each sample measured. Equation (10.7-34) can then be written for each sample, and the two resulting equations may then be solved numerically for the two unknown quantities, τ_p and s. A third method is to use only a single sample, and to adjust the values of τ_p and s in (10.7-30) so as to reproduce not only the observed principal mode time constant τ_0 (which can, after all, be obtained from many different combinations of τ_p and s) but *also* the observed initial higher-mode decay scheme. Although the required mathematical analysis is complicated, this method has the advantage that only a single sample is used, and difficulties arising from inhomogeneities in bulk lifetime and surface recombination velocities are thereby minimized.

10·8 RECOMBINATION MECHANISMS; THE SHOCKLEY-READ THEORY OF RECOMBINATION

We shall now investigate briefly the physical aspects of electron-hole recombination. It is important to understand the physical processes involved in *direct* electron-hole recombination as well as in recombination by *trapping*, although the latter is found to be the dominant mechanism in the majority of situations involving covalent or inter-metallic III-V semiconductors.

In direct recombination an electron in the conduction band and a hole in the valence band recombine without the assistance of any intermediate state. If both conduction and valence bands have energy minima at $k = 0$, then vertical transitions in which a hole near the top of the valence band may recombine with an electron near the bottom of the conduction band are possible. Since momentum must be conserved in such a transition, only electrons and holes whose k-vectors satisfy the relation

$$\mathbf{k}_n + \mathbf{k}_p = 0 \tag{10.8-1}$$

may interact in this way. A photon of frequency ω given by

$$\hbar\omega = \varepsilon_n - \varepsilon_p \cong \varepsilon_c - \varepsilon_v \tag{10.8-2}$$

is emitted in the process.[21] This *direct transition* process is illustrated in Figure 10.16(a). The absorption of a photon, accompanied by the production of an electron-hole pair may take place by a process which is simply the reverse of this one.

In substances such as germanium and silicon, where the minimum of the conduction band occurs for some value of the crystal momentum $\mathbf{k}_0$ other than zero, the recombination of an electron at the conduction band minimum and a hole at the top of the valence band requires (in addition to the production of a photon) the emission of a phonon of propagation constant $\mathbf{k}' = \mathbf{k}_0$ or the absorption of an already present phonon of propagation constant $\mathbf{k}' = -\mathbf{k}_0$ in order that crystal momentum be conserved. Such a process is called an *indirect transition*. Since some energy which would have been carried away by the photon in a direct transition is now carried away (or contributed) by the phonon, the frequency of the radiation associated with such transitions is given by

$$\hbar\omega = \varepsilon_n - \varepsilon_p \pm \varepsilon(\mathbf{k}_0) = \varepsilon_c - \varepsilon_v \pm \varepsilon'(\mathbf{k}_0) \tag{10.8-3}$$

---

[21] In (10.8-1) the momentum carried away by the emitted photon has been neglected. This is possible because the electron wavelength (for an average energy kT) as given by (8.4-1) is much smaller than the photon wavelength for $\omega \cong \Delta\varepsilon/\hbar$. The momenta are, of course, related to the respective wavelength by the de Broglie relation (4.9-7).

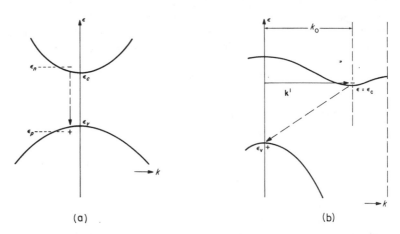

FIGURE 10.16.   Electron-hole recombination via (a) direct transition, (b) indirect transition.

where $\varepsilon'(\mathbf{k}_0)$ is the energy associated with a phonon of wave number $\mathbf{k}_0$ as obtained from the dispersion relation $\omega(\mathbf{k})$ for the lattice vibrations by

$$\varepsilon'(\mathbf{k}_0) = \hbar\omega(\mathbf{k}_0). \tag{10.8-4}$$

The energy $\varepsilon'(\mathbf{k}_0)$, although small compared with the forbidden gap energy $\Delta\varepsilon$, is perceptible through its influence on the form of the *absorption* versus wavelength curve in the neighborhood of the absorption edge. Again, the absorption of a phonon to produce an electron-hole pair may be regarded as the inverse of this process. Actually, since the emitted or absorbed phonon may belong to either acoustical or optical branch, and may be either a transverse or longitudinal vibration, there are *four* separate phonons which can participate in such a process, each with a different frequency and hence a different energy ε' for $k = k_0$. The effects of all four of these phonons can be observed in the absorption spectra of this type of semiconductor near the fundamental absorption edge. Of course, direct transitions can also occur in materials where the conduction band minimum is not at $k = 0$, but if the energy difference at the conduction band edge between the point of minimum energy and $k = 0$ is at all appreciable, the electron population for energies corresponding to $k = 0$ will normally be so small that such transitions will be quite infrequent. Likewise, direct absorptive transitions will require incident photon energies much higher than those required to promote indirect transitions. Indirect transitions are predominant under normal conditions in all the elemental covalent semiconductors and III-V compound semiconductors with the single exception of InSb, in which both band minima are at $k = 0$.

It is possible, in either case, to calculate the rate at which transitions take place from the observed optical absorption spectrum of the material,[22] and therefore to

---

[22] In the steady state, the generation and recombination rate of electron-hole pairs must be equal; for this reason a calculation of the generation rate will also yield the recombination rate. For direct optical excitation of electron-hole pairs, however, the generation rate for any given incident photon wavelength must be proportional to the absorption coefficient for that wavelength. The optical absorption curve and the recombination rate arising from direct recombinative processes are therefore closely related.

infer the recombination rate and the excess carrier lifetime from the absorption spectrum. This calculation was first carried out by van Roosbroeck and Shockley,[23] who found that in germanium, for example, the excess carrier lifetime at room temperature should be about 0.75 sec, independent of donor or acceptor impurity density. This result was contradicted by experimental evidence; observed excess carrier lifetimes rarely exceeded 10^{-3} sec, and were found to be quite sensitive to the donor and acceptor impurity concentration. It was also well established experimentally that excess carrier lifetimes in germanium and silicon could be drastically reduced by the introduction of certain impurities, such as copper, in extremely tiny concentrations, and by the introduction of certain structural imperfections, such as dislocations. In addition, the van Roosbroeck-Shockley calculation predicted a temperature dependence of excess carrier lifetime which was not observed experimentally. It was apparent that in most materials, excess carrier lifetimes were not limited by radiative transitions of the type discussed above, but rather by some other process which was intimately associated with the presence of certain special types of impurity atoms and structural imperfections in small concentrations. These results led to a consideration of recombination by *trapping* of electrons and holes by localized energy levels lying deep within the forbidden energy gap, which were thought to be associated with certain "trapping" impurities and structural defects. This mechanism was first investigated by Shockley and Read.[24]

The Shockley-Read theory of recombination involves a consideration of the statistics of occupation of such trapping levels. We shall consider in detail the behavior of an energy level at an energy ε_t within the forbidden energy region, which is neutral when empty and which may be occupied by an electron, thus acquiring a negative charge. These *traps* may promote electron-hole recombination by capturing electrons from the conduction band and subsequently transferring them to the valence band whenever a hole appears near the trap to recombine with the trapped electron. The net effect is to do away with an electron-hole pair, the trapping level being returned in the end to its original state. Since the trapping center is usually tightly coupled to

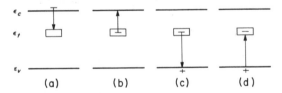

FIGURE 10.17. Four fundamental processes involved in recombination through traps; (a) the capture of a conduction electron by an empty trap, (b) the emission of an electron from the trap to the conduction band, (c) the capture of a hole from the valence band by a trap containing an electron, and (d) the promotion of a valence electron into an initially empty trap.

the lattice, the energy and momentum initially belonging to the electron and hole may be converted into lattice vibrations, with little or no electromagnetic energy being emitted in the process. There are four basic processes involved in electron-hole

[23] W. van Roosbroeck and W. Shockley, *Phys. Rev.*, **94**, 1558 (1954).
[24] W. Shockley and W. T. Read, Jr., *Phys. Rev.*, **87**, 835 (1952).

recombination through trapping centers, as illustrated in Figure 10.17; (a) the capture of an electron from the conduction band by an initially neutral empty trap, (b) the inverse of (a), involving the emission of an electron initially occupying a trapping level to the conduction band, (c) the capture of a hole in the valence band by a trap containing an electron, and (d), the inverse of (c), wherein a valence electron is promoted into an initially empty trap (this can equally well be regarded as the emission of a hole from the trap to the valence band).

Under these circumstances, the rate at which electrons from the conduction band are captured by traps will clearly be proportional to the number of electrons in the conduction band and to the number of *empty* traps available to receive them. We may write this as

$$R_{cn} = C_n N_t (1 - f(\varepsilon_t))n \qquad (10.8\text{-}5)$$

where C_n is a constant and N_t represents the total concentration of trapping centers in the crystal. Likewise, the rate at which electrons are emitted from filled traps to the conduction band will be proportional to the number of filled traps, and can thus be expressed as

$$R_{en} = E_n N_t f(\varepsilon_t) \qquad (10.8\text{-}6)$$

where E_n is another proportionality constant. In equilibrium, in accord with the principle of detailed balancing, these two rates must be equal, whereby

$$E_n = n_0 C_n \frac{1 - f_0(\varepsilon_t)}{f_0(\varepsilon_t)}. \qquad (10.8\text{-}7)$$

By substituting the explicit value of the Fermi function f_0 from (5.5-20) into this expression, we may easily obtain the very simple relation

$$E_n = n_0 C_n e^{(\varepsilon_t - \varepsilon_f)/kT}. \qquad (10.8\text{-}8)$$

Substituting the value given by (9.3-6) for n_0, (10.8-8) may finally be written as

$$E_n = n_1 C_n \qquad (10.8\text{-}9)$$

where
$$n_1 = U_c e^{-(\varepsilon_c - \varepsilon_t)/kT}. \qquad (10.8\text{-}10)$$

We are assuming, in writing (10.8-9) and (10.8-10), that the distribution of electrons in the conduction band is Maxwellian. The quantity n_1 is the electron concentration which would be present in the conduction band if the Fermi level were to coincide with the trap level ε_t. The *net* rate at which electrons are captured from the conduction band is simply the difference between (10.8-5) and (10.8-6), which, using (10.8-9) to express E_n in terms of C_n, is

$$R_n = R_{cn} - R_{en} = C_n N_t [n(1 - f(\varepsilon_t)) - n_1 f(\varepsilon_t)]. \qquad (10.8\text{-}11)$$

In the same way, the rate at which holes are captured from the valence band must be proportional to the concentration of holes and to the number of traps which may receive holes (thus the number which are occupied by electrons) whence

$$R_{cp} = C_p N_t f(\varepsilon_t) p. \tag{10.8-12}$$

The rate at which holes are emitted to the valence band from trapping centers (process (d) in Figure 10.17) is proportional to the number of traps occupied by holes (i.e., *empty*), from which

$$R_{ep} = E_p N_t (1 - f(\varepsilon_t)). \tag{10.8-13}$$

As before, in the equilibrium state, R_{cp} must be equal to R_{ep}, and by reasoning which is exactly analogous to that used in deriving (10.8-9), one may easily show that

$$E_p = p_1 C_p \tag{10.8-14}$$

where
$$p_1 = U_v e^{-(\varepsilon_t - \varepsilon_v)/kT}. \tag{10.8-15}$$

Again, p_1 may be visualized as the concentration of holes which would be found in the valence band if the Fermi level were at the energy of the trapping level. The *net* rate at which holes are captured from the valence band may be expressed as

$$R_p = R_{cp} - R_{ep} = C_p N_t [p f(\varepsilon_t) - p_1 (1 - f(\varepsilon_t))]. \tag{10.8-16}$$

When trapping centers are present it is no longer possible in general to assume that $\delta n = \delta p$, despite the fact that overall electrical neutrality is maintained, because some electrons which would otherwise be present in the conduction band as a part of δn may now be immobilized in traps, unable to function as conduction electrons. This situation causes some difficulty when the density of trapping centers is large, and has been treated in detail by Shockley and Read.[24] If N_t is quite small, however, the approximation $\delta n = \delta p$ is still very good.[25] We shall assume that this is the case, and in the majority of cases of practical interest in the covalent semiconductors this condition is indeed satisfied. In a sample in which excess carriers are recombining, under these conditions, we must have $R_p = R_n$ since every electron recombination is accompanied by the recombination of a hole. The common recombination rate is then related to the lifetime τ by

$$R_n = R_p = \frac{\delta p}{\tau}. \tag{10.8-17}$$

Equating (10.8-11) and (10.8-16) and solving for $f(\varepsilon_t)$, one may obtain

$$f(\varepsilon_t) = \frac{C_p p_1 + C_n n}{C_n(n + n_1) + C_p(p + p_1)}. \tag{10.8-18}$$

Substituting this value into either (10.8-11) or (10.8-16) and noting that $n_1 p_1 = n_i^2$, it

---

[25] It is shown by Shockley and Read that $\delta n \simeq \delta p$ whenever any one of the four quantities n_0, p_0, n_1 and p_1 is large compared with N_t.

is clear that

$$R_n = R_p = \frac{\delta p}{\tau} = \frac{N_t C_n C_p (np - n_i^2)}{C_n(n + n_1) + C_p(p + p_1)}. \tag{10.8-19}$$

If we now write $n = n_0 + \delta p$, $p = p_0 + \delta p$, we may express the above equation in the form

$$\frac{1}{\tau} = \frac{N_t(p_0 + n_0 + \delta p)}{\dfrac{1}{C_p}(n_0 + n_1 + \delta p) + \dfrac{1}{C_n}(p_0 + p_1 + \delta p)}. \tag{10.8-20}$$

In the limit where $n_0 \gg \delta p, p_0, n_1, p_1$ it is evident that (10.8-20) reduces to

$$\tau = \tau_{p0} = \frac{1}{C_p N_t}. \tag{10.8-21}$$

The lifetime of holes in a strongly extrinsic n-type sample having a given trapping center concentration is thus just the reciprocal of $C_p N_t$. Likewise, in the limit where $p_0 \gg \delta p, n_0, n_1, p_1$, (10.8-20) reduces to

$$\tau = \tau_{n0} = \frac{1}{C_n N_t} \tag{10.8-22}$$

showing that in a strongly extrinsic p-type sample the electron lifetime is the reciprocal of $C_n N_t$. We may then write (10.8-20) as

$$\frac{1}{\tau} = \frac{p_0 + n_0 + \delta p}{\tau_{p0}(n_0 + n_1 + \delta p) + \tau_{n0}(p_0 + p_1 + \delta p)} \tag{10.8-23}$$

where (τ_{p0}, τ_{n0}) represent the lifetime of excess (holes, electrons) in highly extrinsic (n-type, p-type) material having the particular trapping characteristics which are being considered. The lifetime for materials intermediate between these two extremes is

FIGURE 10.18. The dependence of lifetime upon the Fermi energy, as predicted by the Shockley-Read theory.

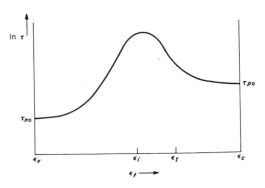

given by (10.8-23). For small values of δp (10.8-23) can be written as

$$\tau = \tau_{p0}\left(\frac{n_0 + n_1}{p_0 + n_0}\right) + \tau_{n0}\left(\frac{p_0 + p_1}{p_0 + n_0}\right).$$ (10.8-24)

From this equation the lifetime can be plotted as a function of the Fermi level, and thus as a function of n_0 and p_0. It is found that the lifetime is a maximum when the Fermi level is at some point within the forbidden gap whose exact location depends upon the values τ_{p0}, τ_{n0} and n_1, as illustrated in Figure 10.18. If τ_{n0} and τ_{p0} are equal, then the maximum will come at the intrinsic condition.

The variation of the lifetime with the injection level δp can be studied by solving (10.8-23) for τ, writing the result in the form

$$\tau = \tau_0 \left[\frac{1 + \dfrac{(\tau_{p0} + \tau_{n0})\delta p}{\tau_{p0}(n_0 + n_1) + \tau_{n0}(p_0 + p_1)}}{1 + \dfrac{\delta p}{n_0 + p_0}}\right]$$ (10.8-25)

where

$$\tau_0 = \frac{\tau_{p0}(n_0 + n_1) + \tau_{n0}(p_0 + p_1)}{n_0 + p_0}.$$ (10.8-26)

From this it is clear that if δp is sufficiently small, the lifetime will have the value τ_0 independent of δp. For larger values of δp, the lifetime depends upon δp, and may either increase or decrease with increasing values of δp, depending upon the relative values of τ_{p0}, τ_{n0}, n_0 and n_1. For large values of δp it is easily seen that (10.8-25) gives

$$\tau = \tau_\infty = \tau_{p0} + \tau_{n0}.$$ (10.8-27)

If τ_∞ is different from τ_0 (as is generally the case), there is a monotonic variation of lifetime between the limits τ_0 and τ_∞ as δp is increased.

It is rather difficult to make precise and meaningful comparisons between the Shockley-Read theory and experiment, due to the difficulty of obtaining samples with precisely controlled concentrations of a single type of trapping center, and due to the number of unknown parameters whose value must be determined from (or adjusted to fit) experimental data.[26] Nevertheless, the theory seems to be consistent with all the experimental results to which it might be expected to apply to date.[27] It explains the marked observed variations of carrier lifetime with donor and acceptor density, injection level, concentration of trapping centers, and temperature. It may also be extended to provide a theory of surface recombination based upon trapping levels associated with the surface, but having properties similar to the bulk centers discussed herein insofar as the statistics of capture and emission are concerned.

Physically, the bulk trapping centers in semiconductor crystals often arise from

---

[26] The difficulty is not so much fitting experimental data, but in having so many adjustable parameters that the data can be explained by many possible sets of values for these parameters.

[27] See, for example, J. A. Burton *et al.*, *J. Phys. Chem.*, **57**, 853 (1953).

the presence of certain "trapping impurities." In the covalent semiconductors silicon and germanium these trapping impurities are often divalent *double* acceptors, such as Cu and Ni, which provide an ordinary acceptor level very close to the valence band and a second level deep within the forbidden region which is much less easily ionized, and which functions as a trapping level. The donor and acceptor levels arising from the presence of substitutional group III and group V impurities are almost completely ionized at all but very low temperatures and therefore do not ordinarily play an important role in recombination processes.[28] The presence of structural imperfections, particularly edge-type dislocations, has also been shown to give rise to acceptor trapping levels within the forbidden region, which may in many instances provide an important contribution to electron-hole recombination.

It sometimes happens that *two* distinct types of trapping centers are present in the same sample. Usually one of these is the normal recombination center as discussed above; the second set of trapping centers, however, quite often has associated with it a very small hole capture probability when filled with electrons. This leads to a very large value for τ_{p0} for that set of levels. Although naturally the Shockley-Read analysis must be repeated, considering both sets of trapping levels from the outset, to describe the details of the resulting situation, it is fairly easy to see intuitively that what must happen initially in a transient photoconductivity measurement such as that discussed in Section 10.7 is that some electrons recombine through the normal recombination levels, while others fall into the second set of traps. Before long, however, the normal recombination process has exhausted all excess electrons except those in the slow second set of traps, which then slowly capture the remaining excess holes. The resulting photoconductivity decay has two separate exponential sections with different time constants, as shown in Figure 10.19. There is an initial "normal"

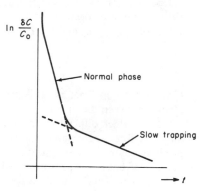

FIGURE 10.19. Transient photoconductive response of a sample which contains a large number of "slow trapping centers."

recombination phase and a long "trapping tail" due to the slow traps. A certain confusion in terminology has arisen around this effect, in that the "normal" initial part of the decay is frequently referred to as "recombination" and the slow final decay is then ascribed to "trapping." Actually *both* phases of the process are governed by trapping phenomena, and it is more accurate to refer to the earlier as due to *normal* trapping and the latter as due to slow trapping effects.

---

[28] Except, of course, insofar as they influence the position of the Fermi level.

EXERCISES

1. Starting with the Boltzmann equation and assuming that the relaxation time approximation holds, show that in a steady state system containing free particles wherein a *gradient of concentration* along the x-direction exists, a diffusion flux $- D(\partial n/\partial x)$ is set up. Show that the diffusion coefficient D is equal to $\lambda \bar{c}/3$ where λ is the mean free path and $\bar{c}$ is the mean thermal speed. You may assume that the mean free path λ is independent of velocity, and that the system obeys Maxwell-Boltzmann statistics in equilibrium. *Hint:* Assume that the distribution function $f(\mathbf{r}, \mathbf{v})$ can be written in the form $n(\mathbf{r})f(\mathbf{v})$; this is reasonable if the departure from the equilibrium state is not too great.

2. The results derived from the calculations of Problem 1 have been applied in the text to nonsteady state systems as well as to systems which are in the steady state. Discuss (qualitatively) under what circumstances this extension should be allowable.

3. Prove that if $f(x, t)$ satisfies the differential equation

$$D \frac{\partial^2 f}{\partial x^2} - \frac{f}{\tau} = \frac{\partial f}{\partial t}$$

then $f(\xi, t)$ ($\xi = x - \mu E_0 t$, with E_0 const) must be a solution to

$$D \frac{\partial^2 f}{\partial x^2} - \mu E_0 \frac{\partial f}{\partial x} - \frac{f}{\tau} = \frac{\partial f}{\partial t}.$$

4. A steady-state excess carrier distribution is created in an extrinsic n-type sample by a plane source of generation at the origin. The resulting diffusion and drift of the carrier distribution can be regarded as one-dimensional, along the x-direction; the extent of the sample along the x-axis is essentially infinite. There is a constant electric field E_0 in the $+ x$-direction The excess carrier density is measured by measuring the reverse saturation current in two collector probes located at two different points, $x = a$ and $x = b$ ($b > a$). It is desired to calculate the diffusion length L_p from the ratio of these two measured excess carrier densities. Show that

$$L_p = \frac{d}{\sqrt{\ln K_0 \left(\ln K_0 + \dfrac{eE_0 d}{kT} \right)}}$$

where $K_0 = \delta p(a)/\delta p(b)$ and $d = b - a$.

5. Starting with the solution (10.3-32) to the continuity equation and defining the transit time as that time when the oscilloscope trace reaches a maximum, derive the results shown in (10.4-3) and (10.4-4).

6. Consider a uniform semiinfinite semiconductor whose surface coincides with the yz-plane. The material has a bulk excess carrier lifetime τ and there is a surface recombination velocity s associated with the surface. There is a constant uniform bulk generation rate of excess carriers g' everywhere within the sample. Assuming that a steady state has been reached, calculate the excess carrier density $\delta p(x)$ at all points within the sample.

7. From the results of Problem 6, show that provided $D/L \ll \bar{c}/2$ ($D = $ diffusion coefficient, $L = (D\tau)^{1/2} = $ diffusion length) it makes little difference in the carrier concentration in the bulk or in the diffusive flux of carriers to the surface, whether s is taken to be $\bar{c}/2$ or infinity. Show that this condition can also be stated in the form $L \gg \frac{2}{3}\lambda$.

8. Prove explicitly that the set of functions $\{\cos \alpha_n a\}$ with α_n defined by (10.7-19) is orthogonal. Find the normalization constant, thus deriving (10.7-22).

9. The observed photoconductive decay constant associated with a very thick sample

of n-type Germanium ($D_p = 45$ cm²/sec, $\mu_p = 1800$ cm²/V-sec) is 500 μsec. The sample is cut into slices 0.1 cm thick, the surfaces are etched chemically, and the photoconductive decay constant is then observed to be 300 μsec. What is the surface recombination velocity associated with the etched surfaces of the thin samples?

10. Show from the results of the Shockley-Read theory for a material with trapping centers for which $\tau_{no} = \tau_{po}$ that the maximum possible lifetime occurs when ε_f is at the intrinsic point, and that under these circumstances the lifetime is given by

$$\tau = \tau_{po}\left[1 + \cosh\frac{\varepsilon_t - \varepsilon_i}{kT}\right]$$

where ε_i is the position of the Fermi level for the intrinsic condition.

11. Discuss the variation of lifetime with the injection level δp for a material in which $\tau_{po} = \tau_{no}$ and in which $\varepsilon_t = \varepsilon_v + \frac{1}{4}\Delta\varepsilon$, in terms of the Shockley-Read theory.

12. Show from the results and techniques developed in Section 10.5 that the numerical constant α in Equations (10.5-8) and (10.5-9) must have the value 2/3.

13. It is sometimes stated that since the equilibrium flux of particles across a plane in a Boltzmann distribution is $p\bar{c}/4$, the maximum obtainable surface recombination velocity (associated with a surface of zero reflection coefficient) should be $s = \bar{c}/4$. This conclusion is contradicted by (10.5-15), which predicts that the maximum obtainable value for s is $\bar{c}/2$. Point out the physical error in this argument, and show that the value $\bar{c}/2$ given by (10.5-15) is really to be expected on physical grounds.

14. Consider a bar of extrinsic n-type germanium, of constant cross-section, extending along the x-axis. A constant current is made to flow along the positive x-direction by connecting a battery to the ends of the sample. A uniform density of excess electron-hole pairs is created at $t = 0$ in the region $x_1 < x < x_2$. This excess carrier distribution drifts with velocity $\mu_p E_0$ in the positive x-direction, according to the results of Section 10.2. Explain, using *physical* arguments only, why the excess carrier distribution moves to the right, along the $+ x$-axis, despite the fact that the excess electrons within the distribution must be moving to the *left* in the applied field. Draw a diagram showing (schematically) the internal field and the space charge density which causes it. You may assume that the bulk excess carrier lifetime is infinite, that the surface recombination velocity is zero, and that the excess carrier density is small compared with the equilibrium majority carrier density. *Hint:* Begin by noting that due to the constancy of current density along the sample and the modulation of conductivity by the excess carriers, the electric field must be *smaller* within the excess carrier pulse than outside it.

GENERAL REFERENCES

A. Many and R. Bray, "Lifetime of Excess Carriers in Semiconductors," in *Progress in Semiconductors*, Heywood and Co., London (1958), Vol. 3, pp. 117–151.

Allen Nussbaum, *Semiconductor Device Physics*, Prentice-Hall, Inc., Englewood Cliffs, N.J. (1962).

W. Shockley, *Electrons and Holes in Semiconductors*, D. Van Nostrand Company, Inc., New York (1950).

R. A. Smith, *Semiconductors*, Cambridge University Press, London (1961).

E. Spenke, *Electronic Semiconductors*, McGraw-Hill Book Co., Inc., New York (1958).

CHAPTER 11

MATERIALS TECHNOLOGY AND THE MEASUREMENT OF BULK PROPERTIES

11·1 PREPARATION OF HIGH-PURITY SEMICONDUCTOR MATERIALS

In the foregoing chapters, very little has been said concerning the actual chemical and metallurgical techniques involved in the preparation of semiconducting substances of high purity. Also, although we have examined the electrical properties of semiconductors in some detail, we have discussed the experimental techniques which are used for measuring these properties hardly at all. In this chapter we shall try to remedy these deficiencies, although, since our main purpose is to achieve an understanding of the physics of semiconductors, we shall have to be content to limit ourselves to a qualitative, or at best semiquantitative discussion of the chemical and metallurgical aspects of materials technology.

Although a very large number of substances have been shown to exhibit the characteristic properties associated with semiconductors, the covalent elemental semiconductors germanium and silicon are by far the best known, most thoroughly understood, and most widely used in device technology. The reason for this is simply that of all semiconducting substances, these are the ones which are most easily prepared in the form of ultra-high purity single crystals, free of dislocations, lattice vacancies and other structural imperfections. The other covalent semiconductors belonging to group IV of the periodic table, diamond and α-(gray) tin, are very difficult to obtain in the form of large high purity single crystals and hence have not been studied quite so thoroughly from a fundamental point of view, nor employed to any extent in electronic device technology.

The intermetallic compounds which are formed between the elements in group III and group V of the periodic table, such as InSb, GaAs, AlP, etc., crystallize in the zincblende structure and form a series of semiconducting materials whose physical properties are very much like those of the covalent group IV elements in many respects. These III–V semiconductors, however, are slightly *ionic* in character, and it is this fact which is responsible for most of the differences in the properties of the III–V compounds and the group IV covalent semiconductors. The effective mass of conduction electrons in III–V compounds is generally quite small, compared to the gravitational mass m_0, while the hole effective mass is not much different from the gravitational electron mass. For this reason, according to (9.7-7) and (9.11-19), the electron mobility in pure samples of these compounds will be very high, as will the mobility ratio b.

In InSb, where this phenomenon is most pronounced, electron mobility values in excess of 50,000 cm^2/V-sec and mobility ratios of nearly 100 are observed at room temperature. Unfortunately, due to a relatively high probability of radiative recombination and to the difficulty of producing crystals of such substances which are free from trapping impurities and structural imperfections, excess carrier lifetimes in these materials are very short (10^{-7} sec or less) and excess carrier effects can be observed only with difficulty. The energy gaps associated with the various III–V intermetallic compounds increase as the atomic masses of the constituent elements decrease, in the same way in which the energy gaps of the group IV semiconductors increase in going from α-Sn through Ge and Si to diamond. Thus InSb, like α-Sn, has quite a small energy gap, while InAs, GaSb, GaAs, GaP, and AlP have successively larger gap widths. In general, any III–V compound will somewhat resemble the group IV semiconductor located in the periodic table halfway along a line joining the two constituents of the III–V compound. The characteristic properties of the III–V compounds have been discussed by Seraphin[1] in terms of an ingenious one-dimensional model (similar to that used in the Kronig-Penney calculation) which permits an exact solution of Schrodinger's equation. In s ite of the disadvantages associated with very low carrier lifetime and with the difficulties involved in the preparation of high-purity single crystals, the III–V semiconducting compounds have been used in a wide range of technical applications, especially where high electron mobility is advantageous.

In addition to these substances, silicon carbide (SiC), certain II–VI compounds such as CdS, CdSe, CdTe, ZnS, PbS, PbSe, PbTe, certain metallic oxides and even a number of organic crystalline substances, such as anthracene, are well-known semiconductors. With the exception of silicon carbide, and, perhaps, cadmium sulfide, however, the technological importance of these materials has been quite small, and we shall therefore restrict our discussions primarily to the covalent group IV elements and to the III–V intermetallic compounds.

The starting material for the preparation of high-purity germanium is chemically pure germanium dioxide, which is obtained as a byproduct in zinc smelting and refining. This material, which is usually supplied in the form of a finely divided white powder, is placed in high-purity graphite crucibles and reduced to germanium by heating to 600–900°C in a stream of hydrogen. The reaction proceeds according to the equation

$$GeO_2 + 2H_2 \rightarrow Ge + 2H_2O.$$

The germanium is obtained, after several hours, as a finely divided black powder which is melted down into a long narrow bar of polycrystalline germanium by raising the temperature above the melting point, which is 936°C. A mold or channel is generally provided for this purpose in the bottom of the reduction crucible. The resulting ingot has a silvery, metallic appearance which is characteristic of germanium. The reduced ingot is then further purified by *zone refining*. In this process, the reduced bar is placed in a long narrow "boat" made from quartz or graphite and pulled through a tube furnace in which several narrow hot zones are created, so as to melt a narrow portion of the ingot. When quartz boats are used, these hot zones may be created by graphite rings, heated by radiofrequency induction or by direct current; with graphite boats a part of the boat itself may act as a susceptor when passed through a narrow

---

[1] B. Seraphin, *Z. Naturforsch.* **9a**, 5 (1954).

rf coil. In any case the effect is to melt a portion of the ingot, and to pass the molten zone continuously along the bar from one end to the other as the ingot is pulled through the hot zone. A hydrogen atmosphere is used to prevent oxidation of the germanium at high temperatures. A diagram of this process is shown in Figure 11.1.

In the thermal equilibrium state, according to Nernst's law of distribution,[2] a dilute solute distributes itself between two solvent phases such that the ratio of the solute concentration in the two solvent phases is a constant, independent of the solute

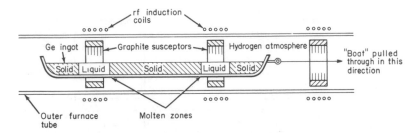

FIGURE 11.1. Schematic diagram of zone refining process for purifying semiconductor materials.

concentrations. If both solvent phases are considered for the moment to be in equilibrium with an amount of the solute in excess of the amount which can be dissolved in equilibrium, and if ideal solutions are formed in both phases, it is clear that the constant ratio of concentrations must be just the ratio of the equilibrium solubilities of the solute in the two solvent phases. In the case of the zone refining process a state of equilibrium may be considered to exist, approximately,[3] between solid and liquid germanium, which may be regarded as two solvent phases, and the impurities in the system which act (hopefully) as ideal solutes. Each solute will then distribute itself between the solid and liquid phases in the ratio of its solubility in the solid to its solubility in the liquid. This ratio is commonly referred to as the *equilibrium K-value* for that specific impurity. Since, for the most part, the impurity elements are much more soluble in the liquid phase than in the solid phase, the equilibrium *K*-values for most impurities are much less than unity. Under these circumstances, as the liquid zone is passed slowly through the solid bar, most of the impurity atoms in the melt are rejected by the solid at the interface where the molten region freezes to form a solid bar once more. When a single molten zone is passed through a long ingot, then, the impurity atoms tend to collect in the liquid and are finally precipitated out only at the very end of the ingot when the last part of the molten zone finally solidifies. This segregation process may be further enhanced by repeating the process and ultimately passing many molten zones through the ingot. In the limit of an infinite number of zone passes, a limiting concentration profile is achieved, and purification beyond this stage cannot be accomplished directly. If further impurity segregation is desired, however, the ingot can be taken out of the furnace, the end portion (which now

---

[2] See, for example, G. N. Lewis and M. Randall, *Thermodynamics*. New York: McGraw-Hill, 1923, p. 234.
[3] Strictly speaking, of course, equilibrium is maintained only in a system at constant temperature, in which the solid-liquid interface would be stationary. The above description thus breaks down when interface velocities and temperature gradients become too large.

contains most of the impurities in the sample) *removed*, and the remaining section subjected to further zone refining. In any case, at the completion of the zone refining process the end section of the ingot must be rejected.

While for very slow interface velocities the ratio of the impurity concentration in the solid to that in the liquid is the ratio of *equilibrium* solubilities, for greater interface velocities impurity atoms which are rejected from the freezing solid pile up near the interface more rapidly than they can be removed by diffusion to the interior of the molten zone, as illustrated in Figure 11.2. The actual impurity concentration c_l near

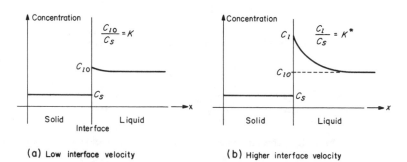

(a) Low interface velocity   (b) Higher interface velocity

FIGURE 11.2. Impurity concentration in solid and liquid phases near a solid-liquid interface in a zone refining apparatus (a) for low interface velocity (b) for higher interface velocity.

the interface is then greater than the equilibrium value c_{l0}. The effect of this is that more impurity atoms are incorporated into the solid than in the low velocity case where constant concentration throughout the liquid zone can be maintained by diffusion. The effect is the same as if the equilibrium K-value were replaced by an *effective* value K^*, where

$$K^* = K \cdot \frac{c_l}{c_{l0}}. \qquad (11.1\text{-}1)$$

Since the ratio c_l/c_{l0} will be affected by the interface velocity and the diffusion coefficient of the impurity atoms in the liquid, K^* must also depend upon these factors. It is clear, however, that if $K < 1$, then K^* must be greater than K. Ordinarily it is advantageous to work under conditions such that K^* may be appreciably different from K, and under such circumstances it is usually necessary to determine K^* by experiment.

It is easily shown[4] that the impurity distribution in an initially uniform ingot after the passage of a single zone of length l is

$$c_s(x) = c_{s0}[1 - (1 - K^*)e^{-K^*x/l}] \qquad (11.1\text{-}2)$$

where c_{s0} is the initial impurity density and x is the distance coordinate along the bar. It can also be shown that the limiting distribution which is approached after a large

[4] W. G. Pfann, *J. Metals*, **4**, 747 (1952).

number of passes is

$$c_s(x) = Ae^{Bx} \qquad (11.1\text{-}3)$$

where

$$A = \frac{c_{s0}BL}{e^{BL} - 1} \qquad (11.1\text{-}4)$$

and

$$K^* = \frac{Bl}{e^{BL} - 1} \qquad (11.1\text{-}5)$$

Here L is the total length of the ingot. For further details on the zone refining process, the reader is referred to Pfann's article[4] or to a recent monograph by Hannay.[5] It is obvious that the process of zone refining may be applied not only to germanium but to a wide range of other materials as well. A list of equilibrium K-values for commonly encountered impurities in germanium and silicon is given in Table 11.1. It should be

TABLE 11.1
Segregation Coefficients for Ge and Si[6,7]

Impurity	K (Ge)	K (Si)
P	0.12	0.04
As	0.04	0.07
Sb	0.003	0.002
B	20.	0.68
Al	0.1	0.0016
Ga	0.1	0.004
In	0.001	0.0003

noted that boron is the only common impurity whose equilibrium K-value in germanium is greater than unity, i.e., which is rejected by the liquid to the solid. This peculiarity results in the segregation of boron in the head end of the ingot rather than the tail.

The preparation of high-purity silicon generally begins with the reduction of silicon tetrachloride ($SiCl_4$) or trichlorosilane ($SiHCl_3$) by hydrogen at about 1100°C. The reaction goes according to the equation

$$SiHCl_3 + H_2 \rightarrow Si + 3HCl.$$

All the reactants are in the gaseous phase at the temperature at which the process is carried out, with the exception of the silicon, which precipitates out upon a suitable substrate. This substrate may be a sample of high-purity single crystal silicon, in

---

[5] N. B. Hannay, *Proceedings of the International School of Physics, "Enrico Fermi."* Course 22 (Semiconductors). New York: Academic Press, 1963, pp. 341–435.

[6] R. N. Hall, *J. Chem. Phys.*, **57**, 836 (1953).

[7] J. A. Burton, *Physica*, **20**, 845 (1954).

which case the underlying lattice structure of the substrate is continued in the precipitated layer, the growth process then being referred to as *epitaxial*. Alternatively the
substrate may be a thin polycrystalline silicon rod which yields a cylindrical ingot
suitable directly for zone refining.

The zone refining of silicon is similar in principle to the process used for germanium, except for the complication that molten silicon wets, reacts with, and is con-
taminated by practically every known substance which can conceivably be used as a
crucible. It is therefore necessary to use a crucibleless technique, the one in most common use being termed the *floating zone* process. In this process a cylindrical silicon
rod is clamped vertically at both ends, and is heated by an rf induction coil coupled
directly to the silicon itself. A narrow region of the ingot can be melted, and the
surface tension of the molten silicon is sufficient to prevent the molten zone from
collapsing and flowing away. The zone may now be passed along the ingot to segregate
impurities toward one end, and the process may be repeated as many times as is
necessary to produce the degree of purification required. The process is illustrated in
Figure 11.3. It is possible to reduce the content of electrically active impurities in both

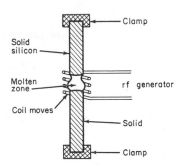

FIGURE 11.3. Schematic representation of the floating zone
process for purifying silicon.

germanium and silicon to as little as one part in 10^{11}. This is sufficient to produce
germanium which is intrinsic at room temperature but not quite good enough to
obtain intrinsic silicon (another factor of 10-100 would be necessary to accomplish
this). Zone-refined silicon and germanium is of sufficient purity to serve as the starting
material for a wide variety of semiconductor devices.

The preparation of high-purity specimens of the III–V intermetallic compounds is
usually achieved by melting together stoichiometric quantities of highly purified
elemental constituents in sealed quartz tubes so as to prevent loss of volatile constituents by vaporization. The resulting ingots may often be purified significantly by
zone refining, but again the sample must be confined within a sealed tube. In general,
efforts to purify the III–V compounds by zone refining or by other techniques have
not been as successful as they are with the group IV materials.

11·2 THE GROWTH OF SINGLE CRYSTAL SAMPLES

In general, the material produced by the zone refining process is polycrystalline,
despite the fact that it is very pure. What is usually desired for experimental investigations or technological purposes are single crystal semiconductor samples which may

have to be intentionally "doped" with some specific impurity element so as to produce material with a given conductivity type and electrical resistivity. There are several ways of growing single crystals, each of which has certain advantages for specific applications.

Single crystals of germanium are often grown by a process in which one end of a zone refined ingot is placed in contact with a single crystal seed in a suitable boat. A molten zone is then formed between the seed crystal and the polycrystalline ingot and slowly passed down the ingot so that the regrowth starts at the seed and continues its crystal structure throughout the ingot. By choosing a seed of proper orientation, any desired crystal orientation within the final ingot may be obtained, and by incorporating small amounts of impurity elements in the molten zone, p- or n-type crystals of any desired impurity content may be made. Due allowance must be made, of course, for impurity segregation at the growth interface when calculating the amount of doping impurity required. Since the necessary quantities of impurity elements which must be added are often microscopically small, it is customary to produce a doping alloy of pure germanium to which perhaps 0.1 or 0.01 per cent of the desired doping impurity has been added, and to add this doping alloy to the melt, rather than attempting to dope the melt directly. This process is capable of producing large crystals of germanium of very high structural perfection, and can also be used with certain of the intermetallic compounds, notably InSb, but cannot be used to produce silicon single crystals since silicon wets or reacts with all materials which can be used to make the boat. High quality silicon single crystals can, however, be grown using a corresponding modification of the floating-zone process.

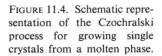

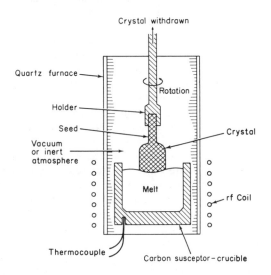

FIGURE 11.4. Schematic representation of the Czochralski process for growing single crystals from a molten phase.

Another very widely used method of growing single crystals is the so-called *Czochralski technique*, in which a single crystal seed of the desired orientation is dipped into a melt prepared from zone refined material, suitably doped. The heat input to the melt is then reduced and the seed crystal is slowly withdrawn, while being rotated at the same time, as shown in Figure 11.4. The diameter of the resulting crystal may be controlled by regulating the rate of withdrawal and the power input to the melt. A graphite crucible is satisfactory for growing germanium crystals, but for silicon a quartz liner is required. Although molten silicon wets and reacts with quartz the liner is attacked only rather slowly and the crystal growth is not seriously impeded. Some oxygen from the quartz liner may be introduced into silicon crystals grown by this technique, but it is not an electrically active impurity and does not seriously affect the quality of the crystal for most experimental or technical purposes. The Czochralski technique can also be used to grow single crystals of the III–V intermetallic compounds, but for these substances a closed system must ordinarily be used to minimize loss of volatile components. The withdrawal of the crystal and the rotation are then often accomplished by magnetic forces acting through the quartz walls of the system upon ferromagnetic armatures inside.

The possibility of growing silicon single crystal regions epitaxially has already been mentioned in connection with the preparation of that element by the vapor phase reduction of the chlorides. This process is especially valuable for growing plane-parallel regions of opposite conductivity type or of abruptly changing impurity density because the doping impurities can be incorporated (for example, as gaseous PCl_3 or BCl_3) in the entering gas stream. Processes have also been developed to adapt this technique for use with germanium.

In addition, controlled *dendritic* crystal growth has been used to produce crystals which are suitable for use in semiconductor device technology. Dendritic growth occurs when a seed crystal is inserted into a highly supercooled melt. Under these conditions a rapid spearlike growth is initiated, the velocity of which is determined by the rate at which the latent heat of fusion at the freezing interface can be conducted away. If the growing crystal is pulled rapidly upward out of the melt a continuous dendritic ribbon of semiconductor can be formed. Growth velocities of the order of 5 cm/sec are typical (compared to velocities of 10^{-3} cm/sec for the Czochralski process, in which the melt is not supercooled). The width and thickness of the dendritic crystal can, within certain limits, be controlled by varying the supercooling of the melt and the velocity with which the ribbon is withdrawn from the melt. It is quite possible, for the case of germanium, to produce dendritic ribbons between 1 and 5 millimeters wide and between 0.1 and 0.5 millimeters thick. The length is limited only by the arrangements which can be made for withdrawing the dendrite. The surfaces of dendritic crystals are optically flat and if certain precautions in growth are taken the crystals may be essentially perfect structurally. The most obvious advantage of the dendritic process is that it yields crystals which may be used directly as starting material for semiconductor devices, without the cutting, lapping, sizing and etching operations which are otherwise required. The thickness of dendritically grown ribbons can be controlled so as to be very uniform, which is of great importance in many device fabrication processes. Very good dendritic crystals of germanium, silicon and InSb are easily produced; dendrites of other intermetallic III–V compounds have also been grown, but with greater difficulty and with rather poorer results.

11·3 MEASUREMENT OF BULK RESISTIVITY

The measurement of resistivity is the most fundamental of the routine measurements which are customarily made on semiconductor crystals. Basically, any such measurement is made by passing an accurately known current through the sample and measuring a voltage drop between two points a known distance apart. A few of the specific techniques, however, are of sufficient importance to warrant extended discussion.

One quite accurate way of measuring the resistivity of a reasonably homogeneous semiconductor specimen is to make a *potential traverse* of the sample, by making a series of potentiometric measurements of the voltage drop between a fixed contact and a movable probe, as the movable probe is moved along the sample by small increments, usually of 0.5 mm or 1.0 mm. A diagram of this measurement is shown in Figure 11.5. If the distance between the probes is increased from x to $x + \Delta x$, the

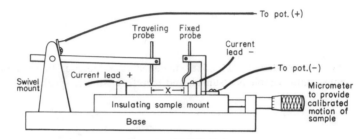

FIGURE 11.5. Diagram of apparatus for determining bulk resistivity by the potential traverse method.

resistance ΔR of the small element of the sample lying between x and $x + \Delta x$ is added to the resistance R of the portion between 0 and x. Then, if the current flowing through the sample is I_0 and if a change ΔV is noted in the potential between the fixed and traveling probe accompanying the change in probe spacing Δx, by the definition of resistivity,

$$\Delta R = \frac{\Delta V}{I_0} = \frac{\rho \Delta x}{A}$$
(11.3-1)

where A is the (constant) cross-sectional area of the sample. Solving for ρ, it is clear that

$$\rho = \frac{A \Delta V}{I_0 \Delta x}$$
(11.3-2)

represents the resistivity of the material lying between x and $x + \Delta x$. By making a whole series of measurements involving constant increments of probe spacing, it is possible to arrive at an experimental determination of the resistivity as a function of distance along the sample. Such a plot is often termed a *resistivity profile* of the sample. It should be noted that since a potentiometer is used to measure the potential between the fixed and moving probe, no current at all flows through these electrodes.

Another very simple but less accurate way of measuring resistivity, which is often used for control of material specifications in device fabrication processes, is by use of the four-probe technique. In this method four equidistant probes are set down upon the sample, as shown in Figure 11.6. The outermost pair of probes are used to pass a steady current through the sample, and the innermost pair are used as potential probes between which a voltage drop can be measured potentiometrically. Using the standard methods of electrostatics and steady current theory one can find the resulting potential distribution and from this solution show that the potential difference between the two inner probes must be given by

$$V = \frac{I}{2\pi\sigma a}$$ (11.3-3)

where I is the steady current flowing through the outer probes and a is the spacing between adjacent probes. The resistivity of the sample is accordingly

$$\rho = \frac{1}{\sigma} = \frac{2\pi a V}{I}.$$ (11.3-4)

The details of deriving (11.3-3) are assigned as an exercise. In obtaining this equation a semi-infinite sample is assumed, and therefore, strictly speaking, (11.3-3) is not valid

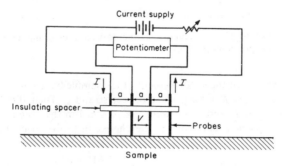

FIGURE 11.6. Schematic diagram of the four-probe resistivity measuring apparatus.

for samples of finite extent. However, if all dimensions of the sample are large compared to the probe spacing a, the error involved in using (11.3-3) or (11.3-4) will be quite small. In both the above methods the conditions of the experiment must be such that effects arising from the injection of minority carriers at the current contacts are avoided. This is most easily arranged by lapping or otherwise abrading the sample surfaces so that injected excess carriers quickly recombine near the current contact, and by keeping the steady current low enough so as to avoid any significant modulation of conductivity by the injected carriers.

    In making resistivity measurements at low temperatures, it is necessary to confine the sample within a vacuum cryostat in order to prevent the condensation of ambient atmospheric moisture upon the sample surfaces. Under these circumstances the resistivity is usually measured by potentiometrically measuring the voltage drop between two *fixed* contacts a known distance apart when a steady current is passed along the sample. The resistivity then follows from (11.3-2) with ΔV the measured voltage drop and Δx the distance between the contacts. All of the methods discussed herein apply primarily to samples of reasonable homogeneity; for specimens in which large resistivity fluctuations over small distances are encountered special techniques must be used.

11·4 MEASUREMENT OF IMPURITY CONTENT AND MOBILITY BY THE HALL EFFECT

The theory of the Hall effect has been treated in Chapters 7 and 9, and only a brief discussion of experimental techniques will be considered here. Large area current contacts are made to the ends of the sample, while tiny probes or soldered contacts are applied to the sides of the sample, midway along the current path, for potential leads. Ultrasonic soldering techniques are often very useful for making these contacts as well as contacts which are used for other experimental purposes. The Hall voltage must be measured potentiometrically, so as to fulfill the condition that the y-component of current be zero. The sample surfaces are usually abraded and minimal values of sample current are generally used so as to avoid as far as possible injection of excess carriers into the sample.

If the potential contacts are not aligned perfectly so as to coincide with opposite ends of the same equipotential line in the absence of the static magnetic field, an ohmic voltage drop proportional to the steady sample current will be measured in addition to the Hall voltage. This situation is illustrated in Figure 11.7. The effect of this

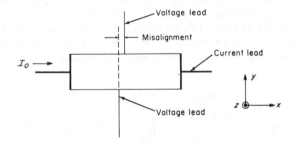

FIGURE 11.7. Geometry of samples used for Hall effect measurements, showing origin of ohmic voltage drops due to contact misalignment.

spurious ohmic drop can be eliminated by making four separate measurements; first with both magnetic field and steady current in the positive direction, then with magnetic field positive and steady current reversed, then with both magnetic field and current reversed, and finally with field reversed and current positive. The Hall voltage is then obtained as

$$V_H = \tfrac{1}{4}[V(H_+,I_+) + V(H_-,I_-) - V(H_+,I_-) - V(H_-,I_+)], \qquad (11.4\text{-}1)$$

where the quantities in parentheses describe the condition of current and field pertaining to the potential measurement. It is evident that the ohmic drop will be eliminated by this procedure, since the Hall voltage changes sign if *either* current or field is reversed, while the ohmic drop changes sign only when the current is reversed. Since the analysis of Hall measurements is quite difficult for intrinsic or near-intrinsic samples, it is often advantageous when dealing with samples which are nonextrinsic at room temperature to conduct the measurements at lower temperatures at which the specimens are definitely extrinsic. This is especially true when the primary objective is to ascertain the net density of impurity atoms in the samples. Under these circumstances the use of a vacuum cryostat is required. It is also necessary to use values of static magnetic field sufficiently low that $\omega_0\tau \ll 1$, where ω_0 is the cyclotron frequency eB_0/m^*c, in order that the analysis of Section 9.8 be valid. The carrier density in the sample may be calculated from the Hall voltage according to methods which have already been developed and described in Sections 7.6 and 9.8. The Hall mobility may be obtained by combining the results of Hall and resistivity measurements; this procedure has been outlined in Section 9.8.

11·5 MEASUREMENT OF EXCESS CARRIER LIFETIME

The measurement of excess carrier lifetime is usually accomplished by the observation of the decay of transient photoconductivity. The theory of this measurement has been discussed in great detail in Section 10.7, and nothing need be added here save a brief description of the apparatus.

A high-pressure xenon arc tube (such as the Edgerton, Germeshausen, and Grier XP-12) may be used to excite photoconductivity in the sample. This tube produces a very intense and abrupt flash of light lasting about 10^{-6} sec. The tube may be used in conjunction with an optical system to concentrate as much light upon the sample as possible. One very good arrangement is to mount the flash tube at one focus of an

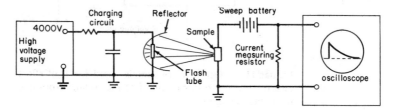

FIGURE 11.8. Schematic representation of apparatus for the determination of excess carrier lifetime by transient photoconductivity measurements.

ellipsoidal reflector and the sample at the other focus. The flash tube is similar to those which are widely used as photographic flash sources. It may be operated as a free-running relaxation oscillator in a circuit such as that shown in Figure 11.8, to obtain repetitive flashes. The photoconductivity decay is observed upon a wide-band oscilloscope, the lifetimes being obtained from the observed decay curves by the methods discussed in Section 10.7. This method is restricted to lifetimes longer than 10^{-6} seconds, since the flashtube illumination does not cut off abruptly, but dies off with a time constant of this order of magnitude. Special spark sources have been developed which can be used to resolve somewhat shorter lifetimes.

11·6 DISLOCATIONS AND OTHER IMPERFECTIONS

Dislocations are lattice defects in crystals which can be understood in terms of partial internal slip. There are two basic types of dislocation structures, *edge-type* dislocations and *screw-type* dislocations. An edge type dislocation is shown in Figure 11.9. This

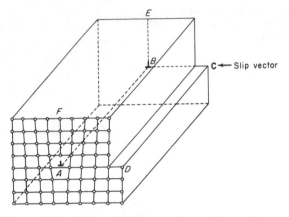

FIGURE 11.9.   Schematic diagram of a crystal containing an edge-type dislocation. The dislocation can be regarded as having arisen by slip amounting to one lattice spacing along the direction *BC*. The crystal may be thought to contain an "extra plane" of atoms *ABEF*.

type of dislocation can be considered to arise as a result of partial internal slip in the *ABCD* plane along the direction *BC*. In the illustration the magnitude of the slip is one interatomic distance. The resulting crystal may be looked upon as containing an *extra* half-plane of atoms (*ABEF* in the diagram), and is clearly in a state of mechanical stress in the region of the dislocation edge *AB*, such that the material is in tension above the dislocation edge and in compression below.

A screw-type dislocation is pictured schematically in Figure 11.10. In this dislocation structure the slip has taken place *parallel* to the line of the dislocation, which is normal to the top surface of the crystal, rather than perpendicular to the line of the dislocation, as in the case of the edge-type dislocation. Actually the edge and screw dislocations may be regarded as different aspects of the same internal slip phenomenon, as illustrated in Figure 11.11, which is a view of the crystal shown in Figure 11.9 with a differently shaped internal slip region. Here the same dislocation line (always denoted by the symbol ⊥) is normal to the slip vector at the front face of the crystal, as in an edge dislocation, and parallel to the slip vector at the side, as in a screw dislocation. It is perhaps more accurate to say that the dislocation has the *edge orientation* at the front and the *screw orientation* at the side, although we shall not always adhere rigorously to this terminology.

Dislocations are formed in the initial stages of crystal growth and are responsible for many of the observed properties of crystals. As a matter of fact, most of the observed characteristics of crystal growth from a liquid melt, from a supersaturated

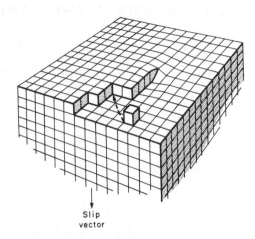

FIGURE 11.10. Schematic representation of a screw-type dislocation.

Slip vector

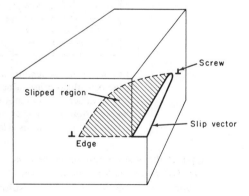

FIGURE 11.11. A crystal containing a region of slip resulting in a dislocation which is partly in the screw orientation and partly in the edge orientation.

Screw

Slipped region

Slip vector

Edge

solution, or from the vapor phase, are governed by dislocations in the growing crystals rather than by any property of the perfect crystal lattice itself. The reason for this is that atoms of the liquid or vapor phases are bound to the growing crystal face more strongly at a "step" where an incomplete layer of atoms ends than at an isolated site on the growth face. This is true because at the step an incoming atom can be bonded to *two* atoms of the crystal lattice rather than only one. The situation is illustrated in Figure 11.10. Growth will therefore proceed much more rapidly (or equally rapidly at a much lower level of supersaturation or undercooling in the liquid or solid phase) at a step face than on a complete and structurally perfect plane surface. The formation of a step upon a structurally perfect crystal by the coalescence of several isolated atoms which are randomly deposited from the liquid or vapor phase, however, is an event of very low probability, because such atoms are not very strongly bound to the crystal face and do not usually stick to the crystal long enough to join up to form a growth step; they are instead usually shaken loose by thermal agitation before this can occur. When a step is formed in this manner, it will then very rapidly spread across the crystal face by the addition of atoms at the edges, where they are strongly bound, until a new atomic layer has been added. Further growth then proceeds only with great difficulty, since now a new step must be formed on the added layer.

The situation is entirely different, however, if the crystal should happen to contain a screw dislocation, as in Figure 11.10. The screw dislocation provides a *built-in* step which can never be removed from the growth face by further addition of atoms. When atoms are added at a constant rate to the step associated with a screw dislocation such as that illustrated in the figure, all parts of the step obtain the same constant linear growth velocity. This results in an *angular* growth velocity about the center which is greater near the dislocation than far from it, and the step tends to assume a spiral form, winding around the central dislocation. Such spiral steplike configurations are frequently observed upon growing crystals, and provide direct evidence that this is indeed an important growth mechanism.[8] In effect, only those tiny crystal nuclei, formed in a slightly undercooled melt or supersaturated solution, which contain dislocations, are *capable* of growing to macroscopic size before being broken up by thermal agitation. The growth processes which we are capable of observing in ordinary crystal samples are then processes which are associated with the presence of screw dislocations.

When stresses are applied to crystals containing dislocations, it is possible to move the dislocations within the crystal. For example, if a shear stress arising from a force directed toward the left applied to the top of the sample and a force toward the right at the bottom is brought to bear on the crystal of Figure 11.9 the dislocation line *AB* may be moved to the left through the crystal. Although a complete discussion of the mechanics of dislocations is beyond the scope of the present treatment, most of the mechanical properties of macroscopic crystals, including elastic moduli, slip and plastic deformation, hardening by alloying and heat treatment, annealing properties, and work hardening, can be explained in terms of the motion of dislocations and the interaction of dislocations with one another and with impurity atoms.[9,10]

The formation of grain boundaries and the regions of structurally imperfect material associated with grain boundaries may also be discussed in terms of dislocations. In particular, the misfit between crystal planes which occurs at a small-angle grain boundary may be resolved by the formation of a row of edge-type dislocations at such a boundary,[11] as illustrated in Figure 11.12. This type of crystal boundary is sometimes called a Burgers lineage boundary.

The presence of edge-type dislocations in semiconductor crystals is easily detected by chemically etching the crystal surface with certain reagents. Since the crystal lattice around the dislocation is already in a state of strain, it takes somewhat less energy to

[8] Of course, in order to be observable even by electron microscopy, the slip vector of the screw dislocation associated with these "growth spirals" must extend over many times the unit cell dimension.

[9] W. T. Read, *Dislocations in Crystals.* New York: McGraw-Hill, 1953.

[10] A. H. Cottrell, *Dislocations and Plastic Flow in Crystals.* New York: Oxford, 1953.

[11] J. M. Burgers, *Proc. Phys. Soc. (London),* **52**, 23 (1940).

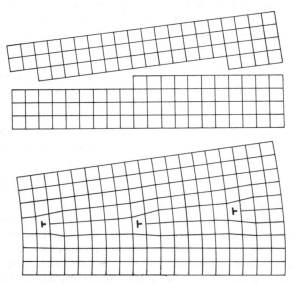

FIGURE 11.12. The resolution of atomic misfit at a small angle grain boundary by the formation of a row of edge-type dislocations.

break up the lattice there than it does elsewhere. As a result of this, the etch attacks the crystal surface more rapidly near the dislocations, forming etch pits, clearly visible under the microscope, which mark the intersection of edge dislocations with the crystal surface. A photograph of etch pits at the dislocations associated with a small angle grain boundary in germanium is shown in Figure 11.13. Such photographs provide striking experimental verification of the picture of a small-angle boundary illustrated in Figure 11.12, since the angle of misfit between the two crystal sections can be calculated from the spacing between etch pits and can be determined experimentally from X-ray diffraction measurements. The two values thus obtained are invariably in excellent agreement.[12] A few etch pits arising from the presence of random dislocations can also be seen in Figure 11.13. In silicon crystals it is possible to view dislocation lines directly by diffusing copper into the sample and then examining the interior of the crystal microscopically using infrared light of wavelength just beyond the absorption edge.[13] An image converter tube is used to make the infrared image visible to the eye. The copper, which is opaque to infrared radiation, precipitates out along the dislocation lines, forming dark traces marking the course of those lines.

The presence of dislocations in semiconductor crystals also gives rise to certain electrical effects. In germanium, for example, edge-type dislocations appear to introduce very deep-lying acceptor levels, located about 0.2 eV below the conduction band

[12] F. L. Vogel et al., Phys. Rev., **90**, 489 (1953).
[13] W. C. Dash, J. Appl. Phys., **27**, 1193 (1956).

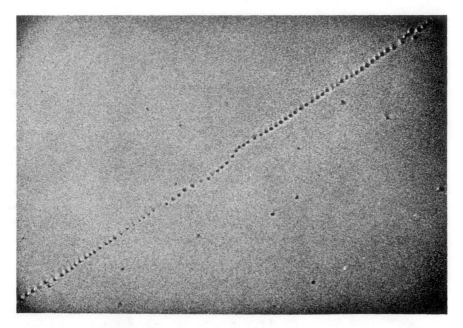

FIGURE 11.13.   Edge-type dislocations at a small angle grain boundary in a germanium crystal as revealed by chemical etching. Magnification about $600\times$. [Photograph by the author.]

edge in energy.[14,15] Furthermore, several independent investigations have shown that edge-type dislocations act as recombination centers in germanium, and, in fact, that the excess carrier lifetime in high-purity germanium is usually dislocation limited.[16,17] The situation in silicon and other semiconductors has been less thoroughly investigated on account of technical difficulties, but is presumably similar.

FIGURE 11.14. (a) Frenkel and (b) Schottky defect in a crystal lattice.

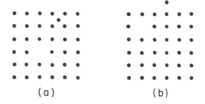

(a)   (b)

Another important example of a structural imperfection is the lattice vacancy. There are two basic types of lattice vacancies, called Frenkel defects and Schottky defects, as illustrated in Figure 11.14. In the formation of a Frenkel defect, an atom of the crystal lattice moves to an interstitial position, leaving a lattice vacancy behind, while in the formation of a Schottky defect an atom in the region near the surface is

[14] G. L. Pearson, W. T. Read, and F. J. Morin, *Phys. Rev.*, **93**, 666 (1954).
[15] W. T. Read, *Phil. Mag.*, **45**, 775 (1954).
[16] J. P. McKelvey, *Phys. Rev.*, **106**, 910 (1957).
[17] G. K. Wertheim and G. L. Pearson, *Phys. Rev.*, **107**, 694 (1957).

removed to a surface site leaving a mobile vacancy behind, which may then diffuse into the interior of the crystal. There is an activation energy associated with the formation of any such defect, and since the thermal energy available for surmounting the activation barrier is greater at high temperature, the equilibrium concentration of vacancies increases with temperature. Defects of the Schottky type are believed to predominate in germanium and silicon. If the crystal is cooled very suddenly from a high temperature a concentration of vacancies far above the equilibrium concentration at ambient temperatures may be "frozen in." The role of lattice vacancies in the diffusion of substitutional impurity atoms from the surface into the interior of a crystal lattice is an important one. The presence of lattice vacancies allows the atoms of the lattice (as well as impurity atoms) to migrate from one lattice site in the crystal to another, and thus permits diffusive transport of surface impurity atoms into the interior. This process is illustrated in Figure 11.15. Most of the important features of the diffusion of substitutional impurities in semiconductor crystals, including the magnitude of the diffusion coefficients and their temperature dependence are consistent with the vacancy mechanism.

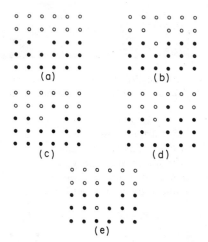

FIGURE 11.15. Interdiffusion at an interface between two different substances resulting from the motion of a lattice vacancy. Successive stages in the diffusion process are shown from (a) through (e).

EXERCISES

1. Derive Equations (11.1-2) through (11.1-5) from fundamental considerations.

2. It is desired to grow a single crystal of p-type germanium of resistivity 2 to 5 Ω-cm by the Czochralski technique. As starting material, 500 g of intrinsic zone refined germanium will be used. How much of a doping alloy consisting of germanium containing 0.1 atomic per cent gallium must be added to the melt so that the room-temperature resistivity of the

part of the crystal which is grown initially will be just 5 Ω-cm? You may assume that the segregation coefficient which is appropriate is the equilibrium segregation coefficient as given by Table 11.1. How will the resistivity of the crystal so obtained vary along its length?

3. Using electrostatics and steady-current theory, and assuming a semi-infinite homogeneous and isotropic sample, derive equation (11.3-3) for the four-probe resistivity apparatus. Show explicitly that your solution satisfies all the necessary boundary conditions. *Hint:* Begin by considering the electrostatic field created by a positive point charge and a negative point charge in free space.

4. A Burgers lineage boundary in a simple cubic crystal creates a line of etch pits spaced 2×10^{-4} cm apart when the crystal is etched. Assuming that the lattice constant of the crystal is 4×10^{-8} cm, find the angle of misfit between the sections of the crystal on opposite sides of the lineage boundary.

5. A metal point probe makes contact with a semi-infinite semiconductor crystal of constant conductivity σ. A known constant current I_0 passes through the probe into the semiconductor. Assuming that the interface between the probe point and the semiconductor is a hemisphere of radius a extending into the semiconductor, its center lying in the plane of the semiconductor surface, and that there are no excess carriers injected by the probe, calculate the impedance (spreading resistance) of the contact.

GENERAL REFERENCES

W. Bardsley, "The Electrical Effects of Dislocations in Semiconductors," in *Progress in Semiconductors*, Vol. 4, Heywood and Co., London (1960), p. 155.

A. H. Cottrell, *Dislocations and Plastic Flow in Crystals*, Oxford University Press, New York (1953).

I. G. Cressel and J. A. Powell, "The Production of High Quality Germanium Single Crystals," in *Progress in Semiconductors*, Vol. 2, Heywood and Co., London (1957), p. 137.

P. Haasen and A. Seeger, "Plastische Verformung von Halbleitern und ihr Einfluss auf die Elektrische Eigenschaften," in *Halbleiterprobleme*, Vol. 4, Friedrich Vieweg and Sohn, Braunschweig (1958), p. 68.

N. B. Hannay, "Semiconductor Chemistry," in *Proceedings of the International School of Physics, "Enrico Fermi."* Course 22 (Semiconductors), Academic Press, Inc., New York (1963), p. 341.

W. G. Pfann, "Techniques of Zone Melting and Crystal Growing," in *Solid State Physics, Advances in Research and Applications*, Vol. 4, Academic Press, Inc., New York (1957), p. 423.

W. G. Pfann, *Zone Melting*, John Wiley & Sons, New York (1958).

W. T. Read, *Dislocations in Crystals*, McGraw-Hill Book Co., Inc., New York (1953).

W. Shockley, J. Holloman, R. Maurer, and F. Seitz (editors), *Imperfections in Nearly Perfect Crystals*, John Wiley & Sons, New York (1952).

R. K. Willardson and H. L. Goering (editors), *Compound Semiconductors*, Vol. 1, Reinhold Publishing Corporation, New York (1962).

CHAPTER 12

THEORY OF SEMICONDUCTOR *p-n* JUNCTIONS

12·1 THE *p-n* JUNCTION

In the preceding chapters, we have dealt exclusively with semiconductors which were assumed to be uniform or homogeneous with regard to the density of impurity atoms contained therein. We shall now, however, turn our attention to the consideration of substances wherein the impurity density may vary from point to point, and in particular to samples which contain an extrinsic *n*-type region and an extrinsic *p*-type region separated by a relatively narrow transition zone. This narrow transition region is called a *p-n* junction, and associated with it are a number of rather remarkable physical properties, which, although of definite interest from a purely fundamental viewpoint, are otherwise of great importance since they form the basis of operation of most semiconductor electronic devices.

Let us then consider a semiconductor crystal containing a *p-n* junction. The transition between the *p*- and *n*-type regions may be *abrupt*, in which case a region containing a more or less constant net concentration of donor impurities adjoins another region in which a more or less constant net concentration of acceptors is found. On the other hand, the junction may be a *graded* one, in which case N_d and N_a are functions of distance along the normal to the junction, N_d gradually decreasing from a large value and N_a gradually increasing from a small one as the junction is approached from the *n*-type side; the two quantities become equal at the junction, and N_a exceeds N_d on the far side of the junction in the *p*-region. These two cases are illustrated in Figure 12.1. We shall for the most part confine our attention to abrupt junctions, since they are much simpler to analyze mathematically, since they illustrate most features of importance about the behavior of *p-n* junctions in general, and since for a large and important class of devices the abrupt junction approximation is a good one. It should be noted, however, that in many cases the *p-n* junctions in semiconductor device structures (particularly those which are produced by solid state diffusion of impurity atoms which produce *n*- or *p*-type conductivity into a substrate of opposite conductivity type) are graded junctions. We shall try to point out, qualitatively, at least, the differences between the physical behavior of abrupt and graded junctions, wherever the occasion to do so arises.

Suppose that an abrupt *p-n* junction is somehow formed instantaneously by joining a uniform *p*-type sample to a uniform *n*-type sample[1] to form a single crystal.

---

[1] The formation of a junction in this manner is merely a thought experiment and cannot be

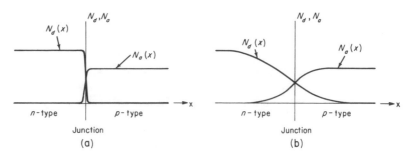

FIGURE 12.1.   Impurity atom concentrations in the neighborhood of (a) an abrupt *p-n* junction, and (b) a graded *p-n* junction.

At the instant of formation there exists a uniform concentration n_{n0} of mobile free electrons and p_{n0} mobile free holes on the *n*-side extending up to the junction and on the *p*-side a uniform concentration p_{p0} of mobile holes and n_{p0} free electrons, also extending right up to the junction. These concentrations are related to the net donor and acceptor densities N_d and N_a by (9.5-23) and (9.5-24), while, of course, on either side the electron and hole densities satisfy the relation

$$n_{n0}p_{n0} = p_{p0}n_{p0} = n_i^2. \qquad (12.1\text{-}1)$$

Since the concentration n_{n0} of electrons on the *n*-side is much larger than the electron concentration n_{p0} on the *p*-side, at the instant of formation there exists an enormous gradient in the concentration of electrons at the junction between the two regions. The same situation exists with respect to hole concentration at the junction. The large initial concentration gradients set up diffusion currents, electrons from the *n*-region and holes from the *p*-region flowing down the respective concentration gradients into the region of opposite conductivity type, and leaving the region near the junction *depleted* of majority carriers. This initial diffusion flow cannot continue indefinitely, however, because as the regions near the junction become depleted of majority carriers, the charges of the fixed donor and acceptor ions near the junction are no longer balanced by the charges of the mobile free carriers which were formerly there, and so an electric field is built up. The direction of this electric field is such as to oppose the flow of electrons out of the *n*-region and the flow of holes out of the *p*-region, and the magnitude of the field builds up to the point where its effect exactly conteracts the tendency for majority carriers to diffuse down the concentration "hill" into the region of opposite conductivity type. A condition of dynamic equilibrium is established in which the region near the junction is depleted of majority carriers and in which strong *space charge layers* containing high electric fields are formed near the junction. The situation is illustrated in Figure 12.2, and from this it is clear that the space charge configuration is an electric dipole layer with uncompensated donor ions on the *n*-side and uncompensated acceptor ions on the *p*-side furnishing the positive and negative charge components, respectively.

---

effected in actual practice. The *p-n* junction structure must exist in a nearly perfect single crystal, and can be made only by properly doping and counterdoping the melt as the crystal is grown or by some special diffusion or alloying process which accomplishes essentially the same result.

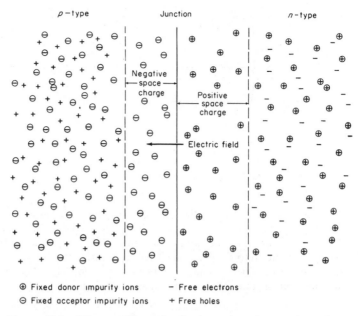

FIGURE 12.2. Diagram illustrating the formation of space charge layers
and internal electric field by diffusion of majority carriers near a junction
to the region of opposite conductivity type.

It is a well-known result of electrostatics that there is a difference in potential
between the two extreme limits of an electric dipole layer which is related to the strength
(dipole moment per unit area) of the layer by

$$\phi_2 - \phi_1 = 4\pi\Delta \tag{12.1-2}$$

where ϕ_1 and ϕ_2 are the edge potentials and Δ is the dipole layer strength. Beyond the

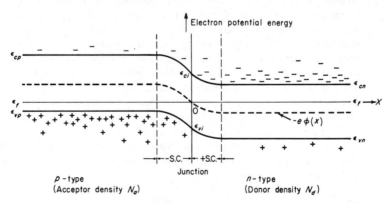

FIGURE 12.3. Potential energy diagram showing energy band configuration
near a *p-n* junction in the absence of applied voltage.

space charge regions, of course, there is no charge density, and therefore the electrostatic potential is constant. According to (12.1-2), the potential energy of an electron "at rest" at the bottom of the conduction band is lower by $4\pi e \Delta$ on one side of the junction than the other. The same statement is true of a hole "at rest" at the top of the valence band. Since the state of the system is one of thermal equilibrium, however, the Fermi energy ε_f must be the same throughout the system. These considerations enable one to depict the energy bands of the semiconductor in the neighborhood of an abrupt junction as shown in Figure 12.3. There is clearly an "internal contact potential" $e(\phi_2 - \phi_1) = e\phi_0$ built up between the two regions. The counteracting tendencies of the concentration gradients and electric fields are easily seen in Figure 12.3; the electrons in the conduction band on the n-side would obviously diffuse down the concentration gradient to the p-side were it not for the electrical potential "hill" they must surmount to do so.

12·2 THE EQUILIBRIUM INTERNAL CONTACT POTENTIAL

The value of the internal contact potential ϕ_0 is easily determined from the values of the equilibrium electron (or hole) density on either side of the junction far from the space charge regions. According to (9.3-6) we may write

$$n_{n0} = U_c e^{-(\varepsilon_{cn} - \varepsilon_f)/kT} \tag{12.2-1}$$

for the density of electrons on the n-side, and for the density on the p-side,

$$n_{p0} = U_c e^{-(\varepsilon_{cp} - \varepsilon_f)/kT}. \tag{12.2-2}$$

These equations may be solved for ε_{cn} and ε_{cp} and the internal potential energy difference $e\phi_0$, which, it is easily seen from Figure 12.3, is simply the difference between the two, is then given by

$$e\phi_0 = \varepsilon_{cp} - \varepsilon_{cn} = kT \ln \frac{n_{n0}}{n_{p0}}. \tag{12.2-3}$$

This may be written in a slightly different way, using (12.1-1) to express n_{p0} in terms of p_{p0}, as

$$\phi_0 = \frac{kT}{e} \ln \frac{n_{n0}p_{p0}}{n_i^2}. \tag{21.2-4}$$

If all donors and acceptors in the n- and p-regions are ionized, as they will be at all but very low temperatures, and if both n- and p-regions are strongly extrinsic, then, from (9.5-25) and (9.5-26), $n_{n0} \cong N_d$ and $p_{p0} \cong N_a$, whereby (12.2-4) becomes

$$\phi_0 = \frac{kT}{e} \ln \frac{N_d N_a}{n_i^2}. \tag{12.2-5}$$

The applicability of these formulas is clearly limited to situations where Boltzmann statistics can be used to describe the equilibrium distribution of carriers in both *n*- and *p*-regions.

Let us now examine what happens when an external voltage is applied to a sample containing a *p-n* junction. It is clear that since the space charge regions on either side of the junction are depleted of carriers, these regions have a much higher resistivity than any other part of the crystal. This means that when an external voltage source is applied to a sample in which there is a *p-n* junction as in Figure 12.4, most of the

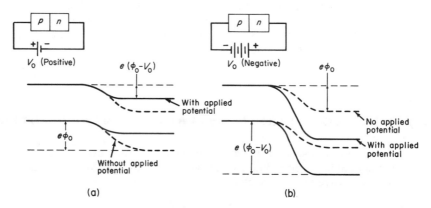

FIGURE 12.4. Potential energy diagrams showing the effect of (a) forward bias and (b) reverse bias voltage upon the energy band configuration at a *p-n* junction.

voltage drop will occur across these regions. Under practically all conditions except that in which a large voltage is applied with the "forward" polarity shown in Figure 12.4(a) the voltage drops across the uniform *n*- and *p*-regions outside the space charge layers near the junction are negligible in comparison with the drop across the junction region itself. If the *p-n* junction is used as a rectifier, it is found that a low impedance condition is obtained when the *n*-region is connected to the negative terminal and the *p*-region to the positive terminal of the external voltage source as illustrated in Figure 12.4(a). This polarity is called the *forward bias* state. It is conventional to regard the sign of the externally applied potential to be *positive* in this case. It is clear from Figure 12.3 that when the external voltage is applied in this way, the effect is to *reduce* the height of the potential barrier, as shown in Figure 12.4(a). If the voltage drop across the bulk *n*- and *p*-regions exterior to the space charge layer is negligible compared to that which occurs at the junction, then the internal barrier height will become $e(\phi_0 - V_0)$ where V_0 is the external voltage. If the polarity of the external potential source is reversed (in which case its sign is regarded as negative), it is again easily seen from Figure 12.3 that the height of the internal potential barrier is now *increased*, as illustrated in Figure 12.4(b). If the drop across the *n*- and *p*-regions exterior to the junction is neglected in comparison to the drop across the junction space charge regions, it is again apparent that the barrier height can be represented as $e(\phi_0 - V_0)$, where V_0 is now a *negative* quantity. Under these circumstances, as we shall see in more detail later, the structure exhibits a very high impedance, and the applied voltage is said to be a *reverse* bias.

12·3 POTENTIALS AND FIELDS IN THE NEIGHBORHOOD OF A *p-n* JUNCTION

Consider now the general model of a *p-n* junction shown in Figure 12.5. As illustrated in that figure there is an applied voltage V_0 (negative in the drawing) as well as the internal potential ϕ_0. Under these circumstances electrical currents will flow, and the state of the system will no longer be one of thermal equilibrium.   For this reason

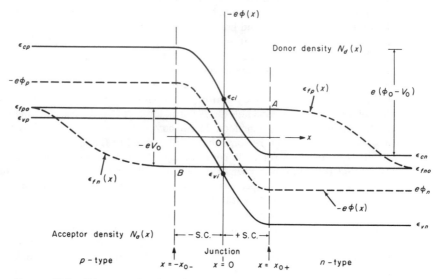

FIGURE 12.5.   Diagram showing details of energy band potentials and quasi-Fermi levels for a *p-n* junction in the presence of applied voltage.

a unique Fermi level for the entire crystal can no longer be defined. We shall, however, assume that far from the junction equilibrium conditions prevail, and that the populations of holes and electrons there can be described by Fermi distributions referred to the two Fermi levels ε_{fn0} and ε_{fp0}. If there were no applied voltage, of course, the system would be at equilibrium and ε_{fn0} and ε_{fp0} would be equal. As it is, it is clear from the figure that they must differ by eV_0. When a voltage is applied to the junction, it turns out, as we shall see more clearly later, that the concentration of *minority* carriers in both *n*- and *p*-regions contiguous to the space charge layers are seriously perturbed from their equilibrium values, while the concentration of *majority* carriers, relative to the equilibrium concentrations, is not much affected. The reason for this can be understood from Figure 12.5, where a junction with reverse bias is shown. The high potential barrier prevents electrons from the *n*-side and holes from the *p*-side from "spilling over" the junction into the opposite side; on the other hand the presence of the barrier *encourages* minority carriers (holes on the *n*-side and electrons on the *p*-side) to slide down the potential hill to the opposite side of the junction where they become majority carriers. This seriously reduces the concentration of minority carriers near the junction. When forward bias is applied the height of the barrier is *reduced* and electrons from the *n*-region and holes from the *p*-region do then spill over the junction, becoming minority carriers on the opposite side. Again, however, the *relative* concentration of minority carriers is much more strongly affected than the majority carrier concentration.[2]

The result of this is that near the junction region, while the electron and hole populations separately are characterized by Fermi distributions, there are two different "effective Fermi levels" or "quasi-Fermi levels"; one for the electron population and one for the hole population. These quasi-Fermi levels are referred to as ε_{f_n} (for electrons) and ε_{f_p} (for holes) in Figure 12.5. Far from the junction in the n-region, the equilibrium concentration of holes is hardly influenced by the effect of holes near the junction sliding down the potential "hill" to the p-region, and therefore the effective Fermi levels for electrons and holes are nearly equal. As the junction is approached, however, the local concentration of holes falls far below the equilibrium value as a result of this effect, while the electron concentration is hardly affected. The effective Fermi level for electrons remains the same, then, while the effective Fermi level for holes rises from the equilibrium value, corresponding to a smaller concentration of holes in the n-region near the junction, as shown in Figure 12.5. A similar effect with the roles of the electron and hole quasi-Fermi levels reversed takes place in the p-region on the other side of the junction. The behavior of the quasi-Fermi levels for holes and electrons must then be somewhat as illustrated in the figure.

It is ordinarily assumed in connection with all the usual calculations which are made that the minority carrier quasi-Fermi level has reached the equilibrium value associated with the same species of carrier as majority carriers in the bulk region on the other side of the junction at the near edge of the junction space charge region. In Figure 12.5, then, the value of ε_{f_p} at A is generally assumed to be equal to the equilibrium value at the far left, and the value of ε_{f_n} at B is assumed to be the same as at the far right. This means simply that the carriers of the same species on opposite edges of the junction space charge region have the same distribution in energy, by virtue of the efficiency of the collision processes which occur in that region. This assumption leads to a boundary condition on carrier concentrations at the junction which we shall find to be useful in the next chapter. Since, for example, for electrons, the distance between the conduction band and the effective Fermi level for electrons on the n-side is less than the corresponding distance on the p-side by an amount $e(\phi_0 - V_0)$, as shown in Figure 12.5, then the ratio of electron concentration at the edge of the space charge region on the n-side to that prevailing at the edge of the space charge region on the p-side must be given by

$$\frac{n_n(x_{0+})}{n_p(-x_{0-})} = e^{e(\phi_0 - V_0)/kT}, \tag{12.3-1}$$

where the coordinates x_{0+} and $-x_{0-}$ refer to the location of the edges of the space charge region on the n- and p-sides of the junction, respectively. By similar reasoning, the hole concentrations on either side must be related by

$$\frac{p_p(-x_{0-})}{p_n(x_{0+})} = e^{e(\phi_0 - V_0)/kT} \tag{12.3-2}$$

In writing these equations it is, of course, assumed that the Boltzmann approximation is valid.

---

[2] The requirement of electrical neutrality in the material away from the junction causes an alteration of the majority carrier density as well. But since the equilibrium majority carrier density is so much larger than the equilibrium minority carrier density, the addition of a given number of electrons and holes may result in increasing the minority carrier density manyfold while increasing the majority carrier density by only a few percent.

The potential and electric field in the space charge regions may be found from Poisson's equation. In the case where the charge density varies only along the x-direction, Poisson's equation may be written as

$$\frac{d^2\phi}{dx^2} = -\frac{4\pi\rho(x)}{\kappa} = -\frac{4\pi e}{\kappa}(p - n + N_d(x) - N_a(x)). \tag{12.3-3}$$

For convenience, we shall choose the origin as shown in Figure 12.5, such that the potential is zero there.[3] Further, the potential function $\phi(x)$, which contains an arbitrary additive constant will be defined relative to the conduction and valence band edges such that it has the position of the intrinsic Fermi level with respect to them. Under these circumstances, we may write

$$n(x) = U_c e^{-(\varepsilon_c(x) - \varepsilon_{fn})/kT} \tag{12.3-4}$$

and

$$p(x) = U_v e^{-(\varepsilon_{fp} - \varepsilon_v(x))/kT}. \tag{12.3-5}$$

Also, $\varepsilon_c(x)$ and $\varepsilon_v(x)$ are related to $\phi(x)$ by

$$\varepsilon_c(x) = -e\phi(x) + \varepsilon_{ci} \tag{12.3-6}$$

and

$$\varepsilon_v(x) = -e\phi(x) + \varepsilon_{vi}. \tag{12.3-7}$$

Substituting these values into Poisson's equation (12.3-3), we obtain

$$\frac{d^2\phi}{dx^2} = -\frac{4\pi e}{\kappa}(U_v e^{-(\varepsilon_{fp} - \varepsilon_{vi} + e\phi(x))/kT} - U_c e^{-(-e\phi(x) + \varepsilon_{ci} - \varepsilon_{fn})/kT}$$

$$+ N_d(x) - N_a(x)). \tag{12.3-8}$$

The solution of this equation under the most general circumstances is a complex and difficult problem. We shall content ourselves with examining several particular cases which may be understood in relatively simple terms. For example, consider a system in which the applied voltage is zero. Then $\varepsilon_{fn} = \varepsilon_{fp} = \varepsilon_f$, and (referring to Figure 12.5) it is clear that in this instance

$$U_v e^{-(\varepsilon_f - \varepsilon_{vi})/kT} = U_c e^{-(\varepsilon_{ci} - \varepsilon_f)/kT} = n_i, \tag{12.3-9}$$

whereupon (12.3-8) may be written as

$$\frac{d^2\phi}{dx^2} = \frac{4\pi e n_i}{\kappa}\left(e^{e\phi(x)/kT} - e^{-e\phi(x)/kT} - \frac{N_d(x)}{n_i} + \frac{N_a(x)}{n_i}\right)$$

$$= \frac{4\pi e n_i}{\kappa}\left(2\sinh\frac{e\phi(x)}{kT} - \frac{N_d(x) - N_a(x)}{n_i}\right). \tag{12.3-10}$$

---

[3] The donor and acceptor densities are assumed for the present to be given functions of distance, although Figure 12.5 is drawn for simplicity as though they were constant.

This is the fundamental differential equation which must be solved to obtain the potential $\phi(x)$. For an *abrupt* junction we have

$$\left.\begin{array}{l} N_a(x) = N_a = \text{const} \\ N_d(x) = 0 \end{array}\right\} \quad (x < 0) \tag{12.3-11}$$

and

$$\left.\begin{array}{l} N_a(x) = 0 \\ N_d(x) = N_d = \text{const} \end{array}\right\} \quad (x > 0) \tag{12.3-12}$$

in which case (12.3-10) yields two equations, one for the potential $\phi_-(x)$ in the region $(x < 0)$, of the form

$$\frac{d^2\phi_-}{dx^2} = \frac{4\pi e n_i}{\kappa}\left(2\sinh\frac{e\phi_-}{kT} + \frac{N_a}{n_i}\right) \quad (x < 0) \tag{12.3-13}$$

and one for the potential $\phi_+(x)$ in the region $(x > 0)$, which may be written

$$\frac{d^2\phi_+}{dx^2} = \frac{4\pi e n_i}{\kappa}\left(2\sinh\frac{e\phi_+}{kT} - \frac{N_d}{n_i}\right). \quad (x > 0) \tag{12.3-14}$$

These equations must be solved for $\phi_+(x)$ and $\phi_-(x)$ using appropriate boundary conditions, which will be discussed in the next section. Unfortunately even these equations cannot be solved exactly by analytical means, and so numerical methods or physical approximations must be introduced. We shall now discuss a simplified model based upon the preceding equations which admits of an exact mathematical solution even when an external voltage is present, and which is a good approximation for the abrupt *p-n* junction under most circumstances of practical importance.

12·4 SIMPLIFIED MATHEMATICAL MODEL OF THE ABRUPT *p-n* JUNCTION

An approximate solution for an *abrupt p-n* junction can, under certain conditions, be obtained by observing from (12.3-4) and (12.3-6) that the carrier concentration varies exponentially with ϕ, so that the carrier density will drop to a value negligible compared to the bulk value far from the junction as the band edge potential energy $-e\phi$ varies by a few times kT. If the total potential drop across the junction, $\phi_0 - V_0$, is much larger than kT, then, a variation of potential energy several times kT will be achieved in a distance quite small compared to the total linear extent of the junction space charge region. Under these circumstances the charge density quickly reaches a constant value $-eN_a$ as the space charge region is entered from the *p*-region and a constant value eN_d as the space charge region is entered from the *n*-side, as shown by the dotted curves in Figure 12.6.

In such a situation it is a good approximation to represent the space charge density by a constant positive value eN_d on the *n*-side from the junction out to the edge of the space charge layer on that side, and zero beyond, and by a constant negative value $-eN_a$ on the p-side from the junction to the edge of the space charge region, and zero thereafter. The charge density, according to this approximation is shown by the solid curve in Figure 12.6. Fortunately, most of the abrupt p-n junctions which are of interest from a practical point of view satisfy the condition $e(\phi_0 - V_0) \gg kT$ under which this approximate representation of the actual situation is a good one.

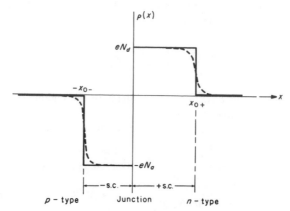

Acceptor density N_a = const   Donor density N_d = const

FIGURE 12.6. The actual charge density (dashed curve) and the charge density assumed in the approximate model of Section 12.4 (solid curve) for an abrupt p-n junction for which $\varphi_0 \gg kT/e$.

If we denote by $\phi_+(x)$ the electrostatic potential for $x > 0$, and by $\phi_-(x)$ the potential in the region $x < 0$, and if we assign the values x_{0+} and $-x_{0-}$ to the coordinates of the edges of the space charge regions on the n- and p-sides, respectively, then Poisson's equation assumes the form

$$\frac{d^2\phi_+}{dx^2} = -\frac{4\pi e N_d}{\kappa} \qquad (0 < x < x_{0+}), \tag{12.4-1}$$

$$\frac{d^2\phi_-}{dx^2} = \frac{4\pi e N_a}{\kappa} \qquad (-x_{0-} < x < 0) \tag{12.4-2}$$

and $\qquad \dfrac{d^2\phi_\pm}{dx^2} = 0 \qquad (x > x_{0+} ; x < -x_{0-}). \tag{12.4-3}$

If we assume that all the voltage applied to the sample appears as a potential drop across the junction region, then, in the regions beyond the boundaries of the space charge layers, the potential must be constant, whereby

$$\phi_+(x) = \phi_+(x_{0+}) = \text{const} \qquad (x > x_{0+}) \tag{12.4-4}$$

$$\phi_-(x) = \phi_-(-x_{0-}) = \text{const} \qquad (x < -x_{0-}) \tag{12.4-5}$$

These solutions clearly satisfy (12.4-3). Since the total fall in potential across the junction region is $\phi_0 - V_0$, we must also have

$$\phi_+(x_{0+}) - \phi_-(-x_{0-}) = \phi_0 - V_0. \tag{12.4-6}$$

In addition, it is easily shown from Gauss's electric flux theorem that the electric field must be continuous everywhere except at a surface bearing a true surface charge distribution. The proof of this statement is recommended to the reader as an exercise. This means that at the junction,

$$-E_+(0) = \left(\frac{d\phi_+}{dx}\right)_0 = \left(\frac{d\phi_-}{dx}\right)_0 = -E_-(0) = -E_0 \tag{12.4-7}$$

where $E_+(x)$ and $E_-(x)$ denote the fields in the regions $(x > 0)$ and $(x < 0)$, and E_0 their common value at the origin.

Equations (12.4-1) and (12.4-2) may be integrated to obtain

$$\frac{d\phi_+}{dx} = -\frac{4\pi e N_d}{\kappa} x + A = -E_+(x) \qquad (0 < x < x_{0+}) \tag{12.4-8}$$

and

$$\frac{d\phi_-}{dx} = \frac{4\pi e N_a}{\kappa} x + B = -E_-(x) \qquad (-x_{0-} < x < 0) \tag{12.4-9}$$

where A and B are arbitrary constants. Beyond the limits x_{0+} and x_{0-} of the space charge region, according to (12.4-4) and (12.4-5), the field is zero, and since the field must be everywhere continuous, (12.4-8) must yield $d\phi_+/dx = 0$ at x_{0+} and (12.4-9) must give $d\phi_-/dx = 0$ at $x = -x_{0-}$, whereby

$$A = \frac{4\pi e N_d}{\kappa} x_{0+} \qquad \text{and} \qquad B = \frac{4\pi e N_a}{\kappa} x_{0-} \tag{12.4-10}$$

and (12.4-8) and (12.4-9) become

$$\frac{d\phi_+}{dx} = \frac{4\pi e N_d}{\kappa} (x_{0+} - x) = -E_+(x) \qquad (0 < x < x_{0+}) \tag{12.4-11}$$

and

$$\frac{d\phi_-}{dx} = \frac{4\pi e N_a}{\kappa} (x + x_{0-}) = -E_-(x) \qquad (-x_{0-} < x < 0) \tag{12.4-12}$$

while, of course,

$$E_\pm(x) = 0 \qquad (x > x_{0+} ; x < -x_{0-}). \tag{12.4-13}$$

If in (12.4-11) and (12.4-12), x is set equal to zero, then, referring to (12.4-7) it is clear that

$$E_0 = \frac{-4\pi e N_d x_{0+}}{\kappa} = -\frac{4\pi e N_a x_{0-}}{\kappa} \tag{12.4-14}$$

and therefore that

$$N_d x_{0+} = N_a x_{0-}. \tag{12.4-15}$$

According to (12.4-15) the ratio of the widths of the two space charge layers is the inverse of the ratio of the respective impurity densities. This implies that the total number of positive and negative charges in each layer is the same.

Integrating (12.4-11) and (12.4-12) once more, the potential is obtained as

$$\phi_+(x) = \frac{-2\pi e N_d}{\kappa} (x_{0+} - x)^2 + C \qquad (0 < x < x_{0+}) \tag{12.4-16}$$

and

$$\phi_-(x) = \frac{2\pi e N_a}{\kappa} (x + x_{0-})^2 + D. \qquad (-x_{0-} < x < 0) \tag{12.4-17}$$

The zero point of the potential may be chosen arbitrarily; for convenience we shall choose it such that $\phi_+(0) = \phi_-(0) = 0$. Substituting this back into (12.4-16) and (12.4-17) it is possible to evaluate both C and D. Combining these results with (12.4-4) and (12.4-5) it is easily seen that the potential must have the form

$$\phi_+(x) = \frac{2\pi e N_d}{\kappa} [x_{0+}^2 - (x_{0+} - x)^2] \qquad (0 < x < x_{0+})$$

$$\tag{12.4-18}$$

$$= \phi_+(x_{0+}) = \frac{2\pi e N_d x_{0+}^2}{\kappa} \qquad (x > x_{0+})$$

$$\phi_-(x) = \frac{2\pi e N_a}{\kappa} [(x + x_{0-})^2 - x_{0-}^2] \qquad (-x_{0-} < x < 0)$$

$$\tag{12.4-19}$$

$$= \phi_-(-x_{0-}) = -\frac{2\pi e N_a x_{0-}^2}{\kappa}. \qquad (x < -x_{0-})$$

The expressions for the field and potential are now complete, except that the distances x_{0+} and x_{0-} are as yet unspecified. To obtain the values of these quantities the values of $\phi_+(x_{0+})$ and $\phi_-(-x_{0-})$ given by (12.4-18) and (12.4-19) may be substituted into the boundary condition (12.4-6), whereby

$$\phi_+(x_{0+}) - \phi_-(x_{0-}) = \frac{2\pi e}{\kappa} (N_d x_{0+}^2 + N_a x_{0-}^2) = \phi_0 - V_0. \tag{12.4-20}$$

In this equation x_{0-} can be expressed in terms of x_{0+} by (12.4-15), and the resulting expression solved for x_{0+} to give

$$x_{0+} = \sqrt{\frac{\kappa(\phi_0 - V_0)}{2\pi e} \frac{N_a}{N_d(N_d + N_a)}}. \tag{12.4-21}$$

From (12.4-15) x_{0-} can be obtained in the form

$$x_{0-} = \sqrt{\frac{\kappa(\phi_0 - V_0)}{2\pi e} \frac{N_d}{N_a(N_d + N_a)}}. \qquad (12.4\text{-}22)$$

The total combined width of the space charge regions is then

$$x_0 = x_{0+} + x_{0-} = \sqrt{\frac{\kappa(\phi_0 - V_0)}{2\pi e} \left(\frac{1}{N_d} + \frac{1}{N_a}\right)} \qquad (12.4\text{-}23)$$

and the field E_0 at the junction, from (12.4-14) can be written as

$$-E_0 = \sqrt{\frac{8\pi e(\phi_0 - V_0)}{\kappa} \left(\frac{1}{N_d} + \frac{1}{N_a}\right)^{-1}} = \frac{2(\phi_0 - V_0)}{x_0} \qquad (12.4\text{-}24)$$

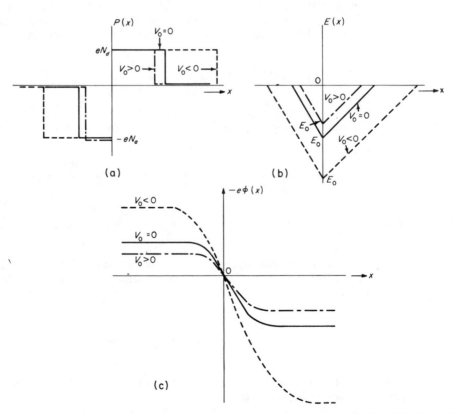

FIGURE 12.7. Schematic representation of (a) charge density (b) electric field and (c) potential in the neighborhood of an abrupt *p-n* junction under various applied voltages, as calculated by the methods of Section 12.4.

It should be noted that x_{0+}, x_{0-}, and E_0 are related to the doping densities N_d and N_a directly, as shown above, and *also* through ϕ_0 which is a function of N_d and N_a as given by (12.2-5), although the logarithmic variation of ϕ_0 with N_d and N_a is much slower than the direct variation.

The space charge density, electric field and electrostatic potential variations as given by the preceding theory are plotted for the zero-bias, forward-bias $(V_0 > 0)$ and reverse-bias $(V_0 < 0)$ conditions in Figure 12.7.

It is clear from this figure, or from the above equations, that

(a) The space charge layer is of greater extent in lightly doped *p-* or *n*-type regions than in regions which are heavily doped. For a given junction the space charge layer extends furthest into the region which is least heavily doped.

(b) The maximum field value E_0 is very high in junctions where both *p-* and *n*-regions are heavily doped and much smaller in junctions where one or both regions are lightly doped.

(c) Under reverse-bias conditions, the space charge regions extend themselves outwards into the crystal; the extent of the space charge regions may become quite large in junctions where either *n-* or *p*-region is lightly doped. For $-V_0 \gg \phi_0$ (large reverse bias) the extent of the space charge regions is proportional to $(-V_0)^{1/2}$, according to (12.4-21) and (12.4-22).

(d) Under reverse-bias conditions the maximum field E_0 becomes quite large; at large reverse biases it is proportional to $(-V_0)^{1/2}$.

These conclusions, while verified for a specific abrupt junction model, are quite generally true for *all* semiconductor *p-n* junctions, whether abrupt or graded (although for graded junctions, the square-root dependence referred to in (c) and (d) is no longer quantitatively correct).

The relations derived in the preceding pages are quite accurate for reverse-bias conditions, but should not be expected to hold very well for high-forward bias. In the latter instance, large currents flow and there are appreciable voltage drops in the

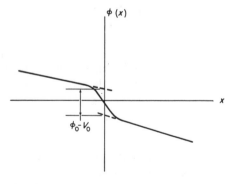

FIGURE 12.8. An approximate representation of the potential near an abrupt *p-n* junction under high forward bias conditions.

bulk regions outside the junction space charge layers which render the junction voltage drop different from the externally applied voltage. The potential then assumes something like the form shown in Figure 12.8. Some numerical values for the properties of the space charge layers associated with typical germanium *p-n* junctions are given in Table 12.1.

12·5 JUNCTION CAPACITANCE; DETERMINATION OF INTERNAL POTENTIAL

The results of the previous section suggest strongly that a *p-n* junction behaves in many ways like a parallel-plate capacitor. There are two space charge layers containing equal and opposite amounts of charge, and the amount of charge belonging to the layers increases as the reverse voltage is increased. The charge associated with unit area of the positive space charge region on the *n*-side of the junction is clearly the product of the charge density eN_d and the volume x_{0+} of space charge region corresponding to unit area of junction. The charge per unit area of negative space charge layer on the *p*-side may be calculated the same way. In either case it is easily seen with the aid of (12.4-21) or (12.4-22) that the charge per unit junction area is

$$|Q| = eN_d x_{0+} = eN_a x_{0-} = \sqrt{\frac{\kappa e(\phi_0 - V_0)}{2\pi} \frac{N_d N_a}{N_d + N_a}}. \tag{12.5-1}$$

A small increment of applied voltage ΔV_0 will be accompanied by a change ΔQ in stored charge. If the differential capacitance per unit area associated with the junction at this value of applied voltage is defined as $|dQ/dV_0|$, then, from the above equation it is evident that

$$C = \left|\frac{dQ}{dV_0}\right| = \left|\frac{d|Q|}{d(\phi_0 - V_0)}\right| = \frac{1}{2}\sqrt{\frac{\kappa e}{2\pi(\phi_0 - V_0)}\frac{N_d N_a}{N_d + N_a}}. \tag{12.5-2}$$

From this we may conclude that junctions in which *both* p- and n- regions are heavily doped will have very high capacitance, while if either p- or n-region, or both, are lightly doped, the capacitance will be much less. For reverse bias voltages much in excess of the internal contact potential ϕ_0, C will be found, according to (12.5-2) to vary nearly proportionally to $(-V_0)^{-1/2}$. Indeed, this variation of capacitance is characteristic of the abrupt junctions which are found in devices made by alloying processes. Graded or nonabrupt junctions, such as are found in many devices made by diffusing *n*-type impurities into a *p*-type crystal, or *vice versa*, exhibit capacitance which varies with reverse voltage more slowly than this. The capacitance of a linearly graded reverse-biased junction, for example, varies with voltage as $(-V_0)^{-1/3}$.

For abrupt-junction rectifiers, capacitance measurements furnish a simple way of determining the internal contact potential ϕ_0. For this purpose, Equation (12.5-2) may be rearranged to read

$$\phi_0 - V_0 = \frac{\kappa e}{8\pi C^2}\frac{N_d N_a}{N_d + N_a}. \tag{12.5-3}$$

Experimentally, the capacitance per unit area C may be measured and a plot of $1/C^2$ *versus* the reverse voltage $-V_0$ made, as shown in Figure 12.9. According to (12.5-3) such a plot should be a straight line. The intercept of this plot on the V_0-axis extrapolated from the measured data gives the value of ϕ_0 as illustrated. From the slope of the experimentally determined line, the quantity $N_d N_a/(N_d + N_a)$ can be determined; if one side of the junction is much more heavily doped than the other,[4] this quantity is approximately equal to the concentration of the doping impurity on the more

---

[4] This is usually the case in junctions made by alloying processes, as we shall see later.

TABLE 12.1.
Properties of Abrupt p-n Junctions in Germanium ($\kappa = 16$)

N_d (cm^{-3})	N_a (cm^{-3})	ϕ_0 (V)	V_0 (V)	x_{0+} (μ)	x_{0-} (μ)	x_0 (μ)	E_0 (kv/cm)	C (μf/cm^2)
10^{14}	10^{14}	0.0702	0	0.788	0.788	1.58	0.891	0.00898
			-1	3.08	3.08	6.15	3.48	0.00229
			-5	6.69	6.69	13.4	7.57	0.00105
			-50	21.0	21.0	42.1	23.8	0.000336
			-500	66.5	66.5	133.0	75.2	0.000106
10^{14}	10^{15}	0.1854	0	0.405	0.405	0.810	4.58	0.0175
	10^{16}	0.3005	0	0.163	0.163	0.326	21.5	0.0434
	10^{17}	0.4156	0	0.0606	0.0606	0.121	68.6	0.117
	10^{18}	0.5308	0	0.0217	0.0217	0.0433	245.2	0.327
	10^{19}	0.6459	0	0.0076	0.0076	0.0151	855.5	0.937
10^{14}	10^{18}	0.3005	0	2.31	0.00023	2.31	2.61	0.00613
			-1	4.80	0.00048	4.80	5.42	0.00295
			-5	9.68	0.00097	9.68	10.95	0.00146
			-50	29.8	0.00298	29.8	33.7	0.000475
			-500	94.1	0.00941	94.1	106.4	0.000151

lightly doped side. The concentration of the doping impurity on the more heavily doped side can be obtained, once ϕ_0 and the other doping impurity concentration are known, from (12.2-5). A somewhat similar procedure can be used to find both

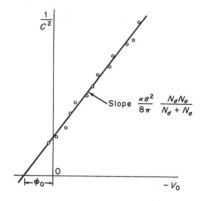

FIGURE 12.9. A plot of $1/C^2$ *versus* bias voltage illustrating the determination of internal contact potential and donor and acceptor impurity concentrations from capacitance measurements.

N_d and N_a even if the doping impurity densities on either side of the junction are comparable, although the mathematical aspects are slightly more involved. The details of this procedure are assigned as an exercise. Some numerical data for capacitance of typical junction structures is shown in Table 12.1.

EXERCISES

1. Show that there is a discontinuity in electrostatic potential of magnitude $4\pi\Delta$, where Δ is the dipole moment per unit area, on opposite sides of an electric dipole layer.

2. An attempt to measure the internal contact potential of a *p-n* junction is made by connecting a high-resistance voltmeter with copper leads to contacts on opposite sides of the crystal containing the junction. The voltmeter reads zero. A potentiometer is substituted for the meter, but the reading is still zero. Explain physically the failure of these methods and discuss the contact potentials between copper and the *n*-type and *p*-type regions.

3. Show, using Gauss's electric flux theorem that the electric field must be continuous everywhere except at a surface bearing a charge distribution.

4. Explain *physically* why the extent of the space charge region of an abrupt *p-n* junction is found to be proportional to the square root rather than the first power of the junction potential drop $\phi_0 - V_0$.

5. Find approximate expressions for the potentials, fields, and extent of space charge layers associated with a linearly graded junction in which

$$N_d = 0, N_a = -\alpha x \qquad (x < 0)$$

$$N_a = 0, N_d = \alpha x \qquad (x > 0)$$

where α is a constant. Assume that the total potential fall across the space charge region is $\phi_0 - V_0$ and that $\phi_0 - V_0 \gg kT$. *Hint:* Use an extension of the methods of Section 12.4.

6. Find the differential capacitance per unit area associated with the linearly graded junction of Exercise 5.

7. Show how the internal contact potential and both N_d and N_a can be obtained from a plot of experimental data such as that shown in Figure 12.9. Derive explicit formulas for N_a and N_d in terms of the slope of the experimental plot and the internal potential ϕ_0.

GENERAL REFERENCES

J. D. Jackson, *Classical Electrodynamics*, John Wiley & Sons, New York (1962), Chapter 1.

Allen Nussbaum, *Semiconductor Device Physics*, Prentice-Hall, Inc., Englewood Cliffs, N.J. (1962), Chapter 4.

W. Shockley, *Electrons and Holes in Semiconductors*, D. van Nostrand Co., Inc., New York (1950).

R. A. Smith, *Semiconductors*, Cambridge University Press, London (1961).

E. Spenke, *Electronic Semiconductors*, McGraw-Hill Book Co., Inc., New York (1958).

p-n JUNCTION RECTIFIERS AND TRANSISTORS

13·1 THEORY OF THE *p-n* JUNCTION RECTIFIER

In the preceding chapter, the physical nature of a semiconductor *p-n* junction was discussed. Certain physical manifestations of the junction structure, such as space charge, electric fields, potentials and capacitance were examined quantitatively on the basis of a simple model of an abrupt junction. In this section we shall turn our attention to the current flow associated with the junction under nonequilibrium conditions and the current-voltage characteristic of a simple *p-n* junction rectifier.

The junction model which will be adopted is shown in the equilibrium state in Figure 13.1. The *p*-type and *n*-type regions are assumed to be homogeneous and to

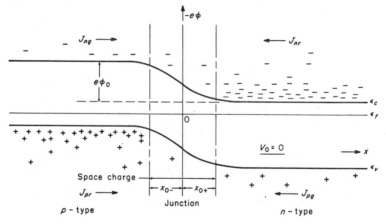

FIGURE 13.1. Potential diagram of *p-n* junction in the equilibrium condition, showing charge carrier distributions and generation and recombination fluxes.

extend to infinity in the $+x$ and $-x$ directions. In the equilibrium state, there will be a certain number of electrons present as majority carriers on the *n*-side which have enough energy to surmount the potential barrier and diffuse over to the *p*-side of the junction. In the *p*-region these electrons are minority carriers and they may, after a time, disappear through recombination with holes. There will therefore, at equilibrium, be a current of electrons from the *n*-region over the junction to the *p*-region where they are lost by recombination. This equilibrium electron flow will be referred to as a recombination flux J_{nr}. Clearly, however, in the equilibrium condition, there can be no *net* current flow; furthermore, according to the principle of detailed balance, at equilibrium any microscopic transport process and its inverse must proceed at equal rates. The inverse of this particular process is the generation of electron-hole pairs on the *p*-side of the junction, and the subsequent diffusion of electrons across the junction to the *n*-side, where they become majority carriers. This inverse process leads to an electron current which flows from the *p*-region to the *n*-region, which will be called a *generation* current. At equilibrium, the resulting particle flux J_{ng} must be exactly equal to $-J_{nr}$. The same argument applies to holes; there is a recombination flux J_{pr}, which arises when holes in the *p*-region wander over the barrier to the *n*-region, where they recombine with electrons. This is accompanied by a generation flux J_{pg} arising from thermal pair generation in the *n*-region followed by diffusion of the thermally generated holes across the junction. At equilibrium we must have $J_{pg} = -J_{pr}$. The total fluxes of holes and of electrons across the junction in the equilibrium state are therefore zero.

The basis for the action of a *p-n* junction as a rectifier, is as follows. Under conditions of reverse bias, the height of the potential barrier at the junction increases by an amount $-eV_0$, where V_0 is the applied voltage. Under these conditions it becomes very difficult for carriers to surmount the barrier by diffusion, and thus J_{nr} and J_{pr} become quite small; the fluxes J_{ng} and J_{pg}, however, depend *only upon the rate of thermal generation of electron-hole pairs* in the respective bulk regions, and are consequently undiminished. As the bias voltage is increased, therefore, the current density across the junction approaches a constant small value $-e(J_{pg} - J_{ng})$, called the *saturation current*, which is limited by the number of thermally generated pairs created in the

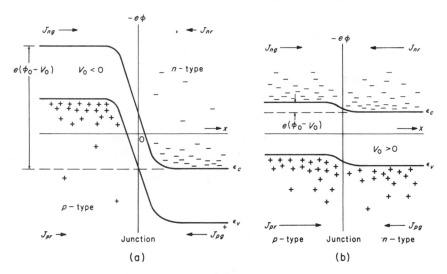

FIGURE 13.2. Potential diagram of *p-n* junction (a) with reverse bias (b) with forward bias, showing how generation and recombination fluxes are modified.

bulk regions near the junction. This current depends only upon material parameters and temperature and cannot increase significantly no matter how large the voltage becomes. The physical situation is illustrated schematically in Figure 13.2(a). When a forward bias voltage is applied, the potential barrier height is reduced, and it then becomes very easy for majority electrons from the n-side and majority holes from the p-side to diffuse over the barrier to the respective opposite sides where they become minority carriers and eventually recombine. In this case the recombination currents J_{pr} and J_{nr} may become very large compared to their equilibrium values. The generation currents J_{ng} and J_{pg}, however, remain the same, since they are limited by the thermal generation rate. As a result of this, a great deal of current may flow, as illustrated in Figure 13.2(b). The junction structure exhibits a very high impedance in the reverse bias direction and very small impedance in the forward direction, thus behaving as a rectifier. The current-voltage relation is shown in Figure 13.3. In the reverse bias

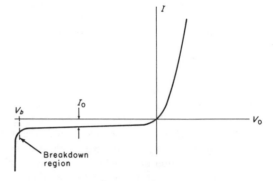

FIGURE 13.3. Current-voltage relation for a p-n junction, showing reverse breakdown region.

condition a saturation current flows, consisting of minority carriers generated thermally in the bulk regions outside of, but adjoining, the junction space charge layers which are *collected* by the junction potential barrier. In the forward direction, the current consists largely of majority carriers which diffuse over the lowered barrier to become minority carriers in the region of opposite conductivity type across the junction; the junction, in effect *injects* a large number of minority carriers into the bulk regions adjoining it on both sides.

Since the number of electrons having energy sufficient to surmount the junction potential barrier is proportional to $e^{-e(\varphi_0 - V_0)/kT} = (\text{const})e^{eV_0/kT}$ (assuming that Maxwell-Boltzmann statistics are valid), we should expect J_{nr} and J_{pr} to be *proportional* to $e^{eV_0/kT}$. But for $V_0 = 0$, $J_{pr} = J_{pg}$ and $J_{nr} = J_{ng}$, and we must have

$$J_{pr} = J_{pg}e^{eV_0/kT} \tag{13.1-1}$$

and

$$J_{nr} = J_{ng}e^{eV_0/kT}. \tag{13.1-2}$$

According to Figure 13.2, then,

$$J_p = J_{pr} - J_{pg} = J_{pg}(e^{eV_0/kT} - 1) \tag{13.1-3}$$

and

$$J_n = J_{ng} - J_{nr} = -J_{ng}(e^{eV_0/kT} - 1). \tag{13.1-4}$$

The relationship between applied voltage and the *electric* current density I would then be given by

$$I = e(J_p - J_n) = e(J_{pg} + J_{ng})(e^{eV_0/kT} - 1) = I_0(e^{eV_0/kT} - 1) \tag{13.1-5}$$

where the saturation current density I_0 is seen to be simply the sum of the generation current densities;

$$I_0 = e(J_{pg} + J_{ng}). \tag{13.1-6}$$

The current-voltage relation shown in Figure 13.3 is essentially a plot of Equation (13.1-5) above. We shall now proceed to derive the result (13.1-5) using a more rigorous method of calculation, which has two advantages; first, that it results in a more explicit expression for the saturation current, and second, that it forms the basis for a calculational technique which can be applied to more complicated junction devices.

The procedure which we shall follow is to solve the continuity equation for excess carriers in both *n*- and *p*-type regions and to apply the proper boundary conditions at the junction and elsewhere in the system. For this purpose it will be assumed that the junction is planar and that the junction plane extends essentially to infinity in the *y*- and *z*-directions, so that only variations of carrier concentration and current flow along the *x*-direction need be considered. It will also be assumed that a steady state has been reached and that all the voltage applied to the system appears across the space charge region associated with the junction. Voltage drops and electric fields in the material outside the junction region are then negligible and transport in those regions is purely diffusive. Under these conditions the continuity equations for electrons in the *p*-region and holes in the *n*-region have the form

$$\frac{d^2(n_p - n_{p0})}{dx^2} - \frac{n_p(x) - n_{p0}}{L_n^2} = 0 \qquad (\text{p-region}, \; x < -x_{0-}) \tag{13.1-7}$$

and

$$\frac{d^2(p_n - p_{n0})}{dx^2} - \frac{p_n(x) - p_{n0}}{L_p^2} = 0 \qquad (\text{n-region}, \; x > x_{0+}) \tag{13.1-8}$$

where

$$L_n = \sqrt{D_n \tau_n} \quad \text{and} \quad L_p = \sqrt{D_p \tau_p}. \tag{13.1-9}$$

As usual, only the continuity equations for minority carriers in each region are treated explicitly, but it must be remembered that there is an excess electron concentration everywhere in the *n*-region just equal to the local excess hole concentration, and an excess hole concentration everywhere in the *p*-region just equal to the local excess electron concentration, in order that the requirement of electrical neutrality be satisfied. The ultimate source of these excess majority carriers is the end contacts of the device (which are far from the junction in the somewhat idealized model we are discussing). Roughly speaking, the injection of a minority carrier distribution results in the immediate formation of an accompanying majority carrier distribution, which depletes the rest of the device of majority carriers, instantaneously creating a field which pulls neutralizing majority carriers in from the end contact. In the steady-state condition approximate neutrality is maintained everywhere except within the junction space charge region.

The boundary conditions to be used are that the excess minority carrier density must approach zero far from the junction, whereby

$$n_p - n_{p0} = 0 \qquad \text{for} \qquad x = -\infty \tag{13.1-10}$$

and
$$p_n - p_{n0} = 0 \quad \text{for} \quad x = +\infty, \tag{13.1-11}$$

while at the junction the boundary conditions (12.3-1) and (12.3-2) developed in the previous chapter must be satisfied. In using those boundary conditions it simplifies the analysis a great deal if one may assume that the excess carrier density is everywhere much smaller than the equilibrium majority carrier density, and it is clear that this condition will always be satisfied in the device at all but rather large forward bias voltages. If this assumption is made, then in (12.3-1) $n_n(x_{0+})$ may be set equal to n_{n0}, while in (12.3-2) $p_p(-x_{0-})$ may be taken[1] to be equal to p_{p0}. These boundary conditions then take the form

$$\frac{n_p(-x_{0-})}{n_{n0}} = e^{-e(\phi_0 - V_0)/kT} \tag{13.1-12}$$

and
$$\frac{p_n(x_{0+})}{p_{p0}} = e^{-e(\phi_0 - V_0)/kT}. \tag{13.1-13}$$

The general solutions of the continuity equations (13.1-7) and (13.1-8) can be expressed as

$$n_p - n_{p0} = A e^{x/L_n} + B e^{-x/L_n} \quad (p\text{-region}) \tag{13.1-14}$$

and
$$p_n - p_{n0} = C e^{x/L_p} + D e^{-x/L_p}, \quad (n\text{-region}) \tag{13.1-15}$$

where A, B, C and D are arbitrary constants. It is easily seen that in order to satisfy boundary conditions (13.1-10) and (13.1-11) we must choose $B = C = 0$. Also, since when the applied voltage V_0 is zero, $n_p(-x_{0-}) = n_{p0}$ and $p_n(x_{0+}) = p_{n0}$, it is apparent from (13.1-12) and (13.1-13) that the quantity $e^{-e\phi_0/kT}$ may be expressed in terms of the ratio of equilibrium hole and electron concentrations on opposite sides of the junction[2] as

$$e^{-e\phi_0/kT} = \frac{n_{p0}}{n_{n0}} = \frac{p_{n0}}{p_{p0}}. \tag{13.1-16}$$

This enables one to write the boundary conditions (13.1-12) and (13.1-13) in the form

$$n_p(-x_{0-}) = n_{p0}e^{eV_0/kT} \tag{13.1-17}$$

and
$$p_n(x_{0+}) = p_{n0}e^{eV_0/kT} \tag{13.1-18}$$

Setting $B = C = 0$ in (13.1-14) and (13.1-15) and applying (13.1-17) and (13.1-18) to evaluate the other two constants one may easily show that

$$n_p - n_{p0} = n_{p0}(e^{eV_0/kT} - 1)e^{(x+x_0-)/L_n} \quad (p\text{-region}) \tag{13.1-19}$$

---

[1] Otherwise, we should have to write $n_n(x_{0+}) = n_{n0} + [p_n(x_{0+}) - p_{n0}]$ and $p_p(x_{0-}) = p_{p0} + [n_p(-x_{0-}) - n_{p0}]$.
[2] This is also readily established from (12.2-3) or (12.2-4).

and $\qquad p_n - p_{n0} = p_{n0}(e^{eV_0/kT} - 1)e^{-(x-x_0+)/L_p}$ (*n*-region)   (13.1-20)

It is clear from these equations that for reverse bias ($V_0 < 0$) the concentration of electrons in the *p*-region and the concentration of holes in the *n*-region is *less* than the equilibrium value; the *excess* minority carrier concentration is *negative* in this case The reverse is true when the bias voltage is applied in the forward direction ($V_0 > 0$). Equations (13.1-19) and (13.1-20) can be used to find the currents of holes and electrons flowing into the space charge regions from the *n*-side and *p*-side of the crystal, respectively. We shall assume that the current of electrons which crosses the junction is the same as the current of minority electrons which flows into the space charge region from the *p*-side, and that the current of holes which crosses the junction is the same as the current of minority holes which flows into the space charge region from the *n*-side. In doing this, we are neglecting any thermal generation or recombination of carriers which may occur *within* the space charge layer itself.[3] This is justified by the fact that the thickness of the space charge layer is in most cases quite small and that the very high fields within the space charge layer ordinarily speed electrons from the *p*-side and holes from the *n*-side through the layer in a time small compared to their lifetime. Making this simplification, the junction current may be evaluated by evaluating the diffusion currents of minority carriers at the boundaries of the space charge region, giving

$$J_n(0) \cong J_n(-x_{0-}) = -D_n\left(\frac{dn_p}{dx}\right)_{-x_{0-}} = -\frac{n_{p0}D_n}{L_n}(e^{eV_0/kT} - 1) \quad (13.1\text{-}21)$$

and $\qquad J_p(0) \cong J_p(x_{0+}) = -D_p\left(\frac{dp_n}{dx}\right)_{x_{0+}} = \frac{p_{n0}D_p}{L_p}(e^{eV_0/kT} - 1).$ (13.1-22)

The electrical current crossing the junction is then simply

$$I = e(J_p(0) - J_n(0)) = e\left(\frac{n_{p0}D_n}{L_n} + \frac{p_{n0}D_p}{L_p}\right)(e^{eV_0/kT} - 1).$$ (13.1-23)

This result has the same form as that obtained previously as (13.1-5); the saturation current is now seen to be

$$I_0 = e(J_{ng} + J_{pg}) = e\left(\frac{n_{p0}D_n}{L_n} + \frac{p_{n0}D_p}{L_p}\right).$$ (13.1-24)

The actual hole and electron concentrations on either side of the junction for both forward and reverse biases are shown for a typical *p-n* junction rectifier of the type described above in Figure 13.4. From the above equation the saturation current is found to involve the equilibrium minority carrier densities n_{p0} and p_{n0}, the diffusion coefficients D_n and D_p, and the excess carrier lifetimes through the diffusion lengths

---

[3] If this approximation were not made the continuity equations (including appropriate field terms) would have to be solved in both space charge layers, and matched to the solutions outside as expressed by (13.1-19) and (13.1-20) at x_{0+} and $-x_{0-}$ with suitable boundary conditions. The calculation would then be very difficult albeit more general.

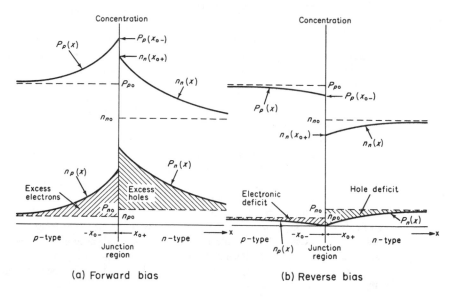

(a) Forward bias

(b) Reverse bias

FIGURE 13.4.   Electron and hole concentrations in the neighborhood of a *p-n* junction (a) with forward bias (b) with reverse bias.

L_n and L_p. It may be expressed in terms of the majority carrier densities by using the relation

$$n_{n0}p_{n0} = p_{p0}n_{p0} = n_i^2,$$ (13.1-25)

whereby (13.1-24) can be put in the form

$$I_0 = en_i^2\left(\frac{D_n}{p_{p0}L_n} + \frac{D_p}{n_{n0}L_p}\right).$$ (13.1-26)

It is apparent that the saturation current may be reduced by the use of relatively heavily doped material for the *n*- and *p*-region, and also by arranging that the excess carrier lifetime be as large as possible in both regions. The saturation current, according to (13.1-26) depends upon temperature through D_n, D_p, L_n, L_p, and n_i. By far the most temperature-sensitive of these parameters is the intrinsic carrier density n_i, and therefore the saturation current must increase rapidly with increasing temperature. Semiconductors having relatively large energy gaps (and correspondingly, relatively small values of n_i) are desirable for making *p-n* junction rectifiers which must handle high current densities and operate at high temperatures. Under such operating conditions the intrinsic carrier density in small-gap semiconductors becomes excessive, and the saturation current becomes quite large, which leads to poor rectification efficiency and excessive heat generation in the reverse-bias region as shown in Figure 13.5. Silicon rectifiers are thus superior to germanium units for high-current or high-temperature application. Unfortunately, the lower saturation current of the large-gap semiconductors at a low temperature results in a higher impedance in the forward bias state, and thus a larger forward voltage drop for a given current. This makes for a greater power loss within the device and greater internal heat generation under small-current conditions. For this reason germanium rectifiers are sometimes preferred where low forward-voltage drop at low current is a requirement and where good performance under high-current, high-temperature conditions is not important.

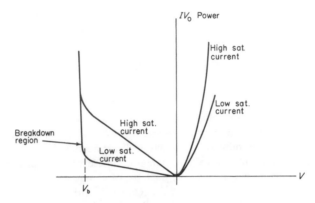

FIGURE 13.5. Power dissipation of a *p-n* junction rectifier *versus* applied voltage.

The fractions of the total junction current carried by electrons and by holes are sometimes referred to as the electron and hole *injection efficiencies*, respectively. From (13.1-21) and (13.1-22) it is clear that these quantities are given by

$$\eta_n = \frac{-eJ_n(0)}{e(J_p(0) - J_n(0))} = \frac{1}{1 + \dfrac{L_n}{bL_p} \dfrac{p_{p0}}{n_{n0}}} \tag{13.1-27}$$

$$\eta_p = \frac{eJ_p(0)}{e(J_p(0) - J_n(0))} = \frac{1}{1 + \dfrac{bL_p}{L_n} \dfrac{n_{n0}}{p_{p0}}}. \tag{13.1-28}$$

In a symmetric junction, where $p_{p0} = n_{n0}$, the electron and hole currents are roughly equal, provided that the mobility ratio b is of the order of unity and that L_n and L_p do not differ greatly. In an asymmetric device, however, in which one region is much more heavily doped than the other, the above expressions show that most of the current flowing across the junction is carried by the carrier which is in the majority in the *heavily* doped region. Physically, this is true because majority carriers in the lightly doped region which could, for example, be injected into the other region if forward bias were applied, are in short supply relative to the majority carriers in the heavily doped region which might under similar conditions be injected into the lightly doped side. We shall see later that the injection efficiency of the emitter junction is an important factor in the gain of a junction transistor.

If the reverse bias voltage applied to a *p-n* junction rectifier is steadily increased, a point will be reached where the reverse current suddenly and abruptly increases, as shown in Figures 13.3 and 13.5. This phenomenon is referred to as *breakdown*, and the voltage at which it occurs is called the *breakdown voltage* of the device. If the device is operated much beyond the breakdown point, a great deal of heat will be generated internally since both I and $|V_0|$ are large, and the device may melt or burn out in a very short time. A brief discussion of the physical mechanisms involved in the breakdown of *p-n* junction rectifiers will be given later in this chapter.

13·2   CURRENTS AND FIELDS IN *p-n* JUNCTION RECTIFIERS

In arriving at the results of the preceding section it was assumed that *all* the charge carrier transport in the bulk *p-* and *n*-regions adjacent to the junction space charge region is purely diffusive. The physical justification for this assumption is that most of the voltage applied to a *p-n* junction device will appear across the junction space charge region, which is largely depleted of charge carriers and is therefore effectively of much higher resistivity than the regions on either side.

It will become apparent, however, that the form of the results obtained thereby actually *contradicts* this assumption; we must now try to determine whether this contradiction is serious enough so that we must abandon this approach and start over from the beginning. We shall see that in most cases it is not. For the moment let us focus our attention upon the *n*-region of the device ($x > x_{0+}$). It is clear that in this section of the device the particle fluxes J_n and J_p can be written as

$$J_n = -D_n \frac{dn_n}{dx} - n_n \mu_n E \tag{13.2-1}$$

and
$$J_p = -D_p \frac{dp_n}{dx} + p_n \mu_p E \tag{13.2-2}$$

where E is the electric field, which was heretofore neglected. Also, since there are no sources nor sinks for electric current within this bulk region we must have, in the steady state, as the equation of continuity for *electrical* current density $\mathbf{I}$,

$$-\frac{\partial \rho}{\partial t} = \nabla \cdot \mathbf{I} = 0 \tag{13.2-3}$$

where ρ is the charge density. In the one-dimensional system at hand, this means simply that

$$dI/dx = 0, \quad \text{or} \quad I(x) = e(J_p(x) - J_n(x)) = \text{const.} \tag{13.2-4}$$

The particle current expressions (13.2-2) may be transformed into a somewhat more useful form by noting that because of the electrical neutrality condition $\delta n = \delta p$, the majority carrier density can be written as

$$n_n = n_{n0} + (p_n - p_{n0}) \tag{13.2-5}$$

whence

$$\frac{dn_n}{dx} = \frac{dp_n}{dx}. \tag{13.2-6}$$

Substituting these relations into (13.2-1) and (13.2-2), inserting the resulting equations into the continuity expression (13.2-4), and solving for the field E, we may obtain

$$E = \frac{I - eD_p(b - 1)\dfrac{dp_n}{dx}}{e\mu_p[bn_{n0} + b(p_n - p_{n0}) + p_n]}. \tag{13.2-7}$$

The results obtained in the preceding section, as Equations (13.1-20) and (13.1-23), under the assumption that the effect of electric fields in the bulk n-region of the device outside the junction space charge region would be negligible in comparison with the effect of diffusive transport, may now be substituted into (13.2-7) to check the consistency of the results with that assumption. In Chapter 10, it was shown that in general an electric field did not alter the diffusive character of the concentration profile, unless the critical value given by (10.2-44) is exceeded. Noting that the appropriate ambipolar transport coefficients are D_p and μ_p, and using (13.1-9) and (10.2-11) this condition may be expressed as

$$|E| \ll \frac{kT}{eL_p}. \tag{13.2-8}$$

If the values of p_n and I given by (13.1-20) and (13.1-23) are substituted into (13.2-7), this equation assumes the form

$$E = \frac{I_0 + \dfrac{ep_{n0}D_p}{L_p}(b - 1)e^{-(x-x_{0+})/L_p}}{\sigma_0 + p_{n0}e\mu_p(b + 1)(e^{eV_0/kT} - 1)e^{-(x-x_{0+})/L_p}} \cdot (e^{eV_0/kT} - 1) \tag{13.2-9}$$

where $\sigma_0 = e\mu_p(p_{n0} + bn_{n0})$ is the equilibrium conductivity of the n-type material and where I_0 is given by (13.1-24). For $V_0 = 0$, the field as given by (13.2-9) vanishes; for sufficiently small bias voltages, either forward or reverse, then, condition (13.2-8) will be satisfied. In general, in the reverse bias range, the factor $(e^{eV_0/kT} - 1)$ must have values between zero and -1. Since the second term in the denominator above is ordinarily small compared to σ_0 for any value of reverse bias, due to the fact that $n_{n0} \gg p_{n0}$, the field magnitude will increase with increasing reverse bias, reaching a *maximum* value for $V_0 \to -\infty$. Under these conditions, neglecting the second term in the denominator for the reason just mentioned, (13.2-9) takes the form

$$E = \frac{I_0}{\sigma_0} + \frac{p_{n0}D_p(b - 1)e^{-(x-x_{0+})/L_p}}{\mu_p(p_{n0} + bn_{n0})}. \tag{13.2-10}$$

Using (13.1-24) and (10.2-11) this can be transformed to

$$E = \frac{kT}{eL_p} \cdot \frac{p_{n0}[1 - (b - 1)e^{-(x-x_{0+})/L_p}] + bn_{p0}\dfrac{L_p}{L_n}}{p_{n0} + bn_{n0}}. \tag{13.2-11}$$

If $L_p \sim L_n$ and $b \sim 1$, condition (13.2-8) must be fulfilled everywhere within the n-region for all values of reverse bias, because if the material in this region is reasonably extrinsic, $n_{n0} \gg p_{n0}$, n_{p0}. These findings are all consistent with the original hypothesis that the field was so small that all but diffusive transport effects could be neglected.

For large values of forward bias $e^{eV_0/kT}$ becomes large and positive. Under these circumstances the second term in the denominator of (13.2-9) may become as large as, or larger than, the first. In the limit of large forward voltage,[4] at the edge of the space charge region ($x = x_{0+}$), (13.2-9) reduces to

$$E = \frac{kT}{eL_p}\frac{b}{b+1}\left(1 + \frac{n_{p0}}{p_{n0}}\frac{L_p}{L_n}\right) \tag{13.2-12}$$

and it is clear that (13.2-8) is no longer satisfied. Under these circumstances the role of the field may be important. We must therefore not expect that the current-voltage law (13.1-23) will be obeyed for large forward bias voltages. Although only the n-region of the device has been investigated in detail, it is clear that similar considerations apply to the p-region as well.

We have verified the essential correctness of the use of the purely diffusive concentration profiles (13.1-19) and (13.1-20) under most circumstances in describing the distribution of excess carriers in the n- and p-regions of a p-n junction rectifier. It is now necessary, to complete the picture, to justify the assumption that the minority carrier *current* flowing into the junction space charge region is purely a diffusive current. This can be accomplished in a somewhat similar way. Again considering the n-region of the device in particular, it is apparent from (13.2-2) that the diffusive hole current component is much larger than the drift component whenever

$$D_p\left(\frac{dp_n}{dx}\right) \gg p_n\mu_p E, \tag{13.2-13}$$

where E is given by (13.2-9). If (13.1-20) and (13.2-9) are substituted into this equation, the result is

$$\frac{D_p}{L_p}e^{-(x-x_{0+})/L_p} \gg \mu_p(1 + (e^{eV_0/kT} - 1)e^{-(x-x_{0+})/L_p})$$

$$\times \left[\frac{I_0 + \dfrac{eD_p p_{n0}}{L_p}(b-1)e^{-(x-x_{0+})/L_p}}{\sigma_0 + p_{n0}e\mu_p(b+1)(e^{eV_0/kT}-1)e^{-(x-x_{0+})/L_p}}\right]. \tag{13.2-14}$$

---

[4] Under these circumstances, of course, the ambipolar transport coefficients (10.2-27) and (10.2-28) should be used in the continuity equations, because the excess carrier concentrations may be greater than the equilibrium majority carrier density. However, for materials where b is of the order of unity the discrepancy involved will not be large enough to invalidate the qualitative conclusions which are derived.

Far from the junction $(x \gg x_{0+})$, the factors $e^{(x-x_{0+})/L_p}$ become extremely small, and (13.2-14) reduces to

$$0 \gg \mu_p \frac{I_0}{\sigma_0};$$

(13.2-15)

and condition (13.2-13) is obviously violated. This is, of course, to be expected on purely physical grounds, since, for example, in the forward bias case, the excess minority carriers which are injected at the junction recombine with majority carriers as they diffuse farther and farther from the junction, and their concentration must for this reason become extremely small far from the junction plane. But if the excess minority carrier concentration becomes small, so also does its gradient and the diffusion current which is proportional thereto. The *total* current, however, must be everywhere the same, according to (13.2-4). The resulting deficit in diffusion current far from the junction is made up by a corresponding increase in *drift* current resulting from the presence of an electric field. In fact, it is clear from (13.2-9), that far from the junction the electric field E approaches the constant value $I_0(e^{eV_0/kT} - 1)/\sigma_0$ giving rise to a purely ohmic drift current free from *any* diffusion component or excess carrier effect in this region. This is just what must be expected intuitively.

Nearer the junction, of course, part of the drift component gives way to diffusion current, and at the junction the relative contribution of the diffusive component may be assessed by setting $x = x_{0+}$ in (13.2-14), whereupon that inequality reduces to

$$\frac{D_p}{L_p} \gg \mu_p e^{eV_0/kT} \left[\frac{I_0 + \dfrac{eD_p p_{n0}}{L_p}(b-1)}{\sigma_0 + p_{n0}e\mu_p(b+1)(e^{eV_0/kT} - 1)} \right]$$

(13.2-16)

Proceeding along a line of reasoning somewhat similar to that which was used previously in connection with (13.2-14), it can be shown that when $L_p \sim L_n$ and $b \sim 1$, (13.2-16) is satisfied for all values of reverse bias voltage and for small forward voltages. As before, also, for large forward bias voltages this condition may be violated. Again (except for large values of forward bias), these results are consistent with the original view that essentially all the minority carrier current flowing into the junction space charge region is diffusion current. The result has been proved explicitly only for the *n*-region, but similar conclusions obviously can be reached for the *p*-region. We are justified in concluding that even though the results of the calculations of the preceding section reveal the presence of electric fields in the bulk regions, in contradiction to the original hypothesis upon which those calculations were based, in the region of reverse bias and small forward bias the fields are so small that no serious violation of the assumption of diffusive minority carrier transport can arise.[5] The calculations under these circumstances are completely self-consistent. For larger values of forward bias voltage the calculated fields become so large as to threaten this self-consistency and we may no longer be confident that the original assumption of purely diffusive minority carrier transport within the bulk *p*- and *n*-regions of the device is at all valid. We shall

---

[5] In some cases, of course, because of ohmic drops within the bulk *p*- and *n*-regions, it will no longer be possible to identify V_0 as the total applied voltage, but only as the drop across the junction region.

discuss the properties of *p-n* rectifiers under these conditions in some detail in the next chapter.

It has already been noted in connection with (13.2-15) that within the bulk regions the diffusive and drift components of the total current are not everywhere the same, in spite of the fact that the total current itself is everywhere constant. The same remark can be made in regard to hole and electron current components. From Figure 13.4(a), for example, it is clear that on the *n*-side near the junction the hole current is much greater (in view of the large diffusive component) than it is far from the junction, where holes have become very scarce and where the hole current is almost exclusively drift current. In view of the requirement that the total current be everywhere the same, it is apparent that the electron current near the junction must be correspondingly smaller than it is far out in the bulk of the crystal. This balance is maintained by the field *E*. given approximately by (13.2-9), which has a much more important effect upon majority carrier currents (due to the much greater *number* of majority carriers present) than upon minority carrier currents, which, as we have shown above, are primarily diffusive in nature.

13.3 JUNCTION RECTIFIERS OF FINITE SIZE; THE EFFECTS OF SURFACES AND "OHMIC" END CONTACTS

The results of the previous section, although derived for a device of infinite extent along the *x*-axis, are clearly applicable to devices of finite size, provided only that the actual *n*- and *p*-regions are many times thicker than the respective diffusion lengths L_p and L_n, and that the quantity V_0 is regarded as the voltage across the junction space charge region rather than the voltage across the entire device. If the thicknesses of one or both bulk regions adjoining the junction space charge layers is of the order of, or smaller than the diffusion length in that region, then the calculations of Section 13.1 must be modified. To see how this may be done, consider first the rectifier shown in Figure 13.6(a), where the device structure is terminated by two free surfaces with surface recombination velocities s_p and s_n. The device is assumed to be large in extent in both *y*- and *z*-directions, so that one-dimensional continuity equations of the form (13.1-7) and (13.1-8) may be written. Although the general form of the solutions as expressed by (13.1-14) and (13.1-15) is still valid, it is slightly more convenient to write the solutions as

$$n_p - n_{p0} = A \cosh \frac{x + x_{0-}}{L_n} + B \sinh \frac{x + x_{0-}}{L_n} \qquad (x < 0) \qquad (13.3\text{-}1)$$

$$p_n - p_{n0} = C \cosh \frac{x - x_{0+}}{L_p} + D \sinh \frac{x - x_{0+}}{L_p} \qquad (x > 0). \qquad (13.3\text{-}2)$$

These solutions can easily be shown to satisfy Equations (13.1-7) and (13.1-8). The boundary conditions (13.1-17) and (13.1-18) are still applicable at the junction edges at x_{0+} and $-x_{0-}$, but now the boundary conditions (13.1-10) and (13.1-11) must be replaced by the surface boundary conditions

$$D_n\left(\frac{dn_p}{dx}\right)_{-d_p} = s_p(n_p(-d_p) - n_{p0}) \qquad (13.3\text{-}3)$$

and

$$-D_p\left(\frac{dp_n}{dx}\right)_{d_n} = s_n(p_n(d_n) - p_{n0}). \qquad (13.3\text{-}4)$$

The above continuity equations may be solved, just as before, with the new boundary conditions, to obtain the excess carrier concentration, and the minority carrier diffusion currents at the edges of the junction space charge regions may be

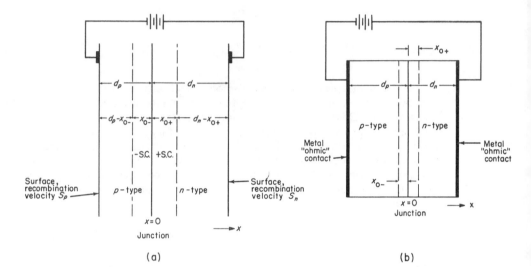

FIGURE 13.6. (a) Geometry of the finite *p-n* rectifier discussed in Section 13.3. (b) Geometry of a typical finite *p-n* rectifier which might be produced by an alloying or diffusion process.

found and combined, exactly as in Section 13.1, to yield the total current flowing across the junction. The same approximations with regard to diffusive carrier transport and neglect of generation and recombination within the space charge region are made in both instances. The results of the calculation using boundary conditions (13.3-3) and (13.3-4) are

$$p_n - p_{n0} = p_{n0}(e^{eV_0/kT} - 1)\left[\cosh\frac{x - x_{0+}}{L_p} - \zeta_n \sinh\frac{x - x_{0+}}{L_p}\right] \quad (n\text{-region}) \quad (13.3\text{-}5)$$

and

$$n_p - n_{p0} = n_{p0}(e^{eV_0/kT} - 1)\left[\cosh\frac{x + x_{0-}}{L_n} + \zeta_p \sinh\frac{x + x_{0+}}{L_n}\right], \quad (p\text{-region}) \quad (13.3\text{-}6)$$

where
$$\zeta_n = \frac{s_n \cosh \dfrac{d_n - x_{0+}}{L_p} + \dfrac{D_p}{L_p} \sinh \dfrac{d_n - x_{0+}}{L_p}}{\dfrac{D_p}{L_p} \cosh \dfrac{d_n - x_{0+}}{L_p} + s_n \sinh \dfrac{d_n - x_{0+}}{L_p}} \qquad (13.3\text{-}7)$$

and
$$\zeta_p = \frac{s_p \cosh \dfrac{d_p - x_{0-}}{L_n} + \dfrac{D_n}{L_n} \sinh \dfrac{d_p - x_{0-}}{L_n}}{\dfrac{D_n}{L_n} \cosh \dfrac{d_p - x_{0-}}{L_n} + s_p \sinh \dfrac{d_p - x_{0-}}{L_n}} \qquad (13.3\text{-}8)$$

while the current equation is found to be

$$I = I_0'(e^{eV_0/kT} - 1) \qquad (13.3\text{-}9)$$

with
$$I_0' = e\left(\frac{n_{p0}D_n}{L_n}\zeta_p + \frac{p_{n0}D_p}{L_p}\zeta_n\right). \qquad (13.3\text{-}10)$$

The current-voltage characteristic so obtained is of the same form as that which was derived for the infinite device, except that the saturation current is modified by the factors ζ_p and ζ_n, which involve the diffusion lengths, recombination velocities and bulk layer thicknesses. If $s_n = D_p/L_p$, it will be noted that $\zeta_n = 1$ and (13.3-5) reduces to (13.1-20). The effect of terminating the n-region of the device with a surface whose recombination velocity is equal to this *characteristic diffusion velocity* D_p/L_p is to leave the charge carrier distribution exactly what it would be if that region of the device were infinite in extent. If, in addition, $s_p = D_n/L_n$, then (13.3-6) likewise reduces to (13.1-19) and

$$I_0' = I_0 \qquad (13.3\text{-}11)$$

where I_0 is the saturation current of the infinite diode, as given by (13.1-24). If $s_n > D_p/L_p$ and $s_p > D_n/L_n$, then both ζ_n and ζ_p will have values greater than unity, and the saturation current will exceed that of the infinite device; if $s_n < D_p/L_p$ and $s_p < D_n/L_n$, ζ_n and ζ_p will be less than unity and the saturation current will be less than that of the infinite device. In the limit where s_n and s_p both approach zero, we may write

$$I_0' = e\left(\frac{n_{p0}D_n}{L_n} \tanh \frac{d_p - x_{0-}}{L_n^n} + \frac{p_{n0}D_p}{L_p} \tanh \frac{d_n - x_{0+}}{L_p}\right), \qquad (s_n = s_p = 0) \quad (13.3\text{-}12)$$

while if s_n and s_p are both very large compared with the respective characteristic diffusion velocity, the saturation current becomes

$$I_0' = e\left(\frac{n_{p0}D_n}{L_n} \operatorname{ctnh} \frac{d_p - x_{0-}}{L_n} + \frac{p_{n0}D_p}{L_p} \operatorname{ctnh} \frac{d_n - x_{0+}}{L_p}\right)$$

$$(s_n \gg D_p/L_p \,;\, s_p \gg D_n/L_n) \quad (13.3\text{-}13)$$

The large area metallic "ohmic" end contacts which are frequently used as end terminals, as illustrated in Figure 13.6(b), usually behave in the same way as surfaces of extremely high recombination velocity. It is reasonable to expect this behavior on physical grounds, for a contact which is electrically ohmic will not inject excess carrier currents, and, in fact, must act in such a way as to maintain the *equilibrium* carrier concentration in the adjacent semiconductor region. But this is just the way in which a surface of very high recombination velocity behaves, according to Equation (10.5-14). Equation (13.3-13) is therefore applicable to the rectifier structure shown in Figure 13.6(b).

As the reverse bias voltage on such a device is increased, the space charge layers adjacent to the junctions extend themselves further and further into the bulk *p*- and *n*-regions. The distances $d_p - x_{0-}$ and $d_n - x_{0+}$ then vary with voltage, in accord with (12.4-21) and (12.4-22), and the saturation current (13.3-10) is seen to depend somewhat upon the applied voltage as well as upon the other parameters which appear there. When d_n and d_p are both large compared with the greatest value which x_{0+} or x_{0-} may attain before the breakdown condition is reached, this variation of saturation current with bias voltage is insignificant, but if x_{0+} or x_{0-} become of the order of d_n or d_p at an experimentally attainable applied voltage, then the effect may be quite important. If the recombination velocity at the end terminal is very high, then, according to (13.3-13), the saturation current may become extremely large as the space charge region approaches the end contact, since $\lim_{x \to 0} \operatorname{ctnh} x = \infty$. An effect somewhat similar to breakdown is then observed, although (as we shall see) the physical causes of true breakdown lie in the bulk physical properties of the semiconductor, while the effect at hand, called *punch-through*, is critically dependent upon the dimensions of the device.

A physical understanding of punch-through may be obtained by recalling that the saturation current of a *p-n* rectifier is simply the current arising from the collection of thermally generated minority carriers by the reverse-biased junction. But according to (10.5-6), the equilibrium flux of thermally generated carriers produced at a surface of high recombination velocity is very large. Therefore, if the edge of the junction space charge region approaches a surface of high recombination velocity (or a metallic ohmic contact, which behaves in very much the same way), the reverse current is greatly increased by this surface thermal generation flux. If the space charge layer is separated from the surface or contact by a bulk region of appreciable thickness, the carriers generated at the surface must diffuse for some distance before being collected; during the time required for them to diffuse to the junction, they may recombine. Under these circumstances, the flux collected by the junction may be much less than the initial surface generation flux, and the reverse current of the junction will be correspondingly smaller. If the junction space charge region is many diffusion lengths from the surface or contact, the increase in saturation current due to surface generation may be negligible.

Ordinarily, punch-through effects may be avoided by making the *p*- and *n*-regions thick enough so that true bulk breakdown takes place before the space charge layers approach the end-contacts. Sometimes, however, the punch-through effect is utilized intentionally to obtain a device exhibiting an abrupt saturation current increase at a definite reverse voltage, which can be used as a voltage regulator element. The current-voltage characteristic of such a punch-through diode is shown in Figure 13.7.

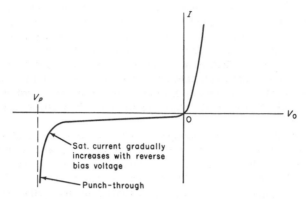

FIGURE 13.7. Current-voltage relation for a *p-n* rectifier of finite size, showing "punch-through" effect.

13·4 PHYSICAL MECHANISMS OF BREAKDOWN IN *p-n* JUNCTIONS

The punch-through phenomenon is an effect resembling breakdown, but its mechanism is related to the structural characteristics of the device rather than the bulk physical properties of the crystal. There are, however, two important mechanisms by which true breakdown of the crystal may occur. These are the *Zener mechanism*[6] and the *avalanche mechanism*.[7]

Zener breakdown is related to the phenomenon of *field emission*, in which electrons are emitted from a cold metallic cathode by the creation of an extremely high field at the cathode surface. The effect may be understood on the basis of the *tunneling* of electrons at or near the Fermi level of the metal through a surface potential barrier, as illustrated in Figure 13.8. In that figure, the surface of an ideal free electron metal is illustrated at (a) in the absence of external fields; the work function of the metal is ϕ_0. If an external voltage is applied such that the metal is negative with respect to an external electrode, there will be a field E between the two electrodes, which will be constant in a plane-parallel geometry. The corresponding potential will be of the form $\phi(x) = -Ex + \text{const}$, as shown at (b). As the externally applied potential difference is increased the slope of the potential becomes greater and greater, until only a very small distance separates electrons near the Fermi level of the metal and allowed states of the same energy outside the metal. Under these circumstances, the free electrons in the metal can *tunnel* through the thin surface potential barrier, creating a field emission current between the two electrodes. Very high external fields are required to produce appreciable field emission currents; at low fields the surface potential barrier is so thick that the associated tunneling probability is negligible.

[6] C. Zener, *Proc. Roy. Soc.* (*London*), **145**, 523 (1934).
[7] K. G. McKay and K. B. McAfee, *Phys. Rev.*, **91**, 1079 (1953).

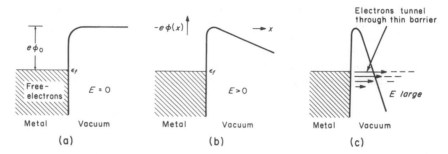

FIGURE 13.8. Potential diagram near a metal surface showing successive stages as an externally applied voltage is gradually increased. At (c) field emission occurs, electrons from within the metal tunneling through the thin surface barrier.

One should note that this is primarily a surface effect, and that the field within the metal is very small, due to the large conductivity.

In an insulator or semiconductor it is possible to attain rather high *internal* electric fields under realizable experimental conditions. Under such circumstances large gradients exist within the crystal, and the potential energy of an electron at the bottom of the conduction band, for example, is lower at one end of the crystal than at the other. The energy bands, when plotted in the usual way as a function of distance along the field direction are then tilted as shown in Figure 13.9. If the applied field, and thus the slope of the bands is sufficiently great, the potential energy barrier separating electrons at the top of the valence band from unfilled states of equal energy at the bottom of the conduction band becomes very narrow, and tunneling may occur in the interior of the crystal, as illustrated in Figure 13.9. This effect is, in actuality *internal* field emission; the mechanism described above was proposed by Zener[6] in 1934 to explain the phenomenon of bulk electrical breakdown in insulators or semiconductors.

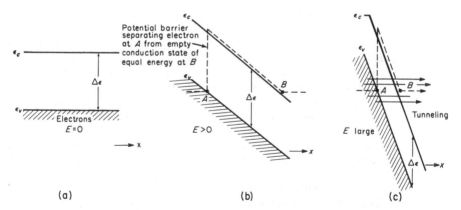

FIGURE 13.9. Potential diagram of a uniform bulk semiconductor showing successive stages as an externally applied voltage is gradually increased. At (c) *internal* field emission occurs, electrons tunneling through the thin potential barrier from valence to conduction band.

In uniform silicon or germanium crystals (except at very low temperatures, when the carriers contributed by donors and acceptors are "frozen out") it is impossible to obtain sufficiently high electric fields to observe bulk Zener breakdown without melting the crystal by ohmic heating. In a sample containing a *p-n* junction, however, there is a built-in electric field at the junction, and the magnitude of the field may be increased enormously by applying a reverse bias voltage to the device, as shown in Figure 13.10.

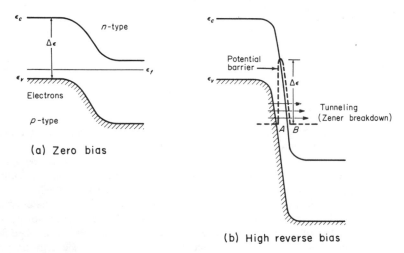

(a) Zero bias

(b) High reverse bias

FIGURE 13.10. Potential diagram of a *p-n* junction (a) at zero bias (b) under high reverse bias. At (b) Zener breakdown occurs when electrons tunnel through the thin barrier from valence to conduction band.

It is clear that the tilt of the energy bands in the reverse biased *p-n* junction gives rise to a situation (analogous to that shown in Figure 13.9(c) for a bulk sample) in which tunneling can take place when the potential barrier separating electrons at the top of the valence band from unoccupied conduction band states of equal energy becomes sufficiently thin. This condition will in general be attained only by the application of more or less reverse bias to the junction, but is more easily arrived at in *abrupt p-n junctions*, wherein the total potential fall is confined to a minimum distance, and in junction structures *wherein both p- and n-regions are quite heavily doped*, resulting (according to Equation (12.4-24) and to the figures of Table 12.1) likewise in very narrow junction space charge regions wherein high fields exist even at zero applied voltage.

In the avalanche mechanism, breakdown of the junction occurs when electrons and holes acquire sufficient energy between collisions from the high electric field in the junction space charge region to create electron-hole pairs by impact ionization of covalent lattice bonds. This effect is similar to Townsend avalanche ionization of a gas, wherein electrons in a gaseous medium acquire enough energy between collisions from an applied field to ionize the atoms with which they collide. The electrons and holes which are thus newly created may in turn acquire enough energy from the field to create still other pairs by impact ionization, initiating a chain reaction which leads to very high-reverse currents.

For an avalanche to occur, an electron or hole must acquire enough energy from the field in traversing a distance of the order of the mean free path to create an electron-hole pair. The condition for avalanche breakdown is then reached when the field in the junction region reaches a critical value E_b such that

$$eE_b\lambda = \Delta\varepsilon$$

or

$$E_b = \frac{\Delta\varepsilon}{e\lambda}. \tag{13.4-1}$$

In germanium and silicon the mean free path at room temperature is of the order of 10^{-5} cm, and avalanche breakdown should thus be expected to occur only when the field intensity in the junction region exceeds about 100 kV/cm. Also, since in all but quite impure crystals, the mean free path is determined by interactions with phonons, and since as the temperature is increased the number of available phonons increases, the mean free path for phonon scattering decreases with rising temperature. Therefore the avalanche breakdown field and the avalanche breakdown voltage *increase* (albeit rather slowly) with rising temperature. This is in sharp contrast with the Zener breakdown field, which is practically independent of temperature, since it depends only upon the energy gap and the thickness of the tunneling barrier.

The actual avalanche breakdown voltage for an abrupt junction can be obtained, approximately, by noting from (13.4-1) and (12.4-24) that if V_b is the breakdown voltage and if $V_b \gg \phi_0$, we may write

$$E_b \cong \frac{\Delta\varepsilon}{e\lambda} = \sqrt{\frac{8\pi eV_b}{\kappa}\left(\frac{1}{N_d}+\frac{1}{N_a}\right)^{-1}}, \tag{13.4-2}$$

whence V_b is given by

$$V_b = \frac{\kappa(\Delta\varepsilon)^2}{8\pi e^3\lambda^2}\left(\frac{1}{N_d}+\frac{1}{N_a}\right). \tag{13.4-3}$$

Again, it is apparent that relatively light doping densities lead to high-breakdown voltages, and *vice versa*. Also, materials with large energy gaps are to be preferred for devices which must resist avalanche breakdown at high reverse voltages.

The type of breakdown that occurs in a given device depends largely upon the material and structural parameters of the device and upon the conditions under which it operates. In general, devices with very abrupt junctions and with quite large doping densities on each side of the junction break down by the Zener mechanism, while for devices with graded junctions or for abrupt junction devices in which one of the regions is only moderately doped, the avalanche mechanism is dominant. Sometimes devices which exhibit Zener breakdown at room temperature will break down by the avalanche mechanism if operated at far lower temperatures, due to the increase of the mean free path. Muss and Greene[8] have found that abrupt germanium *p-n* junctions in which the *p*-region doping concentration is very high (about 10^{18} cm^{-3}) exhibit, at room temperature, Zener breakdown when the donor concentration N_d on the

---

[8] D. R. Muss and R. F. Greene, *J. Appl. Phys.*, **29**, 1534 (1958).

n-side is greater than about 3×10^{15} cm^{-3}, both mechanisms simultaneously for 3×10^{15} cm$^{-3} > N_d > 7.5 \times 10^{14}$ cm^{-3}, and avalanche breakdown for $N_d < 7.5 \times 10^{14}$ cm^{-3}.

Semiconductor p-n junctions, if properly designed and fabricated, may have very sharp breakdown characteristics, breaking down accurately and reproducibly at a given voltage, which may be quite insensitive to temperature. Such devices can be used as voltage reference diodes, and units designed for this specific purpose are often called "Zener diodes," although, despite the nomenclature, the breakdown mechanism which is utilized is frequently the avalanche effect.

13.5 p-n JUNCTION FABRICATION TECHNOLOGY

In this section the more important technological aspects of practical p-n junction fabrication will be reviewed briefly. The discussion will be confined largely to the qualitative aspects of the subject, because to treat the problems which arise rigorously and quantitatively would require the introduction of background material in physical chemistry and metallurgy somewhat beyond the scope of this volume. It is hoped, nevertheless, that this discussion will be sufficient to give the reader a good grasp of the physical principles which are involved.

Experimentally there are three principal ways of preparing p-n junction devices; the grown junction method, the alloying method and the diffusion method. In the *grown junction* method, a crystal of semiconductor is grown, for example, from an initially n-type melt, which, while the crystal is being formed, is counterdoped by adding enough acceptor impurities so that the subsequently grown portion of the crystal is p-type. The resulting junction is usually graded rather than abrupt. Although accurate control can be maintained over the impurity concentrations on either side of the junction by this method, the difficulty of locating the junction in the grown crystal and the difficulty of attaching leads to narrow grown junction regions has prevented widespread adoption of the technique in manufacturing. It is, nevertheless of some historic importance because p-n junction devices were first prepared in this manner.

In the *alloying* method an alloy pellet or foil is melted upon a semiconductor base crystal. The alloy pellet contains impurities which will give rise to the conductivity type opposite to that of the original crystal; after it has melted the temperature is raised to a value high enough to allow the molten alloy to dissolve away the underlying semiconductor crystal to some extent, as shown in Figure 13.11(b). The liquid phase is then a solution of the semiconductor material in the liquid alloy which is approximately saturated at the maximum fabrication temperature. Since the equilibrium solubility of the semiconductor in the liquid metal normally increases with temperature, when the device is slowly cooled the solution will become supersaturated with respect to the solute and some of the dissolved semiconductor atoms will *recrystallize* at the liquid-solid interface, continuing the single-crystal structure of the substrate. Since the liquid phase contains impurities associated with the conductivity type opposite to that of the original crystal, the regrown material will be of the opposite conductivity type, and there will be a p-n junction at the interface between original and

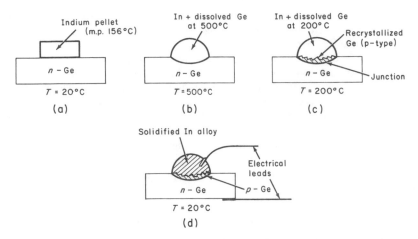

FIGURE 13.11.   Successive stages in the fabrication of an alloyed *p-n* rectifier.

regrown material. Upon further cooling the entire alloy phase solidifies; it is then a simple matter to make electrical connections to the original crystal and to the resolidified alloy pellet. The process is illustrated schematically in Figure 13.11.

   In practice *p-n* junctions can easily be made using *n*-type germanium as the base material and pure indium (m.p. 156°C) as the alloying agent. The recrystallized layer will then be saturated with indium, rendering that region *p*-type, and, in fact, very heavily *p*-type. If the base material is *n*-type germanium of moderate impurity content, then the device is strongly asymmetric, the *p*-region being much more heavily doped than the *n*-region. Even higher impurity concentrations in the *p*-region can be obtained by incorporating a small amount of gallium into the indium, since the equilibrium solid-phase solubility of gallium in germanium is higher than that of indium.[9] Recrystallized *n*-layers can be formed on *p*-type germanium base crystals in a similar manner, using lead-antimony or tin-antimony alloy foils or pellets. On silicon crystals alloyed junctions may be made using aluminium or gold-antimony alloying foils or pellets.

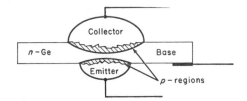

FIGURE 13.12. Cross sectional diagram of an alloyed *p-n-p* transistor.

   The alloying process generally produces junctions that are quite abrupt. It is very well suited to the requirements of manufacturing; in particular, it is quite easy to make two junctions simultaneously on either side of a thin base crystal, thus forming a *p-n-p* or *n-p-n* transistor structure to which it is very simple to attach emitter, base and collector leads (see Figure 13.12). Although it is sometimes difficult to exercise

----

[9] It is relatively simple to estimate the actual *p*-region impurity concentrations which result, using the equilibrium phase diagrams. An exercise on this subject is assigned at the end of this chapter.

the strictest dimensional control over devices made by the alloying process, the ease and simplicity of the technique has led to its widespread use as a production method for many different types of devices.

In the *diffusion* method, an impurity element is diffused at an elevated temperature into a base crystal whose conductivity type is opposite to that which is produced by the presence of the diffusing impurity in the crystal lattice. The depth of the junction below the surface of the crystal and the impurity concentration gradient at the junction can be controlled by controlling the diffusion time and temperature and the surface concentration of the diffusing impurity. If $c(x,t)$ is the concentration of the diffusing impurity at depth x below the surface after a time of diffusing t, and if D is the diffusion coefficient for the diffusion of the impurity within the crystal,[10] then $c(x,t)$ can be found as the solution of the diffusion equation

$$D \frac{\partial^2 c}{\partial x^2} = \frac{\partial c}{\partial t}.$$

(13.5-1)

As boundary conditions, one may assume that the surface concentration $c(0,t)$ is a given constant value c_0 for all t, whence

$$c(0,t) = c_0 = \text{const},$$

(13.5-2)

and also that

$$\lim_{x \to \infty} c(x,t) = 0.$$

(13.5-3)

The solution which satisfies these boundary conditions is

$$c(x,t) = c_0 \, \text{erfc}\left(\frac{x}{\sqrt{4Dt}}\right)$$

(13.5-4)

where

$$\text{erfc } \alpha x = 1 - \text{erf } \alpha x = 1 - \frac{2\alpha}{\sqrt{\pi}} \int_0^x e^{-\alpha^2 x'^2} \, dx'.$$

(13.5-5)

The concentration profile after any given diffusion time may be calculated from (13.5-5). The *p-n* junction will be found at the point where $c(x,t)$, the concentration of diffusing impurity, as given by (13.5-5), just equals the impurity density originally present in the base crystal. A typical family of concentration profiles as given by the above equation is shown in Figure 13.13 for several different values of diffusion time. Successive junction positions for each value of diffusion time are indicated.

Diffused junctions may be either graded or quite abrupt, depending upon the diffusion time, surface concentration and base crystal impurity density. Although

---

[10] The diffusion coefficient D used here is not to be confused with the diffusion coefficients D_n and D_p for electrons and holes. The diffusion coefficient we are discussing here is that associated with diffusion of a substitutional impurity in the lattice by a vacancy mechanism, as discussed in Chapter 11, and is much smaller than either D_n or D_p. Typical values for common substitutional impurities in Ge lie in the range $10^{-12} - 10^{-13}$ cm^2/sec at temperatures of 800°C.

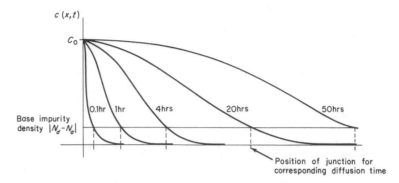

FIGURE 13.13. Schematic representaiton of impurity concentration profile within a diffused rectifier for several different diffusion times.

attachment of leads to diffused layers presents some technical problems, the ease and accuracy with which the thickness of diffused layers can be controlled has resulted in the widespread use of diffusion as a manufacturing technique. The steps in the production of a *n-p-n* diffused transistor are illustrated in Figure 13.14.

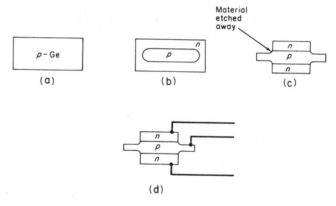

FIGURE 13.14. Successive stages in the fabrication of a diffused *n-p-n* transistor.

13·6 *p-n-p* AND *n-p-n* JUNCTION TRANSISTORS

A semiconductor structure equipped with an emitter electrode to inject excess minority carriers into a semiconductor crystal and a collector electrode to recover them is called a *transistor* and can be made to amplify current or voltage signals or used as an oscillator or switching device. The emitter and collector electrodes may be either point probe contacts or *p-n* junctions. The transistor as first invented by Shockley, Bardeen, and Brattain (1948) utilized point contact emitter and collector elements, but these devices have been entirely superseded on the commercial market by *p-n* junction transistors, which were invented by Shockley (1950). The Haynes-Shockley apparatus for measuring the mobility of injected carriers is actually a transistor, and can be used very conveniently to understand the basic principles of transistor action. A simplified diagram of this device is shown in Figure 13.15.

FIGURE 13.15. Schematic diagram of a point contact transistor and associated external circuitry. The sweep battery connected to the base of the device is not needed for transistor operation but is inserted in this diagram merely to show the relationship of the structure to that of the Haynes-Shockley filamentary transistor of Chapter 10.

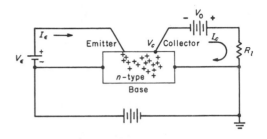

Suppose that a steady emitter current I_ε is made to flow in the forward direction from the emitter electrode in this diagram by a forward bias voltage V_ε. Holes are injected into the semiconductor crystal (here assumed to be *n*-type), drift toward the collector electrode and are there collected by the reverse biased collector probe, causing a collector current I_c to flow through the collector circuit and load resistance to ground. The emitter voltage V_ε required to make the current flow in the emitter circuit may be much smaller than the reverse voltage V_c across the collector. Under these circumstances, if nearly all the emitted holes reach the collector electrode,[11] the collector current will almost equal the emitter current. It is thus evident that a small change in emitter current caused by a small power input at the emitter can cause almost the same change in a collector current flowing through a circuit of much higher impedance, in which higher voltages are maintained, and produce a large change in the power delivered to the load resistance.

This can be shown in a more quantitative way by expressing the emitter current as a function of emitter voltage by the equation[12]

$$I_\varepsilon = I_{\varepsilon 0}(e^{eV_\varepsilon/kT} - 1),$$ (13.6-1)

where $I_{\varepsilon 0}$ is the emitter saturation current. Similarly we may write for the current-voltage relation at the collector

$$I_c = I_{c0}(e^{eV_c/kT} - 1) + \alpha I_\varepsilon$$ (13.6-2)

---

[11] In the Haynes-Shockley experiment, if the object is to measure accurately concentrations of excess carriers and drift times, the fraction of minority carriers which is collected by the collector is intentionally kept very small so as not to disturb the diffusion characteristics of the system which is being investigated. In any such system which is used as a transistor, however, the objective is to collect as large a fraction of the injected minority carriers as possible, so as to maximize the amplification factor associated with the device.

[12] It has been definitely established that a current-voltage relation of this form is to be expected for a *p-n* junction emitter. We shall see in a later chapter, however, that a relation of the same form is to be expected for a rectifying point-probe contact. Similar conclusions may be arrived at in regard to the current-voltage relation for the collector electrode.

where I_{c0} is the saturation current associated with the collector and α represents the fraction of the emitter current which is subsequently recovered as minority carrier current at the collector. In this equation, the first term represents the component of collector current due to the normal collection of minority carriers, which are thermally generated within the semiconductor crystal, by the reverse-biased collector electrode, while the second represents the additional component arising from minority carriers injected at the emitter which are picked up by the collector. In a well-designed transistor the factor α is quite close to unity, but cannot equal or exceed unity because some of the minority carriers injected by the emitter will fail to reach the collector due to recombination in the bulk of the semiconductor crystal or at crystal surfaces. In normal transistor operation, the emitter current is maintained at a value much in excess of $I_{\varepsilon 0}$ and the collector is reverse-biased far into the saturation region; under these circumstances, Equations (13.6-1) and (13.6-2) can be written, approximately, as

$$I_\varepsilon \cong I_{\varepsilon 0} e^{eV\varepsilon/kT} \qquad (I_\varepsilon \gg I_{\varepsilon 0}) \qquad\qquad (13.6\text{-}3)$$

and

$$I_c \cong I_{c0} + \alpha I_\varepsilon \qquad (|V_c| \gg kT/e). \qquad\qquad (13.6\text{-}4)$$

Equation (13.6-3) can be rewritten in the form

$$V_\varepsilon = \frac{kT}{e} \ln \frac{I_\varepsilon}{I_{\varepsilon 0}}, \qquad\qquad (13.6\text{-}5)$$

and from this it is clear that a small variation in emitter current, dI_ε, is related to a variation dV_ε in emitter voltage by

$$dV_\varepsilon = \frac{kT}{e} \frac{dI_\varepsilon}{I_\varepsilon}. \qquad\qquad (13.6\text{-}6)$$

The corresponding change in input power in the emitter circuit is given by

$$dP_\varepsilon = d(I_\varepsilon V_\varepsilon) = I_\varepsilon\, dV_\varepsilon + V_\varepsilon\, dI_\varepsilon = \frac{kT}{e}\left(1 + \ln\frac{I_\varepsilon}{I_{\varepsilon 0}}\right) dI_\varepsilon. \qquad (13.6\text{-}7)$$

In the output or collector circuit, however, from (13.6-4),

$$dI_c = \alpha\, dI_\varepsilon, \qquad\qquad (13.6\text{-}8)$$

and since the power delivered to the load resistor is $P_l = I_c^2 R_l$, we have

$$dP_l = 2I_c R_l\, dI_c = (2\alpha I_{c0} + 2\alpha^2 I_\varepsilon)\, R_l dI_\varepsilon, \qquad (13.6\text{-}9)$$

where I_c and dI_c have been related to I_ε and dI_ε by (13.6-4) and (13.6-8). However, if $\alpha \cong 1$ and $I_\varepsilon \gg I_{c0}$, which is usually the case, the first term in (13.6-9) is much smaller than the second, and may therefore be neglected, whereby this equation reduces to

$$dP_l = 2\alpha^2 I_\varepsilon R_l\, dI_\varepsilon. \qquad\qquad (13.6\text{-}10)$$

If dI_ε is expressed in terms of the input power dP_ε by (13.6-7) the ratio of output power to input power can be expressed as

$$\frac{dP_l}{dP_\varepsilon} = \frac{2\alpha^2 I_\varepsilon R_l}{\dfrac{kT}{e}\left(1 + \ln \dfrac{I_\varepsilon}{I_{\varepsilon 0}}\right)}.$$ (13.6-11)

This equation expresses the small signal dc power gain of the device. In a typical case, we might expect the values of the various parameters to be something like $I_\varepsilon = 10$ ma, $I_{\varepsilon 0} = 10$ μa, $\alpha = 0.99$, $R_l = 1000$ Ω and $kT/e = 1/40$ V at $300°$K. Substituting these values into the above equation, we obtain a power gain of about 100.

The current gain of the device is the ratio dI_c/dI_ε. From (13.6-8) it is clear that this quantity is simply α, and that in general, from (13.6-2) that the current gain may simply be represented as

$$\alpha = \left(\frac{\partial I_c}{\partial I_\varepsilon}\right)_{V_c}.$$ (13.6-12)

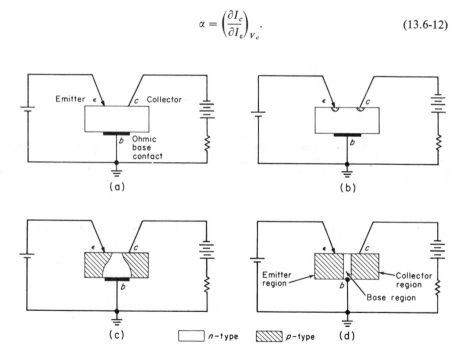

FIGURE 13.16. Successive stages in the "topological transformation" of the Haynes-Shockley filamentary transistor of Chapter 10 into the conventional planar *p-n-p* structure.

The Haynes-Shockley structure, which is shown in Figure 13.16(a) (without the sweep circuit, which is not required in a transistor, where diffusion alone is relied upon to transport excess carriers from emitter to collector) can be converted into a conventional *p-n-p* junction transistor by a "topological transformation" which alters the dimensions and proportions of the various parts of the device, but leaves their basic physical relationships unaltered. This is accomplished as illustrated in Figure 13.16. The point-contact emitter and collector electrodes are first replaced with two small *p*-type regions under the connecting leads, which are now regarded as making *ohmic* contacts to the *p*-regions as shown in Figure 13.16(b). This clearly does not affect the operation of the device, since the current-voltage characteristics of the rectifying point contacts and the *p-n* junctions are essentially the same. The *p*-type regions are then thought of as being extended as at (c) to include a larger volume of the crystal, finally arriving at the conventional *p-n-p* junction transistor configuration shown at (d). This structure, although different in appearance from the Haynes-Shockley filamentary transistor, works in very much the same way. The reason the structure at (d) in Figure 13.16 is adopted is so that the fraction of holes injected at the emitter which survive to be picked up at the collector (and thus the current gain of the device) shall be as large as possible.

The current gain α may differ from unity because of the recombination of excess carriers in the base region of the transistor; it may *also* differ from unity because not all the emitter junction current is accounted for by the injection of holes into the *n*-type base region from the *p*-type emitter. There is also a component of emitter current arising from the injection of electrons from the base region *into the emitter*, which causes no signal current at the collector but which flows through the base lead instead. This amounts to saying that the hole injection efficiency of the emitter junction η_p, as discussed in Section 13.1 is not quite unity, and the electron injection efficiency η_n is not zero. It is obviously advantageous for this reason to design a *p-n-p* transistor in such a way that the hole injection efficiency of the emitter junction is nearly unity and the electron injection efficiency is quite small. In any case, the current gain α can be expressed as the product of the minority carrier injection efficiency of the emitter and the fraction of minority carriers which, once injected into the base region of the device, survive to be picked up by the collector,[13] whereby, if the latter factor is denoted by f,

$$\alpha = \eta_p f. \tag{13.6-13}$$

The current gain of a *p-n-p* planar transistor structure, such as that shown in Figure 13.17, can be calculated by the methods of analysis which were developed in Section 13.1 in connection with the *p-n* junction rectifier. The steady-state continuity equations may be written

$$\frac{d^2(n_p - n_{p0})}{dx^2} - \frac{n_p - n_{p0}}{L_n^2} = 0 \qquad (x < 0 \text{ ; emitter region}) \tag{13.6-13}$$

$$\frac{d^2(p_n - p_{n0})}{dx^2} - \frac{p_n - p_{n0}}{L_p^2} = 0 \qquad (x > 0 \text{ ; base region}) \tag{13.6-14}$$

We assume hereby, as usual, that all voltage drops appear at the junctions and hence that there are no fields in the bulk regions, and also that the injected carrier densities

---

[13] In addition, there is a possibility that the collector efficiency is not unity, whereby some of the collector current will be due to the emission of minority carriers into the base region from the collector region. Under normal conditions of collector bias, however, this effect is so small as to be negligible in comparison with the two effects discussed above. We therefore take explicit account only of the two larger effects.

are everywhere small compared to the equilibrium majority carrier densities. We shall also assume that the effect of carrier generation and recombination within the space charge regions associated with the junctions is negligible, and that the extent of those regions is small in comparison with the thickness of the base region. Under these

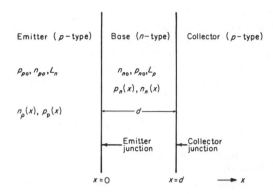

FIGURE 13.17. Geometry and notational usage for the analysis of the planar p-n-p transistor in Section 13.6.

circumstances, we need only apply boundary conditions and evaluate diffusion currents at $x = 0$ and $x = d$, avoiding the complications which the quantities x_{0+} and x_{0-} introduce into the equations of Section 13.1. Clearly, at the expense of a little more computational effort, any effects arising from the finite extent of the space charge layers could be accounted for by the methods used in Sections 13.1 and 13.3. As boundary conditions, we shall demand that

$$n_p(-\infty) - n_{p0} = 0 \tag{13.6-15}$$

(thus that the excess minority carrier density falls to zero far back within the emitter region), that

$$p_n(d) = 0, \tag{13.6-16}$$

whereby the excess minority carrier density is reduced to zero at the edge of the reverse-biased collector junction (this follows from the application of boundary condition (13.1-13) at the collector junction with the collector voltage large and negative), and, further, that at the emitter junction,

$$n_p(0) = n_{p0}e^{eV_e/kT} \tag{13.6-17}$$

$$p_n(0) = p_{n0}e^{eV_e/kT}. \tag{13.6-18}$$

The boundary conditions at the emitter junction correspond fully with those discussed in the case of the p-n rectifier. It is assumed, of course, that the emitter region extends for many diffusion lengths to the left of the emitter junction.

The general solution of these equations can be written in the form

$$n_p - n_{p0} = A e^{x/L_n} + B e^{-x/L_n} \qquad (x < 0) \tag{13.6-19}$$

and

$$p_n - p_{n0} = C \cosh \frac{x}{L_p} + D \sinh \frac{x}{L_p} \qquad (x > 0). \tag{13.6-20}$$

From (13.6-15), we see that $B = 0$; the boundary conditions for electrons and holes at the emitter junction then give

$$A = n_{p0}(e^{eV_e/kT} - 1) \tag{13.6-21}$$

and

$$C = p_{n0}(e^{eV_e/kT} - 1). \tag{13.6-22}$$

The collector boundary condition (13.6-16) may now be utilized to evaluate the one remaining coefficient, the result being

$$D = -p_{n0}\left[(e^{eV_e/kT} - 1) \operatorname{ctnh} \frac{d}{L_p} + \operatorname{csch} \frac{d}{L_p}\right], \tag{13.6-23}$$

whereby, according to (13.6-19) and (13.6-20)

$$n_p - n_{p0} = n_{p0}(e^{eV_e/kT} - 1)e^{x/L_n} \tag{13.6-24}$$

$$p_n - p_{n0} = p_{n0}\left[(e^{eV_e/kT} - 1)\cosh\frac{x}{L_p} - \left\{(e^{eV_e/kT} - 1)\operatorname{ctnh}\frac{d}{L_p} + \operatorname{csch}\frac{d}{L_p}\right\}\sinh\frac{x}{L_p}\right]$$

$$\tag{13.6-25}$$

The electron and hole currents across the emitter junctions are evaluated as before in connection with the *p-n* rectifier to give

$$J_n(0) = -D_n\left(\frac{dn_p}{dx}\right)_0 = -\frac{n_{p0}D_n}{L_n}(e^{eV_e/kT} - 1) \tag{13.6-26}$$

and

$$J_p(0) = -D_p\left(\frac{dp_n}{dx}\right)_0 = \frac{p_{n0}D_p}{L_p}\left[(e^{eV_e/kT} - 1)\operatorname{ctnh}\frac{d}{L_p} + \operatorname{csch}\frac{d}{L_p}\right] \tag{13.6-27}$$

The hole current density across the collector may be evaluated in the same way, the result being

$$J_p(d) = -D_p\left(\frac{dp_n}{dx}\right)_d = \frac{p_{n0}D_p}{L_p}\left[\left\{(e^{eV_e/kT} - 1)\cosh\frac{d}{L_p} + 1\right\}\operatorname{ctnh}\frac{d}{L_p}\right.$$

$$\left. - (e^{eV_e/kT} - 1)\sinh\frac{d}{L_p}\right]. \tag{13.6-28}$$

The hole injection efficiency of the emitter junction is the ratio of hole current to total current across the emitter. This can be calculated at once from (13.6-26) and (13.6-27). In doing so, it may be assumed that in the normal operation of the emitter, the actual emitter current is much larger than its saturation current, so that $e^{eV_e/kT} \gg 1$. The terms in the above equations not containing the exponential factor can then be neglected, and (recalling that $n_{no}p_{no} = p_{po}n_{po} = n_i^2$) it is easily shown that

$$\eta_p = \frac{eJ_p(0)}{e(J_p(0) - J_n(0))} = \frac{1}{1 + \dfrac{bn_{no}L_p}{p_{po}L_n} \tanh \dfrac{d}{L_p}}. \qquad (13.6\text{-}29)$$

The fraction of holes injected at the emitter which survive recombination to be picked up at the collector is, in the steady state,

$$f = \frac{J_p(d)}{J_p(0)} \cong \frac{\operatorname{ctnh} \dfrac{d}{L_p} \cosh \dfrac{d}{L_p} - \sinh \dfrac{d}{L_p}}{\operatorname{ctnh} \dfrac{d}{L_p}} = \operatorname{sech} \frac{d}{L_p}. \qquad (13.6\text{-}30)$$

Here, again, the terms in (13.6-27) and (13.6-28) which do not contain the exponential factor $e^{eV_e/kT}$ have been neglected. The current gain α may now be expressed as

$$\alpha = \eta_p f = \frac{1}{\cosh \dfrac{d}{L_p} + \dfrac{bL_p n_{no}}{L_n P_{po}} \sinh \dfrac{d}{L_p}}. \qquad (13.6\text{-}31)$$

This result has been obtained using the assumption that $V_e \gg kT/e$, but it may be shown by using a slightly more refined method of calculation that (13.6-31) is correct whether this condition is satisfied or not. The details of the argument are assigned as Exercise 13.6 at the end of this chapter. It is clear that in order to make α as large as possible, we must make the width of the base region as small as possible, and the diffusion length L_p (thus the lifetime of holes in the n-type base region) as large as possible; in addition, to provide maximum hole injection efficiency, the emitter doping density p_{po} should be much larger than the impurity density n_{no} in the base region of the device. The latter condition is automatically satisfied when germanium p-n-p transistors are made by the alloying process, using indium pellets to form the recrystallized emitter and collector regions and n-type germanium of moderate impurity density for the base.

Under the assumptions which have been made in this calculation, the current gain is found to be independent of emitter current. For very large values of emitter current, the excess carrier density in the base region near the emitter may equal or exceed the equilibrium majority carrier density there. When this happens, the simple form of the boundary conditions at the junction, as given by (13.1-12) and (13.1-13), can no longer be used, and the more general form (as described in footnote 1 in Section 13.1) must be adopted. It may then be shown that the hole injection efficiency of the emitter junction drops below the value given by (13.6-29), there then being more electron current across the junction than might be expected from the low-level theory.

This decrease in emitter efficiency leads to a decrease in current gain at large values of emitter current. The decrease in injection efficiency is suggested by the form of (13.6-29), wherein a decrease in the value of η_p for large emitter currents is obtained if the actual ratio $(n_n(0)/p_p(0))$ is substituted for the equilibrium ratio n_{n0}/p_{p0}, the actual ratio being much larger than the equilibrium ratio under conditions of large forward bias, due to the fact that the quasi-neutrality condition requires that when large excess concentrations of holes are injected into the *n*-region, they must be accompanied by equal excess electron concentrations.

The preceding analysis has been concerned with finding the current gain as a function of device parameters in the steady-state condition, thus for zero frequency. For sinusoidal signals it is observed that the current gain decreases with increasing frequency. The reason for this effect lies in the nature of the diffusion process. If a sinusoidal signal is applied to the emitter of a transistor, a sinusoidally varying concentration of excess carrier density is generated in the base region. Under these circumstances, carriers will diffuse out of regions of high concentration to "fill in" regions of low concentration, decreasing the signal amplitude, even if the lifetime in the base

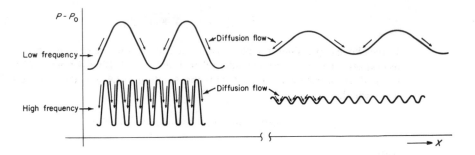

FIGURE 13.18. Excess carrier profiles within the base region of a *p-n-p* transistor associated with high and low frequency ac signals, showing the more drastic effects of diffusion in attenuating high frequency signals.

region is infinite, as illustrated in Figure 13.18. The higher the frequency, the steeper the concentration gradients associated with the local variations in excess carrier density, and the more pronounced the diffusion effects.

The variation of current gain with frequency may be understood quantitatively by considering the time-dependent continuity equation

$$D_p \frac{\partial^2(p_n - p_{n0})}{\partial x^2} - \frac{p_n - p_{n0}}{\tau_p} = \frac{\partial(p_n - p_{n0})}{\partial t}. \tag{13.6-32}$$

If we assume

$$p_n(x,t) - p_{n0} = u(x)e^{i\omega t} \tag{13.6-33}$$

and substitute this into (13.6-32), a differential equation of the form

$$\frac{d^2u}{dx^2} - \frac{1}{L_p^2}(1 + i\omega\tau_p)u(x) = 0 \tag{13.6-34}$$

is obtained for $u(x)$. The general solution to this equation has the form

$$u(x) = A \exp\left(\frac{x}{L_p} \sqrt{1 + i\omega\tau_p}\right) + B \exp\left(-\frac{x}{L_p} \sqrt{1 + i\omega\tau_p}\right)$$

$$= A \exp\frac{(\beta + i\gamma)x}{L_p} + B \exp\frac{-(\beta + i\gamma)x}{L_p}, \tag{13.6-35}$$

where $\qquad \beta = \text{Re}(\sqrt{1 + i\omega\tau_p}) \qquad$ and $\qquad \gamma = \text{Im}(\sqrt{1 + i\omega\tau_p}).$ $\qquad$ (13.6-36)

It is a relatively straightforward matter to calculate the real and imaginary parts referred to in (13.6-36). The results are

$$\beta^2 = \tfrac{1}{2}[1 + \sqrt{1 + \omega^2\tau_p^2}] \tag{13.6-37}$$

and

$$\gamma^2 = \tfrac{1}{2}[-1 + \sqrt{1 + \omega^2\tau_p^2}]. \tag{13.6-38}$$

In the above expressions, the plus sign of the radical must be chosen, since in the limit as ω approaches zero, we must obtain $\beta^2 = 1$ and $\gamma^2 = 0$. If we denote by L_p' the exponential decay constant associated with the solution (13.6-35), then, using (13.6-37), it is clear that

$$u(x) = Ae^{x/L_p'}e^{i\gamma x/L_p} + Be^{-x/L_p'}e^{-i\gamma x/L_p}, \tag{13.6-39}$$

with

$$\frac{1}{L_p'} = \frac{\beta}{L_p} = \frac{1}{\sqrt{2}L_p}\sqrt{1 + \sqrt{1 + \omega^2\tau_p^2}}, \tag{13.6-40}$$

where the value of L_p is just the steady state value given by (13.1-9). From (13.6-39) and (13.6-40) it is easily seen that at low frequencies, the decay constant L_p' associated with the solution of the continuity equation is given by

$$L_p' \cong L_p = \sqrt{D_p\tau_p} \qquad (\omega\tau_p \ll 1) \tag{13.6-41}$$

which is just the steady-state value, while for increasing frequencies, L_p' decreases from this value, approaching in the high-frequency limit the value

$$L_p' \cong L_p\sqrt{\frac{2}{\omega\tau_p}} = \sqrt{\frac{2D_p}{\omega}} \qquad (\omega\tau_p \gg 1). \tag{13.6-42}$$

A plot of L_p' versus frequency is shown in Figure 13.19(a).

From this we see that under ac conditions, the *effective* decay distance which must be used in reckoning the fraction of the input signal which arrives at the collector is L_p' rather than L_p. The value $L_p'(\omega)$ as given by (13.6-40) should thus be substituted into the equation (13.6-30) which gives the decay factor f. The result is

$$\alpha(\omega) = \eta_p f = \frac{\text{sech} \dfrac{d}{L'_p(\omega)}}{1 + \dfrac{bn_{no}L_p}{p_{po}L_n} \tanh \dfrac{d}{L_p}}. \tag{13.6-43}$$

Under the normal circumstances, at zero frequency, $L'_p(0)$ will be much greater than the base width d, and $f = \text{sech } (d/L'_p)$ will be approximately unity. With increasing frequency, a point will be reached where L'_p is comparable in magnitude with d, and

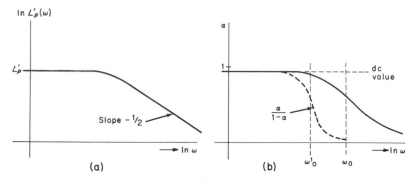

FIGURE 13.19.   (a) A plot of $L'_p(\omega)$ as given by (13.6-40) *versus* ω. (b) A plot of current gain *versus* frequency according to the results of Section 13.6, showing the α- and β- cutoff frequencies ω_0 and ω'_0.

in this region of frequency the value of f, and thus of α will become significantly smaller than unity as shown in Figure 13.19(b). The current gain in this region will begin to be reduced in comparison with its zero frequency value. The "α-cutoff frequency" of the device, ω_0, may be defined by

$$L'_p(\omega_0) = d. \tag{13.6-44}$$

Substituting this value of L'_p into (13.6-40) and solving for the frequency yields

$$\omega_0 = \frac{2\sqrt{D_p}}{d} \sqrt{\frac{D_p \tau_p}{d^2} - 1}. \tag{13.6-45}$$

For frequencies in excess of this value the current gain will be much less than its zero frequency value. It is apparent that to achieve high cutoff frequencies small values of the base width and large values of lifetime in the base region are required.

Before leaving the subject of transistors, we must briefly discuss the grounded emitter circuit and the open-base circuit for these devices. In the circuit of Figure 13.16 (grounded-base circuit) it was assumed that a current increment dI_e was introduced into the emitter electrode. In the grounded emitter circuit, as shown in Figure 13.20, a current signal dI_b is introduced into the *base* lead. It is obvious from the continuity of current that we must always have

$$I_\varepsilon = I_c + I_b \qquad (13.6\text{-}46)$$

whence

$$dI_\varepsilon = dI_c + dI_b. \qquad (13.6\text{-}47)$$

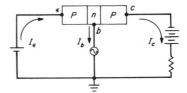

FIGURE 13.20. The grounded emitter circuit for a *p-n-p* transistor.

However, by definition, $dI_\varepsilon = dI_c/\alpha$, whereby

$$dI_c = \frac{\alpha}{1-\alpha} dI_b = \beta dI_b \qquad (13.6\text{-}48)$$

The quantity

$$\beta = \frac{dI_c}{dI_b} = \frac{\alpha}{1-\alpha} \qquad (13.6\text{-}49)$$

is the *grounded emitter current gain* of the transistor. It is clear from this that the grounded emitter circuit configuration can be utilized to provide current amplification factors greater than unity. The grounded emitter configuration is the circuit arrangement which is most commonly used in transistor amplifiers, and is somewhat analogous electrically to the usual grounded cathode vacuum tube amplifier, the emitter corresponding to the cathode, the base to the grid and the collector to the plate of the vacuum tube. The cutoff frequency for the grounded emitter connection is lower than for the same transistor in the grounded base configuration, since if the zero frequency value of α is nearly equal to unity, as is usually the case, only a small reduction of α is required to produce a much larger percentage reduction in $\alpha/(1-\alpha)$. This is illustrated by the dashed curve of Figure 13.19(b), where the β-cutoff frequency ω_0', at which the grounded emitter gain is reduced by the same factor as the grounded base gain at the α-cutoff frequency ω_0, is seen to be considerably lower than the α-cutoff frequency.

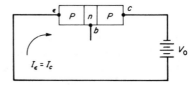

FIGURE 13.21. Self-amplification of a collector current in an open-base circuit.

Another effect which is of some interest in junction transistor operation is the self-amplification of the collector saturation current in an open-base circuit such as that shown in Figure 13.21. Here the emitter and collector currents must be equal, and therefore, provided that the magnitude of the battery voltage is much larger than kT/e, we must have, according to (13.6-2)

$$I_c = I_{c0} + \alpha I_\varepsilon = I_{c0} + \alpha I_c \tag{13.6-50}$$

whence

$$I_c = \frac{I_{c0}}{1 - \alpha}. \tag{13.6-51}$$

In this circuit, then, the collector current is self-amplified by a factor $1/(1 - \alpha)$. This effect provides a rough but very simple way of measuring the dc current gain of a junction transistor. The collector saturation current I_{c0} is first determined by opening the emitter circuit and completing the base circuit and noting the amount of current which flows. The base circuit is then opened and the emitter circuit closed, as shown in Figure 13.21, and the self-amplified collector current in this mode of operation determined. The current gain may then be determined by (13.6-51). This self-amplification of saturation current also provides the basis for an understanding of the operation of four-layer *p-n-p-n* switching devices, which will be discussed in detail in a later chapter.

Although the discussions of this section have been concerned exclusively with *p-n-p* junction transistors, it is obvious that an analysis of *n-p-n* units can be carried out in exactly the same way, leading to equations which are the same as those derived above with the exception that the subscripts *n* and *p* are interchanged, and the quantity *b* is replaced by $1/b$. It should be noted, however, that the bias voltages which must be used in connection with *n-p-n* units are reversed in sign with respect to those which are appropriate with *p-n-p* transistors.

EXERCISES

1. Discuss the limiting form of the current-voltage characteristic of an ideal *p-n* junction rectifier, such as that discussed in Section 13.1, as the temperature approaches zero. What determines the *maximum* temperature of practical operation for the device?

2. Show that the application of the exact boundary conditions (12.3–1) and (12.3–2) to the solutions (13.1–14) and (13.1–15) for the *p-n* junction rectifier of Section 13.1, wherein the variation of *majority* carrier concentration due to the presence of the excess minority carrier distribution is not neglected, leads to a current-voltage characteristic of the form

$$J = \frac{(e^{eV_0/kT} - 1)}{1 - e^{-2e(\phi_0 - V_0)/kT}} \left[\frac{ep_{n0}D_p}{L_p} \left(1 + \frac{n_i^2}{p_{p0}^2} e^{eV_0/kT}\right) \right.$$

$$\left. + \frac{en_{p0}D_n}{L_n} \left(1 + \frac{n_i^2}{n_{n0}^2} e^{eV_0/kT}\right) \right].$$

You may assume that the mobility ratio b is sufficiently close to unity that the concentration profiles (13.1-14) and (13.1-15) are not seriously affected by ambipolar transport phenomena. Note that this equation predicts that $J \to \infty$ as $V_0 \to \phi_0$, which means, in effect, that the internal junction potential barrier is never completely "flattened out" even for extremely large currents. The expression clearly reduces to the previous result for reverse voltages and forward voltages which are small compared with ϕ_0.

Experimentally, under high current conditions, rather than behaving as predicted above, the deviation of the current-voltage curve from the ideal form as given by (13.1-23) is found to be in just the opposite direction, as shown in the accompanying diagram. The reason for this lies not so much in the fact that the above expression does not correctly represent the relation between the junction voltage drop and the current, but rather in the fact that (a) at high currents, large additional ohmic voltage drops occur in the bulk regions on either side of the junction, and (b) under high current conditions an alloyed rectifier structure of finite size may act as a three-region p-(intrinsic)-n structure, where all the carriers injected into the moderately doped n-region are constrained to remain there until they recombine. This region then becomes flooded with electron-hole pairs, the concentration of these pairs rising to a value much in excess of the normal equilibrium majority carrier concentration, whereupon the region behaves very much like an intrinsic semiconductor layer. The metallic contact on the opposite side of this layer, usually made with a solder containing a fairly large con-

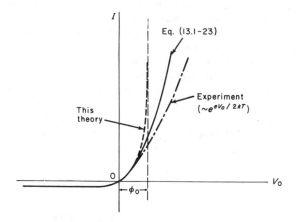

centration of donor impurities, forms a thin, heavily doped, recrystallized n-type layer which terminates the other side of the device. An investigation of the characteristics of a device of this type (such as that set forth in the next chapter) results in a current-voltage relation which is in quite good agreement with those observed in actual rectifiers of this type under such conditions.

3. At 600°C the germanium content of the liquid indium-germanium phase in equilibrium with solid germanium is about 20 atomic percent, as determined from the In-Ge binary phase diagram. Using the equilibrium K-value as stated in Table 11.1, estimate the acceptor concentration in the recrystallized p-layer of an alloyed p-n rectifier made by melting an indium pellet on n-type germanium, raising the temperature to 600°C, and gradually cooling. Perform the same calculation for a rectifier made in the same way using an alloy of indium containing 5 atomic percent gallium. Which alloy would be preferred for forming the emitter of a p-n-p junction transistor?

4. Assume that a small ac signal voltage $V_1 e^{i\omega t}$ is superimposed upon a steady dc voltage V_0 applied to the p-n rectifier of Section 13.1. Assuming that $V_1 \ll kT/e$, find a solution of the time dependent continuity equation for holes in the n-region which reduces to the correct dc value for $\omega = 0$ and which approaches zero in the limit where $x \to \infty$. For simplicity you may neglect the thickness of the junction space charge region in comparison with the diffusion length.

5. Using the solution obtained in Problem 4, find the ac current-voltage characteristic of the *p-n* junction of Section 13.1. Show that the ac impedance of the device is similar to that of a resistance and a capacitor in parallel. Find expressions for the equivalent capacitance in the low- and high-frequency limits, and note that although the equivalent capacitance is in general a function of frequency, in the low-frequency limit it becomes frequency independent just like that of an ordinary condenser. The capacitance so obtained is associated with the storage and diffusion of excess charge carriers in the bulk *p*- and *n*-regions of the device, and is quite distinct from the space-charge capacitance discussed in Section 12.5. This component of the device capacitance is sometimes called the *injection capacitance* or the *diffusion capacitance*.

6. By using the exact expressions (13.6-26), (13.6-27), and (13.6-28) to express the emitter and collector currents and by calculating directly from the definition (13.6-12) the quantity

$$\alpha = \frac{dI_c}{dI_\varepsilon} = \frac{dI_c}{dV_\varepsilon} \cdot \frac{dV_\varepsilon}{dI_\varepsilon}$$

show that the expression (13.6-31) for α is obtained independently of any assumption as to the magnitude of the emitter voltage. The method adopted in the text is used there because it exhibits more clearly the separate roles of the injection efficiency and the transport factor.

GENERAL REFERENCES

J. K. Jonscher, *Principles of Semiconductor Device Operation*, G. Bell and Sons, Ltd., London (1960).

Allen Nussbaum, *Semiconductor Device Physics*, Prentice-Hall, Inc., Englewood Cliffs, N.J. (1964), Chapter 4.

W. Shockley, "The Theory of *p-n* Junction Semiconductors and Junction Transistors," in *Bell System Technical Journal* **28**, 435–489 (1949).

A. van der Ziel, *Solid State Physical Electronics*, Prentice-Hall, Inc., Englewood Cliffs, N.J. (1957), Chapters 13 and 14.

Proc. Inst. Radio Engrs., "Transistor Issues," **30** (December 1952) and **46** (June 1958).

CHAPTER 14

p-n JUNCTIONS AT HIGH CURRENT LEVELS; THE *p-i-n* RECTIFIER

14·1 *p-n* JUNCTIONS AT HIGH CURRENT DENSITIES

The validity of the analysis of the *p-n* junction rectifier given in Section 13.1 is restricted to the region of reverse bias, and to values of forward bias voltage small compared to the equilibrium internal barrier potential ϕ_0. This restriction arises for several reasons, the most important of which are as follows:

(i) In the high-current forward bias region the simple boundary conditions (13.1-12) and (13.1-13) are no longer valid, because the concentration of injected minority carriers in the bulk *p*- and *n*-regions near the junction may be comparable to the equilibrium *majority* carrier density there. Under circumstances such as these, the majority carrier distribution which accompanies the minority carriers to insure quasi-neutrality, is sufficiently dense to seriously perturb the majority carrier concentration near the junction from its equilibrium value. The exact boundary conditions (12.3-1) and (12.3-2) should then be used rather than the approximate ones (13.1-12) and (13.1-13).

(ii) At high forward current levels the injected minority carrier density near the junction may be so large that the total majority and minority carrier densities in that region are of comparable magnitude. The ambipolar transport coefficients D^* and μ^* as given by (10.2-34) and (10.2-35) should then be used in the continuity equation rather than the minority carrier diffusion coefficient and mobility which are applicable only at low excess carrier densities. If the mobility ratio b is large compared to unity (as it may well be in some of the III–V intermetallic semiconductors), this effect may be of great importance, while if b is relatively small (as it is in silicon and germanium) no very large change in device characteristics will be found.

(iii) At high-current levels, the neglect of electric fields arising from concentration gradients and from ohmic voltage drop in the bulk *p*- and *n*-regions is no longer justified.

A correction to the current-voltage characteristic of the *p-n* rectifier of Section 13.1 can be made, taking into account *only* the effect discussed in (i) above. The result is given and discussed in connection with Exercise 13.2; the reader is referred to that discussion for the details of the situation. The conclusion is that in general the effects in (ii) and (iii) above, along with certain other phenomena which will be described shortly, are of such importance that the result so obtained is hardly ever applicable *by itself* in explaining the current-voltage relation of actual devices in the region of substantial forward bias.

In addition to the three effects discussed above, the high-current level characteristics of many actual *p-n* rectifiers (especially those which are made by alloying processes) differ from those expected on the basis of low-current *p-n* junction theory because in the high current region such devices act as structures having *three* distinct regions, *p*-type, intrinsic, and *n*-type, separated by *two* junctions (*p*-intrinsic and intrinsic-*n*). Such a structure is usually referred to as a *p-i-n* rectifier. The reason for this state of affairs can be understood by referring to Figure 14.1, which shows, in cross section,

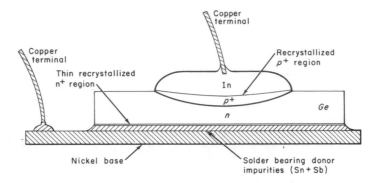

FIGURE 14.1.   Cross-sectional diagram of an alloyed rectifier, showing p^+-n-n^+ structure.

the construction of an actual alloyed *p-n* junction rectifier. A heavily doped *p*-region has been formed by alloying an indium pellet to an *n*-type germanium wafer of moderate donor impurity content,[1] typically of the order of 10^{15} donors per cm³. The heavily doped region is often referred to as the p^+ region; we shall adopt this notation and use it consistently. During the same heating cycle in which the p^+-n junction is formed, the germanium base wafer is usually soldered to a nickel base plate with a tin solder bearing *donor* impurities such as antimony. The objective is to make an "ohmic" contact between the germanium and the nickel; clearly a solder containing acceptor impurities would tend to dissolve some germanium and deposit a recrystallized p^+ layer on cooling to form a second n-p^+ junction whose nature would be decidedly, and undesirably, nonohmic. It is therefore customary to use a solder which contains a small concentration of donor impurities. In this case, a recrystallized n^+-layer is formed in the same way between the germanium and the solder, and there is an n-n^+ junction at the interface between this layer and the original crystal. The resulting structure is a p^+-n-n^+ device, but at low-current densities, such as prevail under reverse bias or small forward bias conditions, the n-n^+ junction acts somewhat like a *p-n* junction having extremely large saturation current and thus rather poor rectification characteristics (see Equation (13.1-24)), thus approximating the behavior of an ohmic contact. At larger values of forward bias voltage, however, a large number of excess holes are injected by the p^+-n junction, and if this excess hole concentration is of the same order of magnitude as, or larger than, the equilibrium electron density in the central moderately doped *n*-region, then the total concentrations of electrons and holes in that region will be comparable,[2] and the region will behave in very much the

---

[1] Larger donor concentrations would result in excessive electric field values in the space charge region, and thus in reduced avalanche breakdown voltages. (See Table 12.1.)

[2] This will be seen to be true *despite* the fact that the excess hole concentration is accompanied by an equal excess electron concentration so as to satisfy the requirements of quasi-neutrality.

same way as an *intrinsic* semiconductor crystal. The high-current representation of the device as an intrinsic layer bounded on the left by a p^+-intrinsic junction and on the right by an intrinsic-n^+ junction, thus as a p^+-i-n^+ rectifier, is easily seen to be justified. The analysis of such a device is illustrative of the general techniques which must be employed for all devices at high-current levels.

14·2 THE ANALYSIS OF THE p^+-i-n^+ RECTIFIER AT HIGH CURRENT LEVELS

In this section we shall examine the properties, in the high current region, of a one-dimensional p^+-i-n^+ device such as that illustrated in Figure 14.2. In this device the equilibrium majority carrier concentrations p_{po} and n_{no} in the p^+- and n^+-regions are assumed to be much larger than the equilibrium majority carrier concentration n_0

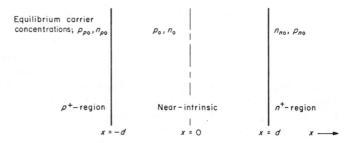

FIGURE 14.2. Geometry and notational usage for the planar p-i-n rectifier discussed in Section 14.2.

in the central moderately doped region. The finite penetration of the junction space charge regions into the central layer can be neglected (for forward bias conditions) in comparison with the total thickness $2d$ of this layer; the recombination of excess electrons and holes within the space charge regions is likewise neglected. It is assumed, however, that the electrical quasi-neutrality associated with ambipolar excess carrier diffusion is maintained everywhere outside the junction space charge layers.

It will also be assumed that *all* the current crossing the junction at $x = -d$ is *hole current* and that *all* the current crossing the junction at $x = +d$ is *electron current*. This amounts to assuming that the hole injection efficiency of the p^+-i junction and the electron injection efficiency of the i-n^+ junction are equal to unity. The hole injection efficiency (as given, in the low level case, by (13.1-28)) is dependent upon the ratio of the concentration of electrons in the n-region to that of holes in the p-region. If this ratio is small the efficiency will be practically unity; if not, it may be significantly less. So far as the p^+-n junction is concerned, at zero current this ratio is small, and will indeed remain small until the excess carrier concentration in the central region is comparable to the equilibrium hole concentration in the p^+-region. The assumption, although still justified for injected excess carrier densities in the central region which are well above the normal equilibrium majority carrier density n_0 *in that region*, will break down when this excess density approaches p_{p0}. Our analysis, then, although extending well into the high-current region, will, on this account, be of doubtful validity beyond a certain point. Similar arguments may be made concerning the electron injection efficiency at the n-n^+ junction, leading to the conclusion that the assumption of unit injection efficiencies is a good one so long as p_{p0} and n_{n0} are both large compared with the largest of the three quantities $(p_0, n_0, \delta p)$, where δp is the excess carrier density in the central region. This assumption also is equivalent to the statement that all the holes injected at the left-hand junction in Figure 14.2 recombine in the region $-d < x < +d$ rather than in the n^+-region at the right, and similarly for electrons injected at the right-hand junction. While it is true that injected electrons and holes are constrained to remain within the central region by the electrostatic potential barriers at the junctions, as illustrated by Figure 14.3, it is clear that for sufficiently

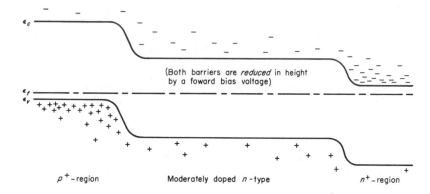

FIGURE 14.3.   Potential diagram for *p-i-n* rectifier.

high forward bias voltages excess carriers may spill over into the end regions before recombining. Nevertheless, for the range of device structures of interest the assumption may be expected to be quite good to values of current far beyond the limits of the low-level theory. For conditions under which it is violated, the physical situation becomes very complex, involving both high- and low-level regions which are not well defined and introducing very complex boundary conditions into the problem.

Under the assumed conditions, the steady-state ambipolar continuity equation (10.2-26) describing the excess carrier density $\delta p(x) = \delta n(x)$ in the central region, which is assumed to behave essentially as an intrinsic semiconductor at the current levels which are involved, is

$$\frac{d^2(\delta p)}{dx^2} - \frac{\delta p}{L_i^2} = 0 \tag{14.2-1}$$

where

$$L_i^2 = D_i^* \tau = \frac{2b}{b+1} D_p \tau. \tag{14.2-2}$$

Here L_i is the ambipolar diffusion length, D_i^* the intrinsic ambipolar diffusion constant, as given by (10.2-36), τ the excess carrier lifetime and b the mobility ratio

$\mu_n/\mu_p = D_n/D_p$. Since the ambipolar mobility μ^* is zero in the intrinsic condition, the field term in (10.2-26) need not be retained. If τ is independent of δp and x, which we shall assume to be the case, the general solution to (14.2-1) can be written in the form

$$\delta p = p - p_0 = n - n_0 = A \sinh \frac{x}{L_i} + B \cosh \frac{x}{L_i}. \qquad (14.2\text{-}3)$$

The equations describing the flow of electrical current in the intrinsic region as a sum of diffusion and drift components are, according to (10.2-8) and (10.2-9),

$$I_p = pe\mu_p E - eD_p \frac{dp}{dx} \qquad (14.2\text{-}4)$$

and

$$I_n = ne\mu_n E + eD_n \frac{dn}{dx}, \qquad (14.2\text{-}5)$$

where E is the electrical field intensity. As a boundary condition it has been assumed that

$$I_n(-d) = I_p(d) = 0 \cdot \qquad (14.2\text{-}6)$$

Setting $I_p = 0$ and $x = d$ in (14.2-4), noting from (14.2-3) that $dp/dx = dn/dx$, solving for E and inserting the value so obtained into (14.2-5), it can be shown that

$$\frac{I}{e} = \frac{I_n(d)}{e} = D_n \frac{dp}{dx}\left(1 + \frac{n}{p}\right) \qquad (\text{at } x = d). \qquad (14.2\text{-}7)$$

Likewise, setting $I_n = 0$ and $x = -d$ in (14.2-5), solving for E and substituting the resulting value into (14.2-4), it is found that

$$\frac{I}{e} = \frac{I_p(-d)}{e} = -D_p \frac{dp}{dx}\left(1 + \frac{p}{n}\right) \qquad (\text{at } x = -d). \qquad (14.2\text{-}8)$$

Of course, the total current density I must be the same at all points, just as in the case of the p-n rectifier of section 13.1, as explained in connection with Equations (13.2-3) and (13.2-4). In the central region, n is given as the sum of the equilibrium density n_0 and the excess electron density δp. Likewise p is equal to $p_0 + \delta p$. But we are investigating the device under conditions of current and forward bias such that δp is much larger than either n_0 or p_0. Under these conditions $p/n \cong n/p \cong 1$, and (14.2-7) and (14.2-8) become

$$\frac{I}{e} = -2D_p(dp/dx)_{-d} = 2bD_p(dp/dx)_d. \qquad (14.2\text{-}9)$$

Now, writing dp/dx as a function of x from (14.2-3), substituting the boundary conditions, (14.2-9) for $x = d$ and $x = -d$, and solving for the constants A and B, we may obtain

$$A = -\frac{IL_i(b-1)}{4ebD_p \cosh \dfrac{d}{L_i}} \quad \text{and} \quad B = \frac{IL_i(b+1)}{4ebD_p \sinh \dfrac{d}{L_i}}, \tag{14.2-10}$$

whereby (14.2-3) may be expressed as

$$\delta p(x) = \frac{IL_i}{4ebD_p}\left[(b+1)\frac{\cosh \dfrac{x}{L_i}}{\sinh \dfrac{d}{L_i}} - (b-1)\frac{\sinh \dfrac{x}{L_i}}{\cosh \dfrac{d}{L_i}}\right]$$

$$= \frac{IL_i}{2ebD_p \sinh \dfrac{2d}{L_i}}\left[(b+1)\cosh \frac{x}{L_i}\cosh \frac{d}{L_i}\right.$$

$$\left. - (b-1)\sinh \frac{x}{L_i}\sinh \frac{d}{L_i}\right]. \tag{14.2-11}$$

In the central region near the junction edges the concentrations of injected excess carriers can be related to the corresponding equilibrium majority carrier concentrations in the p^+- and n^+-regions (assuming δp now to be much smaller than either of the majority carrier concentrations n_{no} and p_{po} in the exterior heavily doped regions) by boundary conditions of the form (13.1-12) and (13.1-13), to give[3]

$$p(-d) \cong \delta p(-d) = p_0 e^{eV_1/kT} \tag{14.2-12}$$

and
$$n(d) \cong \delta p(d) = n_0 e^{eV_2/kT}, \tag{14.2-13}$$

where V_1 and V_2 are the applied voltages appearing across the left- and right-hand junctions in Figure 14.3, respectively. Substituting these values into (14.2-11) and simplifying, it is easily seen that

$$\delta p(-d) = p_0 e^{eV_1/kT} = \frac{IL_i\left(1 + b\cosh \dfrac{2d}{L_i}\right)}{2ebD_p \sinh \dfrac{2d}{L_i}} \tag{14.2-14}$$

---

[3] The use of boundary conditions of the form (13.1–12) and (13.1–13) rather than the more general forms (12.3–1) and (12.3–2), in which effects arising from the disturbance of the majority carrier densities in the heavily doped exterior regions due to the presence of carriers injected into these regions by the junctions would be taken into account, amounts to neglecting the effects discussed in Section 14.1 under (i) and in Exercise 13.2. If the n^+- and p^+-regions are very heavily doped, however, not much error results until very high current levels are attained.

and
$$\delta p(d) = n_0 e^{eV_2/kT} = \frac{IL_i \left(b + \cosh \dfrac{2d}{L_i}\right)}{2ebD_p \sinh \dfrac{2d}{L_i}}. \qquad (14.2\text{-}15)$$

Multiplying the two equations (14.2-14) and (14.2-15) together, solving for I and recalling that $n_0 p_0 = n_i^2$, it is found that

$$I = \frac{2n_i ebD_p \sinh \dfrac{2d}{L_i}}{L_i \sqrt{\left(1 + b \cosh \dfrac{2d}{L_i}\right)\left(b + \cosh \dfrac{2d}{L_i}\right)}} \cdot e^{e(V_1 + V_2)/2kT}. \qquad (14.2\text{-}16)$$

It will be observed that the forward current is related to the total voltage drop $V_1 + V_2$ across the junctions by the exponential factor $e^{e(V_1 + V_2)/2kT}$ rather than the factor $(e^{eV_0/kT} - 1)$ which arises in the low level theory of Section 13.1. The low-current treatment of Section 13.1 thus predicts that a plot of $\ln (I + I_0)$ versus voltage should be a straight line of slope e/kT, while the p-i-n theory predicts a straight line of slope $e/2kT$. Figure 14.4 shows an experimental plot of the actual current-voltage curve of

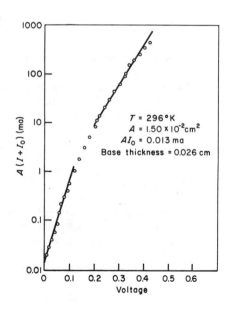

FIGURE 14.4. Current-voltage relation for an actual alloyed rectifier showing separate regions of slope e/kT and $e/2kT$ corresponding to low current level and high current level regions. [Figure courtesy of Dr. H. F. John, Westinghouse Research Laboratories.]

a germanium alloyed p-n rectifier in the forward bias region. In this diagram it is actually possible to distinguish a linear region of slope e/kT at voltages below about 0.1 V, where one might expect the low level theory to be quite good, and another linear region of slope $e/2kT$ at higher voltages (0.2–0.45 V) where the conditions of the p-i-n theory might be expected to apply.

In addition to the voltage across the junctions, there is an additional voltage drop which appears across the central region, arising from ohmic effects and concentration gradients in the near-intrinsic layer. This voltage drop may be determined by calculating the field within the central region and integrating over that portion of the structure. The field may be found by adding the two current equations (14.2-4) and (14.2-5) to obtain an equation for total current, solving for the field, and using the Einstein relations (10.2-11) to obtain

$$E(x) = \frac{1}{nb + p}\left[\frac{I}{e\mu_p} - \frac{kT}{e}(b-1)\frac{dp}{dx}\right]. \tag{14.2-17}$$

Under the conditions prevailing in the central region, $p \cong n \cong \delta p$ and (14.2-17) becomes

$$E(x) = -\frac{d\phi}{dx} = \frac{I}{e\mu_p(b+1)\delta p} - \frac{kT}{e}\frac{b-1}{b+1}\cdot\frac{1}{\delta p}\frac{d(\delta p)}{dx}. \tag{14.2-18}$$

This result is clearly in agreement with (10.2-39). Integrating from $-d$ to d to find the internal voltage drop V_i, one may write

$$\phi(d) - \phi(-d) = -V_i = -\int_{-d}^{d} E(x)\,dx = \frac{-I}{e\mu_p(b+1)}\int_{-d}^{d}\frac{dx}{\delta p(x)} + \frac{kT}{e}\frac{b-1}{b+1}\int_{-d}^{d}d(\ln \delta p) \tag{14.2-19}$$

or

$$V_i = \frac{I}{e\mu_p(b+1)}\int_{-d}^{d}\frac{dx}{\delta p(x)} - \frac{kT}{e}\frac{b-1}{b+1}\ln\frac{\delta p(+d)}{\delta p(-d)}. \tag{14.2-20}$$

Substituting the value of $\delta p(x)$ from (14.2-11) and the value $\delta p(+d)/\delta p(-d)$ from (14.2-14) and (14.2-15), integrating, and doing some rather tedious algebra to simplify the resulting expression, it is found that

$$V_i = \frac{4b}{1+b}\frac{kT}{e}\cdot\frac{\sinh\dfrac{2d}{L_i}\tan^{-1}\left[\sinh\dfrac{d}{L_i}\sqrt{1-\left(\dfrac{b-1}{b+1}\right)^2\tanh^2\dfrac{d}{L_i}}\right]}{\sqrt{1+b^2+2b\cosh\dfrac{2d}{L_i}}}$$

$$-\frac{kT}{e}\frac{b-1}{b+1}\ln\left(\frac{b+\cosh\dfrac{2d}{L_i}}{1+b\cosh\dfrac{2d}{L_i}}\right). \tag{14.2-21}$$

According to this expression, the internal voltage drop V_i is *independent* of current. Physically, this result is obtained because although an increase of I tends to increase V_i ohmically, it also tends to increase δp, which leads to a decrease of V_i; these two opposing effects cancel, leaving V_i independent of I. The total voltage V_0 across the device (neglecting voltage drops at contacts or in the heavily doped p^+- and n^+-layers) is simply the sum of the three voltages discussed previously;

$$V_0 = V_1 + V_2 + V_i. \tag{14.2-22}$$

Adding the expressions for these voltages, as obtained from (14.2-16) and (14.2-21) and simplifying (recalling that $D_i^* = 2bD_p/(1+b)$) one may obtain

$$V_0 = \frac{2kT}{e}\left[\ln\frac{IL_i}{en_iD_i^*} + \Phi_b(d/L_i)\right] \tag{14.2-23}$$

where

$$\Phi_b(\xi) = \ln\left[\frac{\sqrt{(1 + b\cosh 2\xi)(b + \cosh 2\xi)}}{(b+1)\sinh 2\xi}\right.$$

$$+ \frac{2b}{1+b}\frac{\sinh 2\xi \tan^{-1}\left[\sinh\xi\sqrt{1 - \left(\frac{b-1}{b+1}\right)^2\tanh^2\xi}\right]}{\sqrt{1 + b^2 + 2b\cosh 2\xi}}$$

$$- \frac{1}{2}\frac{b-1}{b+1}\ln\left(\frac{b+\cosh 2\xi}{1+b\cosh 2\xi}\right) \tag{14.2-24}$$

The forward voltage drop across the device is thus given as the sum of a term logarithmic in current and a term independent of current. The slope of a plot of $\ln I$ versus V_0 is clearly $e/2kT$. It should be noted that in view of the initial assumptions and approximations which have been made, this expression holds only in the high forward current region and is no longer correct as $I \to 0$. The fact that (14.2-23) yields a nonzero voltage for $I = 0$ is explained by this observation. The use of the high current theory requires that δp be large compared with the equilibrium majority carrier density n_0 of the central region. This can be shown to be equivalent to the requirement that

$$\frac{I}{e(b+1)}\sqrt{\frac{\tau}{D_i^*}} \gg n_0. \tag{14.2-25}$$

The details of the argument are assigned as an exercise. For silicon, where $b \sim 4$ and $D_i^* = 16$, this works out to the requirement that $(3.1 + 10^{14})\, I\sqrt{\tau}$ be large compared with n_0, I being expressed in amp/cm^2, τ in microseconds and n_0 in cm^{-3}.

The function Φ_b is plotted in Figure 14.5 for $b = 1$ and $b = 4$. It will be observed that $\Phi_b(\xi)$ goes to infinity as the argument approaches either zero or infinity, and has a minimum for a value of the argument which depends somewhat on b, but equals about 0.7 for $b = 4$. The minimum value of V_0 at a given constant current, however, does not occur at the minimum value of Φ_b due to the presence of the logarithmic term in (14.2-23). It can be shown, in fact, that for a structure in which the width d is constant, the minimum voltage drop V_0 for a given current density occurs for a value of L_i determined by the condition

$$\xi\frac{d\Phi_b}{d\xi} = 1 \tag{14.2-26}$$

where

$$\xi = d/L_i. \tag{14.2-27}$$

The proof of this assertion is left as an exercise. It is practice quite difficult to arrive

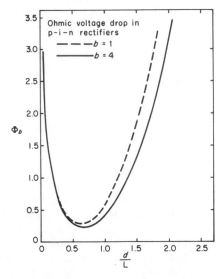

FIGURE 14.5. The function $\Phi_b(d/L)$, as given by (14.2-24), for $b = 1$ (dashed curve) and $b = 4$ (solid curve).

at the value of L_i for minimum forward voltage drop analytically, and therefore the evaluation of this quantity may equally well be done graphically or numerically. When V_0 is plotted as a function of L_i, according to (14.2-23) and (14.2-24), assuming d to be constant, curves such as those illustrated in Figure 14.6 are obtained.

In most *p-n* junction rectifiers the maximum power dissipation and internal heat generation occurs when the device is conducting large currents in the forward direction.

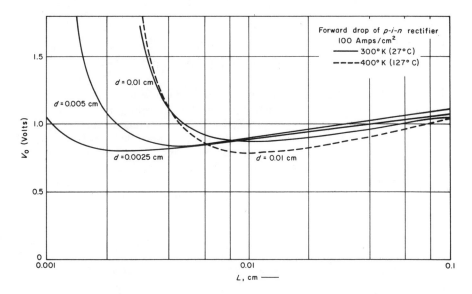

FIGURE 14.6. Forward voltage drop of *p-i-n* rectifier plotted *versus* central region diffusion length L for several central region thicknesses.

This can, for example, easily be demonstrated by comparing the product IV_0 for a typical device such as the one investigated in Figure 14.4 in the reverse bias region at $V_0 = -500$ V and in the forward bias region at a current density of 50 amp/cm^2 (See Exercise 7 at the end of this chapter.) The maximum load current which can be carried without destructive internal heat generation is therefore critically dependent upon the voltage drop across the device at high forward current values. The curves of Figure 14.6, which are calculated for silicon p-i-n rectifiers clearly show that the minimum forward voltage drop is obtained for a certain optimum value of d/L_i, which, for the range of conditions illustrated in that diagram is almost exactly unity. If the lifetime of excess carriers in the central region of the device is so low that L_i is much less than d or so high that L_i substantially exceeds d, an excessively high forward voltage drop will result. Physically this state of affairs arises because the junction voltage drop $V_1 + V_2$ (for a given constant load current) becomes very large when L_i is much larger[4] than d, according to (14.2-16), while the internal drop V_i across the central region becomes very large when L_i is much smaller than d, as shown by (14.2-21). A minimum total voltage must therefore be observed at some value of d which is neither much larger nor much smaller than L_i. This result illustrates the importance of maintaining at least a rough control over the excess carrier lifetime is such structures. One must note, however, that the value of τ which is required in the equations of this section is not necessarily that which would be obtained in a measurement of transient photoconductivity decay involving only a small excess carrier concentration, but the effective value which prevails under the assumed conditions in the central region of the device, where very large excess carrier concentrations are present. The two values may be quite different in view of the concentration dependence of τ predicted by (10.8-24) and (10.8-25).

14·3 FORWARD VOLTAGE DROP IN p-i-n RECTIFIERS AS A FUNCTION OF TEMPERATURE

In Figure 14.6, the dotted curve represents the forward voltage drop at a temperature of 400°K (127°C) for the unit represented by the solid curve for $d = 0.01$ cm, which is plotted for a temperature of 300°K (27°C). It is observed that for some values of L_i the forward voltage drop decreases over this temperature range, while for other values of L_i (below about 0.004 cm for this case), it increases. This rather curious situation is brought about by the fact that the internal voltage V_i has a temperature dependence different from that of the junction voltages. For an "ideal" p-n junction rectifier, such as that treated in Section 13.1, one would expect from (13.1-23) that

$$V_0 = \frac{kT}{e} \ln\left(1 + \frac{I}{I_0}\right) \tag{14.3-1}$$

where I_0 is the saturation current. But if we consider a device in which the n-type

---

[4] This behavior is also predicted by the low-level p-n junction theory of Section 13.1; it is clearly illustrated by equation (13.1-23), in which for a given current density the required junction voltage clearly becomes very large when both L_n and L_p are allowed to approach infinity.

region is much less highly doped than the *p*-region, and in which, therefore, $p_{n0} \gg n_{p0}$, I_0 is simply $ep_{n0}D_p/L_p = en_i^2 D_p/n_{n0}L_p$. From (9.3-21), however, so long as Boltzmann statistics are valid, n_i^2 may be written in the form

$$n_i^2 = CT^3 e^{-\Delta\varepsilon/kT} \qquad (14.3\text{-}2)$$

where $C = U_c U_v/T^3$ is a temperature-independent constant. Substituting this into (14.3-1) and assuming $I/I_0 \gg 1$, which is true for all but very small forward voltages, it is found that

$$V_0 = \frac{\Delta\varepsilon}{e} + \frac{kT}{e} \ln \frac{IL_p n_{n0}}{eCD_p} - \frac{3kT}{e} \ln T. \qquad (14.3\text{-}3)$$

In the usual range of temperatures, concentrations, lifetimes, etc., and for current densities for which this formula is applicable, it is found that the third term predominates over the second, and, as a result, V_0 decreases with temperature. It is assumed here that the diffusion length changes with temperature only through the change in the diffusion coefficient, and thus that the lifetime is *independent* of temperature. Any effect due to a change in lifetime with temperature will, of course, be superimposed upon the results discussed for this simpler case. The same assumption will be made in all subsequent discussions.

In the case of the *p-i-n* rectifier, operating in the high-current forward bias condition, one must proceed initially from (14.2-23) and (14.2-24). To account for the variation of the diffusion coefficient D_i^* with temperature, it is noted that since the diffusion coefficients D_p and D_n are related to the corresponding mobilities by Einstein's relations, then the ambipolar diffusivity D_i^* of (10.2-36) can be written as

$$D_i^* = \frac{2D_n D_p}{D_n + D_p} = \frac{kT}{e} \frac{2b\mu_p}{b+1}. \qquad (14.3\text{-}4)$$

The effective diffusion coefficient is thus kT/e times an expression involving the mobility of electrons and holes. If b is reasonably independent of temperature, the temperature dependence of D_i^* is related to that of the drift mobility of holes by a factor of T. If not, the temperature dependence is more involved, although it is still described accurately by (14.3-4) if mobility data for both holes and electrons are introduced. The temperature dependence of L_i is then obtained from that of D_i^* through (14.2-2).

In actual practice the temperature variation of the mobilities can usually be described by a power law, such as

$$\mu \propto T^{-n} \qquad (14.3\text{-}5)$$

where *n* is a positive constant, as discussed in Chapter 9. If the dominant scattering mechanism which determines the mean free path and thus the mobility is the interaction of electrons and holes with acoustical mode phonons, *n* will be 3/2; if optical mode or intervalley scattering is important, then *n* will be larger. In silicon *n* (as determined experimentally) is not far from 5/2 for both electrons and holes. In germanium, $n = 1.65$ for electrons and 2.3 for holes. Adopting (14.3-5) to express the temperature dependence of mobility and assuming b to be practically independent of temperature

458

SOLID STATE AND SEMICONDUCTOR PHYSICS

SEC. 14.3

one may, according to (14.3-4) express the temperature variation of D_i^* as

$$D_i^*(T) = D_{i0}^* \left(\frac{T}{T_0}\right)^{-n+1} \tag{14.3-6}$$

where D_{i0} is the value of D_i^* at some reference temperature T_0. Likewise, one may write

$$\frac{d}{L_i} = \frac{d}{L_{i0}} \left(\frac{T}{T_0}\right)^{\frac{n-1}{2}} \tag{14.3-7}$$

where

$$L_{i0} = \sqrt{D_{i0}^* \tau}. \tag{14.3-8}$$

Substituting (14.3-5), (14.3-6), (14.3-7) and (14.3-2) into (14.2-23) and simplifying yields the result

$$V_0 = \frac{\Delta\varepsilon}{e} + \frac{kT}{e} \ln \frac{I^2\tau}{e^2 D_{i0}^* C T_0^{n-1}} - (4-n)\frac{kT}{e} \ln T$$

$$+ \frac{2kT}{e} \Phi_b \left\{ \frac{d}{L_{i0}} \left(\frac{T}{T_0}\right)^{\frac{n-1}{2}} \right\}, \tag{14.3-9}$$

which describes the overall variation of V_0 with temperature. Note that the argument of the logarithm in the second term on the right-hand side of the equation is independent of temperature.

Near the *optimum* value of d/L_i, the temperature dependence of V_0 arises largely from the second and third terms on the right-hand side of (14.3-9). For this range of values of d/L_i, the contribution of the last term is quite small, and the overall variation of V_0 is *downward* with increasing temperature, for typical values of the device parameters. When d/L_i becomes much smaller than unity, however, $\Phi_b(d/L_i)$ becomes quite large and the last term on the right-hand side then dominates, the result being that V_0 *increases* with temperature. This term increases with temperature not only because of the coefficient $2kT/e$, but also because of the variation of the argument of Φ_b with temperature. When d/L_i becomes appreciably larger than unity Φ_b again becomes quite large, and an *increase* of V_0 with temperature is once more expected. Particularizing (14.3-9) to the two cases which we shall consider in detail ($n = 3/2$, corresponding to acoustical mode lattice scattering, which approximates the actual state of affairs in germanium, and $n = 5/2$, which is very nearly correct for silicon), one may write

$$V_0 = \frac{\Delta\varepsilon}{e} + \frac{kT}{e} \ln \frac{I^2\tau}{e^2 D_{i0}^* C T_0^{1/2}} - \frac{5}{2}\frac{kT}{e} \ln T + \frac{2kT}{e} \Phi_b \left\{ \frac{d}{L_{i0}} (T/T_0)^{1/4} \right\}$$

$$(n = 3/2) \tag{14.3-10}$$

and $$V_0 = \frac{\Delta\varepsilon}{e} + \frac{kT}{e} \ln \frac{I^2\tau}{e^2 D_{i0}^* C T_0^{3/2}} - \frac{3}{2}\frac{kT}{e} \ln T + \frac{2kT}{e} \Phi_b \left\{ \frac{d}{L_{i0}} (T/T_0)^{3/4} \right\}.$$

$$(n = 5/2) \tag{14.3-11}$$

The results predicted by these equations for a typical set of device parameter values and for a current density of 100 amp/cm² are shown in Figure 14.7. It is clear that the trend of V_0 may be *either* upward or downward with temperature, according to the value of d/L. It is once more obvious that the magnitude of the forward voltage drop

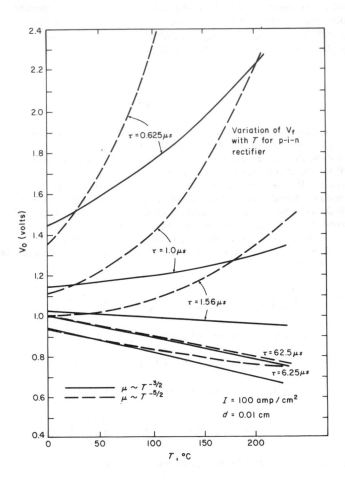

FIGURE 14.7. Forward voltage drop of *p-i-n* rectifier plotted *versus* temperature at constant current density of 100 A/cm² for the two cases $\mu \propto T^{-3/2}$ (solid curves) and $\mu \propto T^{-5/2}$ (dashed curves).

is quite sensitive to the excess carrier lifetime in the central region, and that it rises rapidly with decreasing lifetime, once past the minimum point. These conclusions are generally verified by device characteristics which have been observed experimentally. It should be observed that the temperature dependence of the drift mobility plays a significant role in determining the temperature dependence of the forward voltage drop. Device structures in which the forward voltage drop increases with rising temperature are generally most undesirable, since when overheated due to overload currents, their internal power dissipation rises and tends to overheat them still more, leading to premature burnout. Units in which the forward voltage drop decreases with rising temperature, on the other hand, dissipate *less* power in the forward direction when hot than when cold, are much more stable when carrying high currents, and are in general much more tolerant of transient or permanent overloading. For additional details and discussion of the theory of *p-i-n* rectifiers the original article by Hall[5] and the very thorough extended treatment of the subject in a series of articles by Herlet and Spenke[6] may be consulted.

EXERCISES

1. For the germanium p^+-i-n^+ rectifier discussed in connection with Figure 14.4, find the low-current diffusion length L_p in the central region from the given saturation current, assuming that the central layer is n-type material of 1 ohm-cm equilibrium resistivity at 300°K, and that all the saturation current is hole current from thermal generation in the n-region. If L_i at high-current levels is of the order of L_p, will this be a device with near-optimum internal losses at high current levels? Compare the internal power dissipation of this device in the forward bias region at a current level of 50 amp/cm², as given by the experimental data of Figure 14.4, with the power dissipation in the reverse direction at a voltage $V_0 = -500$ V, as calculated by the low-level theory of Section 13.1.

2. Show explicitly the algebraic steps required to obtain equation (14.2-21) from (14.2-20). Note that

$$\int \frac{dx}{a \cosh \alpha x + b \sinh \alpha x} = \frac{-2}{\alpha \sqrt{a^2 - b^2}} \tan^{-1}\left(e^{-\alpha x}\sqrt{\frac{a-b}{a+b}}\right) + \text{const.}$$

3. Show (a) that in the special case $b = 1$, (14.2-24) reduces to

$$\Phi_1(\xi) = \ln\left((1 + \cosh 2\xi)/2 \sinh 2\xi\right) + \tfrac{1}{2} \sinh 2\xi \tan^{-1}(\sinh \xi)$$

(b) that for $\xi \gg 1$, (14.2-24) may be approximated by

$$\Phi_b(\xi) \cong \frac{\pi}{2} \frac{\sqrt{b}}{1+b} e^{\xi} + \ln \frac{\sqrt{b}}{1+b} + \frac{1}{2}\left(\frac{b-1}{b+1}\right)\ln b$$

(c) that for $\xi \ll 1$, (14.2-24) may be approximated by

$$\Phi_b(\xi) \cong \frac{4b}{(1+b)^2} \xi^2 - \ln 2\xi$$

4. Show that for a given constant forward-current density in the high-current region, the minimum voltage drop is obtained when the ratio $\xi = d/L_i$ satisfies the condition $\xi(d\Phi_b/d\xi) = 1$.

5. Show from the previous calculations of Section 14.2 that the high-current theory developed therein is valid for current densities large enough that the inequality (14.2-24) is satisfied.

[5] R. N. Hall, *Proc. Inst. Radio Engrs.*, **40**, 1512 (1952).
[6] A. Herlet and E. Spenke, *Z. Angew. Phys.* **7**, 99 (1955); **7**, 149 (1955); **7**: 195 (1955).

CHAPTER 15

OTHER SEMICONDUCTOR DEVICES

15·1 THE *p-n* PHOTOVOLTAIC EFFECT AND *p-n* JUNCTION PHOTOVOLTAIC CELLS

If an ordinary *p-n* junction is short-circuited in darkness, as shown in Figure 15.1(a), no steady current will flow in the external circuit in spite of the fact that there is a potential ϕ_0 between the *p*- and *n*-regions of the device. The reason for this is that contact potential differences are developed between the *p*- and *n*-regions of the semiconductor and the metallic leads which precisely cancel out the internal potential at equilibrium. This situation is to be expected, of course, on thermodynamic grounds alone, since otherwise current would flow and work would be done with no energy input, in violation of the first law of thermodynamics.

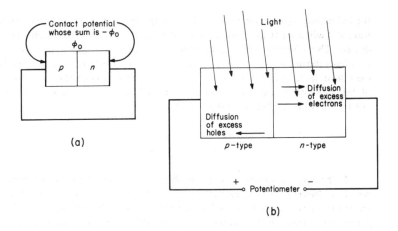

FIGURE 15.1. (a) A "shorted" *p-n* junction in the equilibrium state. (b) An illuminated *p-n* junction exhibiting the photovoltaic effect.

If light is allowed to fall on a *p-n* junction, as illustrated in Figure 15.1(b), the situation is materially changed, and measurable voltages and currents can be observed. Let us first consider the case where the *p-n* junction is connected to an open circuit. The light falling on the *p*- and *n*-regions on either side of the junction creates many excess electron-hole pairs in both regions. Excess electrons created in the *p*-region may diffuse to the junction, and descend the potential barrier to the *n*-side, while excess holes created by optical excitation in the *n*-region may diffuse to the junction and "float" up the barrier to the p-region. The effect of this is to put a net positive charge on the p-side and a net negative charge on the n-side; the presence of these charge densities is such as to *lower* the barrier potential difference from ϕ_0 to some value $\phi_0 - V_0$, as shown in Figure 15.2. The internal potential is now different from

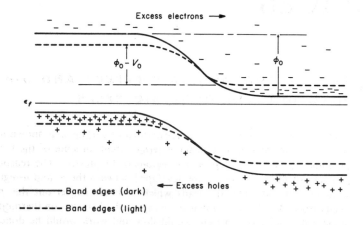

FIGURE 15.2. Potential diagram for a photovoltaic cell.

the balancing contact potentials and a voltage equal to this difference (thus of magnitude V_0) will appear as a measurable potential difference at the circuit terminals of the device. This phenomenon is called the *p-n photovoltaic effect*, and the output voltage is often referred to as a *photovoltage*. The magnitude of the photovoltage can be expressed in terms of the excess minority carrier concentrations at the boundaries of the junction space charge regions by (13.1-12) and (13.1-13), or better still by (13.1-17) and (13.1-18), whereby

$$\frac{n_p(-x_{0-})}{n_{p0}} = \frac{p_n(x_{0+})}{p_{n0}} = e^{eV_0/kT}. \tag{15.-1}$$

If the external circuit connecting the p- and n-regions is closed, electrical current will flow therein so long as there is a diffusion current of optically created excess electrons from the n-region and a current of optically created excess holes from the p-region to maintain the internal barrier height at a value other than ϕ_0. The current will then flow as long as the semiconductor regions are illuminated. It is evident that the energy source which maintains the current flow is the incident illumination, which serves to create and maintain the excess carrier distribution from the outset. In the open-circuit condition, the generation current densities discussed at the beginning of Section 13.1 are increased by illuminating the device, while the recombination currents are initially unaltered. The unbalanced diffusion currents, which flow as a result, cause a charge accumulation whose field lowers the internal barrier height, thus increasing the recombination currents until a balance between generation and recombination current is once more achieved, although at the cost of maintaining a voltage across the terminals of the device. In the closed-circuit condition, the generation currents are always larger than the recombination currents (which, of course, have no inherent dependence upon light intensity), and it is these generation currents which provide the source for the current which flows in the external circuit.

Suppose that a photovoltaic cell is made by diffusing a p-type impurity into an initially n-type semiconductor so that a thin p-type surface layer is formed. The cell is illuminated by *monochromatic light* incident upon the p-type front surface. The physical situation may then be represented by the one-dimensional model illustrated in Figure 15.3, where uniform n- and p-regions are shown separated by an abrupt

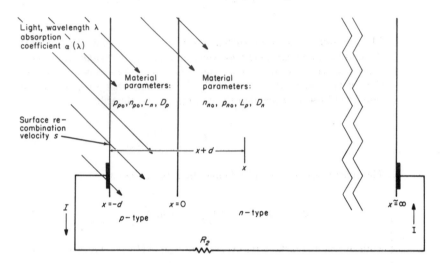

FIGURE 15.3.   Geometry and notational usage for the p-n photovoltaic cell of Section 15.1.

p-n junction.[1] The extent of the space charge region at the junction will be neglected, since under the usual conditions under which the device operates the thickness of both space charge layers is extremely small. Since the incident light is monochromatic, its intensity within the crystal will diminish exponentially with an absorption coefficient α, which, of course, will vary with wavelength, being quite large for photon energies larger than $\Delta\varepsilon$ and very small for photon energies much less than $\Delta\varepsilon$. If the number of photons of wavelength λ entering at the front surface of the device per unit area per unit time is $N_0(\lambda)$, then the flux of photons at any depth x will be

$$N(x) = N_0 e^{-\alpha(x+d)}. \tag{15.1-2}$$

The number of electron-hole pairs per unit area per unit time generated by photon absorption in a distance dx about x will be equal to $-(dN(x)/dx)\,dx$, whereby the excess carrier generation rate $g'(x)$ will be given by

$$g'(x) = \alpha N_0 e^{-\alpha(x+d)}. \tag{15.1-3}$$

---

[1] For a complete treatment of the p-n photovoltaic effect, see R. Wiesner, *Halbleiterprobleme*, Vol. III, Friedrich Vieweg u. Sohn, Braunschweig (1956), pp. 56–71.

Under these circumstances, the continuity equations for excess holes in the n-region and electrons in the p-region may be written as

$$\frac{d^2(n_p - n_{p0})}{dx^2} - \frac{n_p - n_{p0}}{L_n^2} = -\frac{\alpha N_0}{D_n} e^{-\alpha(x+d)} \qquad (x < 0) \qquad (15.1\text{-}4)$$

and

$$\frac{d^2(p_n - p_{n0})}{dx^2} - \frac{p_n - p_{n0}}{L_p^2} = -\frac{\alpha N_0}{D_p} e^{-\alpha(x+d)} \qquad (x > 0). \qquad (15.1\text{-}5)$$

The solutions of equations such as these may be written as the sum of the solution of the corresponding *homogeneous* equation and any *particular* solution to the inhomogeneous equation which can be found. One may, for example, seek a particular solution of (15.1-4) of the form $Ce^{-\alpha(x+d)}$, and find that such a solution will indeed satisfy[2] (15.1-4) *provided* that the constant C has the value

$$C = \frac{\alpha N_0 \tau_n}{1 - \alpha^2 L_n^2}. \qquad (15.1\text{-}6)$$

The general solution to (15.1-4) may then be written as

$$n_p(x) - n_{p0} = A \cosh \frac{x}{L_n} + B \sinh \frac{x}{L_n} + \frac{\alpha N_0 \tau_n}{1 - \alpha^2 L_n^2} e^{-\alpha(x+d)}, \qquad (15.1\text{-}7)$$

while in a similar fashion the general solution to (15.1-5) may be expressed in the form[3]

$$p_n(x) - p_{n0} = Ce^{x/L_p} + De^{-x/L_p} + \frac{\alpha N_0 \tau_p}{1 - \alpha^2 L_p^2} e^{-\alpha(x+d)}. \qquad (15.1\text{-}8)$$

In the model illustrated in Figure 15.3, the device is considered to be of essentially infinite extent in the $+x$-direction. Therefore, since $p_n(x) - p_{n0}$ must decay to zero for large values of x, the arbitrary constant C in (15.1-8) must be zero. The other three arbitrary constants in (15.1-7) and (15.1-8) may be evaluated by imposing the usual boundary conditions (13.1-17) and (13.1-18) at the junction, and the surface boundary condition at $x = -d$. In the situation at hand, these may be written

$$\frac{n_p(0)}{n_{p0}} = \frac{p_n(0)}{p_{n0}} = e^{eV_0/kT} \qquad (15.1\text{-}9)$$

and

$$D_n \left(\frac{d(n_p - n_{p0})}{dx} \right)_{-d} = s \cdot (n_p(-d) - n_{p0}). \qquad (15.1\text{-}10)$$

---

[2] Note that the solution found in this way is not a general solution, since it contains no *arbitrary* constants. The general solution is the sum of the solution of the corresponding homogeneous equation (which contains the two required arbitrary constants) and this *particular* solution.

[3] It is ordinarily somewhat more convenient for algebraic reasons to write the solution of the homogeneous equation in terms of exponential functions in an infinite or semi-infinite region and in terms of hyperbolic functions in a region of finite extent.

Following this procedure it may be shown that

$$D = p_{no}(e^{eV_0/kT} - 1) - \frac{\alpha N_0 \tau_p}{1 - \alpha^2 L_p^2} e^{-\alpha d} \qquad (15.1\text{-}11)$$

$$A = n_{po}(e^{eV_0/kT} - 1) - \frac{\alpha N_0 \tau_n}{1 - \alpha^2 L_n^2} e^{-\alpha d} \qquad (15.1\text{-}12)$$

and

$$B = A \left(\frac{s \cosh \dfrac{d}{L_n} + \dfrac{D_n}{L_n} \sinh \dfrac{d}{L_n}}{\dfrac{D_n}{L_n} \cosh \dfrac{d}{L_n} + s \sinh \dfrac{d}{L_n}} \right) + \frac{\alpha N_0 \tau_n}{1 - \alpha^2 L_n^2} \left(\frac{s + \alpha D_n}{\dfrac{D_n}{L_n} \cosh \dfrac{d}{L_n} + s \sinh \dfrac{d}{L_n}} \right). \quad (15.1\text{-}13)$$

The junction current

$$I = e(J_p(0) - J_n(0)) = e\left(D_n \frac{d(n_p - n_{po})}{dx} - D_p \frac{d(p_n - p_{no})}{dx} \right)_0 \qquad (15.1\text{-}14)$$

may now be evaluated to give the current-voltage equation for the device.[4] **The result may be expressed in the form**

$$I = I_0(e^{eV_0/kT} - 1) - I_g \qquad (15.1\text{-}15)$$

where

$$I_0 = e \left[\frac{p_{no}D_p}{L_p} + \frac{n_{po}D_n}{L_n} \cdot \frac{s \cosh \dfrac{d}{L_n} + \dfrac{D_n}{L_n} \sinh \dfrac{d}{L_n}}{\dfrac{D_n}{L_n} \cosh \dfrac{d}{L_n} + s \sinh \dfrac{d}{L_n}} \right] \qquad (15.1\text{-}16)$$

is a *saturation current* which is a function only of the dimensional and material parameters of the device, *independent* of the level of illumination, and where

$$I_g = \frac{e\alpha N_0 L_p^2 e^{-\alpha d}}{1 - \alpha^2 L_p^2} \left(\frac{1}{L_p} - \alpha \right) + \frac{e\alpha N_0 L_n^2}{1 - \alpha^2 L_n^2}$$

$$\times \left\{ \frac{1}{L_n} \left[\frac{\left(s \cosh \dfrac{d}{L_n} + \dfrac{D_n}{L_n} \sinh \dfrac{d}{L_n} \right) e^{-\alpha d} - s - \alpha D_n}{\dfrac{D_n}{L_n} \cosh \dfrac{d}{L_n} + s \sinh \dfrac{d}{L_n}} \right] - \alpha e^{-\alpha d} \right\} \qquad (15.1\text{-}17)$$

---

[4] In doing this, it is simpler to leave the expressions for excess carrier concentrations in the form of (15.1-7) and (15.1-8), inserting the expressions (15.1-11, 12 and 13) for the arbitrary constants only at the very end.

is a generation current, independent of the junction voltage V_0 but *directly proportional* to the illumination intensity as represented by N_0. The generation current (15.1-17) represents that part of the current of excess electrons and holes created by the incident light which succeeds in diffusing to the junction. It is easily shown that the hole generation current, as represented by the first term in (15.1-17) is *always positive*, whatever the relative magnitudes of α, L_p^{-1} and d may be; the details of the proof are assigned as an exercise. It is much more difficult to show that the same is true of the electron generation current from the p-region, but it is intuitively clear that it must be.

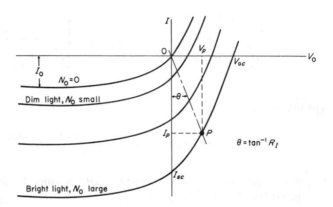

FIGURE 15.4. Current-voltage relation for p-n photovoltaic device under different illumination intensities, showing "load line" construction.

The current-voltage relations given by (15.1-15) are illustrated in Figure 15.4 for several different incident light intensities. It is immediately evident that if $N_0 = 0$, the current-voltage relation is just that of an ideal p-n diode, while as N_0 is increased the generation current increases proportionally, giving rise to the family of curves shown in the diagram. A "load line" corresponding to a given value of circuit resistance is shown; this establishes an operating point corresponding to an output voltage V_p and an output current density I_p. Clearly, if the circuit resistance is infinite, corresponding to open-circuit conditions, I vanishes and the *open-circuit voltage*, according to (15.1-15), is

$$V_{oc} = \frac{kT}{e} \ln\left(1 + \frac{I_g}{I_0}\right). \tag{15.1-18}$$

This is the maximum voltage which can be obtained from the device under a given level of illumination. As indicated by (15.1-18), the open circuit voltage increases logarithmically with I_g (and thus with illumination intensity) when $I_g \gg I_0$. From Figure 15.2 one may easily see that the maximum open-circuit voltage that can ever be obtained is of the order of the internal potential ϕ_0. If the circuit resistance (including the internal resistance of the device itself) is zero, then V_0 becomes negligibly small and a maximum *short-circuit current* current, corresponding to current density

$$I_{sc} = -I_g \tag{15.1-19}$$

flows. In both instances the power delivered to the external circuit as expressed by the product IV_0 is zero. It is clear that the maximum power is delivered to an external circuit under circumstances which are intermediate between the two extremes discussed above, for example, under conditions such as those which obtain at the point P in Figure 15.4. The precise determination of the circuit resistance which results in maximum output power is not difficult, and is recommended to the reader as an exercise.

It is evident from (15.1-15) and Figure 15.4 that the greater the generation current I_g, the greater the power delivered to the external circuit, and the higher the efficiency of conversion of light energy to electrical energy. Although the generation current is in all cases directly proportional to the illumination intensity, there are other factors which play an important role in determining the magnitude of I_g. It is clear from (15.1-17), as well as being evident intuitively, that long lifetimes (hence large diffusion lengths in the n- and p-regions) aid in producing large values of I_g. Likewise, it is obvious that surface recombination can only detract from the diffusion currents which might flow across the junction, and hence a reduction in surface recombination velocity will result invariably in an increased value for the generation current density. In addition, the depth of the junction below the illuminated surface is a critical factor. If it is too deep all the light will be absorbed far in front of the junction, and diffusion currents will have to flow a long way and suffer serious reduction from bulk recombination before reaching the junction. On the other hand, if it is too shallow, then much of the light will penetrate far beyond it before being absorbed, and again the diffusion currents will have a long way to flow. The ideal depth is of the order of the exponential decay distance $1/\alpha$ associated with the absorption of the incident light. Since the maximum open-circuit voltage which can be obtained under strong illumination is of the order of the internal potential ϕ_0, it is advantageous to use fairly heavy doping levels in both n- and p-regions, so as to obtain as large a value as possible for this quantity. Also, since no excess carriers are produced if the energy gap width $\Delta\varepsilon$ is much in excess of the photon energy of the incident light, and since much of the energy of the photons is wasted in heating the crystal if $\Delta\varepsilon$ is much less than the incident photon energy, maximum energy conversion efficiency is obtained when $\Delta\varepsilon = hc/\lambda$.

All the preceding discussion has been based upon a situation in which the photovoltaic cell is illuminated with monochromatic radiation. In most of the actual applications to which such devices are put, the illumination source is sunlight, which, of course, has a broad continuous spectrum of wavelengths. In this case, it is necessary to calculate the actual spatial distribution of excess carrier generation, which will now no longer be the simple exponential function (15.1-3) but a *superposition* of many such exponential functions representing many wavelengths, *each* with a characteristic absorption coefficient $\alpha(\lambda)$ as obtained from the observed optical absorption spectrum of the semiconductor crystal and each with a characteristic intensity factor $N_0(\lambda)$ according to the spectral intensity of sunlight at each wavelength. Ideally, then, $g'(x)$ would be represented as an integral of the quantity $\alpha(\lambda)N_0(\lambda)e^{-\alpha(x+d)}$ over wavelength, and the continuity equations would then become integral equations, which in the case at hand are not at all amenable to analytical solution. The only alternative is to proceed using numerical techniques, and it is clearly beyond the scope of this work to progress any further in this direction. It should, nevertheless, be noted that the general validity of the results obtained in the monochromatic case extends even to this more complex situation. For example, it is clearly advantageous to use a semiconductor in which $\Delta\varepsilon$ is as closely as possible matched to the photon energy at the wavelength of maximum spectral intensity and to locate the junction at a depth corresponding to the mean absorption depth of the wavelength of maximum spectral intensity.

The photon energy associated with the maximum spectral intensity in sunlight is about 2 eV. Unfortunately, the semiconductors having this value of $\Delta\varepsilon$ which are available at present (GaP, SiC, etc.) are very unsatisfactory for making efficient photovoltaic cells because of the very low excess carrier lifetimes which are invariably associated with them. The energy gap width of germanium is too low for best efficiency, and in addition, the output voltages which are obtained are too low. The best photovoltaic cells for solar illumination which have been obtained to date have been made of silicon, although gallium arsenide has also been used with some success. The best silicon photovoltaic devices can convert 15 percent of the incident solar radiation into electrical energy, giving output voltages of 0.55 to 0.6 V and current densities of about 45 ma per cm^2. These devices are used in large numbers as primary power sources for satellite power systems.

15·2 OTHER PHOTO-DEVICES; PHOTOTRANSISTORS, PARTICLE DETECTORS AND INFRARED DETECTORS

In addition to the *p-n* photovoltaic cells which were treated in detail in the preceding section, there are a number of other types of devices which rely upon the generation of excess carrier pairs by optical or elementary particle excitation for their operation.

Photoconductive cells rely on the increase of dc photoconductivity under illumination to provide an electrical signal proportional to the exciting radiation intensity. The operation of devices of this type has been analyzed in some detail in Section 10.6. Cadmium sulfide photoconductive cells are in common use as sensing elements in cameras and photographic exposure meters. Photoconductive cells in which the ionization of deep-lying donor levels at low temperatures by incident radiation gives rise to the photoconductivity are often used as infrared detectors. Gold-doped germanium is frequently utilized for this purpose.

A junction transistor in which excess electron hole-pairs are introduced into the base region by optical excitation rather than by a *p-n* junction emitter is called a *phototransistor*. Such a device is illustrated in Figure 15.5. In the phototransistor, the excess electron-hole pairs created by the incident light diffuse to and are collected by a reverse-biased collector junction. In effect, the excess carriers generated by the light add to the thermal saturation current of the collector junction. The phototransistor is capable of greater sensitivity as a detector of radiation than either the photovoltaic cell or the photoconductive cells, and if properly designed and operated can function very efficiently indeed in detecting low-level light signals. Since electron-hole pairs may be produced by incident beams of X-rays, γ-rays or elementary particles such as electrons, protons or α-particles as well as by light, devices having this structure function admirably as nuclear particle detectors. Since a very large number of electron-hole pairs are created when a nuclear particle of several MeV energy is stopped inside a silicon or germanium crystal (the number can be calculated, approximately, by noting that one electron-hole pair is obtained for about each 3 eV of initial particle energy in

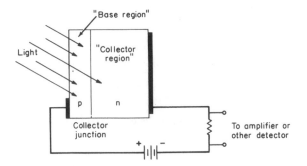

FIGURE 15.5. Schematic diagram of phototransistor or radiation detector.

both silicon and germanium) the detection of a single nuclear particle is easily accomplished. The magnitude of the resulting current pulse is approximately proportional to the incident particle energy, which allows the determination of energy spectra by pulse height analysis.

$15\cdot3$ p-n-p-n CONTROLLED RECTIFIERS

The p-n-p-n controlled rectifier is a four-layer structure, as depicted in Figure 15.6. It is usually designed with much heavier doping impurity densities in the outer layers than in the inner ones, and hence might be more accurately identified as a p^+-n-p-n^+

FIGURE 15.6. Schematic diagram of four layer p-n-p-n switching device.

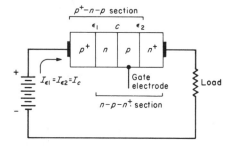

device. The p-n-p-n controlled rectifier may be regarded as two transistors, a p^+-n-p and an n-p-n^+, with a common collector junction, as illustrated. The junction ε_1 is the emitter junction of the p^+-n-p section, while ε_2 is the emitter of the n-p-n^+ section. The junction c is a collector junction common to both sections. When a supply voltage is connected as shown, this will be the only reverse-biased junction in the device. If α_1 is the current gain of the p^+-n-p section and α_2 the current gain of the n-p-n^+ section, then, using the same line of reasoning which was applied in the case of open-base self-amplification of collector current for a p-n-p transistor in Equations (13.6-50) and (13.6-51), we may write

$$I_c = I_{c0} + \alpha_1 I_{\varepsilon_1} + \alpha_2 I_{\varepsilon_2}, \tag{15.3-1}$$

where I_{c0} is the collector saturation current. Since

$$I_{\varepsilon_1} = I_{\varepsilon_2} = I_c, \tag{15.3-2}$$

however, Equation (15.3-1) may be expressed in the form

$$I_c = \frac{I_{c0}}{1 - (\alpha_1 + \alpha_2)}. \tag{15.3-3}$$

We have already seen that under the assumptions which were made in Section 13.6, the current gain of a junction transistor is expected to be independent of emitter current except at large values of emitter current, where it should decrease due to a decrease of injection efficiency at high current levels. Actually, however, the current gain does vary with emitter current also at very low-current levels; it rises from a rather low value at $I_\varepsilon = 0$ quite rapidly until the value predicted by (13.6-31) is attained, then becoming practically independent of I_ε until the decrease at high current density due to degradation of injection efficiency sets in. This state of affairs, which is illustrated in Figure 15.7, may in part be due to surface leakage effects at the junction, or

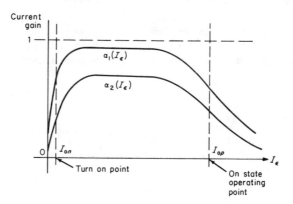

FIGURE 15.7. Current gains α_1 and α_2 of the two transistor sections of a four-layer switching device plotted against current.

to recombination within the junction space charge region. At any rate, as the supply voltage is increased, the collector current remains nearly constant at a rather small value as given by (15.3-3) with α_1 and α_2 referring to values of these current gains somewhere in the low-current region ($I < I_{on}$ in Figure 15.7). In this region, $\alpha_1 + \alpha_2$ is less than unity and a stable operating condition is reached in which the current flow through the device is low and the voltage drop (which appears for the most part across the central collector junction) may be quite large. As the supply voltage is increased, a point is reached where avalanche breakdown starts to occur at the center junction. At this point (which is still well within the pre-breakdown region) the collector current rises somewhat and the current gains α_1 and α_2 increase correspondingly as shown by Figure 15.7 until a critical point ($I_\varepsilon = I_{on}$) is reached where the sum of α_1 and α_2 reaches unity. As predicted by (15.3-3) an instability occurs at this "turn on" point. The current rises rapidly and the device switches from a high-voltage, low-current mode to a low-voltage, high-current state in which strong regenerative amplification of excess carrier currents injected by the emitter junctions maintains large excess carrier densities within the two central regions of the device, producing a condition in which the impedance is very low. A stable high-current, low-impedance operating point is reached when the current becomes so large that due to the decrease of current gain with emitter current in the high-current region the sum of α_1 and α_2 has once more dropped to unity; this point is shown as I_{op} in Figure 15.7. The device continues in the high-current mode until the supply voltage is interrupted, whereupon (after a short-recovery time of the order of the lifetime of excess carriers in the inner regions) the low-current mode may again be resumed. The switching action can be initiated not only by the onset of avalanche breakdown, but also by injecting a small current pulse (of sufficient magnitude to increase the collector current to the point where $\alpha_1 + \alpha_2$ exceeds unity) into a "gate electrode" connected to one of the inner regions. The device may thus be switched on at will.

The current-voltage characteristic of a p^+-n-p-n^+ controlled rectifier is illustrated schematically in Figure 15.8. Devices of this type, made with silicon, can withstand

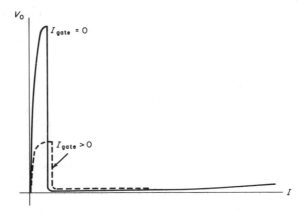

FIGURE 15.8.   Current-voltage relation for four-layer switch.

voltages up to nearly 1000 V in the high-impedance mode and carry load currents of several hundred amperes in the low-impedance state. In the two inner regions of the device the excess carrier concentrations are generally far in excess of the equilibrium values at operating values of load current, and in this mode of operation the device closely resembles a p-i-n rectifier, wherein the role of the i-region is played by the two inner regions. Devices of this type have found widespread technological application as switches, power control devices, inverters and frequency converters.

15·4 TUNNEL DIODES

The tunnel diode is a semiconductor device which is made very much like an ordinary alloyed *p-n* rectifier, except that both *p-* and *n*-regions are as heavily doped as possible, and the thermal cycle associated with the alloying process is completed *very* quickly, so as to provide little or no opportunity for solid-state diffusion of donor impurity atoms from the *n*-region into the *p*-region, and *vice versa*. Under these circumstances a very abrupt transition from p^+- to n^+-type material within the crystal is achieved, and consequently the junction space charge region is very narrow and gives rise to extremely large electric fields. This is illustrated by some of the figures given in Table 12.1; a typical tunnel diode may have doping densities of 10^{18} or 10^{19} cm^{-3} in both *p*- and *n*-regions. In crystals where the impurity content is so large, the Boltzmann approximation is no longer a good one, the electron distribution is *degenerate*, and the Fermi level is actually *within* the conduction or valence band, accordingly as the semiconductor is *n*-type or *p*-type, respectively. The equilibrium energy band diagram for a tunnel diode is therefore as depicted in Figure 15.9(a).

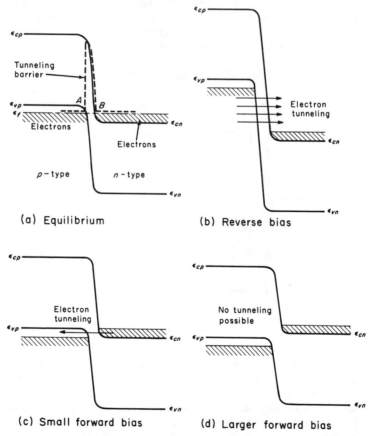

FIGURE 15.9.   Potential diagrams illustrating the operation of the tunnel diode.

In the tunnel diode the junction is so abrupt and the space charge region so extremely thin that the potential energy barrier (shown in Figure 15.9(a)) which separates electrons in the valence band from states of equal energy in the conduction band is narrow enough even in the absence of a reverse bias to permit quantum tunneling of electrons through the barrier to take place. In the equilibrium state, with no applied bias voltage, the tunneling of electrons proceeds with equal probability in both directions so that there is no net flow of current.

If a reverse bias voltage is applied, as at (b) in Figure 15.9, the electron distributions on either side are displaced vertically with respect to one another by an amount equal to the bias voltage. In this condition, empty states on the n-side are brought opposite filled states on the p-side, and electrons from these states tunnel through the potential barrier to supply large net currents. The device does not withstand any voltage at all in the reverse direction, but can be considered to break down by the Zener mechanism at infinitesimally small values of applied voltage.

In the forward direction, tunnel current starts to flow at the beginning by this same mechanism, as at (c) in Figure 15.9, but after the forward bias voltage is increased beyond a certain point, the bands "uncross" as shown in (d) and no more final states for tunneling are then available, so the tunneling current drops sharply. Upon the application of still higher forward bias voltages, the usual forward diffusion or injection current associated with any p-n rectifier begins to flow *over* the potential barrier in significant quantity and the current again increases.

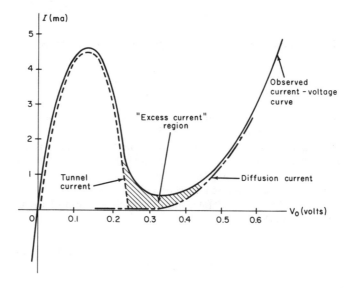

FIGURE 15.10. Current-voltage relation for a typical tunnel diode (solid curve) compared with predicted values for tunneling and diffusion currents (dashed curves).

The result of these successive processes is a current-voltage characteristic curve as illustrated in Figure 15.10, containing a negative resistance region. The negative resistance characteristic of the device can be utilized to make the unit function as an amplifier, an oscillator or a switching device. Since the tunneling process (in contrast to diffusion) is practically instantaneous, tunnel diodes can be used at very high frequencies (up to about 10,000 mc in practice) as amplifiers or oscillators. In germanium tunnel diodes, the ratio of maximum to minimum currents may be as large as 15, with maximum currents of a fraction of a milliampere to perhaps an ampere being obtainable through suitable modifications of fabrication procedure. Tunnel diodes have also been successfully produced from silicon and GaAs. The tunneling current density observed at the peak of the current-voltage curve is strongly dependent upon the impurity density in the n- and p-regions of the device. With high impurity concentrations, the space charge region is very narrow, the tunneling barrier extremely thin and the maximum tunneling current density may be very large. In germanium, with impurity densities approaching 10^{20} cm^{-3}, tunneling current densities of nearly 2000 amp cm^{-2} have been observed. With smaller impurity concentrations the tunneling current density decreases, and the effect is barely observable at impurity levels of the order of 10^{18} cm^{-3}.

The tunneling current may be calculated provided the valence and conduction band density of states functions and the tunneling probability associated with the potential barrier is known. Since the transition rates between initial and final tunneling states are proportional to the density of initial states and the density of final states, one may write for the tunneling current from conduction to valence band[5]

$$I_{c\to v} = A \int_{\varepsilon_{cn}}^{\varepsilon_{vp}} f_c(\varepsilon)g_c(\varepsilon)(1 - f_v(\varepsilon))g_v(\varepsilon)Z \, d\varepsilon \qquad (15.4\text{-}1)$$

while the tunneling current in the opposite direction may be written

$$I_{v\to c} = -A \int_{\varepsilon_{cn}}^{\varepsilon_{vp}} f_v(\varepsilon)g_v(\varepsilon)(1 - f_c(\varepsilon))g_c(\varepsilon)Z \, d\varepsilon. \qquad (15.4\text{-}2)$$

In these equations $f_c(\varepsilon)$ and $f_v(\varepsilon)$ are the electron energy distribution functions in the conduction band on the n-side and the valence band on the p-side, respectively, Z is the tunneling probability associated with the barrier, and A is a constant. The tunneling probability may be obtained as the transmission coefficient of the potential barrier for electrons by the solution of a quantum mechanical barrier problem somewhat similar to the one discussed in Section 9.11 in connection with acoustical mode lattice scattering. The total net tunneling current is the sum of the two currents given by (15.4-1) and (15.4-2), or

$$I_t = I_{c\to v} + I_{v\to c} = A \int_{\varepsilon_{cn}}^{\varepsilon_{vp}} (f_c(\varepsilon) - f_v(\varepsilon))g_c(\varepsilon)g_v(\varepsilon)Z \, d\varepsilon. \qquad (15.4\text{-}3)$$

In the equilibrium condition, with no applied voltage, $f_c(\varepsilon) = f_v(\varepsilon)$ and (15.4-3) reduces to $I_t = 0$, as required.

The observed current-voltage characteristic for a typical germanium tunnel diode is shown in Figure 15.10. The tunneling current contribution, as given by (15.4-3) and the ordinary forward diffusion current component, as given by (13.1-23) are

---

[5] See, for example, the original article about the tunnel diode, by L. Esaki, *Phys. Rev.*, **109**, 603 (1958).

shown as dashed curves. The sum of these two currents reproduces the observed current quite well in the tunneling region and the region of large forward voltage, but near the current minimum the observed current density is invariably much larger than that predicted by the simple theory outlined above. This "excess current" near the minimum has been ascribed to the presence of energy levels in the forbidden region (due to trapping impurities or structural imperfections), and to a modification of the actual density of states curve in the conduction and valence band by the introduction of donor and acceptor impurities in concentrations so large that the donor and acceptor states form bands of levels which may be broad enough under the prevailing circumstances to overlap the conduction and valence bands.[6]

15·5 UNIPOLAR OR FIELD EFFECT TRANSISTORS

In the field effect transistor (or *unipolar* transistor), amplification of an input voltage is accomplished by applying the signal to an electrode which controls the width of a narrow current-carrying channel by varying the width of the space-charge region associated with a *p-n* junction, as shown in Figure 15.11. The load current (sometimes

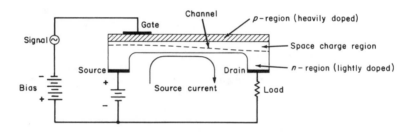

FIGURE 15.11.   Schematic diagram of a field effect transistor and associated circuitry.

referred to as the *source current*) in general must flow through a narrow channel which is bounded on one side, at least, by the space charge region associated with a *p-n* junction. As the reverse voltage across this junction varies, the extent of the space charge region and hence the conductance of the channel changes correspondingly.

The channel current input terminal of a field effect transistor is usually called the *source* electrode, the channel current return terminal the *drain* electrode and the signal input terminal the *gate* electrode. Although we shall not attempt to give a complete analysis of the behavior of such devices here,[7] it may be noted that their electrical characteristics are closely analogous to those of the ordinary high-vacuum pentode.[8]

---

[6] See, for example, T. P. Brody, *J. Appl. Phys.*, **32**, 746 (1961).

[7] An analysis of this type of device has been made by W. Shockley, *Proc. I.R.E.*, **40**, 1365 (1952).

[8] Although it is able to perform all the electrical functions normally associated with the vacuum triode, the circuit behavior of the ordinary junction transistor differs qualitatively from that of the vacuum tube because it depends upon the presence of *two* species of carriers rather than one. It is therefore sometimes referred to as a *bipolar* device.

Both the field effect transistor and the tunnel diode are somewhat unusual among semi-conductor devices in that they are not dependent upon the injection or transport of minority carriers for their operation, but rather utilize essentially only the transport of majority carriers to perform the required device function.

This discussion of semiconductor devices is by no means comprehensive, but has included only a few of the more important examples, and certain others whose operation illustrates important physical phenomena or lends insight into analytical techniques of general applicability. A systematic treatment of all the numerous semiconductor devices now in current technological use is not possible in a work of the scope or size of the present one. Such devices include microwave detectors, p-n junction lasers, thermistors, analogue multipliers and magnetic-field measuring probes utilizing the Hall effect, thermoelectric refrigerators and power generators, and "integrated circuit" modules of various types. It is hoped that the examples which have been discussed in detail herein will provide sufficient familiarity with the behavior of p-n junctions and with the mathematical techniques which are necessary for semiconductor device analysis to enable the reader to proceed independently into the literature. Further details concerning the devices which have been treated herein, and material related to other semiconductor devices may be found in some of the works listed as references at the end of this chapter.

EXERCISES

1. Verify algebraically that the expressions (15.1-15), (15.1-16) and (15.1-17) follow from the preceding results, thus filling in the formal details of the derivation of these equations.

2. Show that the generation current of holes in a p-n photovoltaic cell, as given by the first term in Equation (15.1-17), is always positive, regardless of the relative magnitudes of α, L_p^{-1} and d.

3. For what value of output voltage is the maximum power output of a p-n photovoltaic cell obtained? What is the value of circuit resistance which is required to obtain this maximum power output?

4. A reverse bias voltage of -100 V is maintained across the p-n junction of the phototransistor of Figure 15.5. An electron of energy 500 keV collides with the front surface of the device, and its energy is absorbed entirely within the space-charge region associated with the junction. Assuming that on the average 3 eV of incident particle energy are dissipated for each electron-hole pair, and that all the electron-hole pairs are created instantaneously, find the magnitude of the current pulse which is observed in the external circuit.

5. It is clear from Figure 15.2 that the maximum open circuit voltage obtainable from a p-n photovoltaic cell is ϕ_0. According to Equation (15.1-18), however, the open circuit voltage can be *arbitrarily* large, provided that the light intensity (and hence I_g) is made sufficiently large. What is the explanation for this apparent paradox, and which statement is correct?

GENERAL REFERENCES

J. K. Jonscher, *Principles of Semiconductor Device Operation*, G. Bell and Sons, Ltd., London (1960).

M. J. Morant, *Introduction to Semiconductor Devices*, Addison-Wesley Publishing Co., Reading, Mass. (1964).

A. Nussbaum, *Semiconductor Device Physics*, Prentice-Hall, Inc., Englewood Cliffs, N.J. (1964).

E. Spenke, *Electronic Semiconductors*, McGraw-Hill Book Co., Inc., New York (1958).

R. Wiesner, *p-n Photoeffect*, in *Halbleiterprobleme*, Friedrich Vieweg und Sohn, Braunschweig (1956), Vol. III, pp. 56–71.

A. van der Ziel, *Solid State Physical Electronics*, Prentice-Hall, Inc., Englewood Cliffs, N.J. (1957).

"Transistor Issues," *Proc. Inst. Radio Engrs.*, **30** (December 1952) and **46** (June 1958).

CHAPTER 16

METAL-SEMICONDUCTOR CONTACTS AND SEMICONDUCTOR SURFACES

16·1 METAL-SEMICONDUCTOR CONTACTS IN EQUILIBRIUM

When two metals having different work functions are brought into contact with one another, a brief transient current flow takes place which transfers electrons from the metal with the larger Fermi energy to the one with the smaller Fermi energy, thereby generating an equilibrium *contact potential* difference between the two. This process is shown schematically in Figure 16.1. At (a) two metals having different work

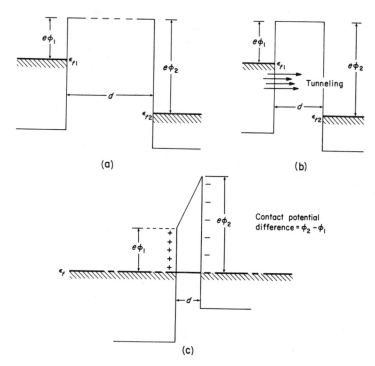

FIGURE 16.1. Successive stages in the establishment of equilibrium between two metals with different work functions.

478

functions are separated by a distance d which is assumed to be quite large, possibly of the order of a centimeter. Under these circumstances the Fermi levels of the two metals do *not* coincide, and the system is *not* in equilibrium. To achieve equilibrium between the two metals, electrons would have to *tunnel* through the potential barrier from the metal having the higher Fermi energy to the one with the lower Fermi energy until a condition of equilibrium in which the two Fermi levels coincide is attained. This cannot be accomplished while the metals are widely separated, as at (a), because the potential barrier is so high and thick that the tunneling probability is negligible. Equilibrium might only be established after millions of years under these conditions.

As the two metals are brought closer together, as in (b), the thickness of the tunneling barrier becomes less and less, until finally the tunneling probability becomes sufficiently large that electrons penetrate the barrier, flowing from left to right in the diagram. The physical separation d which is necessary to bring this state of affairs about is a few times the interatomic distance. The current of electrons creates a negative charge in the metal on the right, and the excess of positive ions which are left behind on the left gives rise to a positive charge in that region. Accordingly, a field is set up which raises the potential energy of the electrons on the right with respect to those on the left until at length the Fermi levels of the two metals coincide. Under these circumstances the system is in equilibrium and the net tunneling current drops to zero, since now tunneling in either direction is equally probable. The transfer of the electrons which flowed from left to right, however, has established a potential difference of magnitude $\phi_2 - \phi_1$ between the two metals. This voltage is called the contact potential difference between the two metals. The number of electrons which are transferred from one metal to the other in order to establish this potential difference is so small compared with the total number of free electrons in either substance that the relative values of free electron population are practically unaffected.

Somewhat the same line of reasoning may be used to infer the properties of a metal-semiconductor contact. The semiconductor differs from a metallic substance, however, in that an electric field may exist within the interior of a semiconductor. For this reason, the contact potential drop between the metal and semiconductor may take place *within* the semiconductor rather than at the contact interface. In the simplest possible instance, what may happen is illustrated in Figure 16.2 for the case of a contact between a metal and an n-type semiconductor crystal, where the work function of the metal, ϕ_M, is larger than the work function ϕ_{SC} associated with the semiconductor. As in Figure 16.1, a process is envisioned wherein the distance d between the two substances is decreased until at length this distance is small enough so as to permit electrons to tunnel freely through the barrier. The field which arises due to the contact potential difference exists now largely within the semiconductor; the potential energy of an electron at rest at the bottom of the conduction band in the interior of the crystal thus differs from the potential energy of such an electron at the surface by the amount $e(\phi_M - \phi_{SC})$, and as a result the conduction and valence band edges are shifted with respect to the Fermi level as illustrated. The positive space charge density in the surface region, due to the excess concentration of ionized donor atoms over the electron population (in conjunction with the electrons which tunneled through to the metal) is just such as to produce a field sufficient to sustain the potential difference $\phi_M - \phi_{SC}$ between the two materials. The situation within the semiconductor is not essentially different from that which was discussed in connection with space charge layers of a p-n junction.

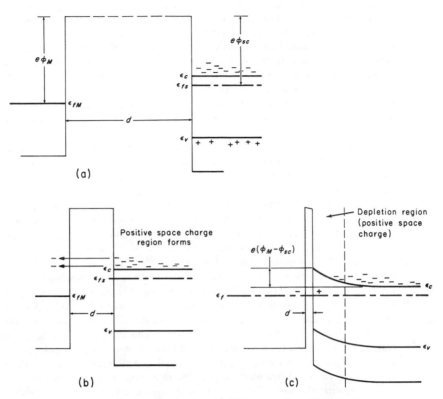

FIGURE 16.2. Successive stages in the establishment of equilibrium between an *n*-type semiconductor and a metal having a greater work function.

In this example, the net carrier density near the surface of the semiconductor is reduced from its bulk equilibrium value, and the surface layer is referred to as a *depletion* region. If $\phi_M - \phi_{SC}$ is sufficiently large the bands may be shifted with respect to the Fermi level to such an extent that next to the surface the valence band is nearer the Fermi level than the conduction band, the material just adjacent to the surface then becoming in effect *p-type*. The surface is then said to be inverted in conductivity type, and the surface *p*-type region is called an *inversion* region.

It is important to note that in the present example a potential barrier of height $e(\phi_M - \phi_{SC})$ is formed at the surface. The formation of this potential barrier and the existence of the depletion region are the basis of the explanation of how a metal-semiconductor contact rectifier operates, as we shall see later. One should observe that in general the depletion region is quite thick. As a matter of fact the thickness of the surface space charge region is essentially the same as that of a *p-n* junction space charge region having an internal potential difference $\phi_M - \phi_{SC}$. There is therefore no possibility (except under extreme circumstances) for electrons ever to be able to tunnel through such a barrier. On the other hand, the very narrow barrier of thickness d, as shown in Figures 16.1(c), 16.2(e) and 16.3, is envisioned as being ultimately so thin as to permit electrons to tunnel freely through it. In two materials which are joined in intimate contact, d is zero, and the only barrier which remains is the surface potential barrier of the semiconductor.

In Figure 16.4 the situation which arises at the contact interface between a metal and an n-type semiconductor where ϕ_{SC} is greater than ϕ_M is shown. Here the semiconductor must acquire a negative charge, the metal a positive one, and the bands

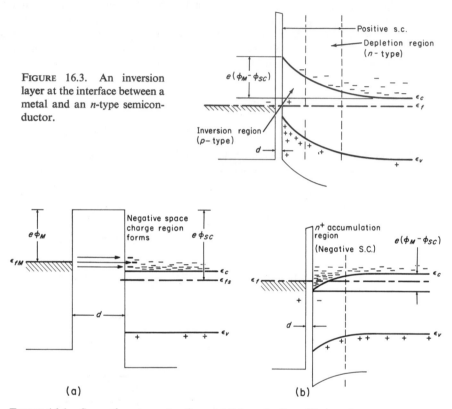

FIGURE 16.3. An inversion layer at the interface between a metal and an n-type semiconductor.

FIGURE 16.4. Successive stages in the establishment of equilibrium between an n-type semiconductor and a metal having a smaller work function.

shift downward at the surface. As a result, instead of a potential barrier, an *accumulation region* in which the electron concentration is *greater* than the concentration of ionized donor atoms is formed. The excess electron concentration in the surface accumulation region gives rise to the negative space charge necessary to support the contact potential difference between the two substances. The contact effects which arise between metals and p-type semiconductors may be discussed in very much the same way as those associated with n-type semiconductors. It may thus be shown that a depletion or inversion region, with an accompanying potential barrier, is formed at the interface between a p-type semiconductor and a metal whose work function is smaller than that of the semiconductor, and an accumulation layer is formed between a p-type semiconductor and a metal having a larger work function. The verification of these conclusions is assigned as an exercise.

16·2 METAL-SEMICONDUCTOR CONTACT RECTIFICATION

The potential barrier at a metal-semiconductor interface where a depletion or inversion layer is present behaves as a rectifier in somewhat the same way as does a *p-n* junction in the interior of a semiconductor crystal. Since the concentration of carriers within the depletion region is much smaller than in the other parts of the system, any externally applied voltage will tend to appear primarily across this high-resistivity layer. The applied voltage will thus tend either to increase or reduce the effective barrier height, as shown in Figure 16.5 for the case of an *n*-type semiconductor and a metal

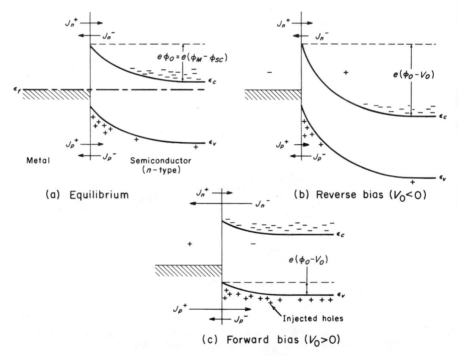

FIGURE 16.5. Potential diagrams illustrating the operation of a metal-semiconductor contact rectifier (a) in the equilibrium state (b) under reverse bias, and (c) under forward bias.

having a larger work function. At (a), in equilibrium, a certain fraction of the electrons in the conduction band of the semiconductor will have sufficient energy to surmount the surface potential barrier. These electrons will cause an electron current J_n^- to flow to the left. An electron flux J_n arising from the small fraction of electrons in the metal which can surmount the barrier will also flow to the right from the metal to the semiconductor. According to the principle of detailed balance, these two fluxes must be *equal* in magnitude in the equilibrium state; under these circumstances there is no net electron current. In a similar way, there are two hole fluxes J_p^+ and J_p^-. The former consists of holes which are generated at the semiconductor surface (by electrons from the valence band occupying occasional empty electronic states in the electron distribution of the metal) having sufficient energy to overcome the potential barrier and enter the interior of the semiconductor. The latter is made up of holes from the interior of the semiconductor which diffuse to the surface and disappear there by extracting an electron from the Fermi distribution of the metal. Again, at equilibrium, these two fluxes are equal and opposite and no net hole current flows.

When an external voltage is applied to the contact, the situation is as shown in Figure 16.5(b) or (c). The two fluxes J_n^- and J_p^+ are clearly governed by the number of carriers present in the semiconductor possessing enough energy to get over the surface potential barrier, which, when an external voltage is applied, is of height $e(\phi_0 - V_0)$, where

$$\phi_0 = \phi_M - \phi_{SC} \qquad (16.2\text{-}1)$$

is the difference between the work functions. This number is simply proportional to $e^{-e(\phi_0 - V_0)/kT}$. The other two fluxes, J_p^- and J_n^+, are dependent only upon the number of thermally generated holes available in the interior of the semiconductor, and the number of electrons from the metal which can surmount the barrier from the left (where the barrier height is fixed), respectively. These two fluxes are functions only of temperature and the material parameters of the two substances, and are independent of the barrier height $e(\phi_0 - V_0)$. We may thus express J_n^- and J_p^+ in the form

$$J_n^- = -J_{n0}e^{-e(\phi_0 - V_0)/kT} \qquad (16.2\text{-}2)$$

and

$$J_p^+ = J_{p0}e^{-e(\phi_0 - V_0)/kT}, \qquad (16.2\text{-}3)$$

where J_{n0} and J_{p0} are constants expressing the *gross* fluxes of electrons and holes, of whatever energy, which are initially incident at equilibrium upon the surface barrier. Since J_n^+ and J_p^- must equal $-J_n^-$ and $-J_p^+$, respectively, when there is no applied voltage, we may write these fluxes as

$$J_n^+ = J_{n0}e^{-e\phi_0/kT} \qquad (16.2\text{-}4)$$

and

$$J_p^- = -J_{p0}e^{-e\phi_0/kT} \qquad (16.2\text{-}5)$$

The total fluxes of electrons and holes are then

$$J_n = J_n^+ + J_n^- = -J_{n0}e^{-e\phi_0/kT}(e^{eV_0/kT} - 1) \qquad (16.2\text{-}6)$$

and

$$J_p = J_p^+ + J_p^- = J_{p0}e^{-e\phi_0/kT}(e^{eV_0/kT} - 1), \qquad (16.2\text{-}7)$$

while the total electrical current density is simply

$$I = e(J_p - J_n) = I_0(e^{eV_0/kT} - 1), \qquad (16.2\text{-}8)$$

where

$$I_0 = e(J_{p0} + J_{n0})e^{-e\phi_0/kT} \qquad (16.2\text{-}9)$$

The metal-semiconductor rectifier thus has the same current-voltage dependence as the *p-n* junction, except that the saturation current is different. The gross fluxes of electrons and holes approaching the barrier in equilibrium from the right and left, respectively, are easily expressed[1] as

$$J_{n0} = \frac{n_0 \bar{c}}{4} \qquad (16.2\text{-}10)$$

and

$$J_{p0} = \frac{p_s \bar{c}}{4} = \frac{p_0 \bar{c}}{4} e^{e\phi_0/kT}, \qquad (16.2\text{-}11)$$

where p_s refers to the concentration of holes in the valence band at the metal surface. Using these results, the saturation current may now be written in the form

$$I_0 = \frac{e\bar{c}}{4} (p_0 + n_0 e^{-e\phi_0/kT}). \qquad (16.2\text{-}12)$$

In the reverse bias condition, the current flow is small, approaching I_0 at reverse voltages large compared with kT/e. Under these circumstances, the depletion region increases in thickness with increasing reverse bias just as does the space charge region of a *p-n* junction. The physical processes which govern the depletion layer thickness are essentially the same in both cases. The current which flows consists almost entirely of holes "collected" from the interior of the semiconductor by the contact and the few electrons from the metal which get over the barrier from the left in Figure 16.5(b). Under foward bias, the barrier height is lowered, and the number of electrons in the conduction band which can overcome the barrier is now much larger than in the equilibrium or reverse bias conditions. A large electron flow from semiconductor to metal results. In addition to this flow of electrons, a flow of holes *into* the interior of the semiconductor will be observed, because now the potential barrier to the holes which are generated at the surface when electrons in the valence band of the semiconductor fall into unfilled electronic states in the Fermi distribution of the metal is *also* lowered. This amounts simply to *injection* of minority carriers by the metal contact in the forward-bias condition.

The above calculations have been made for the case of a contact between an *n*-type semiconductor and a metal having a larger work function. The case of a *p*-type semiconductor and a metal having a smaller work function can be treated in essentially the same way. In this case, however, it is found that the forward direction of rectification is obtained when the semiconductor is positive with respect to the metal, which is just the opposite of the result obtained above. Under these conditions *electrons* are injected into the *p*-type semiconductor when it is positive with respect to the metal. The verification of these results is assigned as an exercise. Due to the absence of a potential barrier, no rectification is obtained at a contact where an accumulation layer is formed. An essentially ohmic contact will therefore be obtained at an interface

---

[1] It should be noted in this connection that there can be no fields nor concentration gradients nor generation and recombination of electrons and holes as such in the metal, and for this reason the situation differs in certain essentials from that discussed in Section 13.1 in connection with the *p-n* junction. Also, it is correct to use the expressions (16.2-10) and (16.2-11) only when the depletion region is thin enough so that essentially no diffusion takes place within it. Since the surface barrier thickness is ordinarily about the same as the electron and hole mean free paths, this condition may sometimes not be fulfilled in practice.

where there is no barrier (i.e., $\phi_M = \phi_{SC}$) or where an accumulation layer is present. The results obtained above serve to verify the observations about the properties of point-contact rectifiers which were made in connection with the discussion of the Haynes-Shockley experiment in Section 10.4.

The predictions of this theory of metal-semiconductor contacts are not, unfortunately, in all respects in agreement with experiment.[2] In particular, these calculations indicate that the rectification properties of metal-semiconductor contacts should be extremely sensitive to the difference in work functions between the metal and the semiconductor, while, in fact they are found to be nearly independent of this quantity. This discrepancy was explained by Bardeen[3] in 1947, who introduced the notion of *surface states* at the contact interface. This subject will be discussed in the next section. In other respects, the theory was more successful. The predicted current-voltage characteristic is in fairly good agreement with experiment, and, in particular, the effects associated with electron and hole injection (as illustrated, for example, by the Haynes-Shockley experiment) are in complete accord with experimental observations.

16·3 SURFACE STATES AND THE INDEPENDENCE OF RECTIFYING PROPERTIES OF WORK FUNCTION

According to the results of the preceding section, the characteristics of metal-semiconductor rectifying contacts should be critically dependent upon the difference in the work functions of the semiconductor and the metal. In particular, if the work functions of the two substances are such as to produce an accumulation layer, no rectification at all would be expected. The experimental results of Meyerhof,[2] on the contrary, showed that in the case of silicon-metal contacts the rectification properties of the contact were practically *independent* of the work function difference, and, in fact, were much the same for all metals. This situation was resolved by Bardeen,[3] who assumed that there could be localized electronic states associated with the surface lying in the forbidden energy region between the conduction and valence bands of the semiconductor.

The fundamental justification for the existence of these *surface states* goes back to the work of Tamm,[4] who showed that if a periodic square-well potential such as that associated with the Kronig-Penney crystal model were terminated on one side by a surface potential barrier such as that illustrated in Figure 16.6, there would be (in addition to the usual allowed energy bands of the Kronig-Penney model) discrete allowed levels within the forbidden energy regions corresponding to wave functions which are localized near the surface. These surface levels were studied in much more detail in a later article by Shockley.[5] According to the calculations there should be one surface state for each surface atom. In addition to this, localized surface levels, discrete or continuously distributed, may be expected to arise from impurity atoms, oxide layers and structural imperfections at the surface. The surface properties of an

[2] See, for example, W. E. Meyerhof, *Phys. Rev.*, **71**, 727 (1947).
[3] J. Bardeen, *Phys. Rev.*, **71**, 717 (1947).
[4] I. Tamm, *Physik. Z. Sowjetunion*, **1**, 733 (1932).
[5] W. Shockley, *Phys. Rev.*, **56**, 317 (1939).

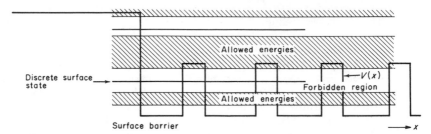

FIGURE 16.6. The formation of localized Tamm states in the forbidden energy region at the surface of a one-dimensional crystal.

individual crystal will then depend upon the density of these surface levels and upon their distribution in energy.

The presence of a considerable density of surface states will result in the formation of a surface depletion layer or accumulation layer within the semiconductor even in the absence of an external metallic contact. To understand how this may come about, let us examine in detail a specific model consisting of an n-type semiconductor crystal having a substantial number of acceptor-type surface states (which are charged when occupied and neutral when empty) whose density and distribution in energy within the forbidden region are as illustrated in Figure 16.7. In the absence of any metallic

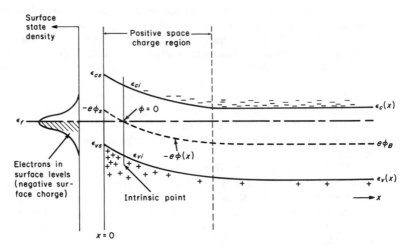

FIGURE 16.7. The formation of a surface inversion layer at a free semi-conductor surface through the interaction of surface states with the interior of the crystal.

contact at the surface, the surface states will be filled up to the Fermi level[6] creating a layer of negative charge at the surface which repels electrons within the conduction

---

[6] This statement refers, of course, to the state of affairs at absolute zero. At higher temperatures, some states above ε_f will be occupied and some states below empty according to the Fermi distribution law.

band of the semiconductor from the surface and which therefore leaves a positive surface space charge layer arising from uncompensated donor ions. In this manner a depletion region and a surface potential barrier are set up by the equilibrium between the bulk semiconductor and the surface states, quite independently of any external contact. If the crystal as a whole is electrically neutral, the total charge contained within the space charge region in the semiconductor must equal in magnitude the total charge associated with electrons in surface states. This fact enables one, as we shall see later, to calculate the value of the surface potential ϕ_S within the semiconductor just adjacent to the surface, and, in fact, the entire potential distribution within the semiconductor surface layer. It is clear that if the surface states are primarily of the donor type (charged when empty and neutral when occupied by electrons) an accumulation layer rather than a depletion layer will be formed on an n-type crystal.

In order to understand the precise effect of the surface states, consider now the effects which take place when a metallic substance is allowed to approach such a semiconductor surface region. We shall discuss specifically the case where the metal is one having a smaller work function than the n-type semiconductor, where a non-rectifying *accumulation* layer was formed in the absence of surface states. The situation is illustrated in Figure 16.8. At (a) the distance d between the metal and semiconductor

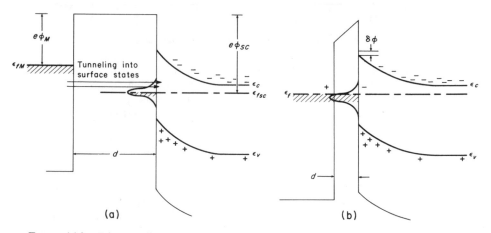

FIGURE 16.8. Diagram illustrating successive stages in the establishment of equilibrium between a metal and a semiconductor with surface states. If the surface state density is reasonably large the interior space charge distribution is barely affected.

has been decreased to the point where tunneling of electrons may begin. The electrons now tunnel from the metal to the surface states, and if the density of surface levels is large enough, the surface states themselves will accommodate all the electrons which are necessary to set up an electric field large enough to equalize the electron potential energies at the respective Fermi levels, the final state of the system being as at (b). The situation inside the semiconductor is essentially the same as at (a), except that the Fermi level is a bit higher with respect to the surface state distribution, and this has resulted in a change of surface barrier height $\delta\phi$ which, however, may be quite small if the surface state density is appreciable. The depletion layer and surface potential barrier still remain, and the rectifying properties of the contact, which, after all, are governed primarily by just these characteristics, are practically invariant. It is an easy matter to verify the fact that, given a situation which is the same except that the work function of the metal exceeds that of the semiconductor, the barrier height change $\delta\phi$ is of opposite sign, the tunneling of electrons from surface states to the metal having lowered the Fermi level with respect to the surface state distribution.[7] Again, the rectifying properties are virtually unaffected. The rectification characteristics are thus seen to be practically independent of the relative values of the two work functions, in agreement with what is observed experimentally.

Another experimental proof of the existence of surface states is afforded by the so-called field effect experiment, in which a semiconductor is made to function as one of the plates of a parallel plate condenser, as shown in Figure 16.9. In this drawing

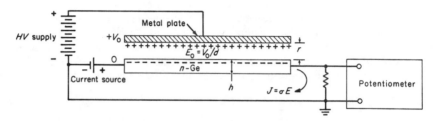

FIGURE 16.9.   Schematic diagram of the field effect experiment and associated circuitry.

a sample of n-type germanium is used as the negative condenser plate, and as large a negative surface charge as possible is induced by connecting a high voltage supply between the plates. Since the capacitance of a parallel plate condenser is just $A/4\pi r$, where A is the area of the plates and r their separation, a charge of magnitude

$$q = CV_0 = \frac{V_0 A}{4\pi r} \tag{16.3-1}$$

will be induced in the semiconductor when a potential difference V_0 is applied. If there were no surface states, this charge would consist solely of mobile free electrons in the conduction band at the surface. The number of induced charge carriers, ΔN is simply q/e, or,

$$\Delta N = \frac{V_0 A}{4\pi e r}. \tag{16.3-2}$$

This change in the number of mobile free electrons would cause a change in sample conductivity $\Delta\sigma$ which would amount to

$$\Delta\sigma = e\mu_n \Delta n = e\mu_n \frac{\Delta N}{V} \tag{16.3-3}$$

where V is the volume of the sample. Writing the volume V as the product of the

---

[7] If the surface states are all at the same energy (i.e., if their distribution is essentially a δ-function) there will be no shift in barrier height at all.

area A and the sample thickness h, and substituting (16.3-2) for ΔN, one may obtain, finally,

$$\Delta\sigma = \frac{V_0\mu_n}{4\pi hr} \tag{16.3-4}$$

as the change in measured conductivity of the sample due to the presence of induced surface charge.

When the experiment was performed, however, it was found that the observed conductivity change was only about one-tenth the value predicted by this expression.[8] The result of the experiment was interpreted by Bardeen to indicate that about 90 percent of the excess induced charge goes into the surface states rather than into the conduction band of the semiconductor, and that electrons in the surface states are *immobile* rather than free.

16·4 POTENTIAL, FIELD AND CHARGE WITHIN A SEMICONDUCTOR SURFACE LAYER

Let us now consider in detail the fields and potentials within the surface region which are set up when surface states are present. We shall assume, in general, that there is an applied field E_0 outside the semiconductor surface, which is generated by some experimental field-effect apparatus such as that shown in Figure 16.9. The situation within the crystal will then be as represented in Figure 16.7. Note that the origin of of the potential function $\phi(x)$, which can be chosen arbitrarily, is defined such that at the point where the material is intrinsic (i.e., where $p_0 = n_0 = n_i$)ϕ is zero. This is the same choice which was made in the discussion of the p-n junction in Section 12.3, and indeed there are many points of similarity between the two examples. Under these circumstances the variation of the conduction and valence band edge energies with distance are as given by (12.3-6) and (12.3-7), while the carrier densities at each point within the band are represented by (12.3-4) and (12.3-5). The carrier densities may be expressed as

$$n(x) = U_c e^{-(\varepsilon_c(x)-\varepsilon_f)/kT} = U_c e^{-(\varepsilon_{ci}-\varepsilon_f)/kT} e^{e\phi(x)/kT} \tag{16.4-1}$$

and
$$p(x) = U_v e^{-(\varepsilon_f-\varepsilon_v(x))/kT} = U_v e^{-(\varepsilon_f-\varepsilon_{vi})/kT} e^{-e\phi(x)/kT}. \tag{16.4-2}$$

However, since
$$U_c e^{-(\varepsilon_{ci}-\varepsilon_f)/kT} = n_i = U_v e^{-(\varepsilon_f-\varepsilon_{vi})/kT},$$

according to (9.3-6) and (9.3-15), these may be written simply as

$$n(x) = n_i e^{e\phi(x)/kT} \tag{16.4-3}$$

$$p(x) = n_i e^{-e\phi(x)/kT}. \tag{16.4-4}$$

---

[8] W. Shockley and G. L. Pearson, *Phys. Rev.*, **74**: 223 (1948).

If all the donors and acceptors are assumed to be ionized, Poisson's equation takes the form

$$\frac{d^2\phi}{dx^2} = -\frac{4\pi e}{\kappa}(p(x) - n(x) + N_d - N_a)$$

$$= \frac{4\pi e n_i}{\kappa}\left(2\sinh\frac{e\phi(x)}{kT} - \frac{N_d - N_a}{n_i}\right). \tag{16.4-5}$$

It is hereby assumed, of course, that the symmetry of the system is such that the potential varies only along the x-direction. If the surface is a plane and if the sample is a uniform one this condition will be realized in practice. If now, we represent the potential by a dimensionless variable u, where

$$u(x) = e\phi(x)/kT, \tag{16.4-6}$$

then (16.4-5) becomes

$$\frac{d^2u}{dx^2} = \frac{2}{L_{Di}^2}\left(\sinh u(x) - \frac{N_d - N_a}{2n_i}\right), \tag{16.4-7}$$

where L_{Di} is the intrinsic Debye length as defined by (10.2-49). This can be written in a form which is even simpler by noting that far in the interior of the semiconductor, at very large values of x,

$$\lim_{x\to\infty} n(x) = n_0 = n_i e^{e\phi_B/kT} \equiv n_i e^{u_B}$$

whereby

$$u_B = \ln\frac{n_0}{n_i}. \tag{16.4-8}$$

Here ϕ_B (and thus u_B) refer to the electrostatic potential far in the interior of the crystal, as shown in Figure 16.7. Expressing the quantity $u_B = \ln(n_0/n_i)$ in terms of $N_d - N_a$ by (9.5-23) and recalling that $\ln(x + \sqrt{1 + x^2}) = \sinh^{-1} x$, we may finally write (16.4-7) as

$$\frac{d^2u}{dx^2} = \frac{2}{L_{Di}^2}(\sinh u - \sinh u_B). \tag{16.4-9}$$

It should also be noted that since $n_0 p_0 = n_i^2$,

$$u_B = \ln\frac{n_0}{n_i} = \ln\frac{n_i}{p_0} = -\ln\frac{p_0}{n_i}. \tag{16.4-10}$$

It is possible to integrate (16.4-9) once by letting

$$F = -\frac{du}{dx} = -\frac{e}{kT}\frac{d\phi}{dx} = -\frac{eE}{kT} \tag{16.4-11}$$

where $E(x) = -d\phi/dx$ is the electric field. Then

$$\frac{d^2u}{dx^2} = -\frac{dF}{dx} = -\frac{dF}{du}\frac{du}{dx} = F\frac{dF}{du}. \tag{16.4-12}$$

Substituting this into (16.4-9) and integrating from $F = 0$ (corresponding to $x = \infty$, far in the interior, where clearly the field must vanish) to an arbitrary interior point, we may obtain the relation

$$F = -\frac{du}{dx} = \frac{2}{L_{Di}}\sqrt{\cosh u - \cosh u_B - (u - u_B)\sinh u_B}, \tag{16.4-13}$$

which connects the value of the field at any point to the value of the potential. This can be integrated once again, this time between the limits $x = 0$ (corresponding to the surface, where $u = u_S = e\phi_S/kT$) and an arbitrary interior point to give

$$x = -\frac{L_{Di}}{2}\int_{u_S}^{u}\frac{du}{\sqrt{\cosh u - \cosh u_B - (u - u_B)\sinh u_B}}. \tag{16.4-14}$$

The relationship between the field and potential in the space charge layer just at the surface can be obtained by substituting $u = u_S$ into (16.4-13), to give

$$F_s = -\left(\frac{du}{dx}\right)_s = -\frac{eE_s}{kT} = \frac{2}{L_{Di}}\sqrt{\cosh u_s - \cosh u_B - (u_s - u_B)\sinh u_B}. \tag{16.4-15}$$

Unfortunately, Equation (16.4-14) cannot in general be integrated in closed form to give an explicit relation describing u as a function of x. There is, however, one case in which an exact analytic solution can be obtained, corresponding to a material in which $N_d - N_a = 0$, whereby $n_0 = n_i$ and $u_B = 0$. The bulk semiconductor crystal is therefore intrinsic far in the interior, and this particular situation will be referred to as the *bulk-intrinsic* case. Setting $u_B = 0$ in (16.4-13) it is easy to show that for this particular case

$$F = -\frac{du}{dx} = \frac{2\sqrt{2}}{L_{Di}}\sinh\frac{u}{2}. \tag{16.4-16}$$

this can now be integrated as described above, and the resulting equation solved for u to give

$$u(x) = 4\tanh^{-1}\left(e^{-\frac{x\sqrt{2}}{L_{Di}}}\tanh\frac{u_S}{4}\right). \tag{16.4-17}$$

The relation between field and potential in the space charge layer at the surface of the crystal is, for this example, clearly

$$F_s = -\frac{eE_s}{kT} = \frac{2\sqrt{2}}{L_{Di}}\sinh\frac{u_S}{2}. \tag{16.4-18}$$

The surface fields may be related to the surface state charge density, the interior space charge per unit surface area and an external field E_0, which one may envision as being applied as shown in Figure 16.9, by applying Gauss's electric flux theorem to cylindrical volumes whose axes are normal to the semiconductor surface, as shown in

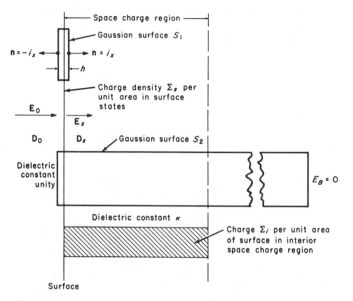

FIGURE 16.10. Diagram illustrating the construction of Gaussian surfaces and notational usage for the calculations of Section 16.4.

Figure 16.10. Considering at first the Gaussian surface S_1 at the top in that diagram, Gauss's theorem[9] requires that

$$\int_{S_1} \mathbf{D} \cdot \mathbf{n} \, da = 4\pi q \qquad (16.4\text{-}19)$$

where $\mathbf{D}$ is the electric displacement vector and $\mathbf{n}$ is the outward normal. Clearly on the lateral surfaces of the cylindrical volume $\mathbf{n}$ and $\mathbf{D}$ are mutually perpendicular, and the dot product vanishes. On the ends of the cylinder, we obtain

$$\int_{S} [\mathbf{D}_S \cdot \mathbf{i}_x + \mathbf{D}_0 \cdot (-\mathbf{i}_x)] \, da = (D_S - D_0)A_1 = 4\pi\Sigma_S A_1 \qquad (16.4\text{-}20)$$

where A_1 is the area of the end of the cylindrical surface and Σ_S the charge density per unit area in surface states. Since the height of the cylinder may be assumed to be infinitesimally small, the amount of charge within the surface contributed by that part

[9] See, for example, J. R. Reitz and F. J. Milford, *Foundations of Electromagnetic Theory*, Addison-Wesley Publishing Co., Reading, Mass. (1960), p. 77.

of the interior volume space charge which is included therein may be neglected. Since $D_0 = E_0$ and $D_S = \kappa E_S$, the result may be written as

$$\kappa E_S - E_0 = 4\pi\Sigma_S. \tag{16.4-21}$$

Gauss's theorem may now be applied in the same way to surface S_2; since the field far in the interior vanishes, the result will be

$$-E_0 = 4\pi(\Sigma_S + \Sigma_i) = 4\pi\Sigma_t \tag{16.4-22}$$

where Σ_i is the charge per unit surface area in the *interior* space charge region and Σ_t represents the *total* charge per unit surface area. By substituting the value given by (16.4-21) for Σ_S into (16.4-22), it is easily seen that the relation between the interior field strength E_S at the surface and the total interior space charge density Σ_i is just

$$E_S = -\frac{4\pi\Sigma_i}{\kappa}. \tag{16.4-23}$$

If the density and distribution of surface states is known, and if the externally applied field E_0 is given, the value of the surface potential u_S may be calculated. Once this is known, the boundary conditions on the interior potential are established and the interior potential and field and the interior space charge density may be determined unambiguously. To illustrate this more clearly, let us consider the bulk-intrinsic case where there are N_S acceptor-type surface states per unit surface area, all at energy ε_S. The surface state density (cm^{-2}) may then be represented as

$$g_S(\varepsilon)\, d\varepsilon = N_S\delta(\varepsilon - \varepsilon_S)\, d\varepsilon. \tag{16.4-24}$$

The energy level associated with the surface states is located at some value which is fixed relative to the conduction and valence band edges at the surface. Since the latter vary with the value of the surface potential so also does the surface state energy ε_S; in fact, referring to Figure 16.11, it is clear that

$$\varepsilon_S = \varepsilon_{S0} - e\phi_S, \tag{16.4-25}$$

where ε_{S0} is the energy difference between the surface level and the intrinsic point in the forbidden gap. Then, since in the bulk-intrinsic case $\varepsilon_f = 0$, we may write n_S, the number of electrons in surface states per unit surface area as

$$n_S = \int_{\varepsilon_c}^{\varepsilon_v} f_0(\varepsilon)g_S(\varepsilon)\, d\varepsilon = N_S \int_{\varepsilon_c}^{\varepsilon_v} \frac{\delta(\varepsilon - (\varepsilon_{S0} - e\phi_S))}{1 + e^{(\varepsilon - \varepsilon_f)/kT}}\, d\varepsilon = \frac{N_S}{1 + e^{(\varepsilon_{S0} - e\phi_S)/kT}}. \tag{16.4-26}$$

The surface charge density Σ_S is then

$$\Sigma_S = -en_S = -eN_S f_0(\varepsilon_{S0} - e\phi_S), \tag{16.4-27}$$

and from (16.4-21) and (16.4-18) an equation for the interior surface field E_S can be

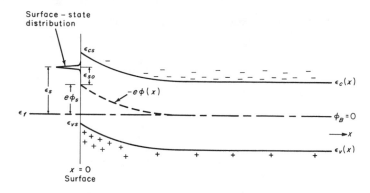

FIGURE 16.11. Potential diagram and notational usage for the calculation of surface potential as a function of applied field discussed in Section 16.4.

written in the form

$$E_S = \frac{4\pi\Sigma_S}{\kappa} + \frac{E_0}{\kappa} = \frac{-4neN_S}{\kappa}\frac{1}{1 + e^{\varepsilon_{so}/kT}e^{-u_S}} + \frac{E_0}{\kappa} = \frac{2\sqrt{2kT}}{eL_{Di}}\sinh\frac{u_S}{2}.$$

(16.4-28)

whence [recalling (10.2-49)],

$$\frac{1}{1 + e^{\varepsilon_{so}/kT}e^{-u_S}} = \frac{E_0}{4\pi eN_S} - \frac{2\sqrt{2}n_iL_{Di}}{N_S}\sinh\frac{u_S}{2}.$$

(16.4-29)

In principle this equation may be solved for u_S to determine the surface potential. Actually, however, the algebraic difficulties in arriving at an analytical solution are so great[10] that for our purposes a graphical solution, as illustrated in Figure 16.12 is more informative.

In that figure the expressions on the left-hand side of (16.4-29) and on the right-hand side of that equation are shown plotted separately as functions of u_S. At the intersection of the two curves, the functions are equal, and the u_S coordinate of that point represents the real root of (16.4-29). The variation of the value of u_S as a function of applied field may be studied by noting that as E_0 is varied, the *sinh* curve is translated up or down parallel to the $f(u_S)$-axis. For $E_0 = 0$, u_S is invariably negative, corresponding to a positive value for $-e\phi_S$. This is to be expected, since the electrons in the surface states create a positive surface space charge layer, hence a region from which electrons are depleted, as in Figure 16.11, leading to a positive value for $-e\phi_S$. With the surface state densities usually associated with germanium and silicon surfaces, it is found that the surface potential energy usually may be varied through several kT units on either side of the zero-field value by the application of experimentally realizable values of E_0 in a field-effect experiment such as that shown in Figure 16.9. Once u_S is determined as a function of the known quantities N_S, ε_{so} and E_0 by this

---

[10] By substituting $x = e^{-u_S}$ into (16.4-29) this equation may be shown to reduce to a cubic equation in x, which has only one real root.

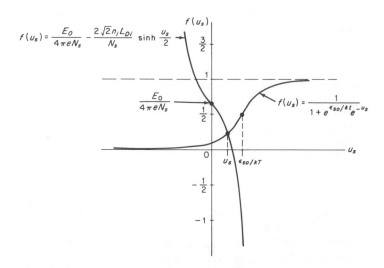

FIGURE 16.12.   Diagram illustrating the graphical solution of equation (16.4-29) and the variation of surface potential with applied field and other variables.

method, the precise form of the interior potential follows from (16.4-17), the interior field E_S at the surface from (16.4-28), the surface state charge density from (16.4-27) and the interior charge density from (16.4-23). The same line of approach may be used with other than bulk-intrinsic samples, although the mathematical work is much more involved.

In actual practice, unfortunately, one ordinarily wishes to reverse this procedure, starting with experimentally measured values of surface and interior charge density as a function of applied field, deriving the corresponding values for E_S and u_S as functions of field, and *inferring* from these data the density of surface states and their distribution in energy. This problem is much more difficult, and the answer can usually be obtained only by trying various *assumed* surface state models, modifying them as needed until a satisfactory fit to the experimental data is obtained when the calculation described above is carried out. Some of the subtleties involved in these field-effect measurements and the conclusions which have been drawn from the experimental results will be discussed in the next section.

16·5   SURFACE CONDUCTIVITY, FIELD EFFECT, AND SURFACE MOBILITY; PROPERTIES OF ACTUAL SEMICONDUCTOR SURFACES

In the field effect experiment of Figure 16.9, the surface conductance arising from an increase or decrease of the carrier density within the interior space charge region may be measured as a function of applied field. The total induced charge density Σ_t may be inferred from (16.3-1) to be $V_0/(4\pi r)$. From these data one may calculate the variation of the surface potential with applied field, and thus (using the procedures outlined in the preceding section) obtain information about the density and distribution in energy of the surface states.

Using the bulk-intrinsic case again as an example, it is clear that if ϕ_S is zero there will be no surface space charge layer and the bulk value of conductivity will prevail even in the surface region. If ϕ_S is positive (thus $-e\phi_S$ negative) the bands as shown in Figure 16.11 will be bent downward near the surface, and electrons which are thermally generated in the interior will slide down the potential energy barrier and accumulate near the surface, while holes will be repelled from the surface region. In effect, the surface will be n-type. The electron concentration near the surface, as a function of ϕ, will be given by (16.4-3), the hole concentration by (16.4-4). It is clear that if ϕ_S is reasonably large the surface electron concentration may be many times what it is in the bulk, and hence the electrical conductivity in the surface layer will be much larger than in the bulk. An increase in sample conductance will therefore result. Likewise, if ϕ_S is negative, the bands will be bent upward, and the resulting accumulation of holes at the surface may lead to a surface layer conductivity which is much larger than the bulk value. It is apparent that the conductance of the sample will reach a minimum near[11] $\phi_S = 0$ and increase as ϕ_S departs from this value in either direction. The exact values of $\phi(x)$ and thus $n(x)$ and $p(x)$ and the conductivity can be calculated as functions of ϕ_S by the methods outlined in Section 16.4, and therefore an exact plot of sample conductance *versus* ϕ_S (or *versus* Σ_i, which is related to ϕ_S through (16.4-23) and (16.4-18)) may be constructed. In making these calculations, it must be recognized that carriers in a surface layer are subject to a scattering process which does not affect carriers in the bulk, namely surface scattering. The effect of surface scattering in space charge layers has been considered by Schrieffer,[12] who adopted a simple linear potential model to approximate the actual surface potential. It was found that the surface mobility was reduced from the bulk value by a factor depending on the surface potential. In pure germanium, this factor ranges from unity for $\phi_S = 0$ to about 0.4 at the largest values of ϕ_S which are realizable experimentally. The value of the factor also depends somewhat upon the bulk carrier density.[13]

Such a plot is shown in Figure 16.13 for a near-intrinsic sample of germanium.[14] In the same diagram is plotted the experimentally determined value of conductance change as a function of the total induced charge Σ_t. Consider first some particular observed value of conductance change. The corresponding values of Σ_t and Σ_i can be read off the curves, and since the conductance change is produced entirely by the

---

[11] The minimum is not exactly at the point $\phi_S = 0$ due to the fact that the electron and hole mobilities are not equal.

[12] J. R. Schrieffer, in *Semiconductor Surface Physics* (R. H. Kingston, Ed.) University of Pennsylvania Press, Philadelphia (1957), p. 55.

[13] This can be understood by recalling that in samples where the bulk impurity concentration is large the thickness of the surface space charge region is small, and *vice versa*. In relatively pure crystals, the surface space charge layer thickness will generally be much greater than the mean free path for scattering in this region, and the effect of the surface scattering mechanism upon the total surface layer conductance will be insignificant. As the impurity density increases, the entire thickness of the space charge region may become of the order of, or less than, the mean free path. Under these circumstances the contribution of surface scattering to the overall mobility associated with surface layer conductance may be very important.

[14] This particular example is discussed by W. L. Brown, W. H. Brattain, C. G. B. Garrett and H. C. Montgomery in *Semiconductor Surface Physics* (R. H. Kingston, Editor), University of Pennsylvania Press, Philadelphia (1957), p. 115.

space charge region, the value of ϕ_S on the computed curve corresponding to this value of conductance change is the actual value of the surface potential. The horizontal distance between the two curves corresponds to the surface state charge density. The variation of ϕ_S with the surface state charge density Σ_S is thus directly obtainable, and a surface state model which reproduces this observed variation may be sought.

It is found experimentally that the gaseous ambient in which the sample is immersed has a profound effect upon the surface state structure. The zero-field value of the surface potential may, in fact, be varied over a wide range simply by changing the nature of the atmosphere surrounding the sample. For example, it has been found[12] that the presence of ozone in the gaseous ambient produces, in the case of germanium, a large negative value of ϕ_S and thus a highly p-type surface, while a moist air ambient produces a large positive value of ϕ_S, hence a strongly n-type surface.

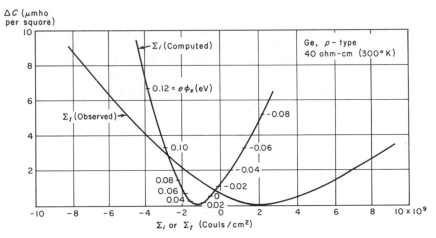

FIGURE 16.13. Observed and computed values of surface charge as a function of surface potential and applied field. [After W. L. Brown, W. H. Brattain, C. G. B. Garrett, and H. C. Montgomery in *Semiconductor Surface Physics*, R. H. Kingston, Editor, University of Pennsylvania Press, Philadelphia (1957), p. 115].

Another readily observed phenomenon is the slow relaxation of the field effect. When a voltage is applied in the experiment shown in Figure 16.9 a change in surface potential and thus a change in interior space charge density is observed immediately, but the surface potential gradually (over a period which, depending upon the surface treatment and temperature, ranges from seconds to hours or days) decays to a value not far removed from the zero-field value. This effect has been explained by assuming that there are two distinct classes of surface states, a group of "fast" states which interact with the interior space charge region very quickly, and a group of "slow" states which may interact only very slowly with the interior of the crystal. The fast states are usually regarded as residing at the surface of the semiconductor crystal, while the slow states are thought to be associated with the outside surface of the oxide layer which is ordinarily present on germanium or silicon samples. This view of the situation is supported by the fact that the relaxation time for the field effect becomes very long when thick oxide layers are present. The magnitude of the observed "fast" field effects, and the fact that the surface potential almost always relaxes to a value not far from the zero-field value after a long time suggests that the density of fast states on germanium is typically of the order of 10^{11} or 10^{12} cm^{-2}, while the slow state density is much higher, perhaps nearly 10^{15} cm^{-2}. The effects associated with changes in gaseous ambient are thought to involve primarily the slow surface states, leaving the fast states practically unaffected. The fact that the observed density of fast surface states is much less than the value predicted by the theories of Tamm and Shockley leads one to the conclusion that the fast states cannot be directly identified with Tamm states, at least in the simple form envisioned in those theoretical investigations.

It is apparent also from the preceding results that the surface states and the surface potential distribution must be largely responsible for determining the recombination probability for excess carriers incident upon the surface, and thus for establishing the surface reflection coefficient and the surface recombination velocity. In particular, it is easily seen that the measured values of the surface reflection coefficient do not represent the "real" reflection coefficient of the surface, but rather express the *combined* effect of the surface itself and the associated space charge region upon the recombination of excess carriers in the surface region. It is to be expected, therefore, that the surface recombination velocity should be affected by anything (for example, applied fields, temperature, gaseous ambient) which can change the surface potential, as well as anything which may change the density and distribution of surface levels themselves. A detailed treatment of surface recombination based upon the Shockley-Read recombination theory has been undertaken by Stevenson and Keyes.[15] This theory assumes that the surface states may act as Shockley-Read traps, and predicts a relation between surface recombination velocity and surface potential which is in general agreement with the results of experimental measurements. The results appear to indicate that it is the fast surface states rather than the slow ones which are of primary importance in determining the surface recombination velocity.

EXERCISES

1. Using the line of reasoning developed in Section 16.1 discuss in detail the nature of an ideal contact, involving no surface states, between a metal and a p-type semiconductor.

2. Derive the current-voltage relation for a rectifying contact between a metal and a p-type semiconductor ($\phi_{sc} > \phi_M$). Assume that there are no surface states.

3. Discuss the nature of the surface space charge layer which is formed when donor-type surface states (neutral when occupied by electrons, positively charged when empty) are present at the surface of a semiconductor.

4. Show explicitly, using the line of approach adopted in Section 16.3, that the rectifying properties of a metal-semiconductor contact for which $\phi_M > \phi_{sc}$ are essentially the same as those which are observed when $\phi_M < \phi_{sc}$ whenever a surface state distribution resulting in a depletion layer is present at the semiconductor surface.

5. Compute the percentage conductivity change to be expected, if no surface states are present, when a potential difference of 1000 V is applied across an air gap of 0.1 mm to a sample of n-type germanium containing 10^{14} electrons per cm^3 at equilibrium which is 0.1 mm thick.

---

[15] D. T. Stevenson and J. R. Keyes, *Physica*, **20**, 1941 (1954).

6. Reproduce the mathematics of the derivation of the electric field, potential and integrated interior space charge density associated with the bulk-intrinsic case.

GENERAL REFERENCES

W. H. Brattain and J. Bardeen, "Surface Properties of Germanium," *Bell Syst. Tech. J*, **32**, 1 (1953).

D. R. Frankl, *Electrical Properties of Semiconductor Surfaces*, Pergamon Press, London (to be published 1966).

R. H. Kingston (editor) *Semiconductor Surface Physics*, University of Pennsylvania Press, Philadelphia (1957).

R. H. Kingston, "Review of Germanium Surface Phenomena," *J. Appl. Phys.*, **27**, 101 (1956).

C. G. B. Garrett and W. H. Brattain, "Physical Theory of Semiconductor Surfaces," *Phys. Rev.*, **99**, 376 (1955).

T. B. Watkins, *The Electrical Properties of Semiconductor Surfaces*, in *Progress in Semiconductors*, Heywood and Co., London (1960), Vol. 5, p. 7.

J. N. Zemel (editor) *Semiconductor Surfaces*, Pergamon Press, London (1960).

APPENDIX A:
THE DIRAC δ-FUNCTION

The Dirac δ-function may be regarded as a function having the following properties:

$$\delta(x - x_0) = 0, \qquad (x \neq x_0) \tag{A1}$$

and
$$\int_{x_0-\varepsilon_1}^{x_0+\varepsilon_2} \delta(x - x_0)\, dx = 1, \tag{A2}$$

where ε_1, ε_2 are *any* positive numbers, which *may* (but need not) be arbitrarily small. Clearly, since according to (A1) $\delta(x - x_0)$ is zero everywhere but at $x = x_0$, in order that (A2) be true, the function must be infinite at $x = x_0$. The Dirac δ-function may then be visualized as the "spike" function[1] shown in Figure A1.

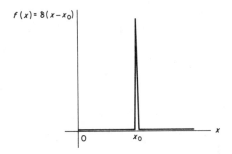

$f(x) = \delta(x - x_0)$

FIGURE A1. Rough visual picture of the Dirac δ-function $\delta(x - x_0)$ as a "spike" of infinite height at the point $x = x_0$.

The most important mathematical property of the Dirac δ-function is expressed by the equation

$$\int_{x_0-\varepsilon_1}^{x_0+\varepsilon_2} f(x)\delta(x - x_0)\, dx = f(x_0), \tag{A3}$$

where $f(x)$ is any function of x which is defined at $x = x_0$. It is possible to arrive at this result simply by noting that if the δ-function were multiplied by a constant A,

---

[1] Strictly speaking, the Dirac δ-function does not satisfy the rigorous mathematical requirements which define a function, and is in the strict sense of the word not a function at all. It may be regarded, however, as the *limit* of a set of functions all of which have unit area and which become progressively thinner and higher about a single point $x = x_0$. The functions (10.3–23) with $x' = x_0$, $E_0 = 0$ and $t' = 0$, in fact, form just such a set in the limit as $t \to 0$. This set is illustrated (for the case $x_0 = 0$) in Figure 10.7(a).

then from (A2) we could write

$$\int_{x_0-\varepsilon_1}^{x_0+\varepsilon_2} A\delta(x - x_0)\, dx = A \int_{x_0-\varepsilon_1}^{x_0+\varepsilon_2} \delta(x - x_0)\, dx = A. \tag{A4}$$

If the integrand, instead of being a constant multiplied by the δ-function is a given function of x multiplied by the δ-function, one may evaluate the integral by initially assuming ε_1 and ε_2 to be extremely small. Then, if $f(x)$ is well-behaved in the neighborhood of $x = x_0$, the function $f(x)$ may be regarded as having the constant value $f(x_0)$ over the tiny interval $(x_0 - \varepsilon_1 < x_0 < x_0 + \varepsilon_2)$, whereby

$$\int_{x_0-\varepsilon_1}^{x_0+\varepsilon_2} f(x)\delta(x - x_0)\, dx = f(x_0) \int_{x_0-\varepsilon_1}^{x_0+\varepsilon_2} \delta(x - x_0)\, dx = f(x_0). \tag{A5}$$

But, clearly, in view of (A1) and (A2), the value of the integral is *independent* of what value is chosen for ε_1 and ε_2. The result (A3) is thus proved.

The Dirac δ-function will be used in the present work to evaluate certain integrals which arise in connection with Fermi-Dirac statistics of free-electron systems, and to describe the density of states function for systems having discrete energy levels using the formalism developed in connection with systems having a quasi-continuous density of quantum states. It has been found to be useful, however, in a variety of situations which arise in other branches of mathematical physics.

APPENDIX B:
TENSOR ANALYSIS

A tensor may be regarded as arising from a linear juxtaposition of two or more vector operations. For example, suppose that we inquire what the *gradient of a vector* might be. We could proceed by applying the gradient operator $(i_x \partial/\partial x + i_y \partial/\partial y + i_z \partial/\partial z)$ to a vector **A**. We would obtain

$$\nabla A = \left(i_x \frac{\partial}{\partial x} + i_y \frac{\partial}{\partial y} + i_z \frac{\partial}{\partial z}\right)(i_x A_x + i_y A_y + i_z A_z)$$

$$= i_x i_x \frac{\partial A_x}{\partial x} + i_x i_y \frac{\partial A_y}{\partial x} + i_x i_z \frac{\partial A_z}{\partial x}$$

$$+ i_y i_x \frac{\partial A_x}{\partial y} + i_y i_y \frac{\partial A_y}{\partial y} + i_y i_z \frac{\partial A_z}{\partial y} \tag{B1}$$

$$+ i_z i_x \frac{\partial A_x}{\partial z} + i_z i_y \frac{\partial A_y}{\partial z} + i_z i_z \frac{\partial A_z}{\partial z}.$$

The gradient of a vector would appear to have *nine* components of the form $\partial A_\alpha/\partial x_\beta$, (where $\alpha = x,y,z$ and $x_\beta = x,y,z$). There are *two* directions associated with any such component, in this example the direction along which the vector component is taken and the direction along which its rate of change is being discussed. The entities $i_\alpha i_\beta$ ($\alpha,\beta = x,y,z$) are clearly analogs of the unit vectors; they are called *unit dyadics*. The sum (B1) is called a dyadic, and the nine components $\partial A_\alpha/\partial A_\beta$ taken together are called the components of a second-rank tensor. A second-rank tensor may be thought of as a collection of nine components which are subject to certain laws of transformation in going from one coordinate system to another, just as a vector may be defined by three components which obey certain coordinate transformation rules. The unit dyadics may be regarded as obeying the usual dot and cross-product laws for unit vectors; for example

$$i_\alpha i_\beta \cdot i_\gamma = i_\alpha (i_\beta \cdot i_\gamma) = i_\alpha \delta_{\beta\gamma}$$

or

$$i_\gamma \cdot i_\alpha i_\beta = (i_\gamma \cdot i_\alpha) i_\beta = \delta_{\gamma\alpha} i_\beta \tag{B2}$$

where $\delta_{\alpha\beta}$ is the Kronecker δ. Under these circumstances, dot and cross-product operations between a tensor and a vector and between two tensors may be defined by the usual rules for those operations applied to the unit vectors. For example, consider a second-rank tensor[1] having nine components $T_{\alpha\beta}$, which we may write as

$$\underline{\mathbf{T}} = \sum_\alpha \sum_\beta T_{\alpha\beta} \mathbf{i}_\alpha \mathbf{i}_\beta, \tag{B3}$$

and a vector

$$\mathbf{A} = \sum_\mu A_\mu \mathbf{i}_\mu. \tag{B4}$$

Then, according to (B2),

$$\underline{\mathbf{T}} \cdot \mathbf{A} = \sum_\alpha \sum_\beta \sum_\mu T_{\alpha\beta} A_\mu \mathbf{i}_\alpha \mathbf{i}_\beta \cdot \mathbf{i}_\mu = \sum_\alpha \sum_\beta \sum_\mu T_{\alpha\beta} A_\mu \mathbf{i}_\alpha \delta_{\beta\mu}.$$

Summing over μ, and noting that $\delta_{\beta\mu} = 0$ except when $\mu = \beta$, we may finally obtain

$$\underline{\mathbf{T}} \cdot \mathbf{A} = \sum_\alpha \sum_\beta T_{\alpha\beta} A_\beta \mathbf{i}_\alpha = (T_{xx}A_x + T_{xy}A_y + T_{xz}A_z)\mathbf{i}_x$$

$$+ (T_{yx}A_x + T_{yy}A_y + T_{yz}A_z)\mathbf{i}_y$$

$$+ (T_{zx}A_x + T_{zy}A_y + T_{zz}A_z)\mathbf{i}_z. \tag{B5}$$

The dot product of a tensor and a vector is thus a *vector*. In the same fashion it is easily shown that

$$\mathbf{A} \cdot \underline{\mathbf{T}} = \sum_\alpha \sum_\beta T_{\alpha\beta} A_\alpha \mathbf{i}_\beta = (T_{xx}A_x + T_{yx}A_y + T_{zx}A_z)\mathbf{i}_x$$

$$+ (T_{xy}A_x + T_{yy}A_y + T_{zy}A_z)\mathbf{i}_y$$

$$+ (T_{xz}A_x + T_{yz}A_y + T_{zz}A_z)\mathbf{i}_z. \tag{B6}$$

From these results it is clear that the two dot products (B5) and (B6) are the same *only* if $\underline{\mathbf{T}}$ is a *symmetric* tensor for which $T_{\alpha\beta} = T_{\beta\alpha}$.

The dot product operation can be readily shown to be equivalent to matrix multiplication. Thus, suppose that the components of the tensor are regarded as the components of a 3 by 3 square matrix and the components of the vector are written as a *column* matrix. Then, by the rules of matrix multiplication[2]

$$\begin{bmatrix} T_{xx} T_{xy} T_{xz} \\ T_{yx} T_{yy} T_{yz} \\ T_{zx} T_{zy} Y_{zz} \end{bmatrix} \begin{bmatrix} A_x \\ A_y \\ A_z \end{bmatrix} = \begin{bmatrix} T_{xx}A_x + T_{xy}A_y + T_{xz}A_z \\ T_{yx}A_x + T_{yy}A_y + T_{yz}A_z \\ T_{zx}A_x + T_{zy}A_y + T_{zz}A_z \end{bmatrix} = \underline{\mathbf{T}} \cdot \mathbf{A}. \tag{B7}$$

The dot product $\mathbf{A} \cdot \underline{\mathbf{T}}$ may be worked out in a similar manner, except that now the vector $\mathbf{A}$ must be written as *row* matrix in order to perform the multiplication;

$$\mathbf{A} \cdot \underline{\mathbf{T}} = \begin{bmatrix} A_x A_y A_z \end{bmatrix} \begin{bmatrix} T_{xx} T_{xy} T_{xz} \\ T_{yx} T_{yy} T_{yz} \\ T_{zx} T_{zy} T_{zz} \end{bmatrix}. \tag{B8}$$

---

[1] The tensor will be denoted by a bold-face letter which is underlined, thus: $\underline{\mathbf{T}}$.

[2] For this, and for a more extended discussion of tensors, the reader is referred to D. H. Menzel, *Mathematical Physics*, Dover Publications, New York (1961), p. 86 ff.

The result is exactly as given by (B6).

The *tensor* product of two tensors, written as $\underline{TU}$, may be defined as the matrix product of the two component matrices, whereby

$$\underline{TU} = \begin{bmatrix} T_{xx}T_{xy}T_{xz} \\ T_{yx}T_{yy}T_{yz} \\ T_{zx}T_{zy}T_{zz} \end{bmatrix} \begin{bmatrix} U_{xx}U_{xy}U_{xz} \\ U_{yx}U_{yy}U_{yz} \\ U_{zx}U_{zy}U_{zz} \end{bmatrix} = \begin{bmatrix} (TU)_{xx}(TU)_{xy}(TU)_{xz} \\ (TU)_{yx}(TU)_{yy}(TU)_{yz} \\ (TU)_{zx}(TU)_{zy}(TU)_{zz} \end{bmatrix} \quad (B9)$$

where

$$(TU)_{\alpha\beta} = \sum_{\mu} T_{\alpha\mu}U_{\mu\beta}. \quad (B10)$$

Likewise, the double product $A \cdot \underline{T} \cdot B = (A \cdot \underline{T}) \cdot B = A \cdot (\underline{T} \cdot B)$ is easily shown from (B5) and (B6) to be a *scalar* quantity whose value is

$$A \cdot \underline{T} \cdot B = \sum_{\alpha}\sum_{\beta} T_{\alpha\beta}A_{\alpha}B_{\beta}. \quad (B11)$$

The tensor whose elements are given by $T_{\alpha\beta} = \delta_{\alpha\beta}$ is referred to as the *unit* tensor, or *idemtensor* $\underline{1}$. Thus

$$\underline{1} = \begin{bmatrix} 1 & 0 & 0 \\ 0 & 1 & 0 \\ 0 & 0 & 1 \end{bmatrix}.$$

From (B5) and (B6), or from (B7) and (B8) it is clear that

$$A \cdot \underline{1} = \underline{1} \cdot A = A, \quad (B12)$$

while from (B9) or (B10) one may easily see that

$$\underline{T1} = \underline{1T} = \underline{T}, \quad (B13)$$

and from (B11), that

$$A \cdot \underline{1} \cdot B = A \cdot B. \quad (B14)$$

Tensor cross-product expressions may be evaluated in the same way using the basic unit vector rules; for example,

$$i_x i_y \times i_z = i_x(i_y \times i_z) = i_x i_x, \quad (B15)$$

and so on. The cross-product of a tensor and a vector is obviously a *tensor*. Tensor cross-products are not used in the present work, and need not be pursued further for our immediate purposes.

INDEXES

INDEX OF NAMES

507

INDEX OF SUBJECTS